20 Springer Series in Chemical Physics
Edited by Robert Gomer

Springer Series in Chemical Physics

Editors: V. I. Goldanskii R. Gomer F. P. Schäfer J. P. Toennies

Chemistry and Physics of Solid Surfaces IV

Editors: R. Vanselow and R. Howe

With 247 Figures

Springer-Verlag Berlin Heidelberg New York 1982

Professor Ralf Vanselow
Professor Russel Howe

Department of Chemistry and Laboratory for Surface Studies,
The University of Wisconsin-Milwaukee,
Milwaukee, WI 53201, USA

Series Editors

Professor Vitalii I. Goldanskii

Institute of Chemical Physics
Academy of Sciences
Vorobyevskoye Chaussee 2-b
Moscow V-334, USSR

Professor Robert Gomer

The James Franck Institute
The University of Chicago
5640 Ellis Avenue
Chicago, IL 60637, USA

Professor Dr. Fritz Peter Schäfer

Max-Planck-Institut für
Biophysikalische Chemie
D-3400 Göttingen-Nikolausberg
Fed. Rep. of Germany

Professor Dr. J. Peter Toennies

Max-Planck-Institut für Strömungsforschung
Böttingerstraße 6–8
D-3400 Göttingen
Fed. Rep. of Germany

ISBN-13: 978-3-642-47499-6 e-ISBN-13: 978-3-642-47495-8
DOI: 10.1007/978-3-642-47495-8

Library of Congress Cataloging in Publication Data. Main entry under title: Chemistry and physics of solid surfaces. At head of title, v. 1: CRC. Vol. 4: editors, R. Vanselow and R. Howe. Heidelberg ; New York : Springer-Verlag. (Springer series in chemical physics ; v. 20). Includes bibliographical references and indexes. 1. Surface chemistry. 2. Solid state chemistry. I. Vanselow, Ralf. II. Tong, S. Y. III. Howe, R. (Russel), 1948–. IV. Chemical Rubber Company. V. Series: Springer series in chemical physics. QD508.C48 541.3'453 77-25890

Softcover reprint of the hardcover 1st edition 1982

Offset printing: Beltz Offsetdruck, 6944 Hemsbach/Bergstr. Bookbinding: J. Schäffer OHG, 6718 Grünstadt
2153/3130-543210

Preface

At the International Summer Institute in Surface Science (ISISS), which is held bienially on the Campus of the University of Wisconsin-Milwaukee, invited speakers present tutorial review lectures during the course of one week. The majority of the presentations deal with the gas-solid interface, but now and then relevant reviews concerning liquid-solid or solid-solid interfaces are included. The goal of ISISS was outlined in the first ISISS publication:

"We recognize that the International Summer Institute in Surface Science should foster mutual understanding and interaction among theorists and experimentalists in the various areas of surface science. Progress can be achieved only when we occasionally peek over the fence into neighboring areas, not so much to amuse ourselves that the grass is greener on the other side as to learn from their progress and, perhaps equally fruitfully, from their limitations and setbacks. In addition, it is an important task in any field of science to assess, take count of what is done and, what is more important, to point in future directions."

Since the foundation of ISISS in 1973, the invited speakers – internationally recognized experts in their area of specialization – have been asked to write review articles too. We wanted in this way to ensure that the largest possible group of scientists could benefit from the special review concept. The collected articles from past Institutes were published by the CRC Press, Boca Raton, Florida, USA, under the following titles:

Surface Science: Recent Progress and Perspectives, Critical Reviews in Solid-State Sciences *4*, 124-559 (1974), and

Chemistry and Physics of Solid Surfaces, Vol. I (1976), Vol. II (1979), Vol. III (1982).

It is not intended that any one volume will offer a complete state-of-the-art report on the whole field of gas-solid interactions, an impossible task given the rapidly expanding scope of the field. Instead, recent developments in selected areas are presented, providing up-to-date supplements to existing textbooks on surface science. Each volume also contains an extensive subject index, vital for the usefulness of a review series.

It has become a tradition to start each volume with a historical review. Distinguished scientists such as P.H. Emmet, E.W. Müller, F.C. Tomkins, and G.-M. Schwab have provided their services in the past. The authors of this volume are led by William E. Spicer, who was recently named "Scientist of the Year" by the journal "Industrial Research and Development."

It is no easy decision to change the publisher for an established book series. Technical reasons, however, did not leave us any choice. We are convinced that the long experience of the Springer-Verlag and its expertise in the area of surface science will ensure continuing success of "Chemistry and Physics of Solid Surfaces." We hope that this fourth volume, with its new format, will be as well received as its forerunners.

Finally, we would like to thank our faithful sponsors: The Air Force Office of Scientific Research and the Office of Naval Research (Grant No. N00014-81-G-0059), as well as the College of Letters and Science and the Graduate School at UWM.

Milwaukee, January 1982

R. Vanselow
R. Howe

Contents

List of Contributors

Block, J.H.
Fritz-Haber-Institut der Max-Planck-Gesellschaft, D-1000 Berlin-Dahlem, Fed. Rep. of Germany

Bootsma, Gosse A.
Department of Applied Physics, Twente University of Technology, NL-7500 AE Enschede, The Netherlands

Briant, C.L.
General Electric Co., Corporate Research and Development Center, Schenectady, New York 12301, USA

Buck, T.M.
Bell Laboratories, Murray Hill, New Jersey 07974, USA

Cardillo, Mark J.
Bell Laboratories, Murray Hill, New Jersey 07974, USA

Delgass, W. Nicholas
School of Chemical Engineering, Purdue University, West Lafayette, IN 47907, USA

Einstein, Theodore L.
Department of Physics and Astronomy, University of Maryland, College Park, MD 20742, USA

Fain, Jr., Samuel C.
Physics Department, University of Washington, Seattle, WA 98195, USA

Gijzeman, Onno L.J.
Van't Hoff Laboratory, University of Utrecht, Padualaan 8, NL-3584 CH Utrecht, The Netherlands

Gruen, Dieter M.
Chemistry Division, Argonne Natl. Laboratory, Argonne, IL 60439, USA

Gunnarson, O.
Max-Planck-Institut für Festkörperforschung, D-7000 Stuttgart 80, Fed. Rep. of Germany

Hanekamp, Lambertus J.
Department of Applied Physics, Twente University of Technology, NL-7500 AE Enschede, The Netherlands

Krauss, Alan R.
Chemistry Division, Argonne Natl. Laboratory, Argonne, IL 60439, USA

Lagally, M.
Department of Metallurgical and Mineral Engineering, Materials Science Center, University of Wisconsin, Madison, Wisconsin 53706, USA

Lauderback, L.L.
School of Chemical Engineering, Purdue University, West Lafayette, IN 47907, USA

Menzel, Dietrich
Physik Department E 20, Technische Universität München, D-8046 Garching, Fed. Rep. of Germany

Messmer, P.
General Electric Co., Corporate Research and Development Center, Schenectady, NY 12301, USA

Mills, D.L.
Department of Physics, University of California, Irvine, CA 92717, USA

Pellin, Michael, J.
Chemistry Division, Argonne Natl. Laboratory, Argonne, IL 60439, USA

Ramaker, David E.
Chemistry Department, George Washington University, Washington, D.C. 20052, USA

Roelofs, Lyle D.
Department of Physics, Brown University, Providence, RI 02912, USA

Ross, Jr., Philip N.
Lawrence Berkeley Laboratory, Materials and Molecular Research Division, University of California, Berkeley, CA 94720, USA

Schönhammer, K.
Max-Planck-Institut für Festkörperforschung, D-7000 Stuttgart 80, Fed. Rep. of Germany

Spicer, William E.
Department of Electrical Engineering, Stanford University, Stanford, CA 94305, USA

Taylor, D.G.
School of Chemical Engineering, Purdue University, West Lafayette, IN 47907, USA

Tong, S.Y.
Department of Physics and Laboratory for Surface Studies, Wisconsin-Milwaukee, Milwaukee, Wisconsin 53201, USA

Venables, J.A.
School of Mathematical and Physical Science, University of Sussex, Brighton, BN1 9QH, Sussex, United Kingdom

Wright, Randy B.
Chemistry Division, Argonne Natl. Laboratory, Argonne, IL 60439, USA

1. Development of Photoemission as a Tool for Surface Science: 1900–1980

W. E. Spicer

With 4 Figures

1.1 Introduction

Since surface science, and particularly the use of photoemission in surface science, has grown so explosively in the past decade, most of you reading this will nave a clear memory of this field which goes back only a decade or less. But this is the decade of "Roots". The decade in which we are rediscovering the value of knowing how we got where we are today – of discovering our foundations. If we gain such perspective, it can help us in going forward and in knowing how to make the best use of future opportunities.

In this field, as in any human endeavor, it is the sum total of the successes and the failures of the individuals and the groups of human beings which determines progress or the lack thereof. We, in science, have the advantage of seeking a provable truth.

Without a proper historical perspective of our field, we will expect too much too soon, and simultaneously too little in the long run. That is, *too much* by expecting to understand all the details of the scientific phenomena under study, and *too little* in the sense of not setting our sights upon the real mountains before us, and thus falling into a quagmire of details.

Therefore, I will try to take you through this history in the hopes of giving you:

1) perspective as to how this one tool, photoemission spectroscopy, was developed;
2) a glimpse of the development of the field of surface science; and
3) perspective, in general, as to how complex scientific tools are developed and deployed.

In all of this you will see the interactions between a wide range of factors, including theory, physical models, experimental techniques, homemade and commercially available equipment, and the critical interaction between practical applications and fundamental advances.

How do I propose to do this? First by going back to the beginnings — to Einstein's 1905 paper on photoelectric emission, to show how this induced a controversy which lasted three decades and focused the attention of the photoemission community on the Einstein photoelectric equation,

$$KE)_{max} = h\nu - e\varphi \tag{1.1}$$

where $KE)_{max}$ is the maximum kinetic energy of the emitted electrons, $h\nu$ is the energy of the incident photon (it was Einstein's postulate of quantized light which made this equation so controversial), and $e\varphi$ is the potential barrier at the surface (the work function, if the solid is a metal). This controversy was not subdued by Milliken's brilliantly capable experiments in the early 1910's, or by the award of two Nobel Prizes: Einstein in 1921 and Milliken in 1923. Why not?

The answer lies in the imperfect experimental equipment and techniques of that time, e.g., there was no available vacuum gauge capable of measuring the ultrahigh-vacuum (UHV, pressure $\approx 10^{-11}$ torr) pressures necessary to insure the maintenance of controlled surface conditions during an experiment. Likewise, there was no convenient tool for detecting surface impurities. The net result of all this was that when experimental results conflicted, there was no easy way to determine if the experiments had been flawed by contaminated surfaces. As we will see, it has only been relatively recently that these problems have been definitively solved (Alpert vacuum gauge — ca. 1950, Auger surface analysis — ca. 1965, etc.).

Evidence of the confusion which characterized photoemission up until the 1940's is the avoidance of it by Irving Langmuir, the father of surface science. Part of his genius was in his concentration on techniques such as thermionic emission in which (because of the hot cathode) effects of contamination could be minimized.

To give some order to this presentation, we will divide the period 1900 - 1980 into subperiods. Thus, the period from 1900 to 1930 will be called "the Einstein era".

The period from 1930 to 1945 was characterized by ill-founded attempts to apply quantum mechanics to photoemission. This will be called "the period of misguided quantum mechanics". In this period theory got off on the wrong foot and stayed there for some time.

Next we will see how a ray of understanding came in 1945 from an obscure university in war-ravaged China. Though little noticed at the time, this begang the true understanding of the process of photoemission from solids. For the next decade and a half an understanding was slowly built which culminated in a simple but fundamentally sound model for photoemission. This model is, in general, the foundation for our current understanding of photoemission and its application to surface science. We will find that the dominant effect, bulk photoemission, had to be understood before photoemission could be used as a surface tool. This period, from 1945 to 1960, will be characterized as "the development of the understanding of photoemission as a bulk phenomenon."

We will, in order to give perspective, backtrack and look at the major experimental developments in vacuum and measurement technology, and see how these developments laid the foundation for the use of photoemission as a scientific tool. We will also see how the development of practical visible photoemitters was essential to the development of a scientific understanding of photoemission. Likewise, we will examine the development of radiation sources, from the early arc and hot tungsten sources to the synchrotron radiation sources of today.

We will then attempt to show how all of these things came together in the late 1960's and early 1970's to make possible the revolution in the use of photoemission spectroscopy as a tool for the study of surfaces. Finally, we will recognize the dangers inherent in such an explosion of activity. These are quite closely related to the dangers specified earlier: the dangers that individual workers become imbedded in the details of their own small "peer" group, and that they become so concerned with these details that they once again lose sight of the important "mountains" and ask detailed questions which are neither of critical long-range importance nor appropriate to our present state of knowledge. There is always concern that theory may drift in the wrong direction, as it did from 1930 to 1945; it is the responsibility of the experimentalists to see that this does not happen. Thus, the experimenter must always question and, where necessary, critically test the theoretical results—always remembering that even the most elegant theory and mathematics must start with approximations. This is because solids and their surfaces are many-body problems which cannot be solved exactly. In the same way, it is essential that the theorists keep the experimentalists from going too far astray. We are emphasizing here a very special and dynamic interaction between theory and experiment, in which each must be aware of the strengths and weaknesses of the other. Lastly, the reader must be warned that today's refereeing system of peer review which determines publications and dominates the funding decisions of the NSF and other agencies, is likely to push the scientist towards excellent mediocrity rather than the true innovation on which the success of science depends. Thus, each scientist must struggle to be "his own man."

Another "trap" the reader is warned of is the historic division between fundamental and applied work. As we will try to illustrate in this paper, fundamental work tends to be sterile and meaningless if it cuts ties with the applied community, just as the applied community makes its job unbearably difficult if it cuts itself off from fundamental research. Thus we should first discard the terms "fundamental" and "applied". They should be replaced by the words scientific and empirical. Fundamental studies must by their nature be scientific, but applied work can also be scientific. The application of established scientific knowledge to practical problems in a process which can often develop new scientific insights and knowledge. A wonderful example from the surface science community is the story of Irving Langmuir who not only won a Nobel Prize and laid the foundations for surface science, but perfected the practical light bulb and helped the General Electric Company to be preemptive in that field for many years following his work. Would Langmuir have been more successful if he had cut himself off from all practical applications? I think not. One reason is that practical work always contains more of the empirical than the fundamental, and in empirical work an immense body of knowledge is built up which can provide an effective resource for the scientist having sufficient strength and ability to find the key elements and systematics, and use these to aid his scientific work.

For the aid of the reader let me close this introduction by listing in a formal way the periods into which this discussion will be divided:

<u>1900 - 1930.</u> The Einstein Era: Emphasis on testing the Einstein equation.
<u>1930 - 1945.</u> The Period of Misguided Quantum Mechanics.
<u>1945 - 1960.</u> The Development of the Correct Fundamental Understanding of the Photoemission Process.
<u>1960 - 1970.</u> The Development of Photoemission Spectroscopy.
<u>1970 - 1980.</u> The Explosive Era in Which Photoemission Spectroscopy was Successfully Applied to the Study of Surfaces.

However, all of this would not have occurred if modern vacuum technology, other areas of surface science, and modern radiation sources had not been developed and successfully combined with photoemission. Thus, in parallel with the scientific eras listed above, we will attempt to mention the critical technological developments without which we would not have modern surface science or even photoemission spectroscopy for the study of surfaces. In doing this we will attempt to identify the key scientific and empirical elements involved.

1.2 The Einstein Era: 1900 - 1930

In 1905 EINSTEIN published his famous paper [1.1] [see (1.1)]. The critical contribution of this paper was assigning a quantization (Energy = $h\nu$) to electromagnetic radiation. Planck had introduced quantization to his theory of blackbody radiation, but he had attributed this to the standing waves characteristic of the blackbody, not to an intrinsic characteristic of electromagnetic radiation. EINSTEIN's concept was hard to swallow because of the prior success of the Maxwell equations which explained electromagnetic radiation in terms of waves with a continuous distribution of energy. EINSTEIN's quantization of the photon forced an intellectually very difficult wave-particle duality on electromagnetic radiation.

EINSTEIN's genius is clear when one examines the crude nature of the experimental data available to him. The combination of 1) the intellectual difficulty of accepting a wave-particle duality, and 2) the unsatisfactory nature of the photoemission experiments (e.g., poor vacuum, ill-defined "dirty" surfaces, the difficulty of reproduction from laboratory to laboratory, etc.) made it easy to argue against the Einstein equation, and kept this controversy alive for almost thirty years.

One can best get perspective on this era and the state is left photoemission in by examining the book *Photoelectric Phenomena* [1.2] by Arthur HUGHES and Lee Alvin DUBRIDGE, published in 1932. (This is the same Lee Alvin DuBridge who became President of Cal. Tech. and Scientific Advisor to the President of the United States at the end of his career, almost 40 years later.) To appreciate the importance and influence of this book, one must emphasize that up until about 1950 there was only one prestigious U.S. series in which books related to physics were published: the green-backed McGraw-Hill series. To give the flavor of this series, we give in Fig.1.1 a reproduction of the listing of these books from ca. 1948. As you can see, the list is short—less than thirty books—and contains such landmark books as those authored or coauthored by Goudsmidt, Morse, Slater, and Seitz, and of course Hughes and DuBridge. Its fame did not die in the thirties. The copy I found in the Stanford library was purchased by L. Schiff after World War II. When I became involved with photoemission in 1955, this book was constantly recommended to me (and I found it basically useless).

First, let us examine the greatest success reported in HUGHES and DUBRIDGE. This was Robert MILLIKEN's work on photoemission, finished in about 1916 [1.3], which provided one of the two citations for his 1923 Nobel Prize in Physics. (Einstein received his Nobel Prize in 1921, the sole specific

INTERNATIONAL SERIES IN
PURE AND APPLIED PHYSICS
G. P. HARNWELL, *Consulting Editor*

BACHER AND GOUDSMIT—ATOMIC ENERGY STATES
BITTER—INTRODUCTION TO FERROMAGNETISM
BRILLOUIN—WAVE PROPAGATION IN PERIODIC STRUCTURES
CADY—PIEZOELECTRICITY
CLARK—APPLIED X-RAYS
CURTIS—ELECTRICAL MEASUREMENTS
EDWARDS—ANALYTIC AND VECTOR MECHANICS
GURNEY—INTRODUCTION TO STATISTICAL MECHANICS
HARDY AND PERRIN—THE PRINCIPLES OF OPTICS
HARNWELL—ELECTRICITY AND ELECTROMAGNETISM
HARNELL AND LIVINGOOD—EXPERIMENTAL ATOMIC PHYSICS
HOUSTON—PRINCIPLES OF MATHEMATICAL PHYSICS
HUGHES AND DUBRIDGE—PHOTOELECTRIC PHENOMENA
HUND—HIGH-FREQUENCY MEASUREMENTS
PHENOMENA IN HIGH-FREQUENCY SYSTEMS
INGERSOLL, ZOBEL, AND INGERSOLL—HEAT CONDUCTION
KEMBLE—PRINCIPLES OF QUANTUM MECHANICS
KENNARD—KINETIC THEORY OF GASES
KOLLER—THE PHYSICS OF ELECTRON TUBES
MORSE—VIBRATION AND SOUND
PAULING AND GOUDSMIT—THE STRUCTURE OF LINE SPECTRA
RICHTMYER AND KENNARD—INTRODUCTION TO MODERN PHYSICS
RUARK AND UREY—ATOMS, MOLECULES AND QUANTA
SCHIFF—QUANTUM MECHANICS
SEITZ—THE MODERN THEORY OF SOLIDS
SLATER—INTRODUCTION TO CHEMICAL PHYSICS
MICROWAVE TRANSMISSION
SLATER AND FRANK—ELECTROMAGNETISM
INTRODUCTION TO THEORETICAL PHYSICS
MECHANICS
SMYTHE—STATIC AND DYNAMIC ELECTRICITY
STRATTON—ELECTROMAGNETIC THEORY
WHITE—INTRODUCTION TO ATOMIC SPECTRA
WILLIAMS—MAGNETIC PHENOMENA

Dr. Lee A. DuBridge was consulting editor of the series from 1939 to 1946.

Fig.1.1. A complete list of books in the prestigious McGraw-Hill "green-backed" series up to about 1948

citation being for "the discovery of the law of the photoelectric effect"). MILLIKEN's photoemission work was described by HUGHES and DUBRIDGE as an example of a "...remarkably accurate study of the photoelectric effect and its relation to the Einstein equation." It is interesting to examine what made this work so unique among the large body of contemporary photoemission work. First, MILLIKEN understood the requirements of his radiation source: it had to be continuously tunable in order to test the equation

$$KE)_{max} = h\nu - e\varphi \qquad (1.1)$$

for a large range of values of $h\nu$. Thus, he made good use of tungsten sources, and he concentrated on emission from the alkali metals which have photoelectron emission thresholds in the visible. In order to determine $e\varphi$ independently, he placed his samples (long "sticks" of alkali metals) on a rotatable mount so they could be moved from one position for photoemission to another for contact potential measurements. The samples could be turned into a third position where a clean surface could be formed by scraping the surface with a sharp edge. Several different alkali metals could be mounted on the "turntable" and measurements made without breaking vacuum. The most interesting and unknown quantity in these experiments was the vacuum. One can estimate that it was better than 10^{-9} torr to explain the excellent results obtained; however, no one will ever know for sure because there were no gauges capable of measuring these low pressures. Until the advent of the Alpert vacuum gauge in 1950, vacuums were truly an empirical art with MILLIKEN their master. How many of us would dare trust scraping in ultrahigh vacuum as a means of obtaining a clean surface? Yet it is still being successfully utilized on some materials today [1.4]. On others, however, it may give misleading results due to damage of the sample created by the scraping [1.5]. Again, we have here an empirical art which has survived until today.

With all of his uncontrollable experimental problems, MILLIKEN designed his experiment so that there was a strong internal check on his data. Knowing both $h\nu$ and $e\varphi$, he could determine h/e for a large range of $h\nu$ for various alkali metals. The value he obtained was quoted by HUGHES and DUBRIDGE as being within 0.5% of the accepted 1932 value. This demonstrates a key point (particularly where critical experimental variables, such as pressure, are not known): the need for the best possible internal experimental checks. Only with these can one be sure that such an experiment is free from major errors.

Unfortunately, in his day, MILLIKEN was the exception and not the rule. With the uncertainties introduced by unknown vacuum pressures, surface conditions, and other key factors, it was possible to do much bad work during the Einstein era, 1900 - 1930. A feeling for this can be obtained from some of the section titles in HUGHES and DUBRIDGE:

Section 2.6: "Recent Tests of Einstein's Equations." (Remember that Milliken's work was complete over 15 years earlier).
Section 3.11: "Is the Presence of Gas Necessary for the Photoelectric Effect?"
Section 6.16: "Is Thermionic Emission Really Photoelectric Emission?" (Richardson had received the Nobel Prize for his correct theory of thermionic emission a decade earlier).
Section 14.6: "Are Positive Rays Produced by Ultra-Violet Light?"

I am happy to say that HUGHES and DUBRIDGE were able to definitively answer each of the questions above correctly. However, the fact that they felt it necessary to answer them, and the fact that tremendous efforts were spent in disproving bad experiments, give some idea of the flavor of those times and illustrates the waste of energy and talent which results from poorly controlled experimental conditions.

Additional evidence of the poor state of photoemission at this time was Irving Langmuir's careful avoidance of it during his epic career. (He received the Nobel Prize in 1933 for "...discoveries and investigations in surface chemistry.") HUGHES and DUBRIDGE have a number of references to Langmuir, but none involving photoemission. I checked in *Science Abstracts* for a five-year period around 1930 and found no publications by Langmuir based on the photoemission technique. Thus, in the Einstein era we find, despite a relatively large amount of work on photoemission, that photoemission was completely avoided by the father of surface science. Clearly, the time was not ripe for a marriage between surface science and photoemission. As we continue this narrative, we will attempt to discover why this marriage did not occur until the 1970's.

1.3 The Period of Misguided Quantum Mechanics: 1930 - 1945

In Sect.1.1 we emphasized that solid-state theory, being concerned with a basically insolvable many-body problem, must start with approximations. If these approximations are well chosen, the results of theory should be important; if not, they can lead to serious confusion and wasted efforts. One must depend on experiment to test theoretical work and, ultimately, to sort good theory from bad. From 1930 to 1945 the theorists got off on the wrong foot, and stayed there. Unfortunately, the experimental side of photoemission was too weak to straighten this out.

How did the theorists get off on the wrong foot? It started with the Sommerfeld model for a metal (Fig.1.2). The Sommerfeld model approximates the interior of a metal by a constant positive potential. Proper quantized states in that potential are filled according to the Pauli exclusion principle; this leads to Fermi-Dirac statistics. This model includes a surface potential (the work function $e\varphi$) determined by the energy difference between the highest filled state at the absolute zero of temperature (i.e., the Fermi level) and the vacuum level. This model immediately had immense success in correctly explaining a range of phenomena in metals where classical models had failed, including the small electronic contribution to the specific heat and the electrical conductivity (with the mean electron scattering length).

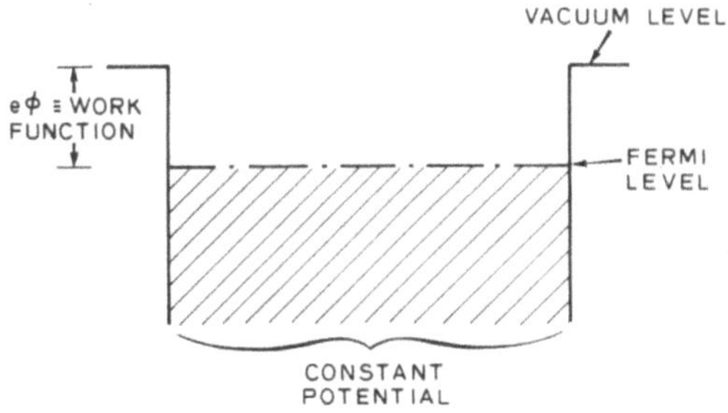

Fig.1.2. The Sommerfeld model of a metal

However, the Sommerfeld model includes a fatal flaw for photoemission, related to conservation of energy and momentum during the optical excitation of an electron. The momentum of a visible or ultra-violet phonon is negligible compared to the change in electron momentum during photoemission (momentum = h/λ, where λ = photon wavelength; electrons $\Delta p \approx h/1$ Å). How, then is the total momentum conserved? Due to the constant potential in the interior of the metal (in the Sommerfeld model), there is no mechanism for momentum conservation. Because of the sharp increase in the potential at the surface, there is only a sufficiently strong interaction between the electrons and the metal to conserve energy. (If one uses standing-wave solutions for the electrons in the Sommerfeld metal, momentum may be conserved in the bulk, but both electrical conductivity and photoemission cannot be explained).

This led to work by Tam and Schuben and others (see HUGHES and DUBRIDGE for details) who explained all photoemission in terms of surface interactions conserving momentum. This gave rise to the "surface photoeffect" (sometimes called the "vector photoeffect"). In the late 1930's MITCHEL [1.6] formulated this problem correctly using quantum mechanics. The only problem was that he and his contemporaries considered this to be the only photoemission mechanism; we now know that, on the average, this is an insignificant contribution. (In the view of this author, it was only in the 1970's that the surface photoeffect was unambiguously isolated and confirmed).

However, this period did have one bright aspect. During it, the first practically important photocathodes were discovered (by completely empirical approaches). As we will see in the next section, the availability of these photocathodes, and the challenge to understand them, played a key role in the development of an understanding of photoemission in general, and in developing the key "three-step model" of photoemission in particular.

1.4 The Development of the Correct Fundamental Understanding of the Photoemission Process: 1945 - 1960

In 1945 a landmark paper [1.7] was published in the *Physical Review*. It made use of the Bloch approximation,

$$\psi(\bar{r},E) = u_{\bar{k},E}(\bar{r}) \exp[-i(\bar{k} \cdot \bar{r} - \frac{E}{h} t)] \quad , \tag{1.2}$$

to show that momentum can be conserved in a bulk optical transition by scattering the photo-excited electron off the periodic potential of the lattice (Bragg reflections).

It is striking that H.Y. FAN, the author of this landmark paper, lived in Kunming, China; a city in the hinterland of China, but strategically placed in war-torn free China during World War II. Kunming is near the terminus of the "Burma Road", and was the heatquarters for the "Flying Tigers" of World War II fame. Soon after this paper was written FAN became Professor of Physics at Purdue University, where he continued a distinguished career developing the field of semiconductor physics.

FAN's paper appeared just before the explosion of work in semiconductor physics stimulated by the invention of the transistor, and was little appreciated at the time. For example, in a graduate physical electronics course (involving principally the emission of electrons into vacuum) which I took from the leader in that field, Professor Wayne B. Nottingham of MIT, in 1950, only the surface photoeffect was taught and there was no mention of FAN's paper. When I began work on photoemission in 1955 no one mentioned FAN's paper to me, whereas I was repeatedly referred to the book by HUGHES and DUBRIDGE.

However, new photoemission experiments were performed following World War II, and through these (and, perhaps, due to FAN's article) the concept that photoemission was a bulk effect gradually took hold. Among the workers who made major contributions here were:

(b) APKER et al. [1.8] (G.E. Labs), 1948-1962,
(a) BORZAK et al. [1.9] (Russia), 1946-1968,
(a) BURTON [1.10] (Bell Labs), 1946-1948,
(b) THOMAS [1.11] (Germany), 1954-1958,
(a) SOMMER [1.12] (RCA Labs), 1936-1978,
(a) SPICER [1.13] (RCA Labs), 1955-1962; (Stanford), 1962-Present.

Those names preceded by an (a) did the majority of their fundamental work on practical photocathodes (principally alkali-antimonide compounds such as Cs_3Sb). Those preceded by a (b) did significant work on other materials. As you can see, a majority of the key scientific work on photoemission during

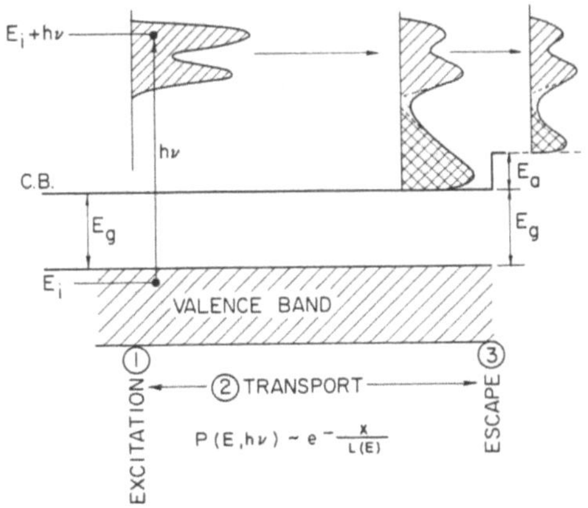

Fig.1.3. The three-step model for photoemission

this period was done on these cathodes, all of which had been developed empirically. The majority had been "discovered" by Sommer, who brought a strong knowledge of chemistry into this field.

There are important reasons why so much of the fundamental work was done on these materials. These reasons all relate to the problems with vacuum and radiation sources which were just beginning to be solved during this period. Of necessity, an empirical technology had been built up in which stable photocathodes were made under relatively poor (10^{-7} torr) vacuum conditions. In order to be of practical usefulness, these cathodes had to be sensitive in the visible. Thus, fundamental work could be done with existing light sources in the visible and near ultraviolet. Above threshold all these cathodes had a high quantum efficiency (~ 0.25 electrons/photon) so that it was easy to show that the photoemission could only be explained by bulk (i.e., volume) and not surface photoemission. They were all semiconductors, enabling much of the rapidly growing body of scientific knowledge concerning semiconductor physics to be applied to them.

In 1958 [1.13] a three-step model for photoemission was suggested by this author which has provided the foundation for the subsequent advances in photoemission in general, and for the development of photoemission spectroscopy (PS) in particular (see the articles in [1.14]). Much important theory has been done since, but the model has retained its usefulness. This model was based on a simple concept which divided the photoemission process into three steps, each of which could be treated in a simple way. First was the optical excitation. Second was the movement of the electron through the solid with energy losses due to scattering processes giving a distance-dependent probability of reaching the surface with sufficient energy to escape. This probability goes as exp[-x/L(E)], where x is the depth from the surface at which the electron is excited and L(E) is a characteristic scattering length dependent on the electron energy E. Third, was the electron escape over the surface barrier. These steps are illustrated in Fig.1.3. This was not the first suggestion of a three-step model; however, it was the first to treat the scattering property and to be applicable to all solids.

1.5 The Development of Photoemission Spectroscopy: 1960 - 1970

Thus, by 1960 the scientific stage was set for the development of photoemission spectroscopy (PS). The work of BERGLUND and SPICER extended both experimental technique and theory [1.15]. In their work, the d bands of Cu and Ag were seen in their full glory for the first time. In addition, electron transport theory was formalized so that the inelastically scattered electron distribution could be fit well with theory, and electron-electron scattering established as the key scattering mechanism in the ultraviolet.

During the 1960's there was steady work which showed the power of PS. For example, using the spectral region below $h\nu = 12$ eV, the rigid band model of alloys was definitively disproved by PS experiments [1.16]. Studies of amorphous Ge and Si established that bulk photoemission did not depend on the crystallinity of materials, but was just as strong (albeit with different optical selection rules) in amorphous materials [1.17]. In the context of the earlier confusion regarding the conservation of electron momentum (Sect.1.3), this was important because it showed that electron momentum could be conserved even with out the Bragg reflections introduced by FAN [1.17].

In Sweden, SIEGBAHN and his co-workers [1.18] began reporting results in which X-ray sources ($h\nu \approx 1500$ eV) were being used in order to study core as well as valence levels. Since this work revealed so much about chemical bonding, SIEGBAHN named this technique electron spectroscopy for chemical analysis (ESCA), but it is now known as X-ray photoemission spectroscopy (XPS).

In England, TURNER and his co-workers [1.19] had developed techniques for photoemission from gases. In this process they developed the helium arc line source [1.19], which had a striking impact on photoemission from solids and surfaces in the 1970's.

Later in this section we will outline in more detail the development of vacuum techniques, radiation sources, and electronics which were essential to the explosion of applications of photoemission spectroscopy to surface

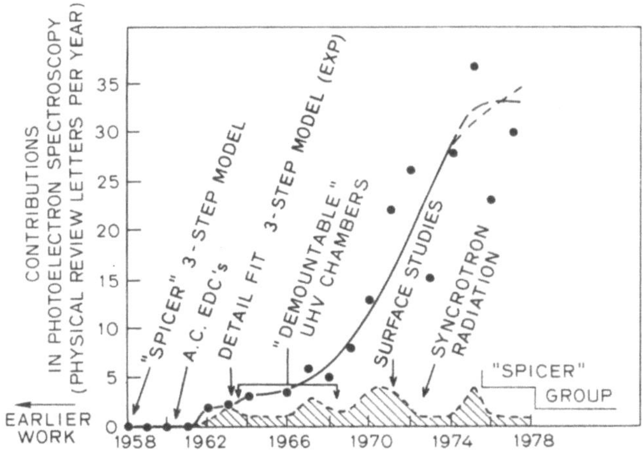

Fig.1.4. The number of photoemission papers published annually in the *Physical Review Letters*

studies which occurred in the 1970's. But first let us attempt to demonstrate that such an explosion did take place.

In Fig.1.4 we plot the number of *Physical Review Letters* published on photoemission spectroscopy as a function of time. At a glance, one can see the increase that took place during the decade following 1969. Note that, although the *Physical Review Letters* started publication in 1958, it was 1962 before the first paper on photoemission appeared. Note the constant rate of two to four publications per year between 1962 and 1967. This was the gestation period for photoemission. This lasted so long because the technique of photoemission spectroscopy are personally demanding and require a large capital investment. People had to be well convinced of its usefulness before they would adopt the technique.

The work during the 1960's was concentrated almost entirely on understanding and making use of the bulk photoemission process. The application of PS to surfaces began in the early seventies and contributed to the explosion shown in Fig.1.4. The success of photoemission spectroscopy in studying surfaces followed its success in studying bulk materials for two reasons: one, because it was necessary that the majority photoemission event, that of the bulk, be understood before the minority surface component could be successfully distinguished and studied; and two, because truly successful surface studies require photon energies in the range $12 < h\nu < 1000$ eV, and these only became available in the seventies.

By the early 1960's a strong scientific understanding of the photoemission process was being developed; however, a number of developments in experimental techniques were necessary before that understanding could be successfully applied to surface studies and before the explosion which occurred during the seventies could take place. Foremost among these was the development of a viable vacuum technology (e.g., [1.20]). The key elements in this were the development by Alpert of his ion gauge capable of measuring pressures to 10^{-11} torr. It should be emphasized that this gauge was the result of "scientific engineering" once Nottingham identified the limiting factor in ionization gauges to be photoemission due to soft X-rays which strike the ion collector. Other gauges followed which had comparable or better low-pressure sensitivity. Once he had a proper ion gauge in hand, Alpert started developments which contributed to the current state of vacuum technology.

The next and final step in the development of modern vacuum technology was the development of dry (i.e., vac-ion) pumps and all-metal (stainless steel) vacuum chambers. These developments took almost twenty years to accomplish.

Your are all familiar (and will see much more elsewhere in this volume) with the various tools of surface analysis other than PS. Thus, I will just emphasize that with these we have the opportunity to accurately characterize the surfaces studied by PS in a way not even dreamed of in Langmuir's days.

A major step for PS was the introduction of phase-sensitive detection for the measurement of energy distribution curves (EDC's [1.21]). This reduced the time necessary to take an EDC of adequate signal to noise from hours in Apker's laboratory to minutes today. It has enabled experimenters to do away with double-screened measurement rooms used to cut down on stray electrical noise and it has diminished the need for ultrastable radiation sources.

Light sources have gradually developed to include sources extending further and further into the ultraviolet, and simultaneously to include sources extending from the X-ray end of the spectrum. In this period synchrotron radiation was being introduced. We will see the full impact of this in the next section. All of these things — a scientific understanding and the development of key experimental techniques — came together to produce the explosion of activity in the 1970's, and led to the very advanced state of PS as a tool for surface analysis. Let us just briefly examine what happened during the explosion of the 1970's.

1.6 The Explosive Era in Which Photoemission Spectroscopy Was Successfully Applied to the Study of Surfaces: 1970 - 1980

It is hard to believe that the first clear photoemission spectra of gases adsorbed on a surface appeared only a decade ago [1.22], and that the first spectra of filled surface states on a semiconductor are less than a decade old. We mention this to emphasize the rapid growth of PS in recent years.

The key technique which came into its own in the seventies was the growing use of synchrotron radiation as a light source [1.23]. Not only does it provide an intense continuum of radiation which bridges the visible to X-ray regions, it is also naturally polarized.

By tuning the wavelength, synchrotron radiation can probe the last few atomic layers [1.24] of a solid, giving tremendous surface sensitivity. By using the $h\nu$ dependence of optical matrix elements, synchrotron radiation can be used to separate adsorbates from substrates, or to distinguish various levels of the same material 1.24 . Combining this with angle-resolved PS, one can determine the angle at which an adsorbed atom sits on a substrate [1.24]. New techniques such as photon-stimulated desorption (PSD) and surface extended X-ray absorption fine structure (SEXAFS) give us strong new insights into the bonding of gases, as well as the local arrangement of atoms on surfaces [1.23].

Thus, we have entered a period in which we have an abundance of riches for the use of PS in the study of surfaces. The challenge is to find the best and most efficiency way to use them to increase our scientific knowledge in general and to develop knowledge which can open doors in understanding a wide range of practical subjects ranging from semiconductor interfaces to catalysts.

1.7 Conclusions

My hope is that, by examining the slow development of PS as a tool for surface studies, I may have stimulated you to develop insights which will help in the field of surface studies. I hope all the difficulties and successes (and the reasons for each) recounted here will help you to find your own mountains. It is also my hope that it will help you to see the importance of developing work which stretches from the fundamental to the practical: to forget the class distinctions of "basic" versus "applied", and think in the new terms of "scientific" versus "empirical." Of course, this would all be meaningless if it just led to a new type of class distinction; but if we realize that at any instant in time each "scientific" and "empirical" approach has its place, and if we remember that each can learn from the other, we may reach a new plateau of scientific and technological productivity and support.

References

1.1 A. Einstein: Ann. Phys. Leipzig *17*, 132 (1905)
1.2 A. Hughes, L. DuBridge: *Photoelectric Phenomena* (McGraw-Hill, New York 1932)
1.3 R.A. Milliken: Phys. Rev. *7*, 362 (1916); *34*, 68 (1912); *35*, 74 (1912); *1*, 73 (1913);
R.A. Milliken, J.R. Wright: Phys. Rev. *30*, 287 (1910); *34*, 68 (1912)
1.4 J.W. Allen, S.-J. Oh, I. Lindau, J.M. Lawrence, L.I. Johansson, S.B. Hagström: Phys. Rev. Lett. *46*, 1100 (1981);
Y. Baer, R. Hauger, C. Zücher, M. Campogna, G.K. Wetheim: Phys. Rev. B*18*, 4433 (1978)
1.5 H.M. Nitz, O. Ganschow, U. Kaiser, L. Wiedmann, A. Benninghoven: Surf. Sci. *104*, 365 (1981);
P. Morgen, J. Silberman, I. Lindau, W.E. Spicer, J.A. Wilson: J. Cryst. Growth (in press)
1.6 K. Mitchell: Proc. R. Soc. London A*146*, 442 (1934); Proc. Cambridge Philos. Soc. *31*, 416 (1935); Proc. R. Soc. London A*153*, 513 (1937);
J.C. Endriz: Phys. Rev. B*7*, 3464 (1973)
1.7 H.Y. Fan: Phys. Rev. *68*, 43 (1945)
1.8 L. Apker, E. Taft, J. Dickey: Phys. Rev. *117*, 550 (1959); *73*, 46 (1948); *74*, 1462 (1948); *76*, 270 (1949); *84*, 508 (1951); J. Opt. Soc. Am. *43*, 78 (1953);
E.A. Taft, H.R. Philipp, L. Apker: Phys. Rev. *113*, 156 (1959);
E.A. Taft, H.R. Philipp: Phys. Rev. *115*, 1583 (1959);
H.R. Philipp, E. Taft, L. Apker: Phys. Rev. *120*, 49 (1960) [and references therein]
1.9 P.G. Borzak: C. R. Dokl. Acad. Sci. URSS *31*, 546 (1941); Zh. Tekh. Fiz. *20*, 923 (1950); Tr. Inst. Fiz. Akad. Nauk. Ukr. SSR *4*, 1924 (1952); *4*, 28 (1953);
P.G. Borzak, V.F. Bibik, G.S. Kramerenko: Bull. Acad. Sci. USSR Phys. Ser. *20*, 939 (1956);
P.G. Borzak, L.S. Miroshnichenko, R.D. Fedoravich: Sov. Phys. Solid State *3*, 1291 (1961) [and references therein]
1.10 J.A. Burton: Phys. Rev. *72*, 531 (1947)
1.11 H. Thomas: Z. Phys. *147*, 395 (1957)
1.12 A.H. Sommer: *Photoemissive Materials* (Wiley, New York 1968) reprinted by Krieger, Huntington, New York 1980
1.13 W.E. Spicer: Phys. Rev. *112*, 114 (1958); RCA Review *19*, 555 (1958); J. Appl. Phys. *31*, 2077 (1960); J. Phys. Chem. Solids *22*, 365 (1961); Phys. Rev. *125*, 1297 (1962);
W.E. Spicer, F. Wooten: Proc. IEEE *51*, 1127 (1963);
W.E. Spicer, A.H. Sommer, J.G. White: Phys. Rev. *115*, 57 (1959)
1.14 Natl. Bur. Stand U.S. Spec. Publ. *323* (1971) [L.H. Bennett (eds.): Electronic Density of States]
1.15 C.N. Berglund, W.E. Spicer: Phys. Rev. *136*, 1030A, 1044A (1964)
1.16 D.H. Sieb, W.E. Spicer: Phys. Rev. Lett. *22*, 1176 (1969); Phys. Rev. B*2*, 1676, 1694 (1970)

1.17 W.E. Spicer, T.M. Donovan: Phys. Rev. Lett. *24*, 595 (1970); D.T. Pierce, W.E. Spicer: Phys. Rev. B*5*, 3017 (1972); W.E. Spicer: In *Proc. of 5th Int. Conf. on Amorphous and Liquid Semiconductors*, ed. by J. Stuke, W. Brenig (Taylor and Francis, London 1974) p.499
1.18 K. Siegbahn, C. Nordling, A. Fahlman, H. Nordberg, K. Hamrin, J. Hedman, R. Nordberg, K. Hamrin, J. Hedman, G. Johansson, T. Bergmark, S.-E. Karlsson, I. Lingren, B. Lindberg: *ESCA – Atomic, Molecular, and Solid-State Structure Studied by Means of Electron Spectroscopy* (Almquist and Wiksells, Uppsala 1967)
1.19 D.W. Turner, C. Baker, A.D. Baker, C.R. Brundle: *Molecular Photoelectron Spectroscopy* (Wiley-Interscience, New York 1970)
1.20 G.F. Derbenwick, D.T. Pierce, W.E. Spicer: In *Solid State Physics*, ed. by R.V. Coleman (Academic, New York 1974) p.67; In *Vacuum Ultraviolet Radiation Physics*, ed. by E.-E. Koch, R. Haensel, C. Kunz (Pergamon, Oxford 1974)
1.21 W.E. Spicer, C.N. Berglund: Rev. Sci. Instrum. *35*, 1665 (1964)
1.22 D.E. Eastman, J.K. Cashion: Phys. Rev. Lett. *27*, 1520 (1971)
1.23 Herman Winick, S. Doniach (eds.): *Synchrotron Radiation Research* (Plenum, New York 1980) [and references therein]
1.24 I. Lindau, W.E. Spicer: In [Ref.1.23], Chap.6 [and references therein]; W.E. Spicer: In *Electron and Ion Spectroscopy of Solids*, ed. by L. Fiermans, J. Vennik, D. Dekeyser (Plenum, New York 1978) Chap.2 [and references therein]

2. Auger Spectroscopy as a Probe of Valence Bonds and Bands

D.E. Ramaker

With 19 Figures

2.1 Introduction

For almost a decade now, Auger electron spectroscopy (AES) has been utilized as a technique for elemental identification and trace analysis at the surface of solids [2.1]. Indeed over these years AES has become a widely available and almost indispensable technique for determining surface cleanliness, surface coverage, and even depth profiling. More recently, the Auger lineshape has been recognized as a source of chemical and electronic structure information. Just how this information can be extracted from the Auger lineshape requires a thorough understanding of the factors contributing to the Auger process. This review will examine several of these factors.

It is often true that one decides to write a review (or requests that someone else write a review) when he perceives that significant progress has recently occurred in an area. Thus, from the seven review articles written in the last two years on AES, one might conclude that significant progress has been achieved recently in our understanding of the Auger process. We believe that this is indeed the case. Table 2.1 summarizes these seven review articles and includes the article from CRASEMANN [2.2] (unpublished lecture notes) for completeness. The principle topics reviewed indicate the breadth of the recent progress and also the relatively small overlap in subject matter between these eight articles. HOLLOWAY's review [2.1] of experimental approach and applications also covers the sometimes difficult but important task of extracting a nearly quantitative Auger lineshape or spectrum from the experimental data. This aspect will not be covered here; I refer the reader to the HOLLOWAY article as well as to some of the original work in this area [2.3,4]. The reviews by LARKINS [2.5] and MADDEN [2.6] concentrate on the chemical information contained in the Auger energies; this important topic also will be omitted from this review. The works by MODDEMAN [2.7], AKSELA [2.8], and CRASEMANN [2.2] concentrate on the Auger effect in atoms and molecules in the gas phase, and as such also will not be considered here.

Table 2.1. Recent AES review articles

Author	Principle topics covered
HOLLOWAY [2.1] (1980)	Experimental approach and applications Extracting the lineshape from experimental data
CRASEMANN [2.2] (Unpublished)	Auger effect in atoms Matrix elements, angular momentum coupling, Auger widths, and fluorescence yields
LARKINS [2.5] (1981)	Auger energies – chemical shifts and solid-state shifts Energies of Auger satellites
MADDEN [2.6] (1981)	Chemical information from AES – chemical shifts and lineshapes Applications to insulators and metals
MODDEMAN [2.7] (1981)	Auger effect in atoms and molecules Gas-phase experiments
AKSELA [2.8] (1981)	Auger lineshapes of transition metals in the gas phase Lineshape interpretation in the atomic model
KLEIMAN [2.9] (1981)	Effects of localization and screening on Auger energies Applications to d-band metals and alloys
FUGGLE [2.10] (1981)	Screening and localization effects on the Auger lineshape Applications to metals and insulators
This work	Auger as a probe of valence bands – radial extent, localization, and screening effects on lineshape Applications to metals, insulators, molecules

The reviews by KLEIMAN [2.9] and FUGGLE [2.10] have the greatest potential overlap with this work since screening and localization effects on the Auger lineshape will be considered here also. However, the results and conclusions presented here on screening do not appear in their reviews; the concept of localization is extended beyond the elemental metals and alloys they consider. The emphasis here will be on AES as a probe of the valence density of states (DOS) with applications to metals, insulators, and molecules.

The Auger process is a two-electron process, which makes it unique from the more common one-electron processes such as are involved in X-ray and photoemission (XES and PES). These three spectroscopies are illustrated in Fig.2.1 for the case of an atom involving the K and L shells. The Auger and X-ray emission processes may be initiated by either an incoming electron or photon, the photoemission process of course only by a photon. The Auger and X-ray emission processes compete to fill the core hole; however, the Auger process clearly dominates for core levels with binding energies less than

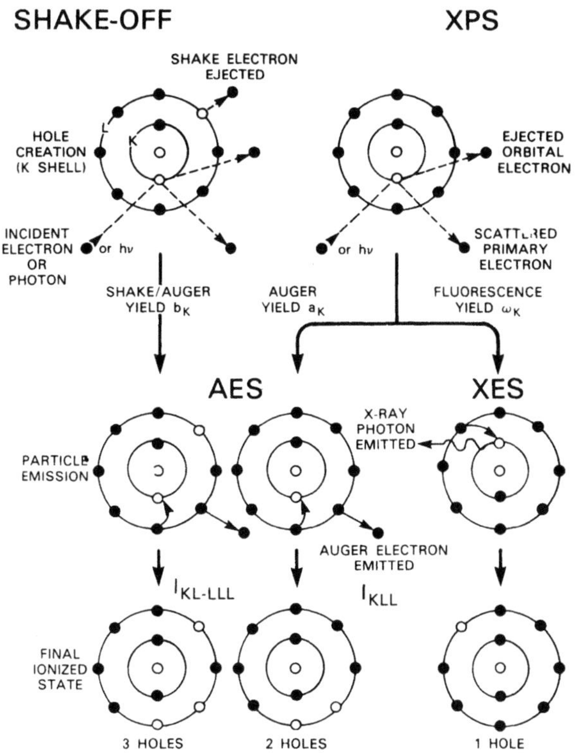

Fig.2.1. An illustration of photoemission (XPS), X-ray emission (XES), and Auger emission (AES) in an atom involving the K and L levels. The shake-off process with a subsequent Auger emission yielding a KL-LLL shake/Auger satellite is also illustrated. (●) electrons, (o) holes in the K and L shells

5000 eV [2.11]. Also illustrated is the shake-off process, a process which always occurs with some probability during an ionization event and results from relaxation of the atom in the presence of the core hole. A subsequent Auger event (a subsequent X-ray emission event may also occur) results in a shake/Auger satellite, i.e., Auger electron contributions generally shifted down in energy from the parent Auger contributions. The parent and satellite contributions are termed in this example the K-LL and KL-LLL transitions on the basis of their initial- and final-state hole configurations.

One may be led to question the appropriateness of AES as a probe of valence-band information. Why bother with AES when XES and PES provide the one-electron DOS, N(E), more directly? The final state of the Auger and shake/Auger events as illustrated in Fig.2.1 leaves the system with 2 or 3 valence holes compared with 1 valence hole in XES and PES. This tends to

Table 2.2. Comparison of PES, XES, and AES features [2.12,13]

Method	Probe input [keV]	Measurement output [eV]	Minimum probe diameter [μm]	Sampling depth atomic layers	Optimum detection sensitivity [ppm atomic]	Sampling region of valence density of states
Electron-excited XES	Monoenergetic electrons 0.01-3	X-rays 10-1500	0.5-100	10^4	1-100	Site specific
Photon-excited XES	X-ray beam ~ 1.5	X-rays 10-1500	10^4	10^5	1-100	Site specific
Electron-excited AES	Monoenergetic electrons 0.01-3	Electrons 10-1500	0.5-100	2-10	10^3-10^4	Site specific
Photon-excited AES	X-ray beam ~ 1.5	Electrons 10-1500	10^4	2-10	10^3-10^4	Site specific
PES[a]	Monoenergetic photon beam	Electrons 0-1500	10^4	2-100	10^4	All sites

[a]PES here includes both X-ray and ultraviolet photoelectron spectroscopy (XPS and UPS)

complicate the spectrum analysis because of the increase in possible final-state hole configurations. As a first approximation, the CVV (core-valence-valence) Auger lineshape A(E) is a fold of N(E) [2.9],

$$A(E) = \int N(E - \varepsilon)N(\varepsilon)d\varepsilon \quad . \tag{2.1}$$

However, Table 2.2, which compares some features of XES and PES with AES, indicates the greater surface sensitivity of AES compared with XES, and the site specificity of the valence DOS probed by AES compared with the all-sites DOS probed by PES [2.12,13]. The lack of surface sensitivity in XES arises from the large escape depth of the observed outgoing photons. AES and XES probe the DOS only near the initial core hole; no initial core hole exists in PES. The combined surface sensitivity and site specificity offer unique advantages for AES. This along with the wide availability of AES instrumentation [2.1] makes it an attractive technique for electronic structure studies.

2.2 Lineshape Description—One-Electron Model

A detailed interpretation of the Auger lineshape requires a quantitative description of each Auger transition. Each transition may be characterized by an Auger energy, width, and intensity. These may be simply expressed in terms of the initial- and final-state holes in a one-electron orbital model [2.14],

$$E_{cxy} = E_c - E_x - E_y - U_{xy} \quad , \tag{2.2}$$

$$\Gamma_{cxy} = \Gamma_c + \Gamma_x + \Gamma_y + 2k_{xy} \quad , \tag{2.3}$$

$$I_{cxy} = 2\pi/\hbar \; |\langle\psi_f|H - E|\psi_i\rangle|^2 \approx \sum_{nm} c_{xn}^2 c_{ym}^2 P_{cnm} \quad , \tag{2.4}$$

where E_c and Γ_c are the binding energy and width of the initial core hole; Γ_c includes the lifetime broadening and spin-orbit splitting contributions. The final valence state holes have the binding energies E_x and E_y and widths Γ_x and Γ_y; the widths in this instance include solid-state banding effects and nuclear (vibrational) effects [2.15]. These one-electron parameters may be obtained from ab initio calculations [2.15], but most often are obtained empirically [2.14]. This is consistent with the desire to make the Auger lineshape interpretation techniques generally practical, available to a large number of experimentalist in the field, and applicable to a large number of interesting systems. Binding energies are best obtained from XPS, the widths from either XES or UPS data [2.14].

The quantities U_{xy} and K_{xy} reflect the presence of two holes in the final state. U_{xy} is the effective hole-hole Coulomb repulsion, and K_{xy} is the singlet-triplet splitting which is assumed to be unresolved in the solid. They may be defined in a LCAO-MO picture (e.g., $X = \sum_n c_{xn} f_n$) by the expressions (utilizing the zero differential overlap (ZDO) approximation)

$$U_{xy} = \langle XY|r_{12}^{-1}|XY\rangle = \sum_{nm} c_{xn}^2 c_{ym}^2 U_{nm} \tag{2.5}$$

$$K_{xy} = \langle XY|r_{12}^{-1}|YX\rangle = \sum_{nm} c_{xn} c_{yn} c_{xm} c_{ym} U_{nm} \quad , \tag{2.6}$$

where the atomic repulsion integrals must reflect the screening and relaxation of the atom,

$$U_{nm} = \langle f_n f_m|r_{12}^{-1}|f_n f_m\rangle = f(n,m) - r_n \delta_{nm} \quad . \tag{2.7}$$

Here f(n,m) are the atomic repulsion integrals and r_n is a "static relaxation energy" [2.16]. The one-center integrals f(n,m) may be obtained from the integral tables of MANN [2.17] and the two-center f(n,m) from some suitable interpolation scheme [2.18]. The quantities r_n have been determined empirically by SHIRLEY [2.16] from atomic Auger spectra.

In the rigid one-electron model, the expression for the Auger intensity reduces to a two-electron matrix element (see Sect.2.4). In the LCAO-MO picture it reduces still further to a sum over terms involving the atomic populations and an atomic Auger matrix element, P_{cnm} (2.4). Assuming P_{cnm} is large only when f_n and f_m are orbitals on the atom with the core hole (see Sect. 2.2.2), the sum in (2.4) often reduces to a single term, $c_x^2\, c_y^2\, P_{c\ell\ell'}$, which involves the square of the local populations (with local angular momentum ℓ or ℓ') on the initial core hole site. The Auger lineshape which involves a sum over the Auger transitions,

$$A(E) = \sum_{\ell\ell} \Big(\sum_{xy} c^2_{x\ell} c^2_{y\ell'}\Big) P_{c\ell\ell'} \quad , \tag{2.8}$$

is then closely approximated by a fold over the DOS (2.1) on the core hole site (site specificity), albeit modulated by the different atomic Auger matrix elements $P_{c\ell\ell'}$, (e.g., P_{css}, P_{csp}, or P_{cpp}; s or p giving the local angular momentum).

Equations (2.2-8) describe the Auger lineshape, but several basic questions remain. What electron density is actually probed by the Auger process? Are the $c^2_{x\ell}$ in (2.8) best given by the local or Mulliken populations? Does the Auger lineshape reflect an initial screened core hole DOS or the final DOS? Are the final-state holes localized on an atom, a molecular cluster, or throughout the solid (i.e., are the DOS probed atomlike, moleculelike, or bandlike)? Finally, what is the best way to determine the Auger matrix elements $P_{c\ell\ell'}$? These questions will be examined in Sects.2.2-4. A review of the outlook for AES will be given in Sect.2.5.

2.2.1 Atomic Auger Matrix Elements

Atomic Auger matrix elements have been calculated for much of the periodic table within a one-electron Hartree-Fock-Slater approximation by McGUIRE [2.19,20] and WALTERS and BHALLA [2.21,22]. Their results are compared with experimental data in Figs.2.2,3 for low atomic number [2.23]. Figure 2.2 compares results for the KLL transitions, Fig.2.3 for L_{23}MM. The matrix element per filled shell, $P_{c\ell\ell'}$ (i.e., s^2, p^6, d^{10}, etc.) is plotted; $P_{c\ell\ell'}$ is defined by the expressions [2.28]

$$P_{c\ell\ell} = \frac{(4\ell + 2)(4\ell + 1)}{(4\ell + 2 - n)(4\ell + 1 - n)} \times (P_{c\ell\ell})_{act.} \tag{2.9}$$

$$P_{c\ell\ell'} = \frac{(4\ell + 2)}{(4\ell + 2 - n)} \times (P_{c\ell\ell'})_{act.} \tag{2.10}$$

where n is the number of holes in the ℓ shell and $(P_{c\ell\ell'})_{act.}$ is the actual Auger matrix element or experimental intensity (it is assumed the ℓ' shell in (2.10) is filled). Since generally one is only interested in the relative

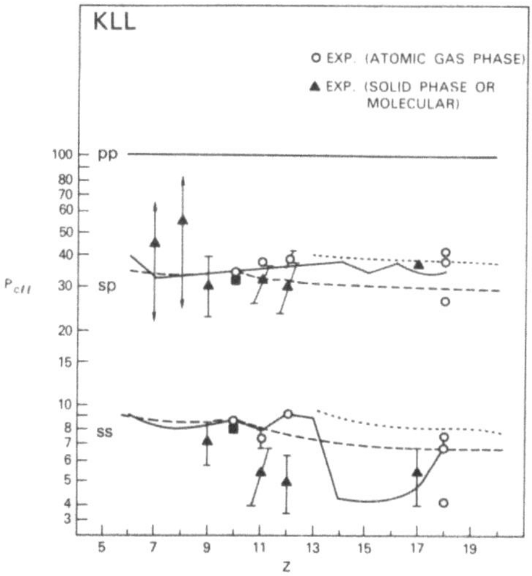

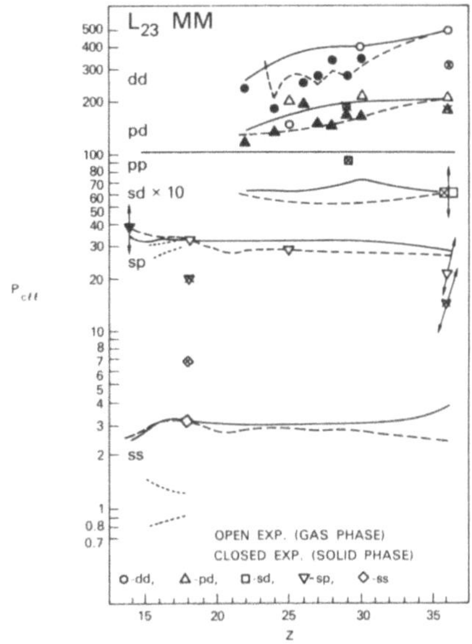

Fig.2.2↑ Fig.2.3→

Fig.2.2. Comparison of the KLL theoretical and experimental atomic Auger matrix elements $P_{c\ell\ell'}$ for atoms in the first two rows of the periodic table as given in [2.23]. Z is the nuclear charge, $P_{c\ell\ell'}$ is the matrix element per filled shell as indicated by (2.9,10). $P_{c\ell\ell'}$ has been normalized for each atom such that P_{cpp} is 100. The theoretical data from [2.19] (——) have been scaled by factors $f_{ss} = 0.65$ and $f_{sp} = 0.81$, that of [2.21] (---) by factors $f_{ss} = 0.59$ and $f_{sp} = 0.78$ to correspond to the experimental data for Ne (Z = 10). The data from [2.24] (...) are unscaled. Sources for the experimental points are given in [2.23]. Results of a one-electron hole calculation for Ne [2.25] are indicated by the solid squares. The brackets indicate stated uncertainties, and arrows indefinite estimated uncertainties in experimental data

Fig.2.3. Comparison of the $L_{23}VV$ theoretical and experimental atomic Auger matrix elements $P_{c\ell\ell'}$ for atoms in the 2nd and 3rd rows of the periodic table as given in [2.23]. $P_{c\ell\ell'}$ is defined and normalized as in Fig.2.1. The theoretical data from [2.20] (——) have been scaled by factors $f_{ss} = 1.97$, $f_{sp} = 1.23$, $f_{sd} = 1.33$, $f_{pd} = 1.27$, and $f_{dd} = 2.06$, that from [2.22] (---) by $f_{ss} = 1.94$, $f_{sp} = 1.03$, $f_{sd} = 1.32$, $f_{pd} = 1.24$, and $f_{dd} = 2.05$, to correspond to the Ar (Z = 18) ss and sp and the Kr(Z = 36) sd, pd, and dd experimental data. The data from [2.26] (···) is unscaled. Sources of the experimental data may be obtained from [2.23]. The crossed symbols for Z = 18 and 36 refer to earlier experimental data, for Z = 29 to a CI calculation for Cu [2.27]

$P_{c\ell\ell'}$ intensities, they have been normalized such that P_{cpp} is 100 for all Z. The theoretical results of McGUIRE and WALTERS and BHALLA have been scaled to the experimental results of the inert gases; this required scale factors (ranging from 0.59 to 2.06) significantly different from 1. These scale factors reflect the magnitude of the errors in the one-electron results.

Electron correlation effects at the atomic level are clearly very important. The configuration interaction (CI) results of CHEN and CRASEMANN [2.24,26] include electron correlation; their unscaled results agree closely with the experimental.

Figures 2.2,3 reveal three very important points. 1) The variation of the relative Auger matrix elements with atomic number is rather small. 2) What variation does exist is predicted remarkably well by the scaled one-electron results. This indicates the large correlation error is rather constant with Z. 3) Both atomic gas phase, molecular phase, and solid phase experimental data are given; no systematic differences between these data exist. This is true even for the dd, pd, and pp matrix elements of the transition metals where electron screening in the solid (Sect.2.4) is expected to cause significant changes. Recently VAYRYNEN [2.29] suggested that the relative atomic matrix elements for solid Mn are very different from those in the gas phase. However, VAYRYNEN's results are inconclusive because of an inadequate background removal procedure in the solid phase. Clearly, more careful work needs to be done to check this important question. For the present, the results of Figs.2.2,3 suggest that gas phase atomic data or the *scaled* one-electron results can be utilized for interpretation of Auger lineshapes in the solid.

2.2.2 Local or Mulliken Populations

The question of whether the Auger process probes the local or Mulliken populations arises from our use of (2.8). Use of (2.4) would automatically weight the DOS in the proper manner. However this question is of general chemical interest in any event. Does the Auger process in any way sample the bonding charge? The local and Mulliken populations are defined by the MO overlap equation,

$$\sum_{nm} c_n c_m S_{nm} = \sum_n \left(c_n^2 + \sum_{m \neq n} c_n c_m S_{nm} \right) = 1 \quad , \tag{2.11}$$

where $S_{nm} = \langle f_n | f_m \rangle$ is the atomic overlap. The Mulliken populations include the bonding charge such that

$$P_n^M = c_n^2 + \frac{1}{2} \sum_{m \neq n} c_n c_m S_{nm} \quad , \tag{2.12}$$

and satisfy the sum rule

$$\sum_n P_n^M = 1 \quad . \tag{2.13}$$

The local populations include only the diagonal terms $P_n^L = c_n^2$.

An analysis of the Auger matrix element by HUTSON et al. [2.30] gives for a two-atom system with atoms a and b (i.e., $X = a\,f_a + bf_b$) with core hole site 'a' the result,

$$I^{Ex} = a^4 M_{caa}^2 + a^3 b(M_{cab} + M_{cba})M_{caa} + \dots \quad . \tag{2.14}$$

It was assumed that $M_{abb} < M_{cab}, M_{cba} < M_{caa}$ since interatomic Auger matrix elements $M_{cab} = \langle\varphi_c\varphi_k|r_{12}^{-1}|f_a f_b\rangle$ are generally much smaller than intraatomic elements, $M_{caa} = \langle\varphi_c\varphi_k|r_{12}^{-1}|f_a f_a\rangle$ [2.31]. The use of local populations,

$$I^L = (P_a^L)^2 M_{caa}^2 = a^4 M_{caa}^2 \quad , \tag{2.15}$$

is valid provided all interatomic elements are indeed negligible. The use of Mulliken populations gives the result

$$I^M = (P_a^M)^2 M_{caa}^2 = (a^2 + abS)^2 M_{caa}^2 \quad , \tag{2.16}$$

$$I^M = a^4 M_{caa}^2 + 2a^3 bSM_{caa}^2 \quad . \tag{2.17}$$

The second terms in (2.14,17) are equivalent if $(M_{cab} + M_{cba})/M_{caa} = 2S$. A semiempirical analysis by MATTHEWS and KOMNINOS [2.31] indicated that $(M_{cab} + M_{cba})/M_{caa} \sim 10^{-2}$. Typical two-center overlaps $S = \langle f_a|f_b\rangle \sim 10^{-1}$, indicating the second terms in general are not equivalent.

The above analysis suggests the result $I^L < I^{Ex} < I^M$ at the bottom of a band when $S > 0$; $I^L > I^{Ex} > I^M$ at the top of a band when $S < 0$. However, one expects some orbital contraction (expansion) at the bottom (top) of a band relative to the free atom. This would have the effect of moving I^{Ex} in the direction of I^M and away from I^L, but no clear answer as to whether I^M or I^L is closer to I^{Ex} comes from this analysis.

Comparison with experiment is equally inconclusive. JENNISON [2.32] first noted for the Si $L_{23}VV$ lineshape of elemental Si that $I^L \approx I^{Exp} \neq I^M$. However, for the Si $L_{23}VV$ lineshape in Pd_4Si [2.33] and at the Pd/Si interface [2.34], it was found that $I^M \approx I^{Exp}$. For the N KVV lineshape of the NO_3^- oxyanion [2.30], a similar result was observed (i.e., $I^M \approx I^L \approx I^{Exp}$). Obviously more work is necessary here. At present it seems reasonable to utilize the more convenient Mulliken populations, particularly in a multi-band lineshape when the bonding charge does not dominate in any region of the spectrum (e.g., such as in the N KVV of NO_3^-). In a single-band lineshape where the bonding charge dominates some part of the spectrum, the local populations may be preferred (e.g., such as in the Si $L_{23}VV$ of elemental Si [2.32]).

2.3 Localization

In the CINI-SAWATZKY [2.35] model of the Auger process in elemental solids, two parameters determine the degree of localization of the CVV two-hole final state. If the effective Coulomb repulsion U^e_{VV} is large compared to the bandwidth ($U^e_{VV} > \Gamma$), the lineshape will be atomiclike, if $\Gamma > U^e_{VV}$ the lineshape will be bandlike. In systems where $U^e_{VV} \approx \Gamma$ both atomic- and bandlike contributions are evident in the lineshape (i.e., correlation effects are important).

The results of CINI and SAWATZKY were obtained from utilizing the Anderson and Hubbard many-body models. The equations which enter are solvable only for the initially filled band or two-hole final state. Some work has been reported for the initially unfilled band [2.36], but the initial conclusions of CINI and SAWATZKY do not appear to be substantialy altered.

The Cini-Sawatzky results can be simply understood by considering a cluster LCAO-MO-CI approach [2.18]. For the moment consider a simple two-orbital system which has two holes present resulting from the Auger process in an initially filled state [2.18,37]. The holes can be described by the one-electron atomic orbitals f_a^2 and f_b^2 with binding energy ε giving the Hamiltonian matrix as follows:

$$\begin{array}{c|ccc|}
 & f_a^2 & f_b^2 & \frac{1}{\sqrt{2}}(f_af_b + f_bf_a) \\
\hline
f_a^2 & 2\varepsilon + U_{aa} & 0 & H_{ab} \\
f_b^2 & 0 & 2\varepsilon + U_{aa} & H_{ab} \\
\frac{1}{\sqrt{2}}(f_af_b + f_bf_a) & H_{ab} & H_{ab} & 2\varepsilon + U_{ab}
\end{array} \quad , \qquad (2.18)$$

where $U_{aa} = \langle f_a^2|r_{12}^{-1}|f_a^2\rangle$, $U_{ab} = \langle f_af_b|r_{12}^{-1}|f_af_b\rangle$, and $H_{ab} = \sqrt{2}\langle f_a|H|f_b\rangle$ ($H_{ab} \propto \Gamma$ in the solid). Clearly if $H_{ab} < U_{aa} - U_{ab}$, very little mixing occurs and the hole states f_a and f_b properly describe the localization of the two holes. In this instance the Auger lineshape is atomic like probing only the two-electron eigenstate f_a^2 (core hole site). If $H_{ab} > U_{aa} - U_{ab}$, the mixing of the configurations is complete and the linear combinations $f_a \pm f_b$ properly describe the localization of the holes. As such, the lineshape is molecular-like (bandlike for the solid) sampling the eigenstates $1/2\,(f_a + f_b)^2$, $1/\sqrt{8}\,\mathscr{A}(f_a + f_b)(f_a - f_b)$, and $1/2\,(f_a - f_b)^2$, and giving relative Auger intensities 1/4, 1/2, and 1/4 respectively. If $H_{ab} \approx U_{aa} - U_{ab}$, of course intermediate mixing occurs giving both contributions.

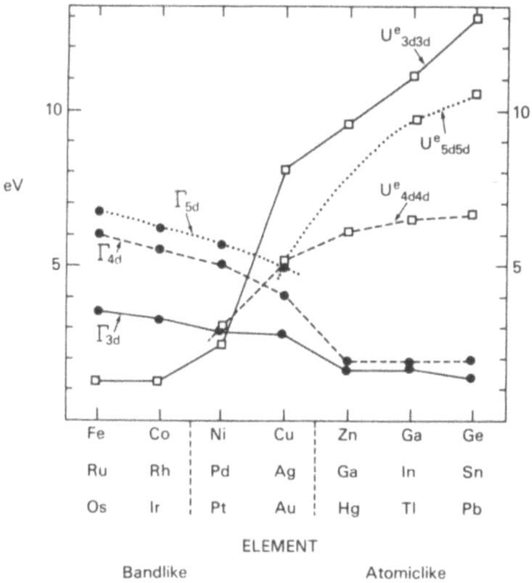

Fig.2.4↑

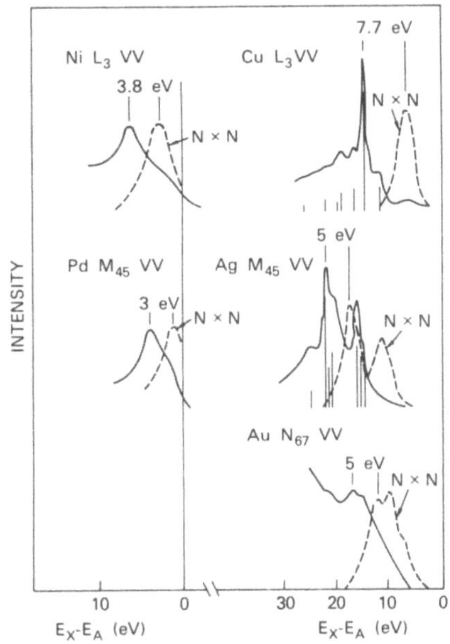

Fig.2.5→

Fig.2.4. Comparison of Γ_{nd} with U^e_{ndnd} for the metals with electron configuration d^6s^2 to $d^{10}s^2p^2$. The 3d and 4d data are from [2.38-41], U^e_{5d5d} from [2.40,42], Γ_{5d} calculated or obtained from [2.40,43]. Metals with bandlike and atomiclike lineshapes are indicated, those between the dashed lines are "transitional"

Fig.2.5. Comparison of the core-(valence d)-(valence d) Auger lineshapes for the 'transitional' metals as indicated in Fig.2.4 with the fold of the one-electron DOS, N(E). Data for Cu, Ag, and Au from [2.40], for Ni and Pd from [2.9]. The vertical bars for Cu and Ag indicate results from an atomic calculation [2.44,45]. E_X is the core binding energy, E_A the Auger electron energy

Figure 2.4 compares the bandwidth Γ_{nd} with U^e_{ndnd} for the metals with electron configurations d^6s^2 to $d^{10}s^2p^2$. For all three rows of this series the Γ and U^e plots cross in the region d^8s^2 to $d^{10}s$. The Auger lineshapes of these "transitional" metals are given in Fig.2.5 and compared with a fold of the DOS, N × N [2.9,40], and in some instances the calculated atomiclike Auger lineshape. The lineshape for Cu is clearly atomiclike [2.38,39,44,46]; that for Ag is also, although some controversy has occurred in the literature concerning this question [2.45,47-50]. The lineshape for Au appears to be bandlike although shifted down in energy by 5 eV [2.40,51]. The Ni [2.9,52] and Pd [2.9,53] lineshapes show atomiclike and bandlike contributions separated by 3 to 4 eV, but no multiplet structure is evident. These lineshapes indicate the rather smooth transition from atomic- to bandlike character.

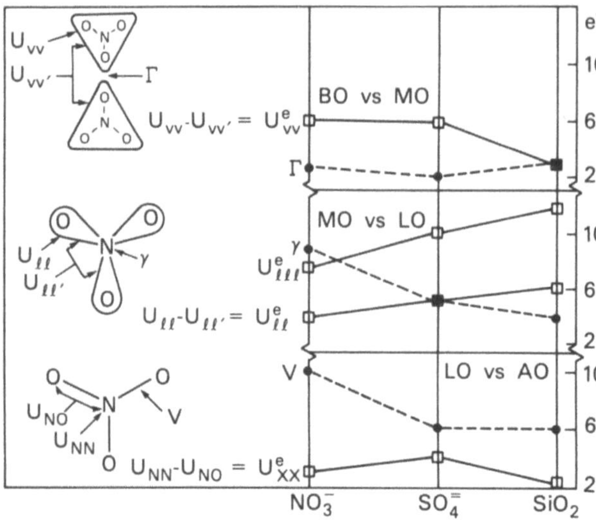

Fig.2.6. Comparison of the effective hole-hole repulsion, U^e_{XX}, $U^e_{\ell\ell}$, or U^e_{VV}, with the corresponding interaction parameter, V, γ, or Γ for NO_3^-, $SO_4^=$, and SiO_2 [2.37]. The definition of each parameter is illustrated on the left for NO_3^-. AO, LO, MO refer to atomic, lobe, or molecular orbitals, also illustrated. $U^e_{\ell\ell\ell}$ is the effective 3-hole lobe orbital repulsion energy [2.37]

The d^2s^2 to d^7s^2 metals are clearly bandlike [2.54,55], the $d^{10}s^2$ to $d^{10}s^2p^2$ metals atomiclike [2.42].

Recently DUNLAP et al. [2.37] showed that intermediate levels of localization can exist, i.e., localization on some subcluster of the system. The systems studied were NO_3^-, $SO_4^=$, and SiO_2. The pertinent parameters are schematically illustrated in Fig.2.6. The degree of localization can best be summarized as follows:

$$\begin{array}{lll}
V < U_{XX} - U_{XO} = U^e_{XX} & & \varphi \rightarrow AO \\
V > U^e_{XX} & \gamma < U_{\ell\ell} - U_{\ell\ell'} = U^e_{\ell\ell} & \varphi \rightarrow LO \\
\gamma > U^e_{\ell\ell} & \Gamma < U_{VV} - U_{VV'} = U^e_{VV} & \varphi \rightarrow MO \\
\Gamma > U^e_{VV} & & \varphi \rightarrow BO
\end{array} \qquad (2.19)$$

where AO, LO, MO, and BO refer to atomic, lobelike, molecular, and bandlike orbitals respectively. The results plotted in Fig.2.6 indicate that the two-hole final state is best represented by NO_3^- and $SO_4^=$ MO's, and a Si-O-Si LO.

Also plotted in Fig.2.6 is $U^e_{\ell\ell\ell'}$, the effective lobe orbital three-hole repulsion energy appropriate for the final state resulting from the shake/Auger process as discussed in Sect.2.1. The three-hole final state is best represented by a NO_3^- MO and S-O and Si-O-Si LO's.

The experimental Auger lineshapes verify these conclusions [2.37]. Figure 2.7 compares the experimental N KVV lineshape for NO_3^- with that predicted theoretically from a semiempirically derived one-electron NO_3^- M.O. DOS [2.30]. The energy separation between the principle peak of the parent and shake/Auger contributions (ΔE, see Sect.2.4) is predicted to be -6 eV utilizing a NO_3^- MO for both the two- and three-hole final states. Experimentally the best fit is obtained with ΔE = -5 eV. Figure 2.8 compares the experimental and theoretical S $L_{23}VV$ lineshapes for $SO_4^=$ [2.37,56], again utilizing a $SO_4^=$ MO DOS to determine the theoretical lineshape. ΔE is predicted to be -8 or -2 eV depending on whether one uses the MO or LO orbitals for the three-hole final state. Experimentally the best fit is obtained with $\Delta E \approx$ -2 eV verifying the appropriateness of the LO three-hole final state of $SO_4^=$.

Figure 2.9 compares the experimental and theoretical (without the shake/Auger contributions in this case) O KVV and Si $L_{23}VV$ lineshapes in SiO_2 [2.14,37,57]. The theoretical Si $L_{23}VV$ lineshape was first determined from a SiO_4^{4-} MO giving the difference curve in Fig.2.9d [2.14,57]. The peaks centered at 55 and 68 eV were attributed to shake/Auger satellites, the peak at 84 eV to beam damage. Later electron correlation was included via a CI calculation allowing for localization on the Si-O-Si LO's [2.18]. These results are indicated by the bars in Fig.2.9c, the dotted bars the Si-O-Si localized contributions, the solid bars the completely delocalized (bandlike) contributions. Notice the localized contributions now account for the intensity originally attributed to the shake/Auger contributions. The shake/Auger satellites may in fact not contribute because the shake hole may not remain localized to the core hole. Recently JENNISON [2.58] suggested electron screening of the final-state holes is important in decreasing $U^e_{\ell\ell}$, thus aiding hole delocalization. However, screening will decrease U^e_{cv} even more than $U^e_{\ell\ell}$ so that localized contributions may still contribute some measure of intensity to the lineshape, but shake/Auger satellites may not.

The O KVV lineshape in SiO_2 was theoretically evaluated assuming initially a totally localized Si-O-Si final state (Fig.2.9a). The difference peaks around 475 and 495 eV have been attributed to shake/Auger satellites, the peak at 510 eV to delocalized contributions dominated by the O 2p nonbonding orbitals. Results of an LCAO-MO-CI calculation on a 16-O-atom cluster in a tetrahedral SiO_2 lattice utilizing a parametrized Hamiltonian [2.18] is indicated

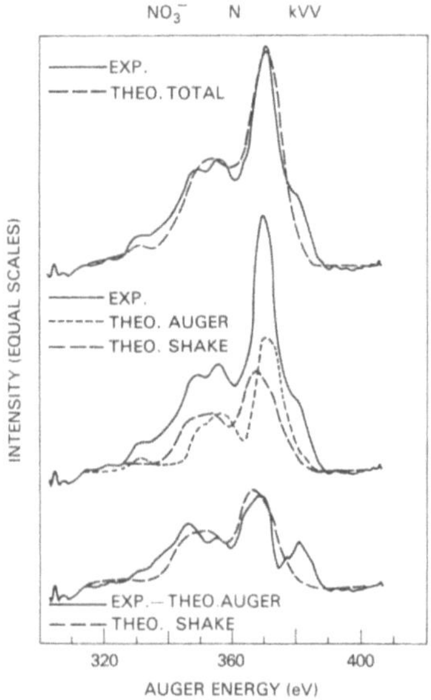

← Fig.2.7

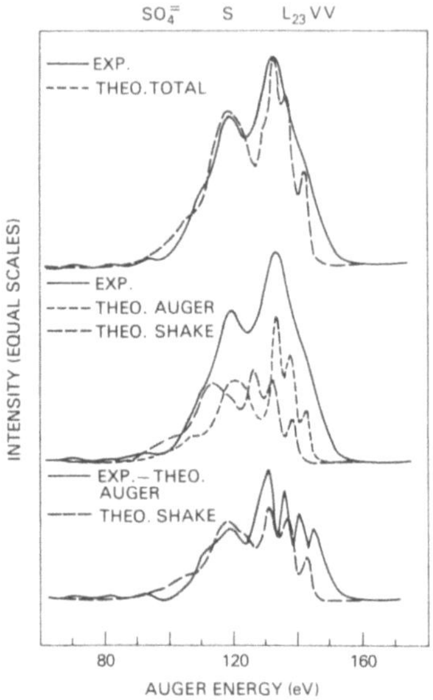

Fig.2.8 →

Fig.2.7. (*Top*) Comparison of the experimental and total theoretical N KVV lineshapes for $NaNO_3$ as obtained utilizing a two- and three-hole final state described by a NO_3^- MO [2.30]. (*Middle*) The parent Auger and shake/Auger contributions compared to the experimental lineshape. (*Bottom*) Comparison of the theoretical shake lineshape with the difference between the experimental and theoretical parent Auger lineshapes

Fig.2.8. Same as Fig.2.7 for Li_2SO_4 except the ΔE between the shake/Auger and parent Auger lineshapes, (2.22), was reduced to -2 eV indicating the preferred 2-hole $SO_4^=$ MO and 3-hole S-O LO final-state description. Data from [2.37,56]

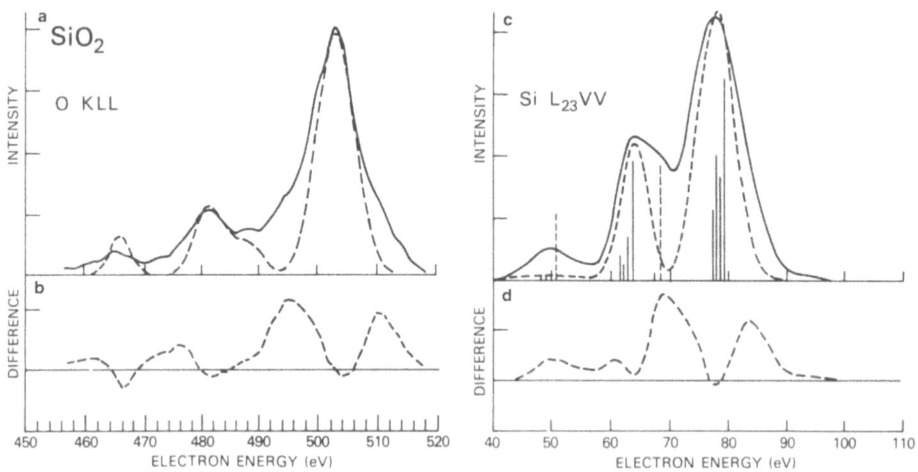

Fig.2.9 (caption see opposite page)

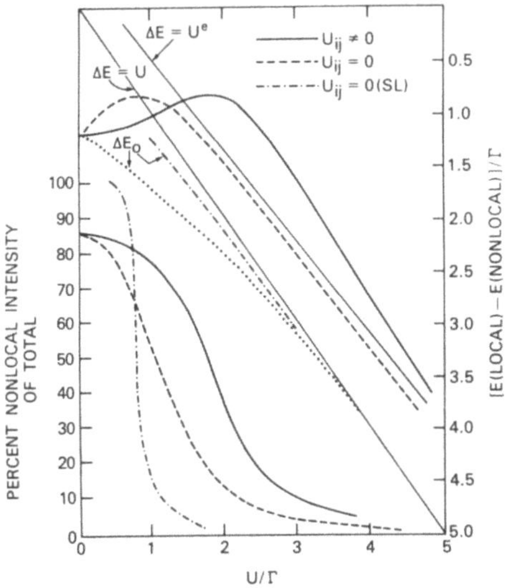

Fig.2.10. (*Lower left*) the percent nonlocal intensity of the total vs U/Γ. U is the one-center hole-hole repulsion, Γ the bandwidth, and $U^e \approx U - U_{ij}$ is an effective Coulomb repulsion. (*Upper right*) the difference in energy between the centroids of the local and nonlocal Auger contributions. ΔE_0 is the shift in energy of the local contribution from the initial band center. The diagonals $E = U$ and U^e are also indicated. The results obtained from a 16-atom cluster in a tetrahedral lattice [2.18] with $U_{ij} = 0$ and $U_{ij} \neq 0$ (U_{ij} is the two-center Coulomb repulsion energy) are compared with the results of [2.59] for an infinite simple cubic lattice with $U_{ij} = 0$

in Fig.2.10. It illustrates the transition from localized to nonlocalized lineshape behavior as a function of U/Γ. A 20% nonlocal intensity separated by ≈7 eV from the local peak as indicated by the O KVV lineshape (Fig.2.9b) gives a U/Γ of 2.5 (Fig.2.10, $U_{ij} \neq 0$). Since the oxygen-atom hole-hole repulsion is 10-15 eV, this corresponds to an oxygen nonbonding bandwidth of 4 to 6 eV in SiO_2, in reasonable agreement with theory and experiment [2.18].

Figure 2.10 clearly shows a point which needs further emphasis here. The transition from the bandlike to the atomiclike lineshape is relatively sharp and occurs at $U/\Gamma \approx 1$, not at $U/\Gamma = 2$, where Γ = the *full* bandwidth. This is in agreement with the results of SAWATZKY and LENSELINK [2.59] utilizing the Hubbard model for an infinite system. There is much confusion in the literature on this point (note Fig.2.4 plots Γ, others have plotted 2Γ [2.5,9,10, 38]), probably because in the original paper CINI [2.35] apparently defined W

Fig.2.9. (a) Comparison of the experimental (——) and theoretical (---) O KLL lineshapes for SiO_2 as obtained utilizing a Si-O-Si LO [2.14,37,57]. (b) The difference between the experimental and theoretical lineshapes. The peaks at 475 and 495 eV have been attributed to shake/Auger satellites, the peak at 510 eV to delocalization [2.18]. (c) Comparison of the experimental (——) and theoretical (---) Si $L_{23}VV$ lineshapes as obtained utilizing a SiO_4^{4-} MO [2.14,37,57]. The vertical bars indicate results of a cluster CI calculation [2.18], the dotted bars denoting Si-O-Si localized holes, the solid bars delocalized holes. (d) The difference between the experimental and theoretical lineshapes. The peaks at 55 and 68 eV may be attributed to the localized hole contributions, that at 84 eV to beam damage [2.18] to be the *half* bandwidth giving the criterion $U > 2W$. SAWATZKY defined W to be the full bandwidth, but still used the criterion $U > 2W$ because his initial paper [2.35] utilized a simplifying approximation to his exact solution valid only for $U > 2W$. The point of steepest slope in the transition occurs at $U/\Gamma = 1.0 \pm 0.2$ [2.59] depending on the crystal structure. The transition becomes less sharp and occurs at different values of U/Γ for smaller systems (e.g., $U/\Gamma \approx 1.2$ for diatomic molecules [2.18,37]).

Another important point concerns the energy separation ΔE between the two contributions. Figure 2.10 shows that the shift ΔE_0 of the atomiclike peak from the zero-point energy (middle of band when $U = 0$) is greater than U in agreement with SAWATZKY and LENSELINK [2.59]. However, the bandlike portion of the spectrum is also skewed to lower energies with increasing U/Γ. Figure 2.10 indicates the shift ΔE between the *centroids* of the two contributions is less than U. It shows results for two cases, assuming two center integrals $U_{ij} = 0$ and $U_{ij} \neq 0$. The calculation of SAWATZKY and LENSELINK within the Hubbard model makes the necessary approximation $U_{ij} = 0$. The effect of $U_{ij} \neq 0$ is to reduce the effective value of U (i.e., $U^e \approx U - U_{ij}$) as indicated by the asymptotic approach of ΔE to U^e and the transition at $U/\Gamma \approx 1.7$ (equivalent to $U^e/\Gamma \approx 1$) in this instance. Thus as shown, the separation between the two contributions slightly *underestimates* U^e and significantly underestimates U.

2.4 Screening

Electron screening of the core hole alters the Auger lineshape in two ways: 1) it introduces shake/Auger satellites as previously discussed, and 2) it changes the parent Auger contribution [2.60]. Consider the core ionization process initiating the Auger process. Its rate r_{ion} is given by the expression [2.61](ignoring CI effects),

$$r_{ion} = 2\pi/\hbar \sum_m |\langle\varphi_c \Psi^G|\hat{T}|\varepsilon_\ell \Psi^i_m\rangle|^2 \tag{2.20}$$

$$\approx 2\pi/\hbar \langle\varphi_c|\hat{T}|\varepsilon_\ell\rangle^2 \langle S_n^2 + \sum_{m=SU} S_m^2 + \sum_{m=SO} S_m^2\rangle \quad ,$$

where $\varphi_c \Psi^G$ is the ground state before the ionization process, $\varepsilon_\ell \Psi^i_m$ is the initial Auger state relaxed in the presence of the core hole, $\hat{T}$ is the appropriate ionization operator, $S_m = \langle\Psi^G|\Psi^i_m\rangle$, and SU and SO denote shake-up and shake-off contributions respectively. The shake-off contributions generally result from atomic relaxation of the core hole site [2.62]; they produce the Auger satellites with lineshape characteristics [2.30,57],

$$I_{cs\text{-}svv'} = S_s^2\left(1 - \sum_s S_s^2\right)^{-1} W_s I_{cvv'} \tag{2.21}$$

$$\Delta E_{cs\text{-}svv'} = U_{cs}^e - U_{sv}^e - U_{sv'}^e \tag{2.22}$$

$$\Gamma_{cs\text{-}svv'} \approx \Gamma_{cvv'} \quad . \tag{2.23}$$

In (2.21-23) s represents the shake electron, I_{cvv} ($\Gamma_{cvv'}$) the parent Auger intensity (width), W_s a statistical weight [2.57], and U_{cs} and U_{sv} the Coulomb repulsion between the initial core-shake and final valence-shake holes respectively. S_s can be determined from the sudden approximation [2.57],

$$S_s^2 = \{1 - [1 - c_{sa}^2 + c_{sa}^2(1 - P_a)^{1/2n}]^{2m}\} \quad , \tag{2.24}$$

where P_a is the atomic shake-off probability [2.63] and c_{sa}^2 is the population of the s MO on the core hole site. In (2.20), the shake-up contributions normally result from charge transfer to the core hole site [2.62]. They generally cause an Auger satellite of smaller intensity and energy shift and thus may be included theoretically with the parent contribution.

Shake/Auger satellites of up to 30-40% of the total intensity are clearly resolved in the gas phase spectra of the inert gases [2.64,65] and in molecules [2.66-68]. In the solid phase, they have been found to significantly alter the experimental Auger lineshape of some metals [2.38,44,69] and insulators [2.70] (see Figs.2.7-9). For quantitative Auger lineshape interpretation, they should not be ignored.

An understanding of the effect of screening on the parent Auger transitions can be gained from the final state (FS) rule. A FS rule was first proposed by von BARTH and GROSSMAN [2.71] and MAHAN [2.72] for the X-ray and photoemission one-electron processes. Later DAVIS and FELDKAMP [2.73] proposed an "orthogonalized" FS (OFS) rule for the same processes. Recently the FS and OFS rules have been extended to the two-electron Auger process [2.61].

Basically the FS and OFS rules are reductions of the many-electron matrix element $\langle\Psi^i|r_{12}^{-1}|\Psi^f\rangle$ (for a single-band system) to different two-electron integrals [2.61],

$$\text{FS:}\ I_{ij} = 2\pi/\hbar|\langle 1s\overline{k\ell}|r_{12}^{-1}|\varphi_i\bar{\varphi}_j\rangle|^2 + O[S_{if}^2 S_{jf}^2(\bar{S}'\rho_0' + \rho_4')^4] \tag{2.25}$$

$$\text{OFS:}\ I_{ij} = 2\pi/\hbar|\langle 1s\overline{k\ell}|r_{12}^{-1}|\tilde{\varphi}_i\tilde{\bar{\varphi}}_j\rangle|^2 + O[S_{if}^2 S_{jf}^2(\bar{S}'\rho_0')^4] \tag{2.26}$$

where φ_i and $\bar{\varphi}_j$ describe the *final* valence holes created by the Auger process (i.e., potential with no core hole), and $\tilde{\varphi}_i$ and $\tilde{\bar{\varphi}}_j$ are final-state orbitals orthogonalized to the existing initial-state valence holes,

$$\tilde{\varphi}_i = \varphi_i - \sum_{\varepsilon_m > \varepsilon_f} S_{mi}\varphi'_m \tag{2.27}$$

(φ'_m are initial orbitals relaxed in the presence of the core hole). The error terms contain $S_{nf} \propto 1/(\varepsilon_n - \varepsilon_f)$ [ε_f is the Fermi energy, $\varepsilon_n = \varepsilon_i$ or ε_j, the valence binding energy of either Auger hole] indicating the increased accuracy of the FS and OFS rules away from the Fermi level. $\bar{S}'$ and $\rho'_0(\rho'_4)$ denote an effective average overlap S_{mi} and the integrated occupied (unoccupied) LDOS respectively. In general, $\bar{S}'\rho'_0 < \rho'_4$ indicating the increased accuracy of the OFS rule near the Fermi level. In the absence of configuration mixing (final-state localization) basic sum rules [2.74] also indicate that the intensity of the individual angular momentum contributions are determined by the initial state. Thus the intensities of the FS and OFS rules are normalized to that given by the initial state (e.g., $I^{\ell\ell'}_{OFS} = I^{\ell\ell'}_{FS} = I^{\ell\ell'}_{initial}$, $\ell\ell'$ = ss, sp, pp, etc.).

The results of the FS and OFS rules can be summarized in three statements: 1) the shape of the individual $\ell\ell'$ Auger contributions are determined by the final DOS, 2) the relative $\ell\ell'$ Auger intensities are determined by the initial state, and 3) the OFS rule is significantly different from the FS rule only near the Fermi level. As stated, the FS rule for the CVV bandlike Auger lineshapes and, as previously stated, for X-ray and photoemission [2.71-73], indicate that XES, UPS, and AES probe an unscreened DOS. Thus they complement each other in electronic structure determinations. The limitation to CVV bandlike Auger lineshapes must be emphasized. A CCV Auger lineshape reflects a *screened* DOS because of the *final*-state core hole [2.75]. An atomic (localized) Auger lineshape does not reflect the bandlike DOS in any case. The Auger energy may however reflect a screened valence hole (e.g., the valence s electrons screen the localized 3d holes in the Cu $L_{23}M_{45}M_{45}$ Auger final state [2.9,76]).

Several applications of the FS rule to experimental data can be given illustrating its validity. Table 2.3 contains data for the O_2 O KVV [2.15], the C_2H_6 C KVV [2.77], and the NO_3 N KVV [2.30,61] lineshapes. It compares relative intensities of various transitions as determined from the initial state (relaxed in the presence of the core hole), from the final state (no core hole), and from either experiment or a more complete theory. (Figure 6.11 also compares these data for NO_3^- [2.30,61]). Since the transitions

Table 2.3. Comparison of actual Auger intensities with those obtained from initial- and final-state populations

O_2 O KVV [2.15]	$\pi_u^{-1}2\sigma_u^{-1}$	$\pi_u^{-1}(\pi_u,3\sigma_g)^{-1}$	$\pi_g^{-1}(\pi_u,3\sigma_g)^{-1}$
X_α initial state	2.0	6	1.6
X_α final state	1.9	6	2.8
Exp.	2.3	6	4
C_2H_6 C KVV [2.77]	$1e_u^{-2}$	$1e_u^{-1}1e_g^{-1}$	$1e_g^{-2}$
RHF-SCF initial state	68	16	1
RHF-SCF final state	0.9	1.9	1
Full theo. matrix element	1.2	3.0	1
NO_3 N KVV [2.30,61]	$2e'^{-2}$	$2e'^{-1}(4e',3e',1a_2'')^{-1}$	$(4e',3e',1a_2'')^{-2}$
X_α initial state	1	7.4	13
X_α final state	1	5	2.5
Empirical fit to data	$\approx$ 1	$\approx$ 5	$\approx$ 2.5

listed in Table 2.3 are dominated by pp contributions, the better agreement with the final-state results indicates the validity of statement 1 above. Statement 2 is best illustrated by examing the Be KVV [2.78], Fig.2.12, and Cu M_1VV [2.79] Auger lineshapes, Fig.2.13. The individual $\ell\ell'$ Auger lineshapes have been determined from the theoretical final DOS, and the relative $\ell\ell'$ intensities determined from the final and initial DOS respectively (assuming complete s electron screening). The s electron screening in metals has been indicated experimentally upon examing the CCV lineshapes in Si, Al, and Mg [2.75].

The utility of the OFS rule near the Fermi level (statement 3) has been illustrated for the intercalated graphite systems C_6Li and C_8Cs. Figure 2.14 compares the DOS as obtained from UPS and from an interpretation of the C KVV Auger lineshape [2.80]. The enhanced intercalant intensity in the Auger-derived DOS can be attributed to electron screening of the initial core hole site within the OFS rule. DUNLAP et al. [2.81] have compared the initial, final, and OFS rule transition ρ_π DOS as obtained from a 109-atom cluster using a parameterized SCF Hamiltonian. The OFS rule transition DOS is essentially the same as the final DOS, however with a much enhanced "intercalant peak", in qualitative agreement with the Auger derived DOS.

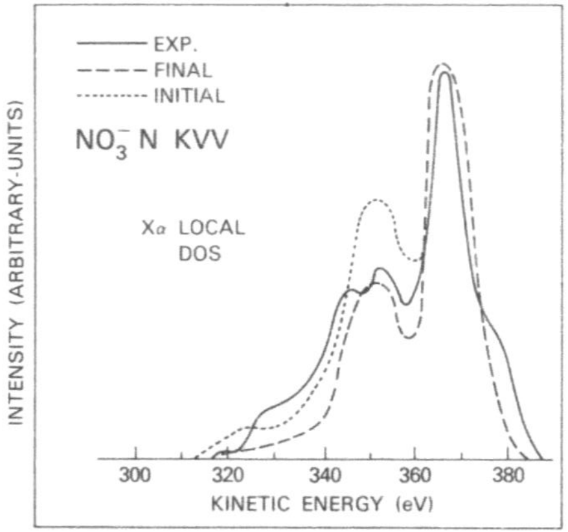

↑ Fig.2.11

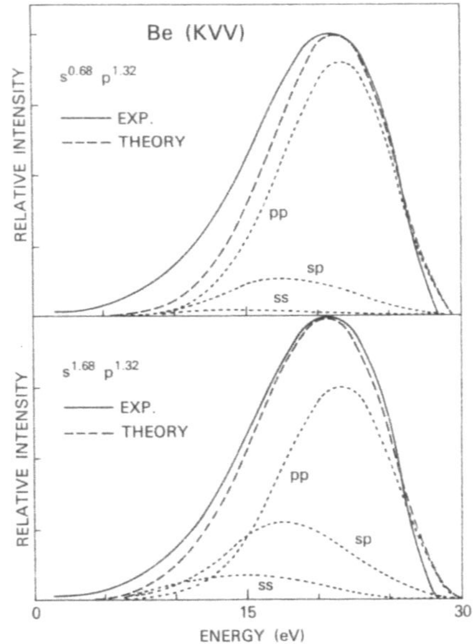

Fig.2.12 →

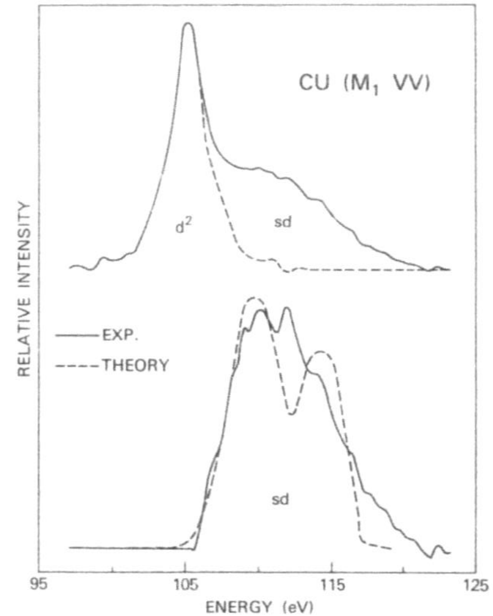

Fig.2.13 ↑

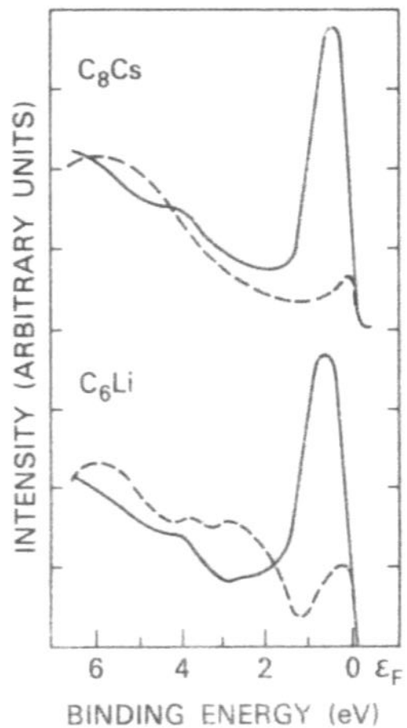

Fig.2.14 ↑

Fig.2.11. Comparison of the experimental N KVV lineshape for $LiNO_3$ with that calculated utilizing an X_α theoretical local DOS from the initial- (core hole) and final- (no core hole) state wave functions [2.30,61]

Fig.2.12. (*Top*) Comparison of the experimental Be KVV Auger lineshape with that calculated utilizing a HF-SCF local DOS from the final state [2.78]. (*Bottom*) Same as the top except the relative ss, sp, and pp areas are determined by the initial state. Complete s electron screening is assumed in the initial state [2.75]

Fig.2.13. (*Top*) The Cu M_1VV experimental Auger spectrum as reported in [2.79]. The narrow localized d^2 contribution is indicated by the dotted line, the difference represents the bandlike sd contribution. (*Bottom*) Comparison of the sd Auger contribution as obtained above (——) with the calculated Auger lineshape (---). The calculated lineshape is obtained from the final DOS but with the relative sd/dd area increased by a factor of 2
(continued on page 39)

consistent with the $d^{10}s^2$ initial-state configuration. Complete s electron screening is again assumed in the initial state [2.75,79]

Fig.2.14. Comparison of the DOS as determined in [2.80] from a quantiative interpretation of the C KVV Auger lineshape (——) with UPS data (---) for the intercalated graphite systems C_8Cs and C_6Li

2.5 Outlook

The outlook for AES as a probe of valence DOS information is obviously good. Three main areas are briefly reviewed here: 1) the utilization of AES in understanding the mechanisms involved in electron- and photon-stimulated desorption (ESD/PSD) in covalent systems, 2) the utility of AES in the gas phase and possibly in the future for chemisorbed systems, and 3) the utilization of AES in gaining valuable electronic structure information in solids and at interfaces.

2.5.1 ESD/PSD

The role of the Auger process for desorption from highly ionic systems is reasonably well understood within the VARLEY mechanism [2.82] as proposed by KNOTEK and FEIBELMAN [2.83]. Although the Auger process has been demonstrated to be of importance in desorption from (or bulk beam damage of) covalent systems, its role is not as clear [2.84-89]. Recently RAMAKER et al. [2.90] proposed and quantified a model for Auger-induced (or for any process creating two valence holes) desorption from covalent (or ionic) systems.

Consider damage to surfaces where the surface atom or cluster is taken to be covalently bonded to a similar neighboring atom or cluster. For definiteness, consider a surface atom S covalently bonded via bonding orbital b_0 to a single bulk atom B with bond energy D. The Auger process resulting from a core hole in S or B can leave the system initially in the repulsive state b_0^{-2}, with excitation energy $2E_b + U$, where E_b is the one-electron binding energy of b_0 and U is the Coulomb interaction energy resulting from two holes in b_0 (see Fig.2.15). Due to this repulsion, the atom S is pushed off the surface initially in state b_0^{-2}. If the holes decay before a critical distance R_c, S is recaptured; otherwise S gains sufficient kinetic energy to be desorbed. Beyond R_c, S can be neutralized leading to the desorption of neutrals (escape along b_n^{-2} where b_n is a bulk bond orbital). The maximum kinetic energy of the outgoing ions is e^2/R_0 (for escape along b_0^{-2}), although the maximum ion intensity will occur at smaller energies if, e.g., escape along $b_n^{-1}b_0^{-1}$ dominates. The most obvious means for decay of b_0^{-2} involves one hole

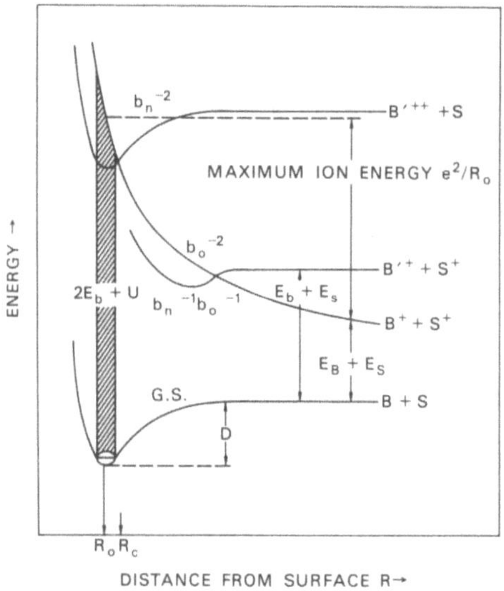

Fig.2.15. Schematic potential-energy curves [2.90] for interaction between a bulk atom B and a surface atom S, before (lower ground-state curve) and after the Auger process. Curves are labeled by the bond orbital hole states and by the separated state of the surface atom. B' denotes bulk atoms not bonded to S; E_b and E_s are the bulk and surface atom ionization potentials; E_B and U are the bond orbital binding energy and hole-hole repulsion energy

hopping into the bulk without appreciable nuclear motion. However this process is blocked if $U > V$ (V = the bond orbital interaction energy) as previously discussed in Sect.2.3. It may not become important until larger R values are reached (Fig.2.15) when $U < V$. Two-hole hopping off the surface bond orbital can occur immediately, but it generally occurs with smaller probability.

To quantify the lifetime aspect of the model [2.90], consider a Hubbard-like single-band Hamiltonian motivated through the bond orbital approximation. Assuming that $U = 0$ and that the bulk band is described by a Bethe lattice of half bandwidth Δ, the probability P of finding two holes, created in the surface bond orbital at time $t = 0$, both in the orbital at a later time t is given by $P \sim \exp(-4\alpha\Delta t/\hbar)$ if $\Delta t \lesssim \hbar$. The factor $\alpha(\sim 1/2)$ arises because we are considering a surface orbital. This expression must be replaced by $\beta(\hbar/\Delta t)^6$ when $\Delta t \gg \hbar$ where $\beta = 4\alpha^4/\pi^2(1-\alpha)^8$. If $U \gtrsim 2\Delta$, the two holes are essentially bound together in which case $P \sim \exp(-2\Delta'\alpha t/\hbar)$ so long as $\Delta' t \lesssim \hbar$ and $P_c \approx \beta^{1/2}(\hbar/\Delta' t)^3$ when $\Delta' t \gg \hbar$, where Δ' is the bound two-hole half bandwidth. Second-order perturbation theory can be used to estimate Δ'; this gives $\Delta' = 2\Delta^2/KU$, where K is the coordination number of the lattice.

The desorption or damage cross section σ_D can be estimated from the expression [2.90],

$$\sigma_D(\text{theor.}) = \sum_{\text{core}} \sigma_I(\text{core}) \quad f_D(\text{core}) \quad P(t_c) \quad , \tag{2.28}$$

which involves two other important factors besides $P(t_c)$ (t_c is the critical desorption time for the atom to move beyond R_c). σ_I (core) is a core ionization cross section, which may include a backscattering factor in the solid, and can be obtained either from theory or experiment [2.91]. f_D is the fraction of core ionization events which result in two holes localized in a bonding orbital via the Auger process. Its magnitude can generally be determined from the Auger lineshape and indicates the importance of AES to estimating ESD cross sections. Utilizing (2.28), the electron beam damage cross section of SiO_2 has been estimated to be 3×10^{-21} cm^2; experimentally it has been determined to be $\approx 2\times10^{-21}$ cm^2 [2.90]. The importance of hole localization under the condition $U > V$ can be realized from the result ($\sim 10^{-39}$ cm^2) obtained when one assumes $U = 0$. In SiO_2, f_D for the ionization thresholds $Si_{2s,2p}$, O_{2s}, and O_{1s} have been estimated to be 0.13, 0, and 0.03 respectively [2.18,90]. These results illustrate an important point not always fully appreciated [2.84-89]: each ionization or Auger yield threshold does not necessarily have a corresponding desorption threshold.

2.5.2 AES in the Gas Phase and Chemisorbed Systems

The utility of AES in the gas phase, with a reduced background and narrower Auger peaks compared to the solid phase, has been clearly demonstrated. Figure 2.16 illustrates the C KVV Auger lineshape for some gas phase molecules. The sensitivity of AES to local hybridization (sp^3, sp^2, sp) is clearly demonstrated by the CH_4, C_2H_4, and C_2H_2 lineshapes [2.92,93]. The insensitivity to substituent effects is demonstrated by the CH_4, CH_3OH, and $(CH_3)_2O$ (all sp^3) lineshapes. The normal alkanes [2.94] show a broadened sp^3 lineshape, the cyclic alkanes [2.95] show an apparent progression from the sp^2 to the sp^3 lineshape as the bond angle strain decreases. The constancy of the principal peak energy in the alkanes, inspite of the increasing size of the molecule, suggests the final-state holes are not completely delocalized about the molecule in these systems.

The Auger lineshapes of small gas phase molecules are rather well understood. Quantitative interpretations have been reported for HF [2.96-98], N_2 [2.68], O_2 [2.15], CO [2.68,99,100], NO [2.68], H_2O [2.67,68,101], NH_3 [2.102-105], CH_4, C_2H_2, C_2H_4, CH_2OH [2.106], CO_2 [2.68,100], C_3O_2 [2.107], and SiH_4 [2.108] to name just some of them.

When molecules or atoms are chemisorbed on the surface of solids, the lineshape often changes little except for the introduction of a peak or shoulder at higher electron energies as illustrated in Fig.2.17. Condensation

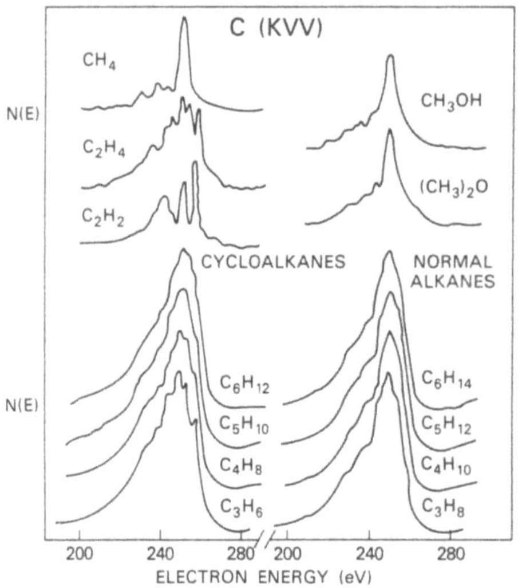

Fig.2.16↑

Fig.2.17→

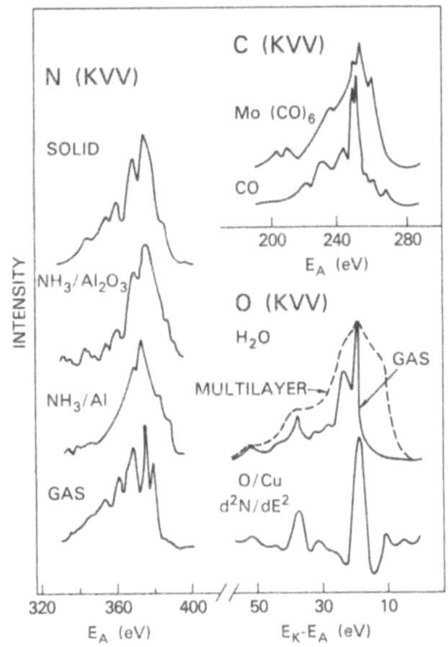

Fig.2.16. The C KVV Auger lineshapes taken in the gas phase for various carbon molecules illustrating the sensitivity to sp^n hydrization [2.92-95]

Fig.2.17. Comparison of some KVV lineshapes of molecules in the gas phase with those in the chemisorbed, condensed, or metal ligand state illustrating the appearance of structure at higher energies. The NH_3 data are taken from [2.109], CO data from [2.110], O/Cu data from [2.111], and H_2O data from [2.92,93]. E_A is the Auger energy, E_K the O 1s binding energy. All spectra are N(E) except for O/Cu

to the solid introduces the same feature as illustrated for NH_3 and H_2O. The source of this high-energy feature is not well understood. At least 6 different mechanisms have been proposed; these are summarized in Table 2.4 along with examples and references.

The first two mechanisms listed in Table 2.4 may be classified as final-state effects, the next two as initial-state effects, the last two as chemical environment effects. In the first two mechanisms the high-energy peak arises when the final-state holes end up on different centers. The interatomic Auger transition model implies the intensity is determined by a two-center matrix element; in the correlation model the intensity is determined by a one-center matrix element. In the second and third mechanisms, the high-energy peak arises because of the relaxation of the system in the presence of the initial core hole. The core hole screening mechanism implies the actual transfer of an electron into the lowest unoccupied orbital of the valence band of the adsorbate; this electron is then capable of contri-

Table 2.4. Summary of mechanisms recently proposed for the existence of high-energy structure above the principle Auger peak

Suggested mechanism	Example	Reference[a]
1) Interatomic Auger transition	MgF_2, NaF	CITRIN [2.112]
	LiF	GALLON and MATTHEW [2.113]
	MgO	BASSETT [2.114]
	S, C, O, N on Ni, Cu, Fe	SALMERON and BARO [2.111]
2) Final-state hole-hole correlation effects	Cu	MADDEN [2.46]
	Cu, Ni, Zn	Antonides [2.38]
	SiO_2	RAMAKER [2.18]
3) Electron transfer or core hole screening	$Cr(CO)_6$, $W(CO)_6$	PLUMMER [2.115]
		JENNISON [2.110]
	CO on Pt, Ru, N_2 on W	NETZER and MATTHEW [2.116]
		FUGGLE and MENZEL [2.117]
4) Shake satellites	Condensed NH_3	LARKINS and LUBENFELD [2.104]
	NH_3 on Al, Al_2O_3	CAMPBELL [2.109]
5) Band splitting due to covalent interaction with substrate or other atoms	FeS_x	MARUSAK and TONGSON[b] [2.118]
	TiC, VC, Cr_3C_2	SMITH and LEVENSON [2.119]
	condensed H_2O	RYE [2.92,93]
	Al, Si, P, S dissolved in Cu	HIRAKI [2.120]
	O, S, on Ni, Mo, Si, Cu	COAD and RIVIERE [2.121], KUNIMORI [2.122], SICKAFUS [2.123, 124], TIBBETTS [2.125]
	C_2H_4, C_2H_2, C_2N_2 On Pt	NETZER and MATTHEW [2.116]
6) Two different chemical environments present; shift of core level	S on Cu-Au alloy	LOSCH [2.126]
	O_2 on Si, Pt	LEGARE [2.127]
	Pt, Sn on Al_2O_3	LEGARE [2.127]

[a] If more than two authors, only the first is listed.
[b] Indicated interatomic Auger transition also possible.

buting to the Auger intensity. The shake-up satellite process implies that the electron removed from the core is deposited into the lowest unoccupied orbital; the one-hole final-state relaxation processes are then significantly different, to produce the high-energy satellite. Finally in the last two mechanisms the high-energy peaks arise from the chemical environment effects on the one-electron DOS, one mechanism concerned with the core level, the other with the valence level. In the band splitting model the chemical environment (covalent interaction) splits off a small electron density to the high-energy side of the main valence peak; in the other mechanism *two different* chemical environments cause a "chemical" shift of the core levels which is reflected in the Auger energy.

It should not be understood that only one mechanism can be responsible for all of the high-energy structure in the lineshapes for the various systems indicated in Table 2.4. The electron screening model as applied to CO on metals is almost certainly correct as there is other evidence, experimental (such as UPS data [2.128]) and theoretical [2.110,129], that the electron transfer does occur. XPS and UPS data [2.118,119,125] also verify that band splitting may indeed occur upon chemisorption to some metals or upon covalent interaction with its neighbors. Nevertheless, in some cases the proper mechanism is not certain. As Table 2.4 indicates, three different mechanisms [2.111,121-126] have been proposed to explain the high-energy structure which arises when O, C, N, or S bonds to Ni, Cu, Si, or Pt. The high-energy shoulder in the O KVV spectrum of MgO and SiO_2 probably results from the same mechanisms, but two different ones [2.18,114] have been proposed. The high-energy peaks which appear in the condensed phases of H_2O and NH_3 (though not in the gas) have been explained differently [2.92,93,109] although they are probably due to the same mechanism. Obviously, further work is needed to determine the proper mechanism for some of these systems.

Important information on chemisorbed systems can also be obtained from the substrate Auger lineshape. Figure 2.18 shows the Si $L_{23}VV$ lineshape for H [2.130] and O [2.131.132] on Si. The structure in the lineshape reveals the surface dangling bonds on clean Si; this structure decreases with exposure to either H or O indicating their use to bond the adsorbed atom. Additional features in the lineshape can also be related to the Si-H bonds or the presence of O_2 or O on Si. Very few substrate lineshapes have been reported, but these two examples indicate their usefulness in providing bonding information.

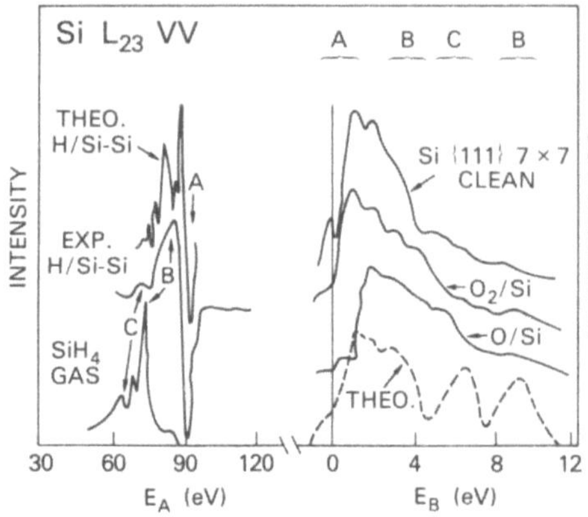

Fig.2.18. The substrate Si $L_{23}VV$ Auger data for either H/Si [2.130] left, or O/Si [2.131], right. (*H/Si*) Comparison of the theoretical and experimental difference spectra (H/Si-clean Si) with the spectrum from gas phase SiH_4. Region A shows the decrease in surface dangling bond states, peaks B and C arise from Si-H bonds as indicated by the SiH_4 spectrum [2.130]. (*O/Si*) Comparison of the Auger transition DOS (self-deconvolution of the Auger lineshape) for clean Si and O/Si. Region A shows the decrease in surface dangling bonds, structure in region B corresponds to O_2 molecular orbitals on Si, structure in region C to O-Si bonds [2.131]. E_A is the Auger energy, E_B the binding energy

2.5.3 AES in the Bulk and at Interfaces

The combined surface sensitivity and DOS site specificity of AES greatly increases the utility of AES for the study of interfaces and the bulk [2.133]. Two examples are given here. Figure 2.19 shows how the Si $L_{23}VV$ lineshape changes with an increasingly thick layer of Pd deposited on the Si surface [2.34]. With a 30-Å Pd layer on the surface, the Si lineshape is still visible. The surface sensitivity of AES indicates the Si must be diffusing into the Pd layer. Two forms of Si must exist in the Pd layer, elemental Si and Pd_2Si [2.34].

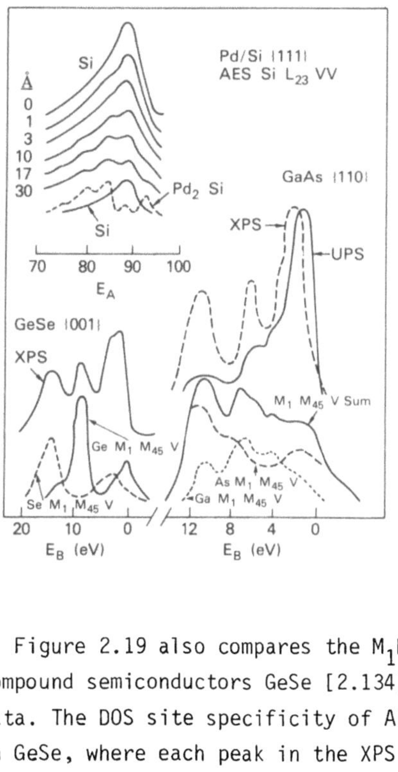

Fig. 2. 19. (*Top*) Comparison of the Si $L_{23}VV$ Auger lineshape with coverage of Pd as indicated in Angströms (Å) [2.34]. (*Bottom*) Comparison of the $M_1M_{45}V$ Auger lineshapes with XPS and UPS data for GeSe [2.134] and GaAs [2.135]. The $M_1M_{45}V$ sum for GaAs corresponds to a direct sum of the Ga and As $M_1M_{45}V$ lineshapes. E_A is the Auger energy, E_B the binding energy

Figure 2.19 also compares the $M_1M_{45}V$ lineshapes of each element in the compound semiconductors GeSe [2.134] and GaAs [2. 135] with UPS and XPS data. The DOS site specificity of AES is clearly illustrated, particularly in GeSe, where each peak in the XPS spectrum can be attributed to either the Ge or Se atom. Although not quite as distinct in GaAs, the separate Auger lineshapes assist in assigning the electron density indicated by the UPS/XPS data to the separate sites. Matrix element effects and final-state core hole screening effects (these are CCV lineshapes, see Sect.2.4) must be included however before quantitative comparisons with the UPS/XPS data or theory [2.136] can be made.

Valence band information is becoming increasingly important for the study and production of new and interesting materials. These include the study of, for example, novel conductors (such as intercalated graphites [2.80,81]) semiconductors, composite materials, and ion-implanted metals. The utility of AES as a source of valence DOS information in the bulk will become more apparent as these studies continue.

2.6 Summary

This review has emphasized the utilization of AES as a probe of valence electron bands and orbitals. A simple and empirical one-electron approach to the interpretation of Auger lineshapes was given to make the interpretative scheme available to a large number of experimentalists in the field. The important effects of localization and screening have been emphasized because they have been only recently identified and understood. Some areas which need further work have been indicated: they include the local versus Mulliken population question, the effects on the atomic matrix elements of introduction into a solid or molecule, and the sorting out of the various mechanisms proposed to interpret the appearence of high-energy structure in the Auger lineshapes of small molecules and atoms chemisorbed on a surface.

A necessarily brief review and optimistic outlook for AES as a source of valence band information was given. Much information can be extracted from the Auger lineshape. However, as indicated in the introduction, Auger spectra are most often utilized for elemental indentification and analysis, thus ignoring this information. It is hoped this review will encourage just a few more experimentalists to utilize the information they already have at their finger tips.

References

2.1 P.H. Holloway: Adv. Electron. Electron Phys. *54*, 241 (1980)

2.2 B. Crasemann: "Auger Effect and Related Topics", Lecture notes, Univ. Pierre and Marie Curie, Paris, France (1974)

2.3 D.E. Ramaker, J.S. Murday, N.H. Turner: J. Electron Spectros. Relat. Phenom. *17*, 45 (1979)

2.4 H.H. Madden, J.E. Houston: J. Appl. Phys. *47*, 3071 (1976); Solid State Commun. *21*, 1081 (1977); J. Vac. Sci. Technol. *14*, 412 (1977)

2.5 F.P. Larkins: Appl. Surf. Sci. (to be published)

2.6 H.H. Madden: J. Vac. Sci. Technol. *18*, 677 (1981)

2.7 W. Moddeman, K. Wittberg: In *Electron Spectroscopy: Theory, Techniques, and Applications,* Vol.4, ed. by C.R. Brundle, A.D. Baker (Academic, New York, forthcoming)
2.8 H. Aksela: Acta Univ. Ouluensis A*89* Phys. 15 (1980) (Dissertation); J. Vayrynen: Acta Univ. Ouluensis, A*99* Phys. 19 (1980) (Dissertation)
2.9 G.G. Kleiman: Appl. Surf. Sci. (to be published)
2.10 J.C. Fuggle: In *Electron Spectroscopy: Theory, Techniques, and Applications,* Vol.4, ed. by C.R. Brundle, A.D. Baker (Academic, New York forthcoming)
2.11 M.O. Krause: J. Phys. Chem. Ref. Data *8*, 307 (1979)
2.12 J.S. Murday: "Review of Surface Physics"; NRL Memorandum Report 3062 (1975); C.C. Chang: J. Vac. Sci. Technol. *18*, 276 (1981)
2.13 J.W. Mayer, A. Turos: Thin Solid Films *19*, 1 (1973); C.A. Evan: Natl. Bur. Stand. U.S. Spec. Publ. *400*, 23 (1976)
2.14 D.E. Ramaker, J.S. Murday, N.H. Turner, C. Moore, M.G. Lagally, J. Houston: Phys. Rev. B*19*, 5375 (1979)
2.15 B.I. Dunlap, P.A. Mills, D.E. Ramaker: J. Chem. Phys. *75*, 300 (1981)
2.16 D.A. Shirley: Chem. Phys. Lett. *17*, 312 (1972); *16*, 220 (1972); Phys. Rev. A*7*, 1520 (1973)
2.17 J.B. Mann: Report No. LASL-3690, Los Alamos Scientific Laboratory (1967)
2.18 D.E. Ramaker: Phys. Rev. B*21*, 4608 (1980)
2.19 E.J. McGuire: Phys. Rev. *185*, 1 (1969); A*2*, 273 (1970)
2.20 E.J. McGuire: Phys. Rev. A*3*, 1801 (1971)
2.21 D.L. Walters, C.P. Bhalla: Phys. Rev. A*3*, 1919 (1971); At. Data *3*, 301 (1971)
2.22 D.L. Walters, C.P. Bhalla: Phys. Rev. A*4*, 2164 (1971)
2.23 D.E. Ramaker: (to be published)
2.24 M.H. Chen, B. Crasemann: Phys. Rev. A*8*, 7 (1973)
2.25 C.P. Bhalla, N.O. Folland, M.A. Hein: Phys. Rev. A*8*, 649 (1973)
2.26 M.H. Chen, B. Crasemann: Phys. Rev. A*16*, 1495 (1977)
2.27 E.J. McGuire: Phys. Rev. A*16*, 2365 (1977)
2.28 E.J. McGuire: Research Report SC-RR-69-137, Sandia Laboratories (1969)
2.29 J. Vayryen: J. Electron Spectros. Relat. Phenom. *22*, 27 (1981)
2.30 F.L. Hutson, D.E. Ramaker, B.I. Dunlap, J.D. Ganjei, J.S. Murday: J. Chem. Phys. (to be published)
2.31 J.A.D. Matthews, Y. Komninos: Surf. Sci. *53*, 716 (1975)
2.32 D.R. Jennison: Phys. Rev. Lett. *40*, 807 (1978)
2.33 S.D. Bader, L. Richter, M.B. Brodsky: Solid State Commun. *37*, 729 (1981)
2.34 G.W. Rubloff, P.S. Ho, J.F. Freeouf, J.E. Lewis: Phys. Rev. B*23*, 4183 (1981); P.S. Ho, G.W. Rubloff, J.E. Lewis, V.L. Moruzzi, A.R. Williams: Proc. Electrochem. Soc. *80*, 85 (1980)
2.35 M. Cini: Solid State Commun. *20*, 605 (1976); Phys. Rev. B*17*, 2788 (1978); G.A. Sawatzky: Phys. Rev. Lett. *39*, 504 (1977)
2.36 M. Cini: Phys. Rev. *87*, 483 (1979); G. Treglia, M.C. Desjonqueres, F. Ducastelle, D. Spanjaard: J. Phys. C *14*, 4347 (1981); P.T. Andrews, T. Collins, P. Weightman: J. Phys. C (to be published)
2.37 B.I. Dunlap, F.L. Hutson, D.E. Ramaker: J. Vac. Sci. Technol. *18*, 556 (1981)
2.38 E. Antonides, E.C. Janse, G.A. Sawatzky: Phys. Rev. B*15*, 1669 (1977); E. Antonides, G.A. Sawatzky: Inst. Phys. Conf. Ser. *39*, 134 (1978)
2.39 F.J. Himpsel, P. Heimann, D.E. Eastman: J. Appl. Phys. *52*, 1658 (1981)
2.40 C.J. Powell: Solid State Commun. *26*, 557 (1978)
2.41 A.C. Parry-Jones, P. Weightman, P.T. Andrews: J. Phys. F*5*, 590 (1975)
2.42 J.F. McGilp, P. Weightman, E.J. McGuire: J. Phys. C*10*, 3445 (1977)

2.43 W.A. Harrison: *Electronic Structure and the Properties of Solids* (Freeman, San Francisco 1980) p.476
2.44 P. Weightman, P.T. Andrews: J. Phys. C*12*, 943 (1979)
2.45 A.C. Parry-Jones, P. Weightman, P.T. Andrews: J. Phys. C*12*, 1587 (1979)
2.46 H.H. Madden, D.M. Zehner, J.R. Nooman: Phys. Rev. B*17*, 3074 (1978)
2.47 P.J. Bassett, T.E. Gallon, J.A.D. Matthew, M. Prutton: Surf. Sci. *35*, 63 (1973)
2.48 J.M. Burkstrand, G.G. Tibbets: Phys. Rev. B*15*, 5481 (1977)
2.49 J.M. Mariot, G. Dufour: J. Electron Spectros. Relat. Phenom. *13*, 403 (1978)
2.50 J.A.D. Matthew: J. Phys. C*11*, L47 (1978)
2.51 L. Oberli, R. Monot, H.J. Mathieu, D. Landolt, J. Buttet: Surf. Sci. *106*, 301 (1981)
2.52 T. Jach, C.J. Powell: Phys. Rev. Lett. *46*, 953 (1981)
2.53 P. Weightman, P.T. Andrews: J. Phys. C*13*, L815 (1980)
2.54 G.A. Sawatzky, D. Post: Phys. Rev. B*20*, 1546 (1979)
2.55 S.P. Withrow, P.E. Luscher, F.M. Propst, W.H. Weinberg: J. Vac. Sci. Technol. *15*, 511 (1978)
2.56 J.S. Murday, D.E. Ramaker, M.K. Bernett, N.H. Turner: Proc. 7th, Int. Vac. Congr. 1977, Vol.2, p.1293
2.57 D.E. Ramaker, J.S. Murday: J. Vac. Sci. Technol. *16*, 510 (1979)
2.58 D.R. Jennison: J. Vac. Sci. Technol. *18*, 466 (1981); Bull. Am. Phys. Soc. *26*, 220 (1981)
2.59 G.A. Sawatzky, A. Lenselink: Phys. Rev. B*21*, 1790 (1980)
2.60 O. Gunnarsson, K. Schonhammer: Phys. Rev. B*22*, 3710 (1980)
2.61 D.E. Ramaker: Phys. Rev. B (to be published)
2.62 W. Domcke, L.S. Cedarbaum, J. Schirmer, W. von Niessen: Phys. Rev. Lett. *42*, 1237 (1979); Chem. Phys. *39*, 149 (1979)
2.63 T.A. Carlson, C.W. Nestor, Jr.: Phys. Rev. A*8*, 2887 (1973)
2.64 H. Korber, W. Mehlhorn: Z. Phys. *191*, 217 (1965);
W. Mehlhorn, D. Stalherm, H. Verbeek: Z. Naturforsch. *23*A, 287 (1968)
2.65 M.O. Krause, T.A. Carlson, W.E. Moddeman: J. Phys. Paris Coll. C4 *32*, 139 (1971);
M.O. Krause, F.A. Stevie, L.J. Lewis, T.A. Carlson, W.E. Moddeman: Phys. Lett. *31*A, 81 (1970)
2.66 W.E. Moddeman, T.A. Carlson, M.O. Krause, B.P. Pullen, W.E. Bull, G.K. Schweitzer: J. Chem. Phys. *55*, 2317 (1971)
2.67 H. Agren, H. Siegbahn: Chem. Phys. Lett. *69*, 424 (1980)
2.68 H. Agren: J. Chem. Phys. *75*, 1267 (1981);
J. Muller, H. Agren, O. Goscinski: Chem. Phys. Lett. *38*, 349 (1979)
2.69 J.N. Schulman, J.D. Dow: Phys. Rev. Lett. *47*, 371 (1981)
2.70 G. van der Laan, C. Westra, C. Haas, G.A. Sawatzky: Phys. Rev. B*23*, 4369 (1981)
2.71 U. von Barth, G. Crossman: Solid State Commun. *32*, 645 (1979); Phys. Scr. *21*, 580 (1980)
2.72 G.D. Mahan: Phys. Rev. B*21*, 1421 (1980)
2.73 L.C. Davis, L.A. Feldkamp: Phys. Rev. B*23*, 4269 (1981)
2.74 V.I. Grebennikov, Y.A. Babanov, O.B. Sokolov: Phys. Status Solidi *80*, 73 (1977)
2.75 R. Lasser, J.C. Fuggle: Phys. Rev. B*22*, 2637 (1980)
2.76 N.D. Lang, A.R. Williams: Phys. Rev. B*16*, 2408 (1977); *20*, 1369 (1979);
A.R. Williams, N.D. Lang: Phys. Rev. Lett. *40*, 954 (1978)
2.77 D.R. Jennison: Phys. Rev. A*23*, 1215 (1981)
2.78 D.R. Jennison, H.H. Madden, D.M. Zehner: Phys. Rev. B*21*, 430 (1980)
2.79 D.R. Jennison: Phys. Rev. B*18*, 6996 (1978)
2.80 J.S. Murday, B.I. Dunlap, F.L. Hutson, II, P. Oelhafen: Phys. Rev. B *24*, 4764 (1981)

2.81 B.I. Dunlap, D.E. Ramaker, J.S. Murday: Phys. Rev. B (to be published)
2.82 J.H.O. Varley: J. Phys. Chem. Solids *23*, 985 (1962)
2.83 M.L. Knotek, P.J. Feibelman: Phys. Rev. Lett. *40*, 964 (1978); Surf. Sci. *90*, 78 (1979); Phys. Rev. B*18*, 653 (1978); P.J. Feibelman: Surf. Sci. *102*, L51 (1981)
2.84 T.E. Madey, R. Stockbauer, S.A. Flodstrom, J.F. van der Veen, F.J. Himpsel, D.E. Eastman: Phys. Rev. B*23*, 6847 (1981)
2.85 G.G. Pantano, T.E. Madey: Appl. Surf. Sci. *7*, 115 (1981)
2.86 F.P. Netzer, T.E. Madey: J. Chem. Phys. *76*, 710 (1982)
2.87 R. Franchy, D. Menzel: Phys. Rev. Lett. *43*, 865 (1979)
2.88 P. Feulner, R. Treichler, D. Menzel: Phys. Rev. B (to be published)
2.89 R. Jaeger, J. Stohr, J. Feldhaus, S. Brennan, D. Menzel: Phys. Rev. B*23*, 2102 (1981)
2.90 D.E. Ramaker, C.T. White, J.S. Murday: J. Vac. Sci. Technol. *18*, 748 (1981);
2.91 E.J. McGuire: Phys. Rev. A*16*, 73 (1977); J.I. Vrakking, F. Meyer: Surf. Sci. *47*, 50 (1975)
2.92 R.R. Rye, T.E. Madey, J.E. Houston, P.H. Holloway: J. Chem. Phys. *69*, 1504 (1978)
2.93 R.R. Rye, J.E. Houston, D.R. Jennison, T.E. Madey, P.H. Holloway: Ind. Eng. Chem. Prod. Res. Dev. *18*, 1 (1979)
2.94 R.R. Rye, D.R. Jennison, J.E. Houston: J. Chem. Phys. *73*, 4867 (1980)
2.95 J.E. Houston, R.R. Rye: J. Chem. Phys. *74*, 71 (1981)
2.96 K. Faegri, Jr., H.P. Kelly: Phys. Rev. A*19*, 1649 (1979); O.M. Kvalheim, K. Faegri: Chem. Phys. Lett. *67*, 127 (1979)
2.97 R.W. Shaw, Jr., T.D. Thomas: Phys. Rev. A*11*, 1491 (1975)
2.98 I.H. Hillier, J. Kendrick: Mol. Phys. *31*, 849 (1976)
2.99 H. Agren, H. Siegbahn: Chem. Phys. Lett. *72*, 498 (1980)
2.100 D.R. Jennison, J.A. Kelber, R.R. Rye: Chem. Phys. Lett. *77*, 604 (1981); J.A. Kelber, D.R.Jennison, R.R. Rye: J. Chem. Phys. *75*, 652 (1981)
2.101 H. Agren, S. Svensson, U.I. Wahlgren: Chem. Phys. Lett. *35*, 330 (1975)
2.102 J.M. White, R.R. Rye, J.E. Houston: Chem. Phys. Lett. *46*, 146 (1977)
2.103 R.N. Shaw, Jr., J.S. Jen, T.D. Thomas: J. Electron Spectros. Relat. Phenom. *11*, 91 (1977)
2.104 F.P. Larkins, A. Lubenfeld: J. Electron Specros. Relat. Phenom. *15*, 137 (1979)
2.105 M.T. Okland, K. Faegri, R. Manne: Chem. Phys. Lett. *40*, 185 (1976)
2.106 D.R. Jennison: Chem. Phys. Lett. *69*, 435 (1980)
2.107 L. Karlson, L.O. Werme, T. Bergmark, K. Siegbahn: J. Electron. Spectros. Relat. Phenom. *3*, 181 (1974)
2.108 F. Maracei, R. Platania, R. Salomone: J. Electron Spectros. Relat. Phenom. *19*, 155 (1980)
2.109 C.T. Campbell, J.W. Rogers, Jr., R.L. Hance, J.M. White: Chem. Phys. Lett. *69*, 430 (1980)
2.110 D.R. Jennison, G.D. Stucky, R.R. Rye, J.A. Kelber: Phys. Rev. Lett. *46*, 911 (1981)
2.111 M. Salmeron, A.M. Baro, J.M. Rojo: Phys. Rev. B*13*, 4348 (1976)
2.112 P.H. Citrin, J.E. Rowe, S.B. Christman: Phys. Rev. B*14*, 2642 (1976)
2.113 T.E. Gallon, J.A.D. Matthew: Phys. Status Solidi *41*, 343 (1970)
2.114 P.J. Bassett, T.E. Gallon, J.A.D. Matthew, M. Prutton: Surf. Sci. *35*, 63 (1973)
2.115 E.W. Plummer, W.R. Selanck, J.S. Miller: Phys. Rev. B*18*, 1673 (1978)
2.116 F.P. Netzer, J.A.D. Matthew: J. Electron Spectros. Relat. Phenom. *16*, 359 (1979)
2.117 J.C. Fuggle, D. Menzel: Surf. Sci. *79*, 1 (1979)
2.118 L.A. Marusak, L.L. Tongson: J. Appl. Phys. *50*, 4350 (1979)
2.119 M.A. Smith, L.L. Levenson: Phys. Rev. B*16*, 1365 (1977)

2.120 A. Hiraki, S. Kim, W. Kammura, M. Iwami: Appl. Phys. Lett. *34*, 194 (1979)
2.121 J.P. Coad, J.C. Riviere: Proc. R. Soc. London A*331*, 403 (1972)
2.122 K. Kinimori, T. Kawai, T. Kondow, T. Onishi, K. Tamaru: Surf. Sci. *46*, 567 (1974)
2.123 E.N. Sickafus, F. Steinrisser: J. Vac. Sci. Technol. *10*, 43 (1973)
2.124 E.N. Sickafus: Phys. Rev. B*7*, 5100 (1973)
2.125 G.G. Tibbetts, J.M. Burkstrand, J.C. Tracy: Phys. Rev. B*15*, 3652 (1979)
2.126 W. Losch: J. Vac. Sci. Technol. *16*, 865 (1979)
2.127 P. Legare, G. Maire, B. Carriere, J.P. Deville: Surf. Sci. *68*, 348 (1977)
2.128 C.F. Bruckner, T.N. Rhodin: Surf. Sci. *86*, 638 (1979)
2.129 K. Hermann, P.S. Bagus: Phys. Rev. B*16*, 4195 (1977)
2.130 H.H. Madden: Surf. Sci. *105*, 129 (1981)
2.131 M.C. Munoz, V. Martinez, J.A. Tagle, J.L. Sacedon: Phys. Rev. Lett. *44*, 814 (1980)
2.132 B. Lang, P. Scholler, B. Carriere: Surf. Sci. *99*, 103 (1980)
2.133 J.A. Tagle, V.M. Saez, J.M. Rojo, M. Salmeron: Surf. Sci. *79*, 77 (1978)
2.134 G.D. Davis, P.E. Viljoen, M.G. Lagally: J. Electron Spectros. Relat. Phenom. *21*, 135 (1980)
2.135 G.D. Davis, M.G. Lagally: J. Vac. Sci. Technol. *15*, 1311 (1978); G.D. Davis, D.E. Savage, M.G. Lagally: J. Electron Spectrosc. Relat. Phenom. *23*, 25 (1981)
2.136 G. De Meyer, R. Hoogewijs, W. Lambrecht, J. Vennik, G. Dalmai: Surf. Sci. *106*, 498 (1981)

3. SIMS of Reactive Surfaces

W. N. Delgass, L. L. Lauderback, and D. G. Taylor

With 16 Figures

3.1 Introduction

The possibility of surface layer analysis by measurement of secondary ions emitted after bombardment of a sample with energetic primary ions has been known for some time [3.1]. Secondary ion mass spectrometry (SIMS) has become a well-established tool in research on both solids and surfaces [3.2]. It is widely used for depth profiling the compositions near interfaces in metals, oxides, and semiconductors [3.3-6]. In 1973, BENNINGHOVEN opened a new avenue of application of SIMS by pointing out that experiments done at low primary ion currents and total doses can probe the uppermost surface layer [3.7]. He termed this mode of the experiment "static" SIMS. At 10^{-9} A/cm^2 with an emission yield of secondary particles of 1 per incident ion, for example, it takes about 1/2 hour to remove 1% of a monolayer. Thus, although each ion impact damages a small region of the surface, the contribution of previously damaged regions to the observed spectrum can be made negligible. To determine the kinds of information that static SIMS can provide for surface science, one must focus on the collision process itself. The central issue is to establish the relations between measured SIMS parameters and the structure and chemistry of the original surface. Significant progress has been made in this area, but the issue is not yet fully resolved. The object of this paper is to examine the state of the art and discuss the promising directions for future development.

The paper is presented in three major sections. The first (Sect.3.2), on single crystal metal surfaces, begins with some general characteristics of the SIMS experiment and its application to metals. In the next segment (Sect.3.2.2), examination of well-studied, ordered Ni and Ru surfaces with adsorbed CO or oxygen shows that SIMS parameters can reflect structure, but that a model is needed to guide interpretation. The classical dynamics calculations of WINOGRAD, GARRISON, and HARRISON, discussed next, provide such a model. This approach also suggests that the greater amount of surface structural information is in the angular-dependent SIMS experiment. The final segment in this section discusses the ability of SIMS to see reactive intermediates during temperature-programmed decomposition reactions. The success of these experiments illustrates the importance of intra-versus intermolecular bonding in determining secondary particle emission characteristics.

When intramolecular bonds are stronger than intermolecular or adatom-surface bonds, then emission of intact molecules becomes significant and the potential for identification of molecular species on surfaces by SIMS rises. SIMS studies of molecular solids have progressed rapidly in recent years and the results are important both for studies of reactions at surfaces and for analysis of catalytic materials. The section on molecular SIMS (Sect.3.3) reviews types of molecular ion formation and the scope of applications achieved so far. A discussion of the discovery by SIMS of the self-hydrogenation of thiophene on silver is presented in Sect.3.3.2. Results on inorganic complexes, interesting analogs to catalyst systems, are presented at the conclusion of this segment.

The last major section (Sect.3.4) concerns the prospects for analysis of the complex surfaces representative of catalytic materials. Results on FeRu alloys show the first steps toward determining the extent to which SIMS can reveal the proximity of one element to another in a multicomponent surface. This section closes with a summary of the prospects for SIMS applications to catalysts, and is followed by a summary of general conclusions.

3.2 Single Crystal Metal Surfaces

3.2.1 Characteristics of SIMS

The SIMS experiment is straightforward. As shown schematically in Fig.3.1, the primary ion beam, typically Ar^+ at 500-5000 eV, impinges on the surface producing emission of positive and negative secondary ions and ion clusters which are detected in a mass spectrometer. A UHV system with an argon ion gun and a mass spectrometer is a candidate for producing SIMS data. Quadrupole mass spectrometers are convenient, but others have been used. In general, energy selection of the secondary ions is necessary for the attainment of high quality data, so a specially designed system is needed [3.8]. Several SIMS systems are available commercially, either complete or as add-ons to existing UHV surface science apparati.

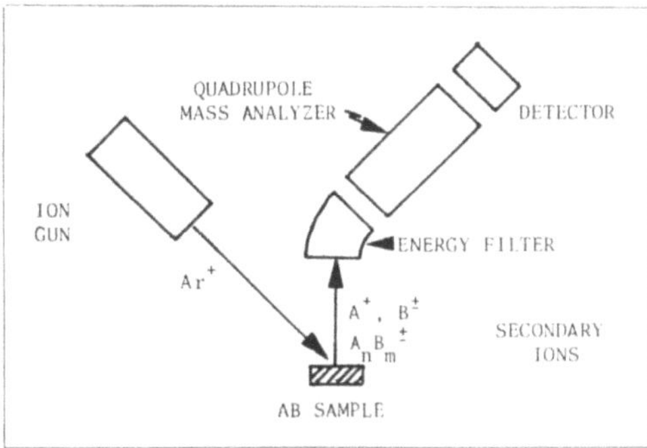

Fig.3.1. Schematic diagram of the SIMS experiment

Perhaps the most striking chemical characteristic in SIMS of metal surfaces is the sensitivity of secondary ion yields to the presence of oxygen [3.7]. Clean surfaces exhibit very low yields of Me_n^+, where Me stands for the metal atom. Oxygen exposure to saturation increases the yields of all positive ions: Me^+ by factors of 10^2 to 10^3 [3.9]. This effect is caused at least in part by the increase in work function accompanying oxygen adsorption. Adsorption of CO and other gases which increase the work function also causes a rise in positive ion yields. Adsorbates, such as Cs, which lower work functions enhance negative ion yields.

Oxygen and carbon monoxide adsorption have been studied on a wide variety of metals [3.10-18]. As an example of the SIMS characteristics in this type of system, we review the results of FLEISCH et al. on Ni{100} [3.17,18]. Figure 3.2 shows the Ni_n^+ yields as a function of oxygen exposure [3.17]. The general trend of the yields follows the work function curve which rises to a maximum at about 10 L, as does the LEED $p(2\times2)$ pattern [3.19]. The plateau in Ni_n^+ yield in the region of 3 L is reproduced in the O^- yield and reflects the decrease in sticking coefficient expected in this exposure region [3.20]. An interesting feature of this figure is the nearly complete loss of Ni_3^+ at saturation exposure. XPS, AES, and other methods show that after such treatment, the surface is covered with 2 layers of NiO [3.19,21]. The loss of Ni_3^+ for this state shows a sensitivity of the high multimer yields to penetration of O into the Ni lattice to interrupt the Ni-Ni bonding. The $Ni_nO_m^+$ ions also show trends that correlate with structure. The Ni_2O^+/NiO^+ ratio is maximum at 30 L, the region where the $c(2\times2)$ structure is reported to be maximized [3.19], while $Ni_2O_3^-$ reaches its highest intensity at saturation, where NiO is well formed.

Results for CO on Ni{100} [3.18] (Fig.3.3) illustrate several additional characteristics of SIMS of adsorbed systems. Most important are the molecular cluster ions Ni_nCO^+. The relatively small amount of NiO^+ and NiC^+ compared

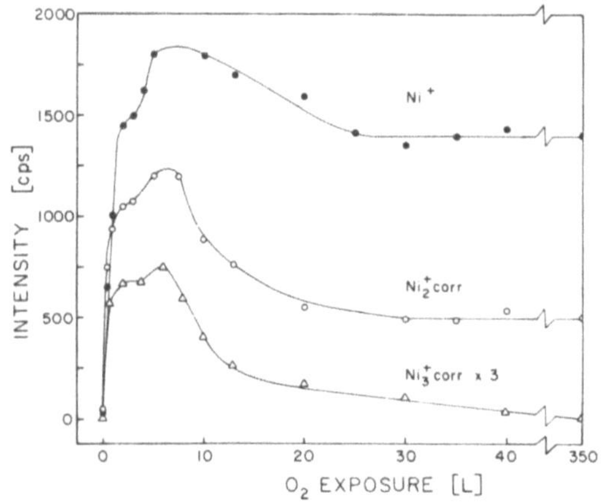

Fig.3.2. Ni^+, Ni_2^+, and Ni_3^+ ion yields versus O_2 exposure. Ni_2^+ and Ni_3^+ intensities are corrected for isotope distribution and quadrupole transmission [3.17] (North-Holland Publishing Company, Amsterdam, 1978)

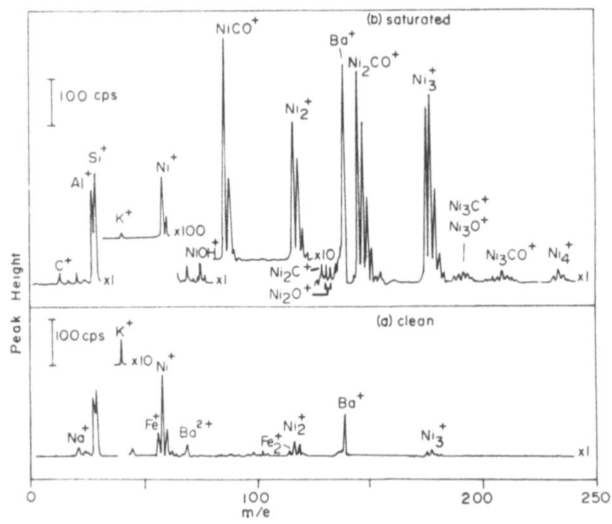

Fig.3.3a,b. Positive ion static SIMS spectra of (a) clean and (b) CO-saturated Ni{100}. Ip = 10^{-9} A or A^+ at 2 kV [3.18] (North-Holland Publishing Company, Amsterdam 1979)

to $NiCO^+$ indicates that the adsorption is nondissociative at room temperature. After exposure of the crystal to CO at 523 K, no $NiCO^+$ was detected, but yields of C^+, C_2^+, Ni_2C^+, Ni_2O^+, NiC^+, and NiO^+ increased. This behavior corresponds to the CO dissociation expected at this temperature [3.22]. Similar results have been found for CO on iron which shows FeC^+ and $FeCO^+$ at room temperature in agreement with the partial disociation reported [3.16], and for associative and dissociative adsorption of NO on Ni {001} [3.23].

The behavior of Ni_3^+ after CO saturation is in sharp contrast to that after O_2 saturation. The strong Ni_3^+ signal in Fig.3.3b shows that when the substrate lattice is not interrupted, the trimer signal can remain high. Figure 3.3 also shows the presence of submonolayer quantities of Ba impurity on the surface, but the concentration is too low to be seen in XPS.

The presence of both $NiCO^+$ and Ni_2CO^+ in the SIMS spectrum of CO on Ni has led to much discussion of whether these ions correlate with structure. The decrease in the $Ni_2CO^+/NiCO^+$ ratio with increasing coverage seems to follow the reported shift from a mixture of linear and bridged CO bonding at low coverage to a high concentration of linear bonding at high coverage [3.24]. In the first SIMS study of this system, BARBER et al. [3.25,26] advanced the concept that Ni_2CO^+ and $NiCO^+$ are formed by intact emission of bridged and linear structures, respectively. FLEISCH et al. [3.18] suggested, on the basis of calculations by GARRISON et al. [3.27], that high Ni_2CO^+ yields could be associated with bridged versus linear bonding but as a result of cluster formation over the surface rather than intact emission. Subsequent work has shown that for the CO/Ni {100} system, the $Ni_2CO^+/NiCO^+$ ratio alone is not sufficient to indicate the CO bonding type. HOPSTER and BRUNDLE [3.28] found that changing adsorption temperature, adsorbate coverage, and CO background pressure to cause large changes in both CO overlayer structure and bonding did not significantly affect the $Ni_2CO^+/NiCO^+$ ratio. DAWSON and TAM [3.29] changed the usual mixed linear and bridged bonding to a state with only bridged bonding by preadsorption of H_2 at 175 K, again with no corresponding changes in the $Ni_2CO^+/NiCO^+$ ratio measured by SIMS. Detailed structural information can be had from this system, but obtaining it requires the more sophisticated angular-dependent measurements described in Sect.3.2.4.

3.2.2 CO and O_2 on Ru {001}

Low-energy electron diffraction (LEED) [3.30,31], electron energy loss spectroscopy (EELS) [3.32], flash desorption spectroscopy (FDS) [3.30,32], and infrared spectroscopy (IR) [3.33] have produced a consistent description of the energetics and dynamics of CO adsorbed on Ru{001}. At relatively low coverages CO molecules are adsorbed at energetically preferred sites on the surface with CO-CO repulsive interactions preventing occupation of nearest-neighbor sites. At coverages above 0.2 monolayers, the long-range attractive potential between CO molecules [3.34] promotes growth of islands of CO in a $(\sqrt{3}\times\sqrt{3})R30^\circ$ structure until the surface coverage reaches 1/3 of a monolayer, corresponding to about 1.8 L at room temperature [3.33]. At this point all next-nearest-neighbor sites are filled and the $(\sqrt{3} \times \sqrt{3})R30^{\circ}$ LEED pattern reaches a maximum. Additional adsorption is accompanied by a shift of some adsorbed molecules to somewhat less preferred adsorption sites. EELS results show that this shift is not accompanied by a change in the Ru-C stretching frequency, suggesting that the surface potential is relatively smooth with preferred adsorption sites only slightly more energetically favored than other sites. Therefore, at higher coverages the adlayer energetics are strongly influenced by repulsive CO-CO interactions which force CO molecules to maintain the maximum possible separation and destroy the registry with the substrate. Saturation at room temperature corresponds to $\Theta = 7/12$, and produces a fuzzy ring LEED pattern. This pattern sharpens to a $(2\sqrt{3} \times 2\sqrt{3})R30^{\circ}$ structure when the sample is cooled to 110 K. In this structure 6 out of 7 molecules are adsorbed in slightly higher binding energy sites than the preferred one [3.31].

In recent work we have used SIMS, AES, and FDS to investigate CO adsorption on Ru{001} [3.34]. We are particularly interested in evaluating the sensitivity of the SIMS cluster ions to the above-mentioned changes in surface structure as a function of coverage and in the question of intact versus above-the-surface recombination mechanisms for CO ejection. The positive ion spectrum for CO on Ru{001} shows Ru^+, RuC^+, RuO^+, $RuCO^+$, Ru_2^+, Ru_2O^+, Ru_2CO^+, and trace amounts of Ru_3^+. The yields of Ru^+, $RuCO^+$, Ru_2^+, and Ru_2CO^+ all increase smoothly with CO exposure and saturate at about 9-10 L, corresponding to saturation coverage at room temperature. The sigmoid shape of these curves and greatest increase in intensity at exposures above 2 L may result from a decreasing heat of adsorption of CO with increasing coverage. The appearance of a second lower temperature peak in the FDS spectrum of CO/Ru{001} at about 1.5 L shows that more weakly bound CO is present (Fig. 3.4). Previous SIMS studies have also noted the high sensitivity of SIMS to decreases in heat of adsorption [3.11,13,29].

The sensitivity of the SIMS cluster ion distribution to the adsorbate structure is illustrated in Fig.3.5 which shows the $(Ru_2CO^+/RuCO^+)/(Ru_2^+/Ru^+)$ ratio as a function of coverage. The minimum in this plot at 1.5 L corresponds to the reported maximum in LEED intensity for the $(\sqrt{3} \times \sqrt{3})R30^{\circ}$ structure. A plot of the $RuCO^+/Ru^+$ ratio versus coverage shows a smooth, exponential-looking increase to saturation in the 9-10 L region. The Ru_2CO^+/Ru_2^+ ratio, on the other hand, shows a minimum at 1.5 L before it begins a smooth increase to saturation. Thus, the structural information in Fig.3.5 appears to be carried in the largest cluster ions. The results for CO on Ni{100} show that careful analysis will be needed before the minimum in Fig.3.5 can

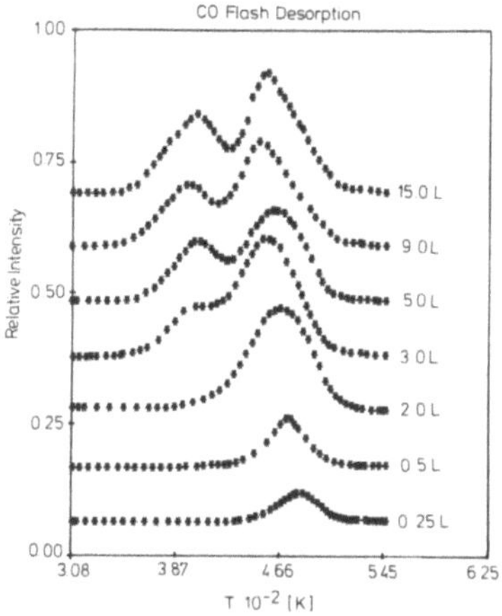

← Fig.3.4. Flash Desorption spectra of CO on Ru{001} at different exposures and a heating rate of 6 K/s

Fig.3.5. Dependence of the $(Ru_2CO^+/RuCO^+)/(Ru_2^+/Ru^+)$ ion yield ratio on exposure of Ru{001} to CO at 300 K
↓

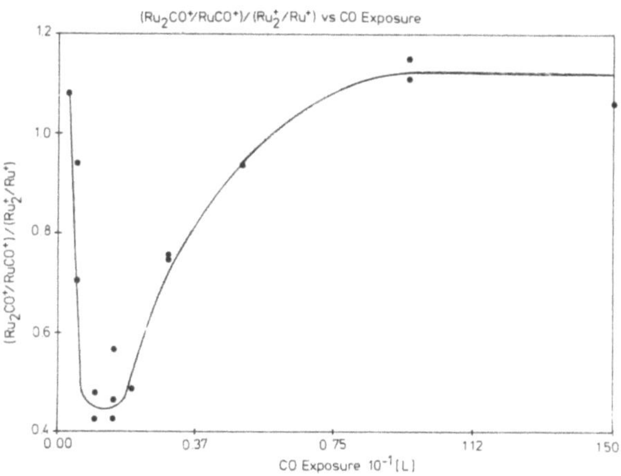

be related to structure. It is noteworthy, however, that a sharp, reproducible minimum appears in a region where changes in the placement of molecules on the surface are known to occur, but are accompanied by only subtle changes in bonding and charge distribution.

The concept that strong intramolecular bonding can preserve intact emission of molecular species has already been advanced above. In the case of CO on Ru{001}, the presence of RuC^+ and RuO^+ ions shows that some C - O bonds are indeed broken either before or during the SIMS process and hence raises the possibility that some $RuCO^+$ and Ru_2CO^+ clusters may contain C and O atoms from different CO molecules on the surface. In order to establish the extent of this type of atomic mixing due to SIMS ejection, the extent of mixing due to reaction was first evaluated by coadsorbing approximately equal amounts of

$C^{18}O$ and ^{13}CO. FDS spectra for $C^{18}O$, ^{13}CO, $^{13}C^{18}O$, and CO show that essentially no isotope mixing occurs by reaction and thus indicate that the adsorption is not dissociative. SIMS spectra of a surface containing approximately equal amounts of $C^{18}O$ and ^{13}CO also show no evidence for any mixed isotopes and about equal intensities of $RuC^{18}O^+$ and $Ru^{13}CO^+$ ions. These data, therefore, support the concept that the CO molecules found in various ion clusters are ejected intact.

Oxygen adsorption on Ru{001} has also received the attention of a variety of surface science techniques, including LEED [3.35,36], EELS [3.32], FDS [3.35], photoelectron spectroscopy [3.37], and work function measurements [3.35]. LEED patterns show a single (2×2) structure at all coverages up to 50 L at room temperature. MADEY et al. [3.35] have observed, however, that the intensity of the (2×2) structure goes through a sharp maximum at 1.5 L and subsequently resumes a smooth increase to saturation. The work function-versus-O_2 exposure plot also shows a sudden increase in slope at 1.5 L [3.35]. These results suggest that some kind of structural change occurs as the O_2 exposure increases through 1.5 L. There is some controversy regarding the nature of this change, however, since the observed (2×2) LEED pattern can result from either a simple (2×2) or three degenerate (1×2) structures. The work of PARROTT et al. [3.36], on a stepped Ru{001} crystal suggests, however, that the structure change results from the completion of a (2×2) structure at 1.5 L. At higher exposures this structure shifts to a (1×2), becoming complete at about 20 L. Electron energy loss results of THOMAS and WEINBERG [3.32] show only a single vibrational mode which varies in frequency from 516 cm^{-1} at low oxygen coverage to 596 cm^{-1} at saturation. This is assumed to be a vibrational mode of atomic oxygen normal to the surface in a highly coordinated site. The appearance of only one vibrational mode suggests that the binding site symmetry does not change significantly as the adlayer shifts from a (2×2) to a (1×2) structure.

There is also evidence that the surface mobility of oxygen is relatively low at room temperature. PARROTT et al. [3.36] have observed a sharpening of the LEED pattern after a crystal exposed to 1.5 L O_2 at 300 K is heated to took and then coded. MADEY et al. [3.35] have also studied the thermal stability of the (2×2) LEED pattern at 1.5 L oxygen exposure. These results show a relatively slow decrease in the LEED intensity as the temperature increases to about 600 K, followed by a rapid decrease between 600 and 700 K. Thus, heating to 700 K increases the surface mobility giving rise to long-range ordering and a sharpening of the LEED pattern upon cooling.

Our SIMS, AES, and FDS studies of oxygen adsorption on Ru{001} [3.34] have the same objective as the CO/Ru{001} work: to evaluate the sensitivity of the ejected cluster distribution in SIMS to surface structural changes. The observed ejected ions include Ru^+, RuO^+, Ru_2^+, Ru_2O^+, and trace amounts of O^+ and O_2^+. The Ru^+ and RuO^+ ion yields increase smoothly with oxygen exposure up to the nominal saturation coverage at about ~10 L, where a plateau occurs. Additional oxygen exposure causes the ion yields to increase further. The Ru_2^+ and Ru_2O^+ ions also increase smoothly with coverage reaching a saturation level corresponding to saturation of oxygen on the surface at about ~10 L. Unlike the Ru^+ and RuO^+ ions, however, these ions maintain essentially constant yields at higher exposures. As already mentioned in the discussion of CO adsorption, the sigmoid shape of all the ion yield curves may result from changes in heat of adsorption with coverage. The additional increase in the Ru^+ and RuO^+ ion yields beyond nominal saturation at ~10 L may reflect additional adsorption of a small amount of more weakly bound oxygen. With AES we have observed a small increase in oxygen coverage with exposure between 10 and 20 L. Evidence for this additional adsorption is also suggested by the work of PARROTT et al. [3.36] in which they observed a doubling in the oxygen surface coverage over that at 10 L upon exposing the surface to 6000 L of oxygen at 10^{-5} torr.

Evidence for the sensitivity of the SIMS cluster ion distribution to structural changes is shown by the $(Ru_2O^+/RuO^+)/(Ru_2^+/Ru^+)$ ratio as a function of oxygen coverage in Fig.3.6. This plot shows a sudden increase in slope at a coverage approximately 70% of saturation, corresponding to an exposure of 1.5 L. Although the mechanistic details giving rise to this sudden change are not yet understood, the break in the curve corresponds to the oxygen exposure associated with the completion of the (2×2) structure and the break in the work function curve. As in the case of CO on Ru{001}, examination of the Ru_2O^+/Ru_2^+ and RuO^+/Ru^+ ratios as a function of coverage shows that most of the structural information is carried by the Ru_2O^+/RU_2^+ ratio which rises rapidly at coverages greater than 75% of saturation.

The effect of temperature on the SIMS spectra as a function of coverage was investigated in two ways. In the first case O_2 was adsorbed at room temperature, the crystal was heated to 700 K for 10 s, and then it was cooled to room temperature before recording the spectra. In the second case, O_2 was adsorbed and the SIMS spectra recorded at 700 K. The results in both cases show no significant differences in either the ion intensities or the ratios of intensities compared to the room temperature data discussed above. These results are consistent with the concept that SIMS reflects short-range rather

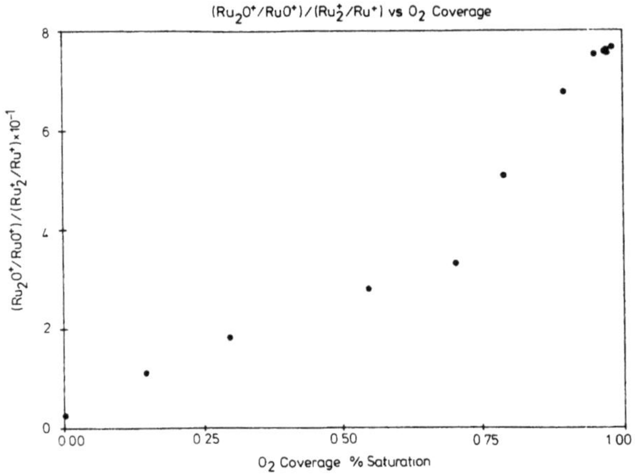

Fig.3.6. Dependence of the $(Ru_2O^+/RuO^+)/(Ru_2^+/Ru^+)$ ion yield ratio on exposure of Ru{001} to O_2 at 300 K

than long-range order. The growth of the (2 × 1) pattern, however, shortens some of the O - O distances on the surface, changes the local environment, and is reflected in Fig.3.6.

3.2.3 Classical Dynamics Modelling

The examples cited above show that SIMS parameters do respond to changes in surface structure and chemical bonding, but that interpretation of SIMS data from new systems will require some guidance from theory. A particularly promising approach to this problem is through the classical dynamics calculations of GARRISON, WINOGRAD, and HARRISON [3.38-41]. In this work, particle ejection resulting from the impact of a primary ion is predicted by integrating the equations of motion to find the positions and momenta of all atoms in a 4-layer crystallite with 60 atoms per layer. All pairwise interactions are summed. The potentials for atoms in the solid are adjusted to be consistent with the properties of the solid and of the adsorbed layer if one is included. Atom-atom potentials in the gas phase are used to predict when ejected atoms will have appropriate positions and momenta to become bound multimers. Examination of the trajectories calculated for a large number of primary ion impact points allows one to predict ratios of Me_n^+ yields, to observe mechanisms of multimer formation, and to depict angular emission patterns for ejected neutral species [3.42]. Although the calculations are sensitive to the potential parameters used and do not account at all for the

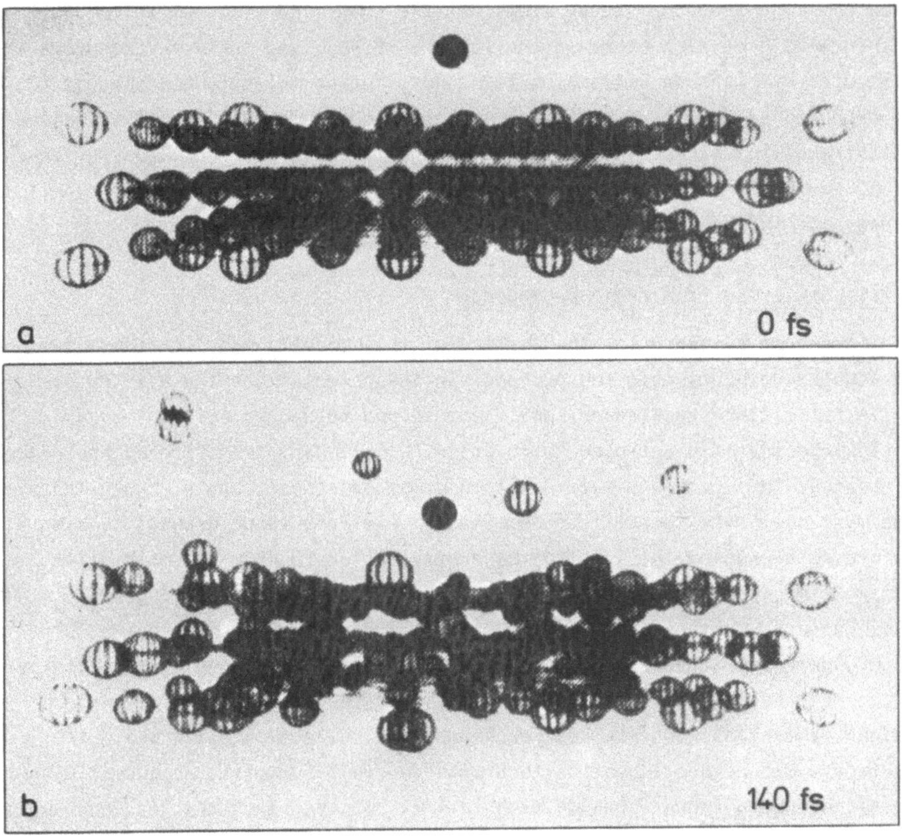

Fig.3.7a.b. Positions of the atoms: (a) before the primary ion (the lone sphere above the solid) has struck; (b) after a single ion impact. The two atoms ejecting to the left form a dimer. For graphical clarity, only a select group of atoms is shown, and their size is arbitrarily drawn for best graphical clarity [3.42]. (Reprinted with permission. Copyright American Chemical Society, 1980)

ionization process, they have been remarkably accurate in illustrating several important aspects of the SIMS process.

Figure 3.7, taken from the work of WINOGRAD and GARRISON [3.42], shows the effect, on a selected group of atoms, of Ar^+ impact on a clean Ni {100} crystal. Many atoms are moved from their starting positions. Of the 6 atoms well above the surface in the lower figure, the pair in the upper left are destined to be a Ni_2 dimer. For this particular trajectory, 140×10^{-15} s was sufficient time to identify all atoms which gained enough energy to leave the crystal. In general the calculations show that for close-packed surfaces nearly all ejected atoms come from the first layer [3.39], that multimers need not come from contiguous atoms but are comprised of atoms which are within 5-10 Å of each other on the surface [3.40], and that, for adsorbed molecules with strong intramolecular bonds, intact molecule emission is to be expected [3.27,43]. One test of the approach has been the ability to predict trends in Me_2^+/Me^+ yields for {110} and {111} versus the {100} surface of FCC metals [3.44]. This result would suggest that ionization contributions cancel out in these particular ratios.

3.2.4 Structure from Angle Dependence

The strongest evidence for the utility of this model comes from the addition of angular resolution to the process. In the calculations the ejected particle velocities are already specified, so one need only select the particles ejected into a chosen solid angle [3.45]. Many trajectories are needed, of course, to accumulate sufficient information at each angle. Experimentally, one must collimate the emitted beam in a field-free space around the sample before accelerating the ions to the energy filter [3.46]. The main effect to be expected from angular resolution is that of simple channelling. On a {100} face-centered surface, for example, atoms with polar emission angles $>0^\circ$ (i.e., away from the surface normal) have a relatively unobstructed path between neighboring atoms, but are scattered if they move directly toward a neighboring atom. Thus for Cu, for example, a collection plate above the surface would record a spot pattern with fourfold symmetry. Such spot patterns have been known for some time [3.47], but the first results under well-controlled surface conditions are those of HOLLAND et al. [3.46], shown in Fig.3.8. The Cu{001} surface was exposed to 1200 L of O_2 at 298 K, a procedure known to produce a $c(2\times2)$ oxygen structure. The polar angle was 45°. The azimuthal angle, defined as zero in the <100> direction, was varied over 360° by rotating the crystal about the surface normal. Both the Cu^+ and O^- yields show the expected fourfold symmetry. The shift of the O^- maximum by 45° with respect to the Cu^+ maximum suggests that the oxygen atoms reside in the fourfold holes in the Cu {001} surface. The lower portion of Fig.3.8 shows the agreement between the fourfold averaged data and the angular variations in Cu and O emission calculated for oxygen in the fourfold holes 1.2 Å above the Cu plane. The calculations also show substantial sensitivity of the angular patterns to placement of adatoms in terms of both registry with the substrate and height above the surface. The most recent angular-dependent results are for the CO/Ni {001} system [3.48]. In that paper, GIBBS et al. showed excellent agreement between experimental polar and azimuthal plots

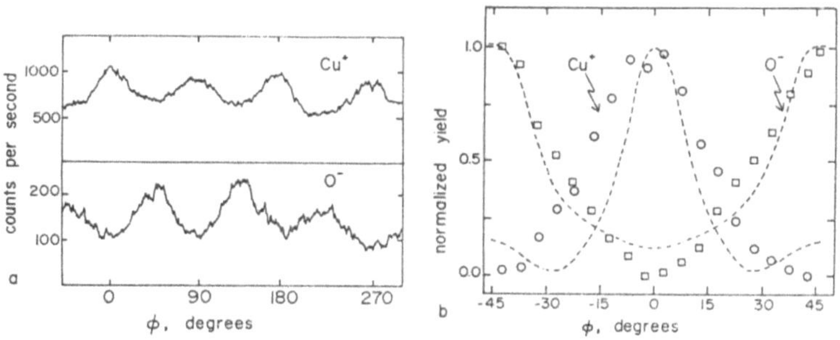

Fig.3.8a,b. Azimuthal angular distributions for oxygen on Cu{001}. (a) The Cu^+ and O^- azimuthal plots recorded from the spectrometer. The primary ion is 1500 eV Ar^+ at a dose of 10^{13} ions/cm^2. (b) A fourfold average of data in (a) with the minimum intensity subtracted from each curve. The dashed lines represent the calculated curve for O placed in a fourfold bridge site 1.2 Å above the Cu plane. The circles represent the Cu^+ intensity and the squares the O^- intensity [3.46]

and calculated plots based on the c(2 × 2) structure with CO atop in the linear bonding position. These new data clearly illustrate two additional important points.

The first has to do with ionization. The success of the model in predicting angular-dependent spectra suggests that ionization contributions in the systems studied so far must not be strongly dependent on angle. One effect of ionization does need to be considered, however. In the experiment, as an ion moves away from the metal surface, its momentum in the normal direction is retarded by the image force. The bending of trajectories by this image force can be significant and must be included to obtain a good match between calculations and experiment [3.48].

The second point has to do with the secondary ion energy. One effect of this energy on angular-dependent data can be appreciated by looking at Fig. 3.7. At the end of the collision cascade, when the slowest atoms are ejected, many of the solid atoms are out of position and the structural relationships are blurred. Thus one would expect a sharpening of angular plots if relatively high-energy (>10 eV) secondary ions were selected. These ions are the ones emitted early in the cascade, when the structure is still preserved. This effect has been demonstrated clearly by GIBBS et al. [3.48]. The secondary ion energy itself can also be a useful property, provided the SIMS prefilter has sufficient energy resolution. Figure 3.9 shows the secondary ion energy distribution for CO on Ni{001} as a function of polar angle. The agreement between experiment and the calculations, which assume linear bonding and include the image charge effect, is quite good.

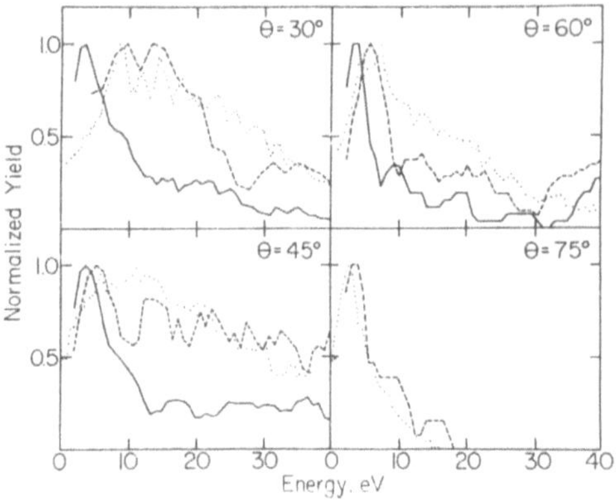

Fig.3.9. Energy distributions for Ni^+ ejected from Ni{001} c(2 × 2)-CO due to bombardment by 1000 eV Ar^+ ions at normal incidence. The three curves in each panel are identified as follows: (——) calculated Ni distribution, (---) calculated Ni^+ distribution with inclusion of the image correction, and (···) experimental points for Ni^+. In each case, $\phi = 0$ [3.48]

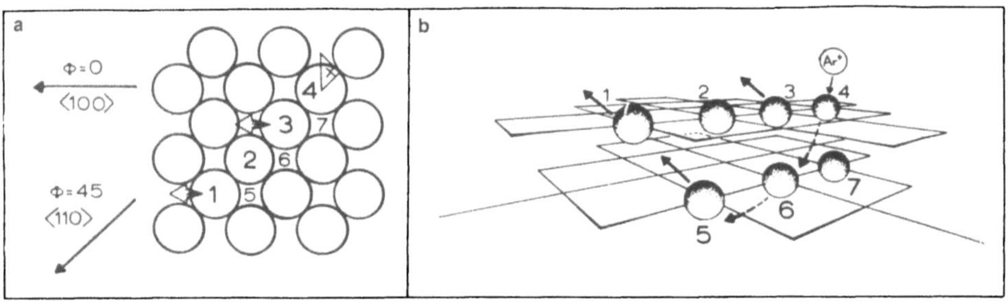

Fig.3.10a,b. Mechanism of formation of the Ni_2 dimer which preferentially ejects in the <100> direction, contributing the majority of intensity to the peak in the angular distribution. (a) Ni{001} showing the surface arrangement of atoms. The numbers are labels while the X denotes the Ar^+ impact point for the mechanism shown in Fig.3.8b. Atoms 1 and 3 eject as indicated by the arrows forming a dimer, which is preferentially moving in a <100> direction. (b) Three-dimensional representation of the Ni_2 dimer formation process [3.49]

A final point concerning angular-dependent experiments is made in Fig.3.10 from HOLLAND et al. [3.49]. The collision cascade shown for formation of the 1,3 dimer accounts for nearly all the high-energy dimers shown by calculations to emit in the <100> ($\phi = 0^{\circ}$) direction. This result suggests that one may be able to choose multimer-forming mechanisms by choosing proper collection angles and energies. This raises most interesting possibilities for the study of ordered alloys.

3.2.5 Reactive Intermediates

The importance of the concept that strong intramolecular bonds can facilitate intact emission of molecules from surfaces is well illustrated in recent studies of reactive intermediates by SIMS. DRECHSLER et al. [3.50] have used

SIMS to identify NH_{ads} as the active intermediate during temperature-programmed decomposition of NH_3 on Fe{110}. They looked first at adsorption of NH_3 at 130 K, where the adsorption is known to be molecular [3.51]. SIMS spectra showed $Fe_n(NH_3)_m^+$ with n and m equal to 1 and 2 and the series NH_x^+ with x = 0 to 4. Data from a series of measurements in the temperature range 130 to 270 K showed that the ratios of yields of NH_y^+/NH_3^+ for y = 0, 1, and 2 were independent of coverage and temperature and surprisingly (although probably accidentally) close to the fragmentation ratios for electron impact ionization of NH_3 in the gas phase in the same spectrometer. This analysis indicates that NH_y fragments were formed by fragmentation of NH_3^+ and that the adsorption was molecular over the full temperature range, as expected. SIMS during thermal decomposition of NH_3 preadsorbed at 130 K or at room temperature showed that all N-containing species were removed from the surface by 350-400 K but that at temperatures above 270 K, NH^+ was no longer proportional to NH_3^+. When the data were normalized by subtracting from measured values the amounts of NH_2^+ and NH^+ expected from the measurements of fragmentation of NH_3^+ at low temperature, the curves in Fig.3.11 were produced. The data show that all the NH_2^+ can be accounted for by fragmentation. Thus, the surface species formed during decomposition at 350 K and identified by UPS as either NH_{ads} or NH_{2ads} is shown unequivocally by SIMS to be NH_{ads}.

A second application of this type is the SIMS-FDS-AES study by MOHRI et al. of formic acid decomposition on Ni [3.52]. Figure 3.12 shows their SIMS data after exposure of a nickel surface to HCOOD at 1×10^{-7} torr for 10 minutes. The ions of special interest are HCO^+ and $HCOO^-$. When DCOOH was used, DCO^+ and $DCOO^-$ were dominant. In both cases the acidic H or D was lost as H_2 or D_2, which desorbed immediately after the room-temperature adsorption. As was the case for NH_3 on Fe{110}, direct fragmentation by ion impact was minor. SIMS spectra obtained during flash desorption showed that the DCO^+ ion from DCOOH on Ni had a maximum at 320 K and disappeared at a temperature of about 390 K. The desorption of CO, on the other hand, was maximum at ~410 K. This lack of correspondence in temperature suggests that the surface decomposition step, $DCO_{ads} \rightarrow D_{ads} + CO_{ads}$, precedes CO desorption.

The NH_3/Fe{110} and DCOOH/Ni studies and others [3.53-56] demonstrate the sensitivity of SIMS to surface molecular species and forecast a role for SIMS in elucidating other reactions on surfaces.

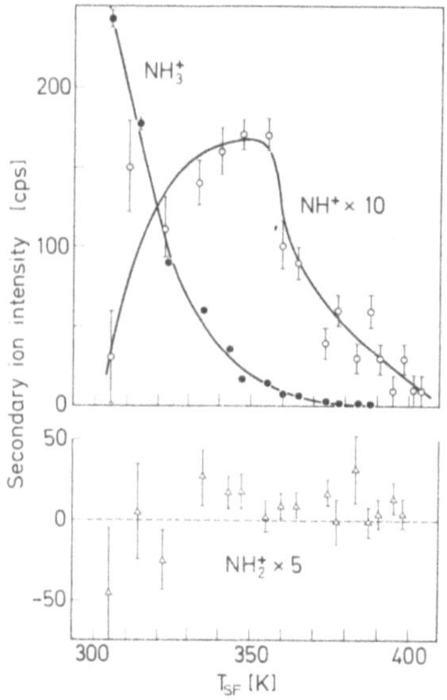

Fig.3.11. Change of secondary ion yields of NH_3^+ and fragmentation-corrected NH^+ and NH_2^+ during heating of a Fe{110} surface after adsorption of NH_3 at room temperature [3.50] (North-Holland Publishing Company, Amsterdam, 1979)

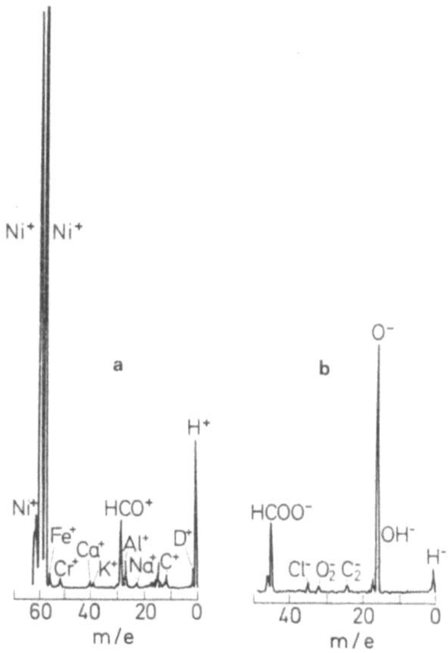

Fig.3.12a,b. SIMS spectra of a nickel surface after exposure to HCOOD at 1×10^{-7} torr for 10 min [3.52] (North-Holland Publishing Company, Amsterdam, 1978)

3.3 Molecular SIMS

3.3.1 SIMS of Molecular Solids

Molecular solids represent a limiting case of the phenomena discussed in the previous section. Here, although the model of single molecules on an ordered surface no longer applies, intramolecular bonds are strong and intermolecular bonds are relatively weak. SIMS experiments by COOKS et al. and others on these materials have been very successful in detecting ions of structural significance [3.57-59]. In many cases large organic molecules (M) are seen in the spectrum in combination with metal cations. More than twenty different alkali, transition, and other metals, for example, have been found to cationize 1,10 phenanthroline [3.60]. Metal-molecule contact can be achieved either by scratching the compound onto a metal foil substrate or by mixing the sample compound with metal salts. As shown in the lower spectrum of Fig.3.13, cationization can also occur when the molecule is adsorbed onto the surface directly from the gas phase [3.61]. In this case a background

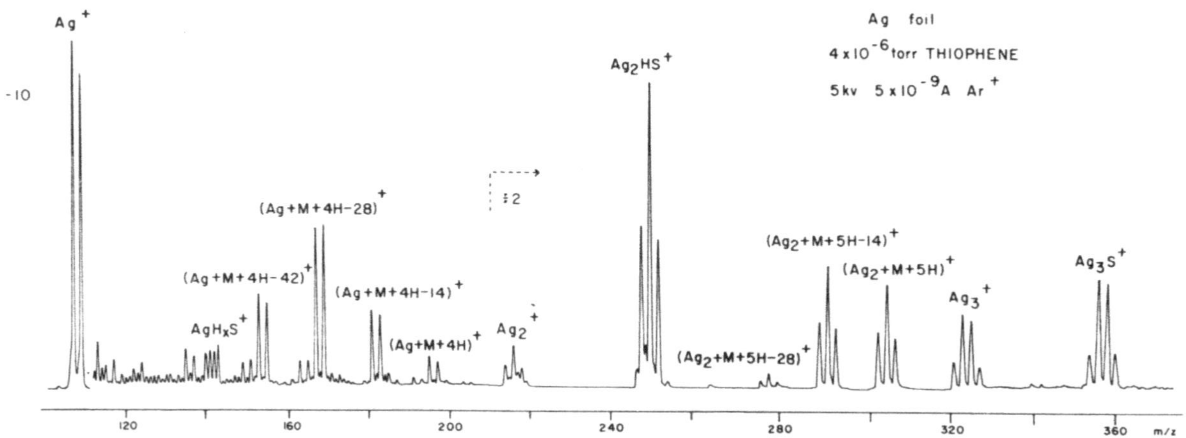

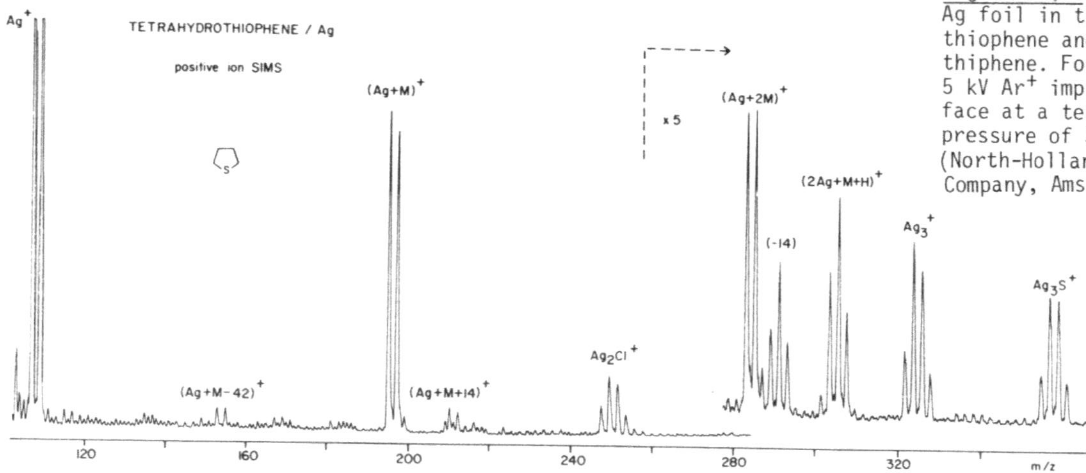

Fig.3.13a,b. SIMS spectra for Ag foil in the presence of (a) thiophene and (b) tetrahydrothiphene. For (b) 5×10^{-9} A of 5 kV Ar^+ impinged on the surface at a tetrahydrothiophene pressure of 3×10^{-6} torr [3.61] (North-Holland Publishing Company, Amsterdam)

pressure of 3×10^{-6} torr of tetrahydrothiophene over a silver foil produces high yields of $(Ag + M)^+$, $(Ag + 2M)^+$, and $(2Ag + M + H)^+$, where M stands for the intact tetrahydrothiophene molecule. Note that the Ag 107, 109 isotope distribution aids in identifying these species.

An especially advantageous case for SIMS of organic solids occurs when preformed ions exist in the matrix. Quaternary ammonium salts, for example, give high ion yields [3.62]. BUSCH et al. have developed this propensity in a derivitization process which maximizes sensitivity by utilizing reagents that create charge centers on the molecules of interest [3.63]. In systems which do not consist of precharged species, and in which cationization does not readily occur, radical molecular ions can sometimes be observed [3.51].

Molecular SIMS is not without complications. Long-range mixing of species components has been observed [3.64]. This phenomenon is thought to be caused by bombardment-assisted surface diffusion [3.64], but its control and analytical implications need further investigation. SIMS of low-temperature condensed phases has also yielded some complexities. SIMS of solid NO, for example, gives NO^+ as the most abundant positive ion but also produces $[NO(N_2O_3)_n]^+$ ions with n from 1-11 [3.65]. The same paper showed that N_2O, N_2O_4, and N_2O_3 form a series of ions of the type $[NO(N_2O_3)_n(N_2O_4)m]^+$. Although the reactions apparent in the low-temperature spectra of the nitrogen oxides need further study, solid H_2O, N_2, and CO give simpler spectra – $H(H_2O)_n^+$ [3.66], $(N_2)_n^+$ and $N(N_2)_n^+$ [3.67] and $C(CO)_n^+$, $C_2(CO)_n^+$, and $(CO)_n^+$ [3.67] – and complex sulfur and nitrogen oxyanions show evidence for intact emission [3.68,69].

3.3.2 Thiophene on Silver

Reactions in the surface layer have also been observed for thiophene on Ag [3.61]. In Fig.3.13a, M represents the thiophene molecule. The appearance of $(Ag + M + 4H)^+$ and $(Ag_2 + M + 5H)^+$ suggests a strong tendency of the thiophene to undergo self-hydrogenation on the silver surface. The spectrum in Fig.3.13b, where M is tetrahydrothiophene, confirms the hydrogenation assignment. To test whether or not the reaction was ion driven, thiophene was passed over a silver powder catalyst in a conventional pulsed microreactor. Chromatographic analysis of the products, showed that tetrahydrothiophene can be formed at 523 K. In this case the reaction of thiophene on the silver surface is stoichiometric rather than catalytic, but it is significant that the reaction was discovered first in a SIMS experiment.

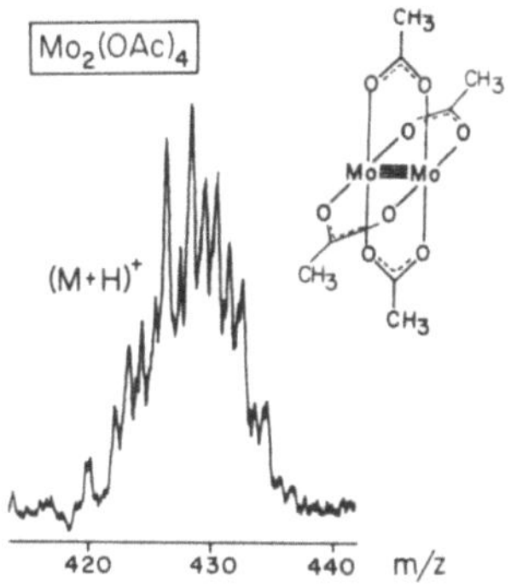

Fig.3.14. Positive ion SIMS spectrum of tetra-μ-acetatodimolybdenum, $Mo_2(OAc)_4$, in the molecular ion region. The isotope distribution is characteristic of the presence of two Mo atoms [3.70]. (Reprinted with permission. Copyright American Chemical Society, 1981)

3.3.3 Inorganic Complexes

Also of special interest for catalysis are applications of SIMS to inorganic complexes. Figure 3.14, from the work of PIERCE et al. [3.70] shows the presence of the $(M + H)^+$ ion for tetra-μ-acetatodiomolybdenum, $[MO_2(OAc)_4]$. The complex isotope distribution clearly indicates the presence of two Mo atoms in this ion. The only fragment ion found for this compound was $Mo_2(OAc)_3^+$. Work on other complexes showed $(RhCl_2Py_4)^+$ from $RhCl_2Py_4Cl$, Ph_3Sn^+ from Ph_3SnCl, $Fe(acac)_2^+$, and $[Ag + Fe(acac)_3]^+$ from $Fe(acac)_3$. In all of these cases, the best spectra were obtained when the inorganic complexes were mixed with NH_4Cl. This matrix has the effect of minimizing intermolecular processes [3.71], but it does not prevent the disadvantageous ligand mixing and cross cationization that also occur in electron impact and chemical ionization of mixtures of β-diketonate complexes such as $Fe(acac)_3$ [3.72]. Although complications such as ligand mixing may ultimately prevent success with some systems, studies of the local chemistry of metal clusters in catalysts through the preservation of metal-ligand bonds in emitted clusters from SIMS of supported transition-metal complexes or even homogeneous catalysts in low-volatility solvents [3.73] are applications to catalysis which may come from extensions of these first studies of inorganic complexes.

3.4 Complex Surfaces

3.4.1 Proximity

An important class of heterogeneous catalysts consists of metals and promoters supported on high-area, porous, insulator supports. Since the distribution of components on the support surface is rarely uniform, large variations in local chemical environment can be expected. The potential of SIMS to probe the proximity of elements on a surface through the formation of

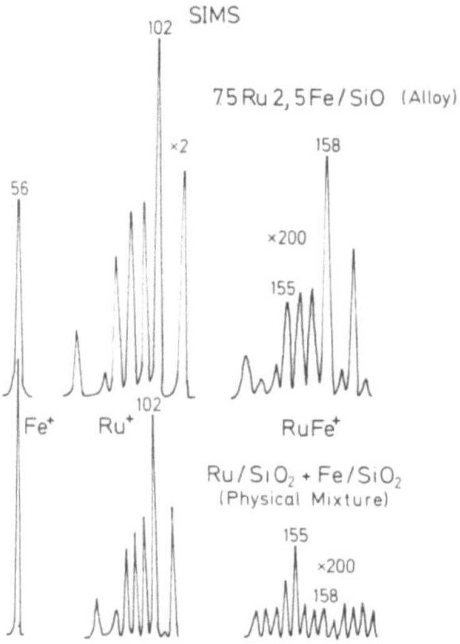

Fig.3.15. Positive ion SIMS spectra of a 7.5 Ru 2.5 Fe/SiO_2 alloy catalyst and a Ru/SiO_2 + Fe/SiO_2 physical mixture.in the Fe^+, Ru^+, and $FeRu^+$ mass ranges. The characteristic isotope distribution shown for Ru^+ allows for easy identification of $RuFe^+$ clusters

secondary cluster ions stands with EXAFS [3.74] as one of few possibilities for local chemical characterization. As a first step in developing this approach we have examined 2.5 Fe 7.0 Ru/SiO_2, where the numbers indicate wt.% of Fe and Ru supported on cabosil EH-5. Independent Mössbauer spectroscopic studies of this catalyst both at Purdue University [3.75] and elsewhere [3.76], have clearly shown that incipient wetness preparation, careful drying, and subsequent reduction at 723 K in flowing H_2 at 1 atmosphere cause all the Fe to be in intimate contact with Ru in small (4-10 nm) metallic particles. SIMS of the catalyst after passivation by slow exposure to O_2 is shown in the upper part of Fig.3.15. The full spectrum includes a variety of metal and metal-oxygen clusters but only Fe^+(m/e = 56 only), Ru^+, and $FeRu^+$ are shown. Mass/charge 102 marks the strongest peak of the Ru isotope distribution, which extends from 96 to 104. Work on unsupported FeRu alloy powders has confirmed our ability to detect $FeRu^+$ ions [3.77] and their appearance in the spectrum of the supported alloy is thus expected from the known proximity of Fe and Ru. To test the specificity of the result, we prepared a physical mixture of Fe/SiO_2 and Ru/SiO_2, having the same overall wt.% of the metals. The Fe/SiO_2 and Ru/SiO_2 were carefully dried separately, mixed, and then reduced and passivated before the SIMS spectrum was taken. The results, in the lower spectra of Fig.3.15, show essentially no $FeRu^+$. A small background is expected in this region from $RuSi_2^+$. Interestingly, a SIMS spectrum of a physical mixture mixed before drying and subsequent reduction did show $FeRu^+$. The Mössbauer effect confirmed, however, that metal mobility during the dehydration step was sufficient to permit nearly half of the iron to alloy with Ru. These experiments are only a beginning, but they suggest that SIMS can extract proximity information from highly heterogeneous surfaces.

3.4.2 Prospects for Catalysis

As a class of materials, catalysts pose some special problems for the SIMS measurement. For example, the insulating character of typical catalyst supports can cause sample charging. This effect can seriously degrade the quality of spectra, but has been adequately compensated for in experiments appearing to date by dilution of the sample in a conducting medium or by the electron flooding techniques familiar in XPS and AES [3.78,79]. Another general problem with porous catalysts is one common to all low-energy particle spectroscopies. The measurement "sees" the outside surface of the sample but the active surface is inside the pores. Thus one must be aware that spectra may not be representative of the active surface. Tests for such problems have been devised for XPS [3.80] and these may well be adaptable to SIMS. A final experimental consideration is the need for a valid pretreatment capability to establish the catalytic chemistry of interest and protect it from contamination until the experiment is completed. Ultrahigh-vacuum transfer of the sample to and from a chamber in which it can receive normal laboratory pretreatments such as reduction in 1 atm of flowing H_2 at 773 K is one way to accomplish this goal [3.81].

Applications of SIMS to catalysts are relatively few so far. Early work on oxide catalysts [3.82,83] has produced an array of cluster ions which are tantalizingly consistent with intact emission assignments, i.e., all ions seen can be ascribed to groups of contiguous atoms in the oxide structure. Similar assignments have been made for SIMS spectra of MICA [3.84]. In light of the clear evidence in metal systems for the formation of clusters by recombination over the surface, however, caution is advised in interpreting data from oxides until a more thorough understanding of emission from these systems is in hand. The results to date and correlations of cluster ion yields with oxidation state [3.85] suggest that this will be an interesting field. Isotope labelling of labile oxygen in partial oxidation catalysts by back titration of a partially reduced catalyst with $^{18}O_2$ is an alternative way to probe the local chemistry of catalytic sites. The appearance of ^{18}O in cluster ions with specific cations would indicate proximity of that cation to the oxidation/reduction site.

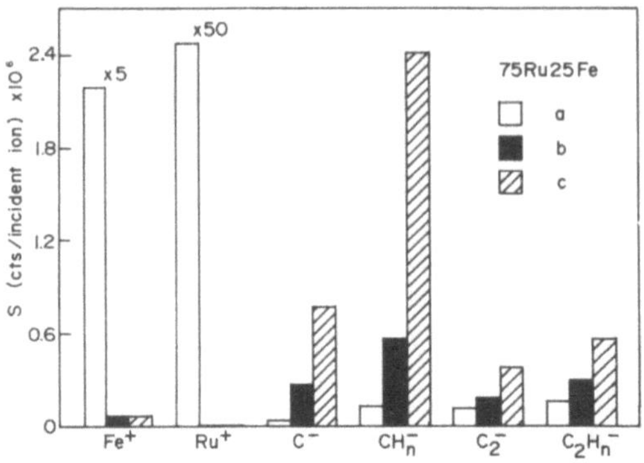

Fig.3.16. SIMS measurement of carbon buildup on 75Ru25Fe alloy powder catalyst during Fischer-Tropsch synthesis, (a) reduced, (b) $3H_2/CO$, 1h, 573 K; (c) $3H_2/CO$, 4 h, 613 K [3.81]

The ability of SIMS to see hydrogen is another important asset. SIMS sensitivity to hydrogen adsorption [3.86] and to the presence of OH groups [3.87] has been well documented and should have important applications in catalyst studies. The adsorption phenomena already discussed for single crystals and molecular SIMS should also carry over to catalyst work. A preliminary study of this kind is shown in Fig.3.16 [3.81]. Unsupported FeRu powders with surface areas of the order of 10 m^2/g were used as Fischer-Tropsch synthesis catalysts in a pretreatment chamber attached to the spectrometer. XPS of the surfaces after reaction showed the growth of thick carbon overlayers which poisoned the catalysts as time on stream increased. The SIMS data in Fig. 3.16 show that the overlayers have significant hydrocarbon content. A detailed analysis of hydrocarbon structure has not yet been undertaken but the possibility of studying coke precursors is demonstrated.

The loss of Fe^+ and Ru^+ as the overlayer thickness increases in Fig.3.16 is a reminder that, in principle, the monomer ion yields carry quantitative information. While such analyses are used routinely in high-current density, depth-profiling experiments [3.3], they are not straightforward at static conditions, in part because of the high sensitivity of the yields to the surface chemical state. In the study of FeRu alloy powders, a plot of Fe^+ and Ru^+ as a function of composition for carefully prereduced samples showed a precipitous drop in Ru^+ after addition of only 3% iron [3.81]. The loss of Ru^+ intensity was not compensated for by a stoichiometric increase in Fe^+, nor was the small Fe enrichment at the surface detected by XPS quantitatively consistent with changes in ion yield ratios. Thus an additional, perhaps electronic, factor has entered. Such effects must be better understood before quantitative analysis can be made reliable. In principle, the Fe_2^+, Ru_2^+, and $FeRu^+$ ion yields should also reflect surface composition, but they were not measured in these experiments. Some of the quantitative complications can be minimized in ion yield ratios, however, as demonstrated by the good quantitative agreement found by BORDOLI et al. for SIMS and conventional measurements of heats of adsorption of CO on Ni [3.16].

3.5 Conclusions

Static conditions have made SIMS a useful surface science tool. Several studies of adsorption on ordered surfaces illustrate a correspondence of SIMS yields with structural characteristics of the surface, but analysis of such data requires a model of the SIMS process. Such a model appears to be forthcoming from classical dynamics calculations which predict many important SIMS features in spite of the fact that they do not yet include the ionization process. Both the model and the data to date suggest that detailed structural information, including adatom placement and bond distances, is best obtained from angular-dependent SIMS measurements. Molecular structure of adlayer species, however, may come from direct emission of such species or their inclusion in cluster ions. This possibility appears to be favored when intramolecular bonds are stronger than intermolecular or molecule-surface bonds. This concept is supported by the success of SIMS in identifying reactive intermediates during thermal decomposition of small molecules on metal surfaces and by the rich store of chemical structure information supplied in applications of SIMS to molecular solids.

The success of SIMS in identifying structurally significant ions from transition-metal complexes provides a bridge to applications in catalysis. Though catalyst studies have barely begun, the potential of SIMS for answering such important questions as the proximity of elements, the molecular structure of adlayers, and the chemistry of complex oxide surfaces is high.

Acknowledgements. This work has been supported by NSF through grants CHE 78-08728, CPE-7911597 and DMR 77-23798 (PURDUE MRL), by the Purdue Schools of Engineering, and by Amoco Oil Company. Our SIMS research has developed in collaboration with N. Winograd and R.G. Cooks. We thank them, K.L. Busch, and R.A. Walton for valuable discussions and access to manuscripts prior to publication.

References

3.1 R.F.K. Herzog, F.P. Viehböck: Phys. Rev. *76*, 855 (1949)
3.2 N. Winograd: In *Progress in Solid-State Chemistry*, ed. by W.L. Worrell, G.M. Rosenblatt (Pergamon, New York 1981)
3.3 J.A. McHugh: In *Methods of Surface Analysis*, ed. by A.W. Czanderna (Elsevier, New York 1975) Chap.6
3.4 H.W. Werner: Surf. Sci. *47*, 301 (1975)
3.5 G. Slodzian: Surf. Sci. *48*, 161 (1975)
3.6 A.G. Koval, V.N. Mel'nikov, Yu.V. Enikov: Nucl. Instrum. Meth. *132*, 467 (1976)
3.7 A. Benninghoven: Surf. Sci. *35*, 427 (1973)
3.8 P.H. Dawson, P.A. Redhead: Rev. Sci. Instrum. *48*, 159 (1977)
3.9 A. Benninghoven: In *Chemistry and Physics of Solid Surfaces*, ed. by R. Vanselow, S.Y. Tong (CRC, Cleveland 1977) p.207
3.10 M. Barber, J.C. Vickerman, J. Wolstenholme: Surf. Sci. *68*, 130 (1977)
3.11 P.H. Dawson: Phys. Rev. B*15*, 5522 (1977)
3.12 P.H. Dawson: Surf. Sci. *65*, 41 (1977)
3.13 P.H. Dawson: Surf. Sci. *71*, 247 (1978)
3.14 A. Benninghoven, O. Ganschow, L. Wiedmann: J. Vac. Sci. Technol. *15*, 506 (1978)
3.15 M.L. Yu: Surf. Sci. *71*, 121 (1978)
3.16 R.S. Bordoli, J.C. Vickerman, J. Wolstenholme: Surf. Sci. *85*, 244 (1979)
3.17 T. Fleisch, N. Winograd, W.N. Delgass: Surf. Sci. *78*, 141 (1978)
3.18 T. Fleisch, G.L. Ott, W.N. Delgass, N. Winograd: Surf. Sci. *81*, 1 (1979)
3.19 P.H. Holloway, J.B. Hudson: Surf. Sci. *43*, 123 (1974)
3.20 A.M. Horgan, D.A. King: Surf. Sci. *23*, 259 (1970)
3.21 P.R. Norton, R.L. Trapping, J.W. Goodale: Surf. Sci. *65*, 13 (1977)
3.22 J.C. Tracy: J. Chem. Phys. *56*, 2736 (1972)
3.23 M.A. Passler, A. Ingatieu, J.A. Schultz, J.W. Rabalais: Chem. Phys. Lett. *82*, 198 (1981)
3.24 S. Andersson: Solid State Commun. *21*, 75 (1977)
3.25 M. Barber, J.C. Vickerman, J. Wolstenholme: J. Chem. Soc. Faraday Trans. 1 *72*, 40 (1976)
3.26 M. Barber, R.S. Bordoli, J.C. Vickerman, J. Wolstenholme: In *Proc. 7th Int. Vacuum Congr. and 3rd Int. Conf. on Solid Surfaces*, Vienna 1977, p.983
3.27 B.J. Garrison, N. Winograd, D.E. Harrison: J. Chem. Phys. *73*, 3473 (1980)
3.28 H. Hopster, C.R. Brundle: J. Vac. Sci. Technol. *16*, 548 (1979)
3.29 P.H. Dawson, W.C. Tam: Surf. Sci. *91*, 153 (1980)
3.30 T.E. Madey, D. Menzel: Jpn. J. Appl. Phys. Suppl. *2*, pt.2, 229 (1974)
3.31 E.D. Williams, W.H. Weinberg: Surf. Sci. *82*, 93 (1979)
3.32 G.E. Thomas, W.H. Weinberg: J. Chem. Phys. *70*, 954 (1979)
3.33 H. Pfnür, D. Menzel, F.M. Hoffmann, A. Ortega, A.M. Bradshaw: Surf. Sci. *93*, 431 (1980)
3.34 L.L. Lauderback, W.N. Delgass: To be published
3.35 T.E. Madey, H.A. Engelhardt, D. Menzel: Surf. Sci. *48*, 304 (1975)
3.36 S.L. Parrott, G. Praline, B.E. Koel, J.M. White, T.N. Taylor: J. Chem. Phys. *71*, 3352 (1979)
3.37 J.C. Fuggle, T.E. Madey, M. Steinkilberg, D. Menzel: Surf. Sci. *52*, 521 (1975)
3.38 D.E. Harrison, Jr., P.W. Kelly, B.J. Garrison, N. Winograd: Surf. Sci. *76*, 311 (1978)
3.39 N. Winograd, D.E. Harrison, Jr., B.J. Garrison: Surf. Sci. *78*, 467 (1978)
3.40 B.J. Garrison, N. Winograd, D.E. Harrison, Jr.: Phys. Rev. B*18*, 6000 (1978)

3.41 B.J. Garrison, N. Winograd, D.E. Harrison, Jr.: J. Chem. Phys. *69*, 1440 (1978)
3.42 N. Winograd, B.J. Garrison: Acc. Chem. Res. *13*, 406 (1980)
3.43 B.J. Garrison: J. Am. Chem. Soc. *102*, 6553 (1980)
3.44 N. Winograd, B.J. Garrison, T. Fleisch, W.N. Delgass, D.E. Harrison, Jr.: J. Vac. Sci. Technol. *16*, 629 (1979)
3.45 N. Winograd, B.J. Garrison, D.E. Harrison, Jr.: Phys. Rev. Lett. *41*, 1120 (1978)
3.46 S.P. Holland, B.J. Garrison, N. Winograd: Phys. Rev. Lett. *43*, 220 (1979)
3.47 G.K. Wehner: Phys. Rev. *102*, 690 (1956)
3.48 R.A. Gibbs, S.P. Holland, K.E. Foley, B.J. Garrison, N. Winograd: J. Chem. Phys. (in press)
3.49 S.P. Holland, B.J. Garrison, N. Winograd: Phys. Rev. Lett. *44*, 756 (1980)
3.50 M. Drechsler, H. Hoinkes, H. Kaarmann, H. Wilsch, G. Ertl, M. Weiss: Appl. Surf. Sci. *3*, 217 (1979)
3.51 M. Weiss, G. Ertl, F. Nitschke: Appl. Surf. Sci. *2*, 614 (1979)
3.52 M. Mohri, H. Kakibayashi, K. Watanabe, T. Yamashina: Appl. Surf. Sci. *1*, 170 (1978)
3.53 M. Barber, J.C. Vickerman, J. Wolstenholme: J. Catal. *42*, 48 (1976)
3.54 P.H. Dawson: J. Vac. Sci. Technol. *14*, 786 (1977)
3.55 M. Mohri, M. Hashiba, K. Watanabe, T. Yamashina: Z. Phys. Chem. *109*, 233 (1978)
3.56 A. Benninghoven, P. Beckmann, D. Greifendorf, M. Schemmer: Appl. Surf. Sci. *6*, 288 (1980)
3.57 R.J. Day, S.E. Unger, R.G. Cooks: Anal. Chem. *52*, 557A (1980)
3.58 A. Benninghoven, W. Sichtermann: Anal. Chem. *50*, 1180 (1978)
3.59 R.J. Colton: J. Vac. Sci. Technol. *18*, 737 (1981)
3.60 R.J. Day, S.E. Unger, R.G. Cooks: Anal. Chem. *52*, 353 (1980)
3.61 S.E. Unger, R.G. Cooks, B.J. Steinmetz, W.N. Delgass: Surf. Sci. (to (in press)
3.62 R.J. Day, S.E. Unger, R.G. Cooks: J. Am. Chem. Soc. *101*, 501 (1979)
3.63 K.L. Busch, S.E. Unger, A. Vincze, R.G. Cooks, T. Keough: J. Am. Chem. Soc. *104* (1982)
3.64 K.L. Busch, S.E. Unger, R.G. Cooks: Conf. on Ion Formation from Organic Solids, Münster F.R.G., 1980
3.65 R.G. Orth, H.T. Jonkman, J. Michl: J. Am. Chem. Soc. *103*, 1564 (1981)
3.66 G.M. Lancaster, F. Honda, Y. Fukuda, J.W. Rabalais: J. Am. Chem. Soc. *101*, 1951 (1979)
3.67 H.T. Jonkman, J. Michl: J. Am. Chem. Soc. *103*, 733 (1981)
3.68 J.D. Ganjei, R.J. Colton, J.S. Murday: Int. J. Mass Spectrom. Ion Phys. *37*, 49 (1981)
3.69 E. dePauw, J. Marien: Int. J. Mass Spectrom. Ion Phys. *38*, 11 (1981)
3.70 J. Pierce, K.L. Busch, R.A. Walton, R.G. Cooks: J. Am. Chem. Soc. *103*, 2583 (1981)
3.71 S.E. Unger, R.J. Day, R.G. Cooks: Int. J. Mass Spectrom. Ion Phys. *39*, 231 (1981)
3.72 J.L. Pierce, K.L. Busch, R.G. Cooks, R.A. Walton: Inorg. Chem. (to be published)
3.73 K.L. Busch, R.G. Cooks: To be published
3.74 G.H. Via, J.H. Sinfelt, F.W. Lytle: J. Chem. Phys. *71*, 690 (1979)
3.75 F. Fortunato: M.S. Thesis, Purdue University (1976)
3.76 M.A. Vannice, Y.L. Lam, R.L. Garten: Am. Chem. Soc. Div. Pet. Chem. Prepr. *23*, 495 (1978)
3.77 G.L. Ott, W.N. Delgass, N. Winograd, W.E. Baitinger: J. Catal. *56*, 174 (1979)
3.78 G. Mueller: Appl. Phys. *10*, 317 (1976)

3.79 K. Wittmaack: J. Appl. Phys. *50*, 493 (1979)
3.80 P.J. Angevine, J.C. Vartuli, W.N. Delgass: Proc. of the 6th Int. Congr. Catal. (The Chem. Soc., London 1977) Vol.2, p.611
3.81 G.L. Ott, T. Fleisch, W.N. Delgass: J. Catal. *65*, 253 (1980)
3.82 R. Buhl, A. Preisinger: Surf. Sci. *47*, 344 (1975)
3.83 M. Barber, P.K. Sharpe, J.C. Vickerman: J. Catal. *41*, 240 (1976)
3.84 M.G. Dowsett, R.M. King, E.H.C. Parker: Surf. Sci. *71*, 541 (1978)
3.85 C. Plog, L. Wiedman, A. Benninghoven: Surf. Sci. *67*, 565 (1977)
3.86 A. Benninghoven, P. Beckmann, D. Greifendorf, K.H. Muller, M. Schemmer: Surf. Sci. *107*, 148 (1981)
3.87 A. Benninghoven, P. Beckmann, K.H. Muller, M. Schemmer: Surf. Sci. *89*, 701 (1979)

4. Chemisorption Investigated by Ellipsometry

G. A. Bootsma, L. J. Hanekamp, and O. L. J. Gijzeman

With 18 Figures

4.1 Introduction

A comparison of the various probes applied in surface investigations indicates that photons in and around the visible region of the spectrum (1-4 eV) are generally less atom specific and less surface sensitive than electrons and ions [4.1]. Moreover the limit of detection of optical techniques is usually larger [10^{-1} to 10^{-2} monolayer (ML)] than in electron or ion spectroscopy. However, in contrast to electrons and ions the influence of photons on the studied samples and processes is generally negligible and in situ measurements at high gas pressures or with liquids are possible.

The microscopic parameters probed by the electromagnetic field are the complex polarizability $\tilde{\alpha}$, which describes instantaneous shifts in the charge density and transitions between filled and empty electronic states. In macroscopic theory the relevant parameters are the dielectric constants $\tilde{\varepsilon}_i$ or indices of refraction $\tilde{n}_i$ of the phases i composing the studied sample. Several theoretical approaches have been proposed for the connection between microscopic and macroscopic quantities. It must be admitted, however, that in many cases optical reflection techniques in the visible region of the spectrum are less attractive from the viewpoint of spectroscopy or qualitative analysis, because parameters like polarizability and dielectric constant are not very discriminatory.

Adsorption measurements with ellipsometry were performed by the founders of the method, DRUDE [4.2] and RAYLEIGH [4.3]. The applicability of ellipsometry to fields like passivity and corrosion of metals, adsorption, heterogeneous catalysis, and surface migration (mentioned in [4.4]) has been considerably improved by the availability of ultrahigh-vacuum (UHV) techniques and by the automation of the measurements. Since the study of oxygen adsorption on a clean silicon surface in UHV by ARCHER and GOBELI [4.5], several ellipsometric investigations of adsorption have been performed, first mainly on semiconductors [4.6,7] and more recently also on metals [4.8].

For physisorption, where the adsorbate is held to the surface by the van der Waals forces that arise primarily from the interaction of fluctuating dipoles, the changes induced in the dielectric constant of the substrate are small. Ellipsometric results for krypton on silicon and germanium [4.7,9] and xenon on graphite [4.10,11] could be interpreted with models assuming (sub)monolayers of noble-gas atoms on an unchanged substrate. To explain the experimental results obtained for chemisorption systems, changes in the optical properties of the substrate induced by the adsorbed layer have to be taken into account; for semiconductors these changes have been related to the removal of surface states [4.6,7,9] and for metals to changes in the density and properties of the free and bound electrons at the surface [4.8]. Notwithstanding the existing problems with the interpretation of the measured changes in the ellipsometric parameters, the method is very suitable for, e.g., the determination of (physical) adsorption isotherms and of chemisorption kinetics or reaction kinetics, especially after calibration against other surface analytical techniques.

Ellipsometry involves the measurement of the change in the state of polarization that occurs when a monochromatic beam of polarized light is reflected from (or transmitted through) an interface. As compared with external reflection spectroscopy, in which only intensity changes are measured, the advantage of ellipsometry is its high sensitivity (down to ~0.01 of a monolayer of adsorbed atoms can be detected) and the fact that two (possibly three) experimental quantities are determined at a given wavelength.

Detailed information on the technique and its applications is found in several reviews [4.12-16] and conference proceedings [4.17]. In the next two sections the principles of ellipsometry and some theoretical models for the interpretation of adsorption effects will be briefly summarized. The review of experimental studies in UHV systems in the remaining sections will mostly concern chemisorption processes on metal surfaces, notably on copper.

4.2 Principles of Ellipsometry

The state of polarization of a polarized light beam is characterized by the phase and amplitude relations of the two components into which the total electric field vector $\underline{E}$ of the light wave can be resolved. These components are chosen parallel with (E_p) and perpendicular to (E_s) the plane of incidence,

$$E_p = A_p \exp[i(\omega t + \delta_p)] \quad , \tag{4.1a}$$

$$E_s = A_s \exp[i(\omega t + \delta_s)] \quad , \tag{4.1b}$$

where ω is the angular frequency ($2\pi\nu$). In general, upon reflection a change will occur in both the phases δ and the amplitudes A of the two components. These changes are described by the two parameters Δ and ψ, defined as

$$\Delta = (\delta_p^r - \delta_s^r) - (\delta_p^i - \delta_s^i) \quad , \tag{4.2}$$

$$\tan\psi = \frac{A_p^r}{A_s^r} \Big/ \frac{A_p^i}{A_s^i} \quad , \tag{4.3}$$

where the superscripts r and i correspond to the reflected and incident beams. The parameters Δ and ψ are measured with the ellipsometer. The total effect upon reflection is summarized in the equation

$$\frac{r_p}{r_s} = \frac{E_p^r}{E_s^r} \Big/ \frac{E_p^i}{E_s^i} = \tan\psi \exp(i\Delta) \quad . \tag{4.4}$$

Here r_p and r_s are the amplitude reflection coefficients, which can be expressed as functions of the wavelength λ, the angle of incidence ϕ_0, and the optical parameters of the reflecting system (see Fig.4.1).

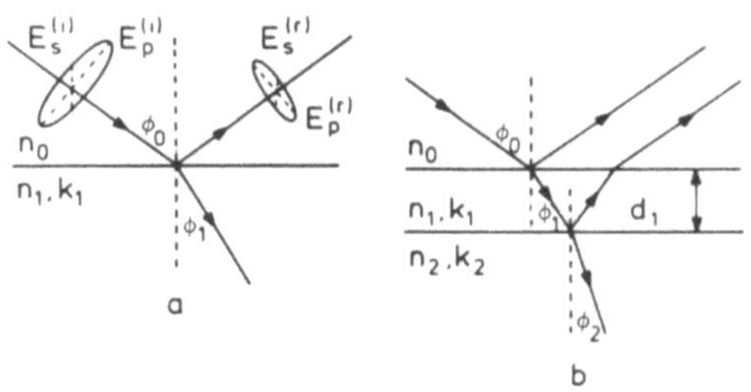

Fig.4.1a,b. Schematic representation of the reflection of polarized light, (a) clean surface, (b) surface covered with one layer

For a clean substrate one has (Fig.4.1a)

$$r_p = \frac{\tilde{n}_1 \cos\phi_0 - n_0 \cos\phi_1}{\tilde{n}_1 \cos\phi_0 + n_0 \cos\phi_1} \quad , \tag{4.5a}$$

$$r_s = \frac{n_0 \cos\phi_0 - \tilde{n}_1 \cos\phi_1}{n_0 \cos\phi_0 + \tilde{n}_1 \cos\phi_1} \quad , \tag{4.5b}$$

where $\tilde{n}_1 = \sqrt{\tilde{\varepsilon}_1} = n - ik$ is the complex refractive index of the substrate, $\tilde{\varepsilon}_1$ its dielectric constant, and n and k are the ordinary refractive index and extinction index. The (complex) angle ϕ_1 is determined by Snell's law

$$n_0 \sin\phi_0 = \tilde{n}_1 \sin\phi_1 \quad . \tag{4.6}$$

In this case a measurement of Δ and ψ yields the optical constants n_1 and k_1, or alternatively, the complex dielectric constant $\tilde{\varepsilon}_1$ if n_0 is known,

$$\tilde{\varepsilon}_1 = \varepsilon_1' - i\varepsilon_1'' \quad , \tag{4.7}$$

$$\varepsilon_1' = n^2 - k^2 \quad , \tag{4.8}$$

$$\varepsilon_1'' = 2\, nk \quad . \tag{4.9}$$

For a substrate covered with a layer of thickness d_1, r_p and r_s are given by (cf. Fig.4.1b)

$$r_p = \frac{r_{Ip} + r_{IIp} \exp(-ix_1)}{1 + r_{Ip}r_{IIp} \exp(-ix_1)} \quad , \tag{4.10a}$$

$$r_s = \frac{r_{Is} + r_{IIs} \exp(-ix_1)}{1 + r_{Is}r_{IIs} \exp(-ix_1)} \quad . \tag{4.10b}$$

Here r_I and r_{II} are the amplitude reflection coefficients at the interface 0-1 and 1-2 respectively, given by the appropriate modification of (4.5a,b) for p and s polarization, and x_1 denotes the phase difference due to one internal reflection within the layer

$$x_1 = \frac{4\pi}{\lambda} \tilde{n}_1\, d_1 \cos\phi_1 \quad . \tag{4.11}$$

So for this system we have, by combining (4.4-6,10,11),

$$\tan\psi \exp(i\Delta) = f(\lambda, \phi_0, n_0, n_1, k_1, d_1, n_2, k_2) \quad . \tag{4.12}$$

The parameters λ, ϕ_0, and n_0 are usually known, as are in many cases the substrate optical constants n_2 and k_2. The changes in Δ and ψ upon formation of a layer are then given by

$$\delta\Delta = \bar{\Delta} - \Delta \quad , \tag{4.13a}$$

$$\delta\psi = \bar{\psi} - \psi \quad , \tag{4.13b}$$

and

$$\delta\Delta, \delta\psi = f(\text{known parameters}, n_1, k_1, d_1) \quad . \tag{4.14}$$

Here $\bar{\Delta}$ and $\bar{\psi}$ refer to the clean substrate and Δ and ψ to the substrate covered with a layer. The problem of deriving three unknowns from two measured parameters has been solved by variation of ϕ_0, n_0, or d_1 (all under the assumption that the other optical parameters remain constant), or by making a suitable choice for the value of one of the unknowns, based on other experiments. In some cases the measurement of a third experimental quantity, the change in intensity, can yield a solution. The change in intensity upon reflection or the reflectivity R is given by

$$R_{p,s} = r_{p,s}\, r^*_{p,s} = |r_{p,s}|^2 \quad , \tag{4.15}$$

where r^* is the complex conjugate of r. At normal incidence ($\phi_0 = 0^\circ$) one gets for reflection at a clean substrate, from (4.10,15),

$$R_p = R_s = \left| \frac{\tilde{n}_1 - n_0}{\tilde{n}_1 + n_0} \right|^2 \quad . \tag{4.16}$$

It should be noted that in the foregoing equations several implicit assumptions have been made. First of all, all boundaries have been chosen to be perfect planes without (local) roughness. Secondly, the overlayer has been attributed a physical thickness d and a dielectric constant $\tilde{\varepsilon}_1$. Since even fractions of a monolayer may still be detected with ellipsometry, both the concept of "thickness" and "dielectric constant" of such an overlayer are somewhat nebulous. A more detailed review of this problem has been given in [4.7,8], where references are also given to the original papers. Surface roughness may also be discussed in terms of an "overlayer with holes" on a chemically identical substrate [4.18-23]. Further problems with the present conventional description of ellipsometry are embodied in the precise meaning of a dielectric constant, even for a pure substance. The Maxwell boundary conditions, used to arrive at (4.5a,b), imply a discontinuity in the electric field vector perpendicular to the surface. In a more refined treatment this discontinuity disappears through the use of a nonlocal dielectric tensor $\underline{\underline{\varepsilon}}(\omega,\underline{q})$, different for longitudinal and transverse electromagnetic waves with wavevector $\underline{q}$ and angular frequency ω. Thus the observed dielectric constant, which is mainly measured in the surface region, may differ from the true bulk value [4.24]. For substrates with an overlayer this phenomenon may be modeled by an anisotropic dielectric constant for the overlayer or by a spatially varying one.

At any rate, this will introduce more unknowns in the expression for the amplitude reflection coefficients [4.8,25,26], whereas the number of observable quantities remains the same.

4.3 (Sub)Monolayer Models

When the layers on the clean substrate are very thin (Fig.4.1b with $d_1 \ll \lambda$), (4.14) may be approximated as

$$\delta\Delta = \zeta(\lambda, \phi_0, n_0, n_1, k_1, n_2, k_2)\, d_1 \quad , \qquad (4.17a)$$

$$\delta\psi = -\eta(\lambda, \phi_0, n_0, n_1, k_1, n_2, k_2)\, d_1 \quad . \qquad (4.17b)$$

Expressions for ζ and η have already been derived by DRUDE [4.2] and are therefore called the Drude approximations. For nonabsorbing isotropic films they are quoted in [4.6], for nonabsorbing uniaxial anisotropic layers in [4.27], and for absorbing isotropic layers in [4.28]. However, because computer calculations with the exact equations (4.2-12) are easily performed – e.g., by using the program of McCRACKIN [4.29] – the use of the Drude approximations is somewhat restricted.

To give an impression of the sensitivity of ellipsometry, the calculated values of ζ and η are plotted in Fig.4.2 as a function of the wavelength for transparent films on a clean copper substrate. Since the changes in the ellipsometric parameters can be measured with an accuracy of $\sim 0.01^\circ$, for such films thicknesses of the order of 0.1 Å can be determined from $\delta\Delta$. However, the changes in ψ are of the order of experimental error.

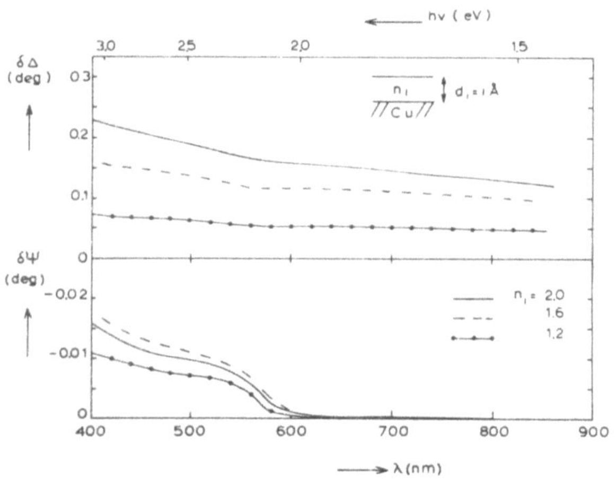

Fig.4.2. Calculated values of $\delta\Delta$ and $\delta\psi$ for nonabsorbing films (n_1, $d_1 = 1$ Å) on a clean copper surface. $n_0 = 1.0$, $\phi_0 = 66^\circ$, $\tilde{n}_2$ from ellipsometric measurements with Cu{110}, $\Omega = 0^\circ$ (see Sect.4.4) [4.30]

The sensitivity of ellipsometry enables one to detect layers with an effective thickness of the order of 1-10% of a monolayer. In the macroscopic description an adsorbed (sub)monolayer is idealized as homogeneous with thickness d_1 and index of refraction n_1 as in Fig.4.2. It has been shown that for *physisorption*, where the interaction between adsorbate and substrate is weak, e.g., krypton on silicon and germanium [4.7,9] and xenon on graphite [4.10,11], the ellipsometric results could be interpreted with a simple extrapolated macroscopic theory in which the relation between n_1 and the microscopic parameters is given by the Lorentz-Lorenz equation (SI units),

$$\frac{n_1^2 - 1}{n_1^2 + 2} = \frac{\alpha\Theta}{3c_m d_m} \quad . \tag{4.18}$$

Here α is the polarizability, c the cross-sectional area, Θ the degree of coverage ($\Theta = c_m/c$, with c_m cross section at $\Theta = 1$), and d_m the diameter of the adsorbed atoms or molecules. In good approximation one gets for (4.17a,b)

$$\delta\Delta = C(\lambda, \phi_0, n_0, n_2, k_2)\,\frac{\alpha\Theta}{c_m} = C'\,\frac{\alpha\Theta}{\lambda} \quad , \tag{4.19a}$$

$$\delta\psi \approx 0 \quad . \tag{4.19b}$$

Expressions (4.18,19) are valid for isotropic films. For the more general case of *anisotropic layers*, of which the anisotropy in n_1 may be due to anisotropy in the molecular polarizability and/or to asymmetry in the shape factor, the extrapolated macroscopic treatment has been summarized in [4.8]. A versatile quasi-microscopic description of anisotropic interfacial layers has been presented by STRACHAN [4.31].

In this approach the interfacial layer is represented by a two-dimensional distribution of Hertzian oscillators, characterized by the scattering indices σ_x and σ_y (both parallel to the phase boundaries) and σ_z (perpendicular to them, plane of incidence xz). By correcting for the field produced by these oscillators on both sides of the interface, first-order expressions may be derived for the amplitude reflection coefficients r_p and r_s [4.6,8,31]. From these expressions the changes in Δ and ψ due to the layer are calculated. For uniaxially anisotropic layers ($\sigma_x = \sigma_y \neq \sigma_z$) the results have been presented in [4.6], and for the general case $\sigma_x \neq \sigma_y \neq \sigma_z$ in [4.8]. The scattering indices $\sigma_j (j = x,y,z)$ may be related to the polarizabilities α_j of the adsorbed atoms or molecules; for absorbing layers the σ_j's and α_j's are taken to be complex,

$$\sigma_j = \sigma_j' - i\sigma_j'' \quad \text{and} \quad \alpha_j = \alpha_j' - i\alpha_j'' \tag{4.20}$$

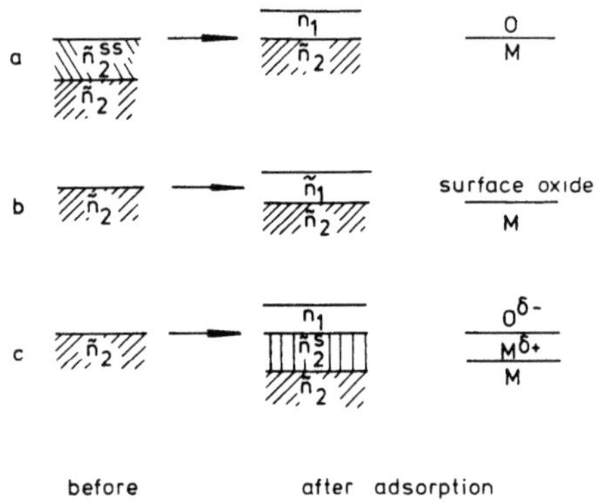

Fig.4.3a-c. Optical models for chemisorption of oxygen

with

$$\sigma_j = \alpha_j/c = \alpha_j\Theta/c_m \quad . \tag{4.21}$$

Microscopic approaches, also applicable to anisotropic layers, have been presented by SIVUKHIN [4.32,33] and by DIGNAM et al. [4.34,35]; for a summary and comparison, see [4.8].

The preceding models have thus far been mainly applied to optical systems with one interfacial layer on an unchanged substrate, such as depicted in Fig.4.1b for the (extrapolated) macroscopic approach. For *chemisorption*, where changes in the electron density distribution of the substrate are to be expected, a simple one-layer description is in general not adequate.

In the model used to interpret the ellipsometric [4.6,7,9] and reflectometric [4.36,37] results obtained for the chemisorption of gases on *semiconductor surfaces* it was assumed that a transition layer at the clean surface, with optical properties different from the bulk ($\tilde{n}_2^{ss}$, associated with surface states), is effectively removed (Fig.4.3a). The measured ellipsometric effects then consist of a contribution from the adsorbed layer itself and a contribution from changes in the substrate surface region,

$$\delta\Delta_{obs} = \delta\Delta_{ad} + \delta\Delta_{ss} \quad , \tag{4.22a}$$

$$\delta\psi_{obs} = \delta\psi_{ad} + \delta\psi_{ss} \quad . \tag{4.22b}$$

The contribution from the adsorbed layer may be estimated in the same way as for physisorbed, nonabsorbing layers, e.g., by applying (4.18,19a,b). Subtraction of these calculated values from the observed values gives the ellipsometric data for the induced substrate change $\delta\Delta_{ss}$ and $\delta\psi_{ss}$.

It appeared that considerable structure was present in the $\delta\Delta$ and $\delta\psi$ curves as a function of wavelength for chemical adsorption on various clean semiconductor surfaces, e.g., Si [4.9,38,39], Ge [4.9,38], GaP [4.40], and

GaAs [4.39,41]. Most of this structure can be ascribed to optical transitions from filled dangling-bond surface states to empty bulk-conduction or surface-states bands [4.7].

By developing the Drude approximation for an absorbing layer on an absorbing substrate under the conditions representative for semiconductors ($n_2^2 \approx n_{ss}^2 \gg 1$, $n_2 > k_2$, $n_{ss} > k_{ss}$), one gets the relation

$$\delta\psi_{ss} \sim (h\nu/n_2)^2 \mathrm{Im}\{\tilde{\varepsilon}_{ss} - \tilde{\varepsilon}_2\} d_{ss} \quad . \tag{4.23}$$

Since the joint density of states of the above- mentioned optical transitions is proportional to $(h\nu)^2$ $\mathrm{Im}\{\tilde{\varepsilon}_{ss} - \tilde{\varepsilon}_2\}$, according to (4.19b,22b,23) the measurement of $\delta\psi$ yields in good approximation the joint-density-of-states curve.

The same type of information can be obtained by differential reflectometry, where the measured quantity is the relative change in reflectivity

$$\frac{\Delta R}{R} = \frac{R(d_1) - R(0)}{R(0)} \quad . \tag{4.24}$$

Here $R(d_1)$ is the reflectivity of the system substrate/layer/ambient (Fig. 4.1b), and R(0) that of the system substrate/ambient (Fig.4.1a). When $d_1 \ll \lambda$ one gets in first-order approximation [4.42,43] at normal incidence on a substrate in vacuum

$$\frac{\Delta R}{R} = \frac{8\pi d_1}{\lambda} \mathrm{Im}\left\{\frac{\tilde{\varepsilon}_1 - \tilde{\varepsilon}_2}{1 - \tilde{\varepsilon}_2}\right\} \quad . \tag{4.25}$$

Applying the model for chemisorption depicted in Fig.4.3a, the differential reflectometric measurements yield directly $\mathrm{Im}\{\tilde{\varepsilon}_{ss} - \tilde{\varepsilon}_2\}$, which is closely related to the above-mentioned joint density of states. In this way WIERENGA et al. [4.36,37] determined the surface-state transitions on the low-index planes of silicon in the energy region of 1 to 3 eV.

On *metals*, where chemisorption prevails for many reactive adsorbates, one rather expects a change in the optical properties of the metal near the surface due to the redistribution of electronic charge. A simple model for this situation (Fig.4.3c) has been outlined by HABRAKEN et al. [4.8] for the chemisorption of oxygen on Cu, Ag, and Ni surfaces. This is basically a three-layer model, adsorbate/modified substrate/unchanged substrate, where the modifications in the substrate are taken into account via a change in free-electron density and collision time (see Sect.4.4).

4.4 Clean Metal Surfaces

The investigation of clean metal surfaces by so-called spectroscopic ellipsometry yields, with the aid of the equations from Sect.4.2, the optical constants n and k, or ε' and ε'' as a function of the wavelength. In the classical treatment of the dielectric constant of a metal two contributions are distinguished, that of free and of bound electrons [4.44]

$$\tilde{\varepsilon} = \tilde{\varepsilon}_f + \tilde{\varepsilon}_b \quad . \tag{4.26}$$

For a free-electron metal with damping $\tilde{\varepsilon}(\nu)$ is given by

$$\tilde{\varepsilon}_f(\nu) = 1 - \nu_1^2/(\nu^2 + \nu_2^2) - i(\nu_2/\nu)\nu_1^2/(\nu^2 + \nu_2^2) \quad . \tag{4.27}$$

Here ν is the frequency of the light, ν_1 the plasma frequency, and ν_2 the electron collision frequency. The plasma frequency is related to the density N_f, the optical mass m^*_f and the charge e of the free electrons,

$$\nu_1^2 = \frac{e^2 N_f}{4\pi^2 \varepsilon_0 m^*} \quad [\text{SI units}] \quad , \tag{4.28}$$

where ε_0 is the permittivity of vacuum. The collision frequency is given by

$$\nu_2 = 1/2\pi\tau = \nu_1^2/2\sigma_0^* \quad . \tag{4.29}$$

Here τ is the collision time and σ_0^* the (optical) conductivity. For metals like copper the collision frequency $\nu_2 \approx 5 \cdot 10^{12} s^{-1}$ [4.44], and thus in the visible region of the spectrum $\nu^2 \gg \nu_2^2$. Neglecting ν_2^2 with respect to ν^2, (4.27) reduces to

$$\tilde{\varepsilon}_f(\nu) = 1 - \nu_1^2/\nu^2 - i(\nu_2/\nu)\nu_1^2/\nu^2 \quad . \tag{4.30}$$

Substituting this into (4.26) yields, with $\nu = c/\lambda$ (c: velocity of light),

$$\varepsilon'(\lambda) = 1 - (\nu_1^2/c^2)\lambda^2 + \varepsilon_b' \quad , \tag{4.31}$$

$$\varepsilon''(\lambda) = (\nu_2\nu_1^2/c^3)\lambda^3 + \varepsilon_b'' \quad . \tag{4.32}$$

In regions of the spectrum where the contributions due to the bound electrons ε_b' and ε_b''/λ are constant, (4.31,32) may be written as

$$\varepsilon'(\lambda) = A - B\lambda^2 \quad , \tag{4.33}$$

$$\varepsilon''(\lambda)/\lambda = C + D\lambda^2 \quad , \tag{4.34}$$

with $A = 1 + \varepsilon_b'$, $B = \nu_1^2/c^2$, $C = \varepsilon_b''/\lambda$, and $D = \nu_2\nu_1^2/c^3$. By plotting ε' and ε''/λ versus λ^2 or, e.g., by linear regression analysis of the experimental data, the "constants" A, B, C, and D may be determined and thus also ν_1, ν_2, ε_b', and ε_b''/λ. In Fig.4.4 the components of $\tilde{\varepsilon}$ for copper, determined at room temperature for samples of different nature, are given in the wavelength region 400-850 nm.

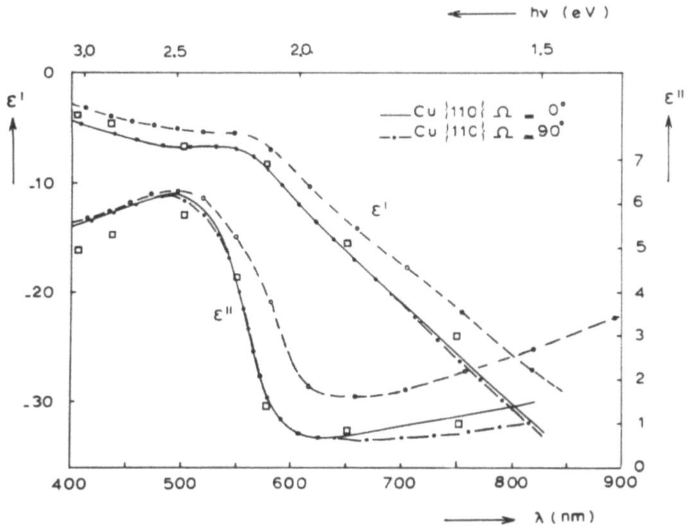

Fig.4.4. The components of the complex dielectric function for copper: Cu{110} calculated from $\bar{\Delta}$, $\bar{\psi}$ data determined at two different orientations of the plane of incidence ($\Omega = 0°$ and $90°$) [4.30]; annealed bulk polycrystal in UHV (□) [4.45]; unannealed thin film (—o—) [4.46]

ROBERTS [4.45] determined the optical properties of annealed and electropolished polycrystalline copper samples by means of ellipsometry. The measurements were done with the sample in a vacuum of about 2×10^{-5} torr after removal of the oxide layer by heating to a temperature of 500 K. The results of JOHNSON and CHRISTY [4.46] were obtained from reflection and transmission measurements on vacuum-evaporated, unannealed films which had been exposed to air. Ellipsometric measurements with single crystals were recently performed with Cu{110} surfaces, whose cleanliness was ascertained by Auger electron spectroscopy [4.30]. The ellipsometric parameters were determined at different values of the azimuth Ω of the plane of incidence, parallel ($\Omega = 0°$) and perpendicular ($\Omega = 90°$) to the <110> direction. After correction for window birefringence the components ε' and ε'' were calculated by applying (4.4-6), valid for isotropic substrates.

In view of the differences in the physical state of the samples and in the cleanliness of sample and bulk, of the topography of the surface, and of errors due to window corrections and angle of incidence, the agreement is satisfactory, especially for well-annealed polycrystalline samples and single crystals in UHV.

From the values of ε' and ε'' at longer wavelengths, outside the absorption-band region, the free- and bound-electron parameters were determined from plots versus λ^2 (4.33,34). The results for room temperature are given in Table 4.1. A significant conclusion from these data, as well as from data obtained at other temperatures [4.30], is that the collision frequency ν_2 and thus, according to (4.29), also the collision time τ and optical conductivity σ_0^* are strongly dependent on the nature and temperature of the sample. Moreover, for Cu{110} a dependence on the orientation of the surface is observed which disappears at higher temperatures.

Table 4.1. Optical parameters of copper at room temperature

Sample	λ range [nm]	$\nu_1 \cdot 10^{-15}[s^{-1}]$	$\nu_2 \cdot 10^{-12}[s^{-1}]$	ε_b'	ε_b''/λ	Ref.
Polycrystalline	650 - 2500	2.18	7.84	5	0.65	[4.45]
Film	700 - 1940	2.11	23.2	5.7	0.63	[4.46]
Cu{110}, $\Omega = 0^\circ$	660 - 820	2.33	13.8	7.3	-0.07	[4.30]
Cu{110}, $\Omega = 90^\circ$	660 - 820	2.36	7.9	8.2	0.17	[4.30]

The last-mentioned observation points to *optical anisotropy* of the clean Cu{110} surface which can in principle be expected because the site symmetry contains no fourfold or higher order symmetry axis. Direct evidence for this anisotropy is presented in Fig.4.5 where the difference in $\bar{\Delta}$ and $\bar{\psi}$ for the two orientations of the plane of incidence is plotted as a function of λ. In view of the errors in absolute measurements not too much value should be attached to the absolute values of $\delta\bar{\Delta}$ and $\delta\bar{\psi}$.

Electroreflectance measurements on Ag{110} [4.47] and Cu{110} [4.48] in electrolyte solution have also revealed optical anisotropy, which was explained in terms of direction-dependent surface conductivities [4.48]. In fact, for Cu{110} in UHV a comparable conclusion could be drawn from Table 4.1: a higher conductivity in the direction perpendicular to the rows (<001> direction) than along the rows <110>.

However, it should be remarked that the treatment used for the determination of the optical parameters in Table 4.1 is only valid for isotropic substrates.

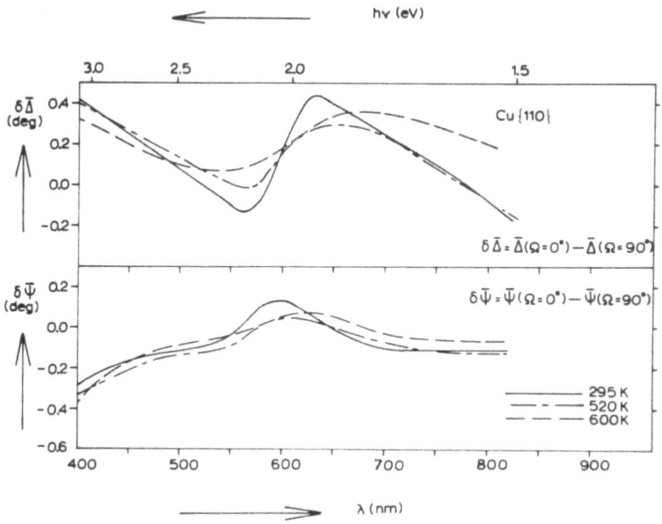

Fig.4.5. Difference in ellipsometric parameters $\bar{\Delta}$ and $\bar{\psi}$ measured for Cu{110} at Ω = 0° and 90°; ϕ_0 = 66° [4.30]

A more appropriate model for clean Cu{110}, e.g., with an isotropic copper substrate ($\tilde{\varepsilon}_2$) covered by a biaxially anisotropic surface layer with thickness d_1, contains more parameters ($\tilde{\varepsilon}_2$, $\tilde{\varepsilon}_{1x}$, $\tilde{\varepsilon}_{1y}$, $\tilde{\varepsilon}_{1z}$, and d_1) than can be determined from the four experimental quantities. Nevertheless, as has been illustrated in [4.8], calculations with the stratified layer model for the system ambient/anisotropic layer/isotropic substrate may give some qualitative insight into the factors that determine the degree of anisotropy. The same insight is obtained in a simpler way with Strachan's approach (Sect.4.3). For an absorbing dipole layer Strachan's expressions for $\delta\Delta$ and $\Delta\psi$ have the general form

$$\delta\Delta = a'_x\sigma'_x + a'_y\sigma'_y + a'_z\sigma'_z + a''_x\sigma''_x + a''_y\sigma''_y + a''_z\sigma''_z \quad , \tag{4.35a}$$

$$\delta\psi = b'_x\sigma'_x + b'_y\sigma'_y + b'_z\sigma'_z + b''_x\sigma''_x + b''_y\sigma''_y + b''_z\sigma''_z \quad . \tag{4.35b}$$

The coefficients a'_j, a''_j, b'_j, and b''_j depend on $\tilde{\varepsilon}_2$, λ, and ϕ_0 according to Ref.4.8, Eqs.(22,23). Since rotation of the plane of incidence over 90° implies only the interchange of x and y components, we get

$$\delta\bar{\Delta} \equiv \bar{\Delta}(\Omega=0^\circ) - \bar{\Delta}(\Omega=90^\circ) = -(a'_x-a'_y)(\sigma'_x-\sigma'_y) - (a''_x-a''_y)(\sigma''_x-\sigma''_y) \quad , \tag{4.36a}$$

$$\delta\bar{\psi} \equiv \bar{\psi}(\Omega=0\ ) - \bar{\psi}(\Omega=90\ ) = -(b'_x-b'_y)(\sigma'_x-\sigma'_y) - (b''_x-b''_y)(\sigma''_x-\sigma''_y) \quad . \tag{4.36b}$$

Table 4.2. Values of coefficients in (4.35,36) for Cu{110}, $\phi_0 = 66^\circ$

λ[nm]	$a'_x-a'_y$	a'_z	$a''_y-a''_x$	a''_z	$b'_x-b'_y$	$-b'_z$	$b''_x-b''_y$	b''_z
460	0.41	2.3	0.23	0.88	0.11	0.42	0.19	1.11
500	0.35	2.2	0.20	0.75	0.09	0.36	0.17	1.06
560	0.36	1.8	0.11	0.43	0.06	0.21	0.17	0.90
600	0.28	1.9	0.02	0.06	0.008	0.03	0.15	0.97
640	0.21	2.0	0.01	0.03	0.004	0.01	0.10	1.03
700	0.15	2.0	0.004	0.02	0.003	0.01	0.07	1.02
760	0.11	2.0	0.004	0.02	0.002	0.008	0.06	0.98

Calculated values of coefficients in (4.35,36) are given in Table 4.2 for a few wavelengths. The coefficients a'_y, a''_x, b'_y, b'_z, and b''_y are negative, and the absolute values of a'_j and b''_j turn out to be considerably larger than those of a''_j and b'_j. From the values of the coefficients it follows that differences in $\delta\Delta$, and also in $\delta\bar{\Delta}$, are primarily associated with differences in σ'_j, and differences in $\delta\psi$ and $\delta\bar{\psi}$ mainly with differences in σ''_j. For example, the large value of $\delta\bar{\Delta}$ at $\lambda \approx 640$ nm (Fig.4.5) implies $\sigma'_x < \sigma'_y$, i.e., a smaller $\sigma'(\alpha')$ in the direction parallel to the rows than perpendicular to them. If the absolute values of $\delta\bar{\Delta}$ and $\delta\bar{\psi}$ were reliable one could even calculate the differences $(\sigma'_x - \sigma'_y)$ and $(\sigma''_x - \sigma''_y)$ with (4.36a,b).

The elements whose optical constants have been determined by spectroscopic ellipsometry have been reviewed in [4.49]. With a few exceptions all experiments were carried out with either film or polycrystalline bulk samples, often in air or in poor vacuum conditions and without techniques that monitor the surface cleanliness and/or topography.

4.5 Spectroscopic Ellipsometry of Overlayers

In spectroscopic ellipsometry of overlayers one essentially measures $\delta\Delta$ and $\delta\psi$ as a function of wavelength. In order to relate these changes to the optical parameters of the overlayers, independent measurements have to be carried out to determine the thickness of the layer. Moreover, one has to make assumptions about the changes induced in the substrate upon formation of the overlayer, e.g., upon adsorption (see Sect.4.3). Not many spectroscopic ellipsometric studies of adsorption processes in UHV systems have been performed to date. Examples are oxygen on silver films [4.50], H_2O/Li films [4.51], CO on copper [4.52], nickel [4.53], silver [4.54] and gold [4.55] films, CO on Ni{100} [4.56] and Ni{110} [4.57], and methanol on Ni{100} [4.58] (for a discussion see [4.49]).

Very recently KÖTZ et al. [4.59] studied the interaction of oxygen with Mg{001} using automatic spectroscopic ellipsometry at wavelengths between 235 and 700 nm. The first stage of oxidation, characterized by a decrease in Δ and ψ, was described in terms of the growth of a surface layer consisting of a mixture of bulk Mg and MgO of which the dielectric constant was calculated with the effective medium approximation (EMA) due to BRUGGEMAN [4.60]. The second stage of the interaction with oxygen, characterized by an increase in ψ, was interpreted in terms of a phase transition, where part of the mixed layer is transformed into three-dimensional MgO film. Further oxidation appeared to proceed via layer growth of pure MgO.

As stated by ASPNES et al. in a series of papers (e.g., [4.61-63]), the various forms of simple effective-medium descriptions of a heterogeneous dielectric mixture can be summarized in the following equation:

$$\frac{\langle\varepsilon\rangle - \varepsilon_h}{\langle\varepsilon\rangle + 2\varepsilon_h} = f_1 \frac{\varepsilon_1 - \varepsilon_h}{\varepsilon_1 + 2\varepsilon_h} + f_2 \frac{\varepsilon_2 - \varepsilon_h}{\varepsilon_2 + 2\varepsilon_h} + \dots \quad . \tag{4.37}$$

Here $\langle\varepsilon\rangle$, ε_h, ε_1, ε_2,... are the (complex) dielectric functions of the effective medium, host medium, and inclusions of the type 1, 2, ... in the host medium, with volume fractions f_1, f_2,... In the Lorentz-Lorenz approximation the host medium is assumed to be vacuum ($\varepsilon_h = 1$), and in the Maxwell-Garnett theory either one of the inclusion materials may represent the host medium ($\varepsilon_h = \varepsilon_1$ or ε_2, or...). In the EMA description the effective medium itself acts as host medium ($\varepsilon_h = \langle\varepsilon\rangle$) and thus self-consistency is achieved.

For a medium consisting of two components the EMA formulation of (4.37) reads

$$f \frac{\varepsilon_1 - \langle\varepsilon\rangle}{\varepsilon_1 + 2\langle\varepsilon\rangle} + (1 - f) \frac{\varepsilon_2 - \langle\varepsilon\rangle}{\varepsilon_2 + 2\langle\varepsilon\rangle} = 0 \quad . \tag{4.38}$$

By comparing their spectroscopic ellipsometric results with calculated curves, ASPNES et al. showed that in several cases the EMA theory gives the best description of the optical properties, e.g., of microscopically rough silicon surfaces [4.61], of the interface between silicon and SiO_2 [4.62], and of gold films [4.63].

Apart from the underlying assumptions of (4.37) (spherical inclusion geometry, only dipole interactions), application of the EMA theory, e.g., (4.38), to a surface layer formed upon chemisorption (Fig.4.3b) presupposes that this layer is a physical mixture of the bulk compounds of which the $\varepsilon(\lambda)$ data are known. This is not always the case: chemical bonds and charge distributions may be considerably different and thus also the (complex) polarizabilities of the microscopic building units and the related macroscopic dielectric functions. In fact, the models of Fig.4.3a,c discussed in Sect.4.3 attempt to take these differences into account.

An interesting example of a chemisorption system is oxygen on Cu{110}. This system has been investigated by HABRAKEN et al. with ellipsometry at one wavelength (λ = 632.8 nm) [4.8,64,65] and by HANEKAMP et al. with spectroscopic ellipsometry ($400 < \lambda < 820$ nm) [4.30], both in combination with AES and LEED. The initial interaction of oxygen with a Cu{110} surface occurs in two stages, characterized by (2×1) and $c(6\times 2)$ overlayer structures [4.65].

The (2×1) structure is obtained at oxygen exposures of about 100 L (1 L = 10^{-6} torr s = 1.33×10^{-4} Pa•s) and is assumed to correspond to 0.5 monolayers (one O atom per two Cu surface atoms) [4.64]. Further oxygen exposure produces the $c(6\times 2)$ structure; at about 10^5 L the oxygen coverage was assumed to be one monolayer [4.65]. HABRAKEN et al. observed that in the first stage the $\delta\Delta$ and $\delta\psi$ values depend on the azimuth of the plane of incidence of the light beam and on the temperature of the crystal. This is confirmed by the spectroscopic ellipsometric curves given in Fig.4.6 for $\Omega = 0^\circ$ and in Fig.4.7 for $\Omega = 90^\circ$. These figures also reveal remarkable differences in the spectral dependences of $\delta\Delta$ and $\delta\psi$ for the two orientations of the plane of incidence, notably in the region of the absorption band of copper. Comparison with Fig.4.2 shows that the measured curves cannot be explained with a model in which a nonabsorbing layer with constant index of refraction is deposited on the clean copper substrate (Fig.4.3b with n_1 real and constant): the sign and order of magnitude of $\delta\psi$ as well as the λ dependences of $\delta\Delta$ and $\delta\psi$ are wrong. Results calculated with the same model but with (absorbing) layers of Cu_2O and of mixtures of Cu and Cu_2O are given in Fig.4.8; the thickness was arbitrarily chosen equal to 2 Å; according to (4.17a,b) another choice does not influence the shape but only the height of the curves. The optical constants of Cu_2O were taken from recent measurements on films grown on copper [4.66] and on reactively sputtered thin films [4.67]. The effective dielectric constants of the mixtures were calculated with (4.38) by taking the values of DROBNY and PULFREY [4.67] for ε_1 and those measured for clean Cu{110} at $\Omega = 0^\circ$ for ε_2. Comparison with the experimental curves of Figs.4.6,7 shows that apparently the optical properties of the surface oxide are quite different from those of "bulk" Cu_2O. Better but not yet satisfactory agreement is found with the EMA results for mixtures of Cu

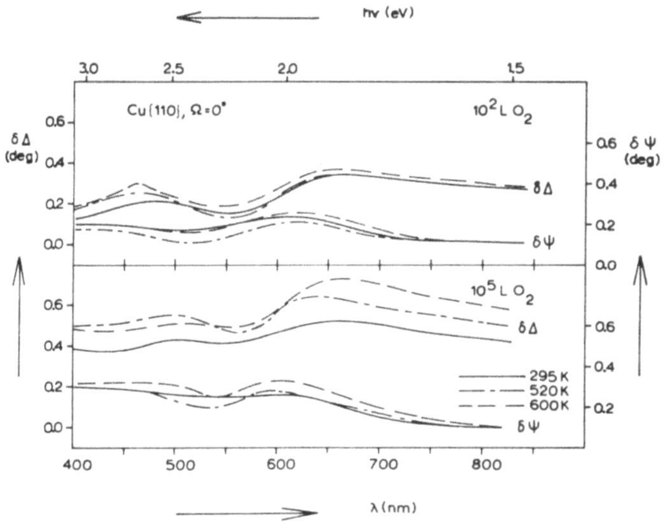

Fig.4.6

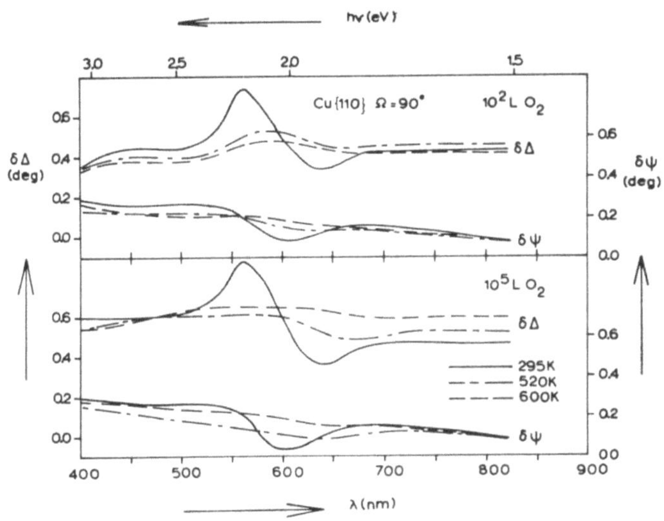

Fig.4.7

Fig.4.6. Ellipsometric effects measured upon adsorption of oxygen on Cu{110}, $\phi_o = 66^o$, plane of incidence parallel to the <110> direction [4.30]

Fig.4.7. Ellipsometric effects measured upon adsorption of oxygen on Cu{110}, $\phi_o = 66^o$, plane of incidence parallel to <001> direction [4.30]

and Cu_2O, at least at $\Omega = 0^o$. However, with this model it is impossible to describe the results at $\Omega = 90^o$ (Fig.4.7), even with ε_2 values measured for clean Cu{110} at $\Omega = 90^o$.

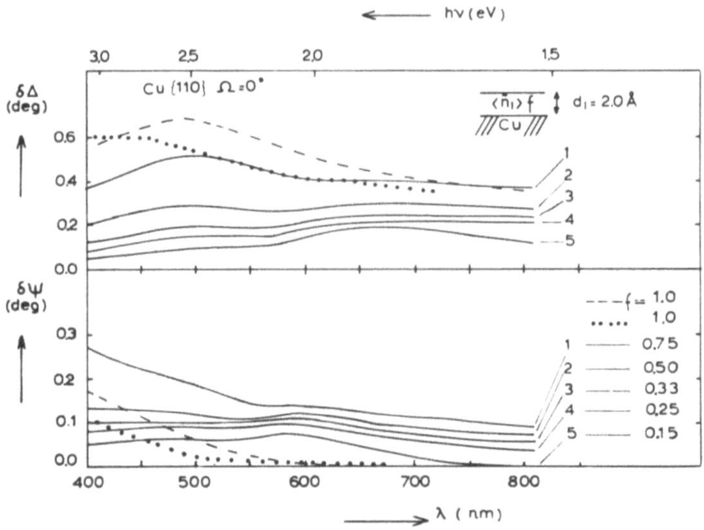

Fig.4.8

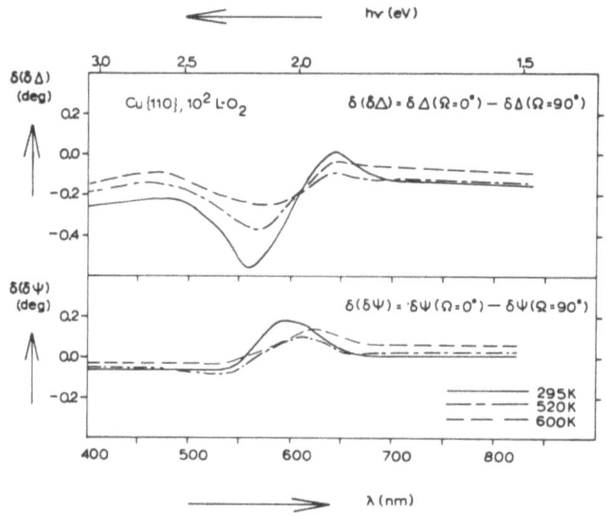

Fig.4.9

Fig.4.8. Ellipsometric effects calculated for layers of Cu_2O {(---) [4.67], (...) [4.66]} and of mixtures of Cu and Cu_2O [(f) volume fraction of Cu_2O] on a clean copper surface. $n_0 = 1.0$, $\phi_0 = 66°$, n_2 from ellipsometric measurements with Cu{110}, $\Omega = 0°$ [4.30]

Fig.4.9. Difference in ellipsometric effects $\delta\Delta$ and $\delta\psi$ measured for Cu{110} at $\Omega = 0°$ (Fig.4.6) and 90° (Fig.4.7) after 100 L oxygen exposure [4.30]

The anisotropy of the $\delta\Delta$ and $\delta\psi$ curves may in principle be due to anisotropy of the clean Cu{110} surface and/or to anisotropy in the distribution of the chemisorbed oxygen atoms or in the copper-oxygen bonds. The spectral and temperature dependence of the differences in $\delta\Delta$ and $\delta\psi$ measured for the two orientations of the plane of incidence, plotted in Fig.4.9, appear to be similar to those for the clean surface in Fig.4.5, notably in the region 500-700 nm. This suggests that the differences in the $\delta\Delta$ and $\delta\psi$ curves may at least partially be ascribed to a disappearance of the anisotropy of clean Cu{110} upon oxygen adsorption. For a further discussion see [4.30].

4.6 Kinetic Studies of Chemisorption

4.6.1 Coverage Calibration

As has been remarked in the introduction, ellipsometry may be used conveniently to study the kinetics of adsorption or chemical reactions on surfaces once a calibration of the observed values of Δ and/or ψ has been obtained.

In the case of the adsorption of oxygen on copper, the calibration has been performed with AES-LEED. The ellipsometric measurements were carried out at one wavelength (λ = 632.8 nm) with $\phi_0 = 69^\circ$. For Cu{111}/O [4.68] and Cu{100}/O [4.69] a linear relation exists between the relative oxygen Auger signal O(515 eV)/Cu(920 eV) and the change in Δ upon adsorption. In the case of Cu{110} this relationship depends on the angle between the plane of incidence and the <110> direction [4.64]. Figure 4.10 gives $\delta\Delta$ as a function of the oxygen Auger signal upon reaction of O_2 with clean Cu{111} and Cu{110} surfaces. At saturation of the chemisorption stage on Cu{110} a (2×1) LEED pattern is observed which is supposed to be due to an oxygen coverage $\Theta = 0.5$. On Cu{100} at the end of the chemisorption stage a $(\sqrt{2}\times\sqrt{2})R45^\circ$ pattern is developed, also attributed to the presence of 0.5 monolayer of oxygen. Comparing the combined results obtained with ellipsometry, AES, and LEED it is concluded that the $\delta\Delta$ per adsorbed oxygen atom is equal for the three copper planes.

For the Ag{110}/O_2 system the calibration has been performed by combining LEED and ellipsometry (λ = 632.8 nm, $\phi_0 = 71^\circ$) [4.70]. The change in Δ was found to be proportional to the coverage 1/n derived from $(n\times1)$ superstructures in LEED, with n = 5, 3, and 2. Oxygen adsorption on clean, annealed Ag{111} practically does not occur, but may be induced by surface roughening via Ar^+ ion bombardment [4.71,72].

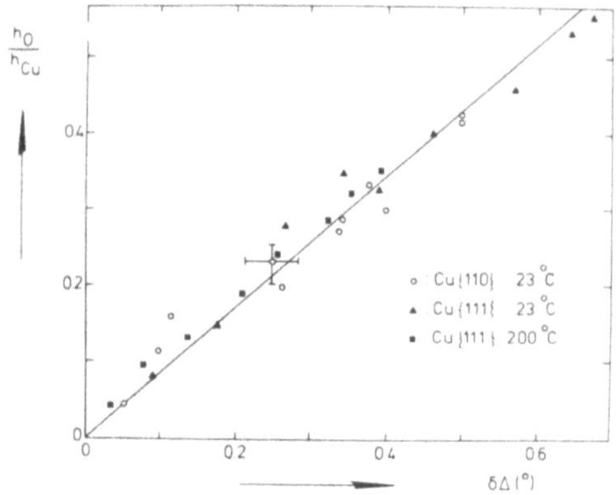

Fig.4.10. Oxygen Auger signal versus $\delta\Delta$ upon oxygen adsorption on Cu{111} (23°C) and on Cu{110} (23°C), and upon removal of oxygen from Cu{111} (T = 200°C) [4.64]

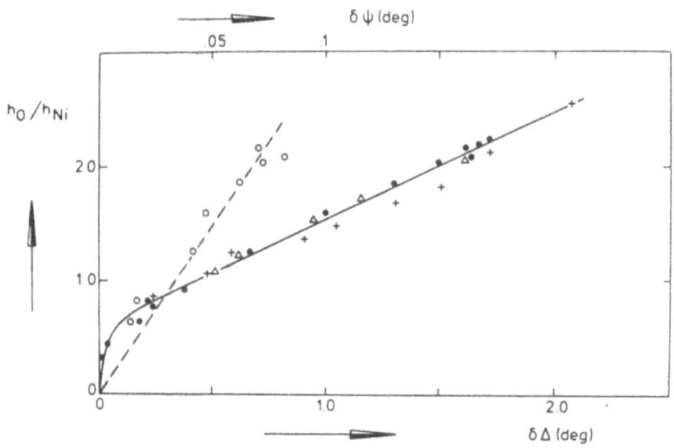

Fig.4.11. Plot of the relative oxygen Auger signal h_O (515 eV)/h_{Ni} (848 eV) as a function of $\delta\Delta$ (lower scale) and $\delta\psi$ (upper scale, open circles). (o) $\delta\psi$ at 25°C; (•) $\delta\Delta$ at 25°C; (Δ) $\delta\Delta$ at 50°C; (+) $\delta\Delta$ at 150°C [4.73]

The system Ni{100}/O_2 has also been studied with ellipsometry (λ = 632.8 nm, ϕ_0 = 68°) and AES-LEED [4.73]. Surprisingly, the calibration is qualitatively different from those shown above (Fig.4.11). Only at higher oxygen coverages (in excess of half a monolayer) is a linear relation obtained. These coverages correspond to the incorporation of oxygen in the nickel lattice leading to a NiO surface layer [4.74].

It thus appears that for oxygen coverages below θ = 0.5 the relation between the observed $\delta\Delta$ and θ is

$\theta = (0.65 \pm 0.06)\ \delta\Delta$ Cu{111}

$\theta = (1.0 \pm 0.1)\ \delta\Delta$ Cu{110}

$\theta = (0.77 \pm 0.08)\ \delta\Delta$ Cu{100}

$\theta = (0.92 \pm 0.05)\ \delta\Delta$ Ag{110}

$\theta \approx 12\ \delta\Delta$ Ni{100}

With an accuracy in $\delta\Delta$ of about 0.02° the ellipsometric sensitivity at λ = 632.8 nm for oxygen on copper and silver amounts to about 0.02 monolayer. For nickel ellipsometry appears rather insensitive in the (sub)monolayer range at least for oxygen and at λ = 632.8 nm.

Linear relationships between $\delta\Delta$ and Auger signals have also been established for other chemisorption systems, e.g., oxygen on Si{111} and Si{100} [4.75] and on Al [4.76].

4.6.2 Initial Stages of Oxidation

The initial stages of oxidation (and the subsequent reduction of the oxide layer) have been studied with the aid of the aforementioned relations between $\delta\Delta$ and the oxygen coverage. Kinetic data are obtained by monitoring $\delta\Delta$ during or after exposure of the surfaces to O_2 or N_2O and CO or H_2 at various gas pressures and at different temperatures of the crystal.

On Cu{111}[4.68], Cu{110} [4.64], and Ag{110} [4.70] a well-defined saturation coverage is obtained ($\Theta \simeq \frac{1}{2}$) at pressures below 10^{-5} torr. For Cu{111} this is illustrated in Fig.4.12. On rough (ion-bombarded) Ag{111} the amount of oxygen adsorbed depended on the amount of surface damage present [4.71,72]. On Cu{100} [4.69] and Ni{100} [4.73] oxidation continued beyond the half-monolayer stage under these conditions. Using the calibration as discussed in the previous section from curves like those in Fig.4.12, sticking coefficients s may be derived from the slopes of the curves and the number of colliding oxygen molecules. For Cu{111} some results are presented in Fig.4.13. In the pressure region considered (10^{-6} to 10^{-5} torr) s appeared to be independent of the pressure and constant over a large coverage region. The solid lines in Fig.4.13 have been calculated with a model in which it is assumed that the adsorption proceeds via a precursor state (probably loosely bound, mobile O_2 molecules) and requires a pair of adjacent free sites for dissociation.

In Table 4.3 initial sticking coefficients s(0) are reported for the copper and silver planes studied, as well as the change in s(0) in the temperature range indicated. In general copper exhibits a larger reactivity

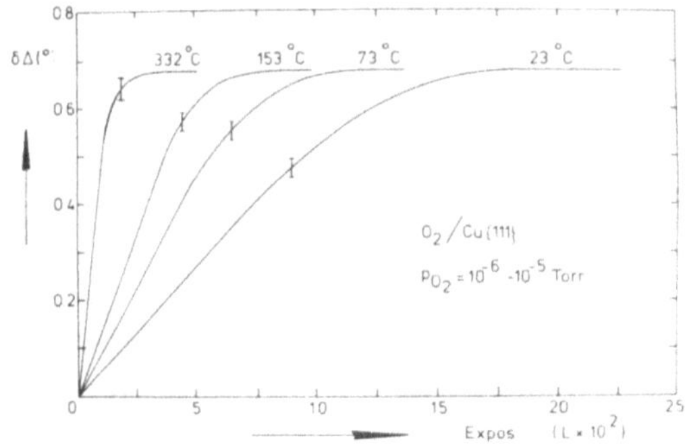

Fig.4.12↑

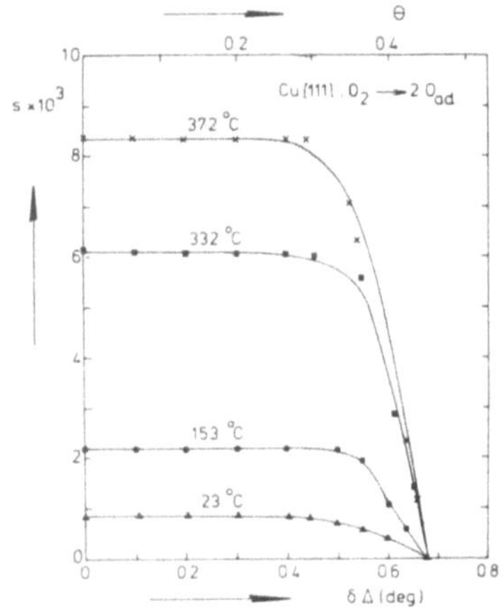

Fig.4.13→

Fig.4.12. δΔ versus exposure of oxygen to a clean and annealed Cu{111} surface (experimental errors indicated) [4.68]

Fig.4.13. Oxygen sticking coefficients s vs δΔ, derived from ellipsometric measurements. The solid lines have been calculated with a precursor model [4.68]

Table 4.3. Initial sticking coefficients of O_2 and N_2O determined by ellipsometry

		T[K]	p[torr]	s(0)	Ref.
O_2	Cu{111}	300-650	10^{-6} - 10^{-4}	1×10^{-3} - 8×10^{-3}	[4.68]
	Cu{100}	300-675	10^{-7} - 10^{-4}	9×10^{-3} - 7×10^{-2}	[4.69]
	Cu{110}	300-650	10^{-8} - 10^{-4}	0.2 - 0.2	[4.64]
	Ag{111}	300	10^{-5} - 10^{-3}	3×10^{-5} - 4×10^{-4}	[4.72]
	Ag{110}	300-475	10^{-6} - 5×10^{-5}	6×10^{-4} - 4×10^{-4}	[4.70]
N_2O	Cu{111}	525-650	10^{-4} - 5×10^{-4}	3×10^{-6} - 2×10^{-4}	[4.68]
	Cu{100}	300-675	10^{-5} - 10^{-3}	5×10^{-5} - 8×10^{-4}	[4.69]
	Cu{110}	300-625	10^{-7} - 10^{-5}	0.15 - 0.02	[4.64]

towards oxygen than silver, and the open {110} planes are more reactive than the closer packed {100} and {111} planes (plane specificity). As is shown in Fig.4.14 the dependence on Θ also differs between the different copper planes (for a detailed discussion see the original papers). A more striking example of plane specificity is the decomposition of N_2O on copper ($N_2O_g \rightarrow N_{2g} + O_{ad}$). The initial decomposition probability s(0) varies over several orders of magnitude at lower temperatures (Table 4.3), but there

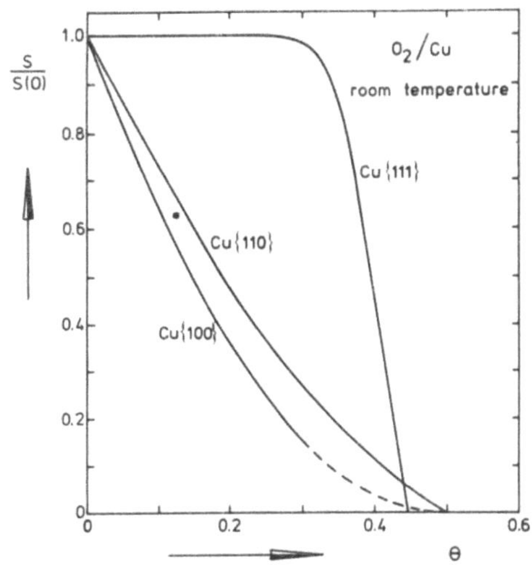

Fig.4.14. Sticking probability of O_2 relative to s(0) as a function of the oxygen coverage θ on Cu{111}, Cu{110}, and Cu{100} [4.77]

appears to be a temperature where its value is the same for all planes and coverages. This phenomenon is known as the compensation effect [4.78].

Beyond the first chemisorption stage generally a further slower uptake of oxygen takes place. For Cu{100} and Ni{100} this occurred spontaneously. Cu{111} and Cu{110} required higher pressures of O_2 or higher temperatures, whereas on Ag{110} above 200°C an apparently reversible adsorption of oxygen was observed. At least for all copper planes N_2O is unable to oxidize the metal beyond the (half-)monolayer stage at $T < 650$ K and pressures below 10^{-3} torr. For Cu and Ni this second stage is accompanied by changes in the LEED pattern and in the work function indicative of a first reconstruction of the metal substrate atoms. For Ni{100}, Cu{110}, and Cu{100} a still greater amount of oxygen could be incorporated in the lattice at the expense of an increase in temperature and/or pressure. The further uptake could only be measured by ellipsometry, as can be seen in Fig.4.15 for Cu{110}. In this third stage (indicated by C in the figure) the other techniques used display a constant reading for the parameter monitored (AES peak height, work function change). At higher oxygen pressures (5×10^{-5} to 5×10^{-4} torr) $\delta\Delta$ rose rapidly by ~0.5° (stage A), less rapidly to $\delta\Delta \simeq 0.7°$ (stage B), and thereafter linearly in stage C.

For Cu{100} this linear increase of $\delta\Delta$ occurred after an induction period at a constant $\delta\Delta \simeq 0.8°$, which was rapidly reached upon admission of the oxygen. In stage C the rates of change of $\delta\Delta$ with oxygen exposure are roughly equal for both planes, indicating that here a non-plane-specific process is rate limiting. The proportionality of the rates to the oxygen pressure suggests

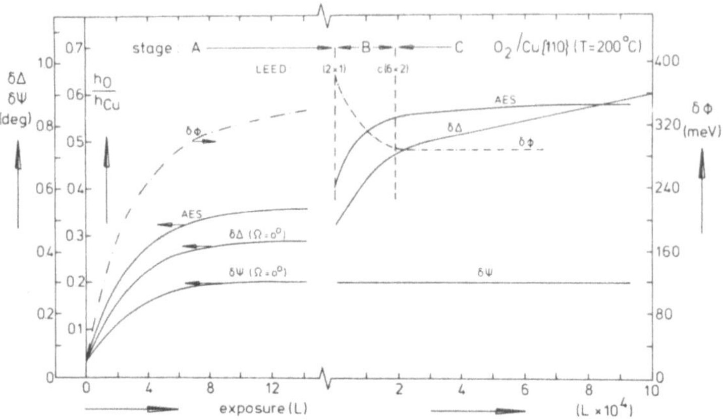

Fig.4.15. Oxygen Auger signal (AES), δΔ, δψ, δϕ, and LEED patterns versus oxygen exposure to Cu{110} at 200°C. The plane of incidence of the light for ellipsometry was parallel to the <110> direction (i.e., Ω = 0°) [4.65]

that the rate-limiting process takes place at the surface. It may be remarked that in the cited study by HABRAKEN et al. the maximal oxygen exposure ($\sim 5 \times 10^5$L) was always lower than the critical exposure reported for the formation and growth of oxide nuclei on Cu{110} [4.79] and Cu{100} [4.80].

In the subsequent stages B and C the ellipsometric response per sorbed oxygen atom is considerably different from that reported in Sect.4.6.1 for the first chemisorption stage. For copper a ratio of 2.3 for the δΔ per oxygen atom adsorbed (stage A) to δΔ per oxygen atom absorbed (stages B,C) was deduced from the experimental data. If one assumes that after the chemisorption stage an oxide layer is formed on top of a copper substrate, with effective $\tilde{n}_2$ calculated from Δ, Ψ at $\Theta \simeq \frac{1}{2}$, one gets for $\tilde{n}_1$, using the model in Fig.4.3b, values that are close to the literature values for bulk Cu_2O [4.8].

Similar results, a first fast chemisorption and incorporation stage followed by a further oxidation have been reported by BRADSHAW et al. for aluminium [4.81-83] and magnesium [4.59,84]. From their figures it can also be concluded that only ellipsometry is capable of monitoring and determining the kinetics of the subsequent stages of the interaction with oxygen. Qualitatively the same picture emerges from work on W{011} [4.85] and Fe{011} [4.86] although no detailed kinetic results have been reported.

4.6.3 Reactions of Adsorbed Oxygen

The reaction of adsorbed oxygen with CO to produce CO_2 has been studied for the three copper planes [4.64,65,68,69], Ag{111} (damaged) [4.72] and Ag{110} [4.70]. Only for damaged Ag{111}/O does a simple, first-order process seem to describe the experimental results,

$$\Theta = \Theta_{max} \exp(-at) \quad , \tag{4.39}$$

where the rate constant a is proportional to the CO pressure, varying between 10^{-7} and 10^{-6} torr. For oxygen adsorbed on Ag{110} the reaction followed a more complex path, the reaction rate increasing with decreasing coverage. This has been explained by assuming lateral interactions within the adlayer. The existence of these interactions is demonstrated by LEED photographs, which show that the original (2×1) superstructure disorders mainly in one direction and even produces a (3×1) superstructure at lower oxygen coverage.

The reduction of oxidized copper surfaces follows an equally complex pathway, also depending upon the amount of preadsorbed oxygen. The nature of the oxygen (whether deposited from N_2O or O_2) did not influence the kinetics. The reduction of preadsorbed oxygen on Cu{110} ($\Theta = 0.5$) is shown in Fig. 4.16. At first the reaction proceeds slowly but after some exposure an increase in the reaction rate occurs. This behavior was explained by assuming that reaction takes place between O_{ad} and a CO molecule, adsorbed on an adjacent free copper site (Langmuir-Hinshelwood mechanism), also taking the lateral interactions between the adsorbed oxygen atoms into account.

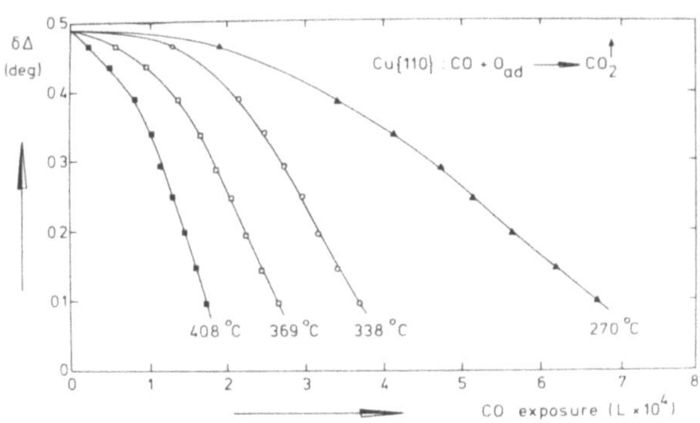

Fig.4.16. δΔ versus exposure of CO to a Cu{110} surface with oxygen, preadsorbed up to $\Theta = 0.5$, at different temperatures. (——) ellipsometric off-null measurements ($\Omega = 0°$). The indicated points were calculated with a model involving interactions between adsorbed oxygen atoms [4.64]

In Fig.4.17 $\delta\Delta$, $\delta\psi$, and the oxygen Auger signal are depicted versus exposure of CO to a Cu{111} surface oxidized up to $\delta\Delta = 1.1^\circ$. From the AES measurements it can be concluded that in the first reduction stage (exposures up to 15×10^4L) the near-surface oxygen concentration remains constant, although, ellipsometrically, the reaction does proceed. This implies that surface oxygen is removed by CO, but immediately replaced by oxygen from the bulk. Once the bulk supply has run out the "normal" removal of surface oxygen takes over, as shown by both AES and ellipsometry. Thus in this interpretation the difference in slope of the $\delta\Delta$ curve represents the difference in ellipsometric effect of subsurface and surface oxygen atoms.

As a measure of the efficiency of the CO reaction we give in Table 4.4 the reaction probability P, defined as the number of removed oxygen atoms per incident CO molecule,

$$P = \frac{N(-d\Theta/dt)}{n_{CO}} , \qquad (4.40)$$

where N is the number of oxygen atoms in a monolayer ($\Theta = 1$) and n_{CO} the collision frequency of CO molecules. In the pressure ranges indicated in the table the reaction rates appeared to be proportional to the CO pressure. The reaction probabilities are temperature dependent, but for all cases given this temperature dependence does not amount to more than an order of magnitude in the range quoted (Fig.4.18).

Table 4.4 also gives the measured activation energies (E_{act}) for the overall reaction $CO + O_{ad} \rightarrow CO_2$ at $\Theta \simeq \frac{1}{4}$. If a Langmuir-Hinshelwood mechanism is operative the "true" activation energy for the reaction $CO_{ad} + O_{ad} \rightarrow CO_2$ must be given by the sum of the heat of adsorption of CO ($\Delta H \simeq 12$ kcal/mol for Cu) and the measured activation energy. For the different copper planes the small differences then almost disappear and it seems that the reaction is not very plane specific.

A different situation prevails in the reduction of oxidized Ni{100} [4.73] where the kinetics seem to be governed by a decrease in the size of nickel-oxide islands. Moreover, CO is not able to reduce the surface. Hydrogen, however, does remove virtually all oxygen from the surface via the reaction $H_{ad} + OH_{ad} \rightarrow H_2O$. Surprisingly, but probably accidentally, the activation energy for this process is also 20 kcal/mol, as for the CO_{ad}/O_{ad} reaction on copper.

A comparison of the ellipsometric results obtained for the interaction with oxygen (Sect.4.6.2) with those for the reaction of adsorbed oxygen

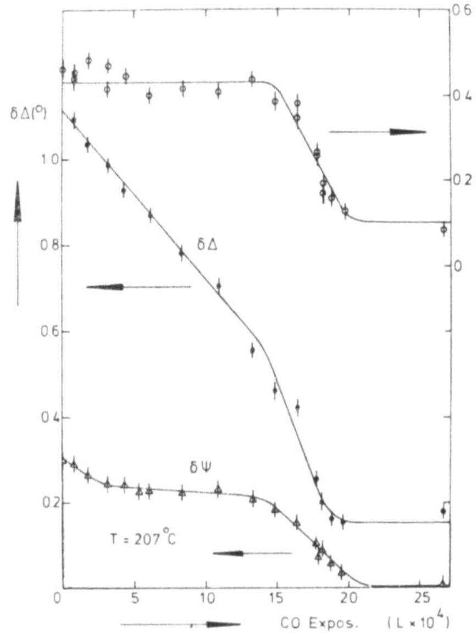

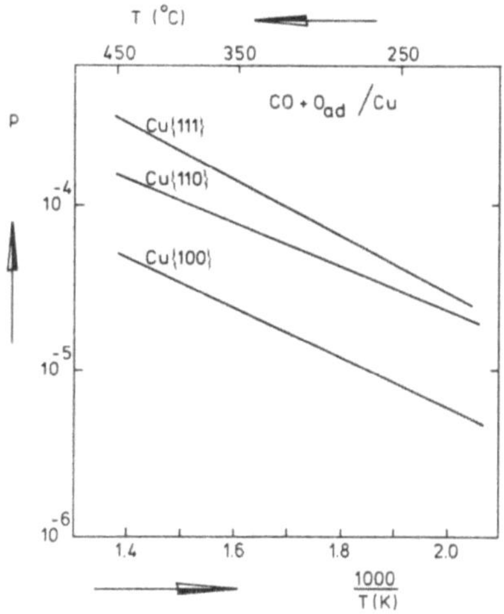

↑Fig.4.18

←Fig.4.17

Fig.4.17. Interaction of CO with oxygen adsorbed at 200°C and 2×10^{-4} torr (1.2×10^{5}L) on Cu{111}. (*Upper curve*) AES measurements; (*other curves*) ellipsometric two-zone measurements. During the measurements the exposure was interrupted [4.68]

Fig.4.18. Arrhenius plot of the reaction probability P of CO with oxygen preadsorbed on Cu{111}, Cu{110}, and Cu{100} at θ = 0.25 [4.77]

Table 4.4. Reaction probability of CO with O_{ad}

	T[K]	p_{CO}[torr]	P(Θ=¼, T=525 K)	E_{act}[kcal/mol]	Ref.
Cu{111}	475-735	5×10^{-5} - 5×10^{-4}	4×10^{-5}	8.0 ± 0.6	[4.68]
Cu{100}	475-675	2×10^{-4} - 8×10^{-4}	8×10^{-6}	6.9 ± 0.4	[4.69]
Cu{110}	500-685	5×10^{-5} - 5×10^{-4}	3×10^{-5}	6.3 ± 0.3	[4.64]
Ag{111}	300	10^{-7} - 10^{-6}	10^{-1}		[4.72]
Ag{110}	300-475	10^{-7} - 4×10^{-6}	5×10^{-2}(300 K)	-2.1 ± 0.4	[4.70]
Ni{100}	475	10^{-4}	0		[4.87]

with CO (this section) shows that the kinetic parameters s and P depend on the oxygen coverage and on the substrate temperature. The ratio P/s strongly increases in the sequence Ni → Cu → Ag and for Cu and Ag also from {110} to {111}.

References

4.1 G.A. Bootsma: Sensors Actuators *1*, 289 (1981)
4.2 P. Drude: Ann. Phys. Chem. *36*, 532, 865 (1889); *39*, 481 (1890)
4.3 Lord Rayleigh: Philos. Mag. *33*, 1 (1892)
4.4 L. Tronstad: Trans. Faraday Soc. *29*, 502 (1933)
4.5 R.J. Archer, G.W. Gobeli: J. Phys. Chem. Solids *26*, 343 (1965)
4.6 G.A. Bootsma, F. Meyer: Surf. Sci. *14*, 52 (1969)
4.7 F. Meyer: Surf. Sci. *56*, 37 (1976)
4.8 F.H.P.M. Habraken, O.L.J. Gijzeman, G.A. Bootsma: Surf. Sci. *96*, 482 (1980)
4.9 F. Meyer, E.E. de Kluizenaar, G.A. Bootsma: Surf. Sci. *27*, 88 (1971)
4.10 G. Quentel, J.M. Rickard, R. Kern: Surf. Sci. *50*, 343 (1975)
4.11 G. Quentel, R. Kern: Surf. Sci. *55*, 545 (1976)
4.12 W. Paik: In *Electrochemistry*, ed. by J.O.M. Bockris (Butterworths, London 1973) p.239
4.13 R.H. Muller: Adv. Electrochem. Electrochem. Eng. *9*, 167 (1973)
4.14 R.M.A. Azzam, N.M. Bashara: *Ellipsometry and Polarized Light* (North-Holland, Amsterdam 1977)
4.15 W.E.J. Neal: Appl. Surf. Sci. *2*, 445 (1979)
4.16 W.E.J. Neal (ed.): *Ellipsometry: Theory and Applications* (Plenum, New York forthcoming)
4.17 E. Passaglia, R.R. Stromberg, J. Kruger (eds.): *Ellipsometry in the Measurement of Surfaces and Thin Films*. Natl. Bur. Stand. U.S. Misc. Publ. *256* (1964)
N.M. Bashara, A.B. Buckman, A.C. Hall (eds.): *Recent Developments in Ellipsometry*. Surf. Sci. *16* (1969)
N.M. Bashara, R.M.A. Azzam (eds.): *Ellipsometry*. Surf. Sci. *56* (1976)
D.E. Aspnes, R.M.A. Azzam, R.H. Muller (eds.): *Ellipsometry*. Surf. Sci. *96* (1980)
4.18 E.C. Chan, J.P. Marton: J. Appl. Phys. *45*, 5004 (1974)
4.19 J.P. Marton, E.C. Chan: J. Appl. Phys. *45*, 5008 (1974)
4.20 I. Ohlidal, F. Lukeš, K. Navrátil: J. Phys. Paris *38*, C5-77 (1977)
4.21 O. Hunderi: J. Phys. Paris *38*, C5-89 (1977)
4.22 O. Hunderi: Surf. Sci. *96*, 1 (1980)
4.23 D.E. Aspnes, J.B. Theeten, F. Hottier: Phys. Rev. B*20*, 3292 (1979)
4.24 K.L. Kliewer: Surf. Sci. *101*, 57 (1980)
4.25 W.J. Plieth, K. Naegele: Surf. Sci. *64*, 484 (1977)
4.26 W.J. Plieth: J. Phys. Paris *38*, C5-215 (1977)
4.27 D. den Engelsen: J. Opt. Soc. Am. *61*, 1460 (1971)
4.28 E.W.M. Visscher: Thesis, Eindhoven University of Technology (1967)
4.29 F.L. McCrackin: Natl. Bur. Stand. U.S. Tech. Note *479* (1969)
4.30 L.J. Hanekamp, W. Lisowski, G.A. Bootsma: To be published
4.31 C. Strachan: Proc. Cambridge Philos. Soc. *29*, 116 (1933)
4.32 D.V. Sivukhin: Zh. Eksp. Teor. Fiz. *18*, 976 (1948); *21*, 367 (1951)
4.33 D.V. Sivukhin: Soviet Phys. - JETP *3*, 269 (1956)
4.34 M.J. Dignam, M. Moskovits: J. Chem. Soc. Faraday Trans. 2 *69*, 56 (1973)
4.35 M.J. Dignam, J. Fedyk: J. Phys. Paris *38*, C5- 57 (1977)
4.36 P.E. Wierenga, A. van Silfhout, M.J. Sparnaay: Surf. Sci. *87*, 43 (1979)
4.37 P.E. Wierenga, M.J. Sparnaay, A. van Silfhout: Surf. Sci. *99*, 59 (1980)
4.38 F. Meyer: Phys. Rev. B*9*, 3622 (1974)
4.39 F. Meyer, A. Kroes: Surf. Sci. *47*, 124 (1975)
4.40 A.E. Morgan: Surf. Sci. *43*, 150 (1974)
4.41 R. Dorn, H. Lüth: Phys. Rev. Lett. *33*, 1024 (1974)
4.42 J.D.E. McIntyre, D.E. Aspnes: Surf. Sci. *24*, 417 (1971)
4.43 J.D.E. McIntyre: In *Optical Properties of Solids; New Developments*, ed. by B.O. Seraphin (North-Holland, Amsterdam 1976) Chap.11

4.44 R.E. Hummel: *Optische Eigenschaften von Metallen und Legierungen, reine und angewandte Metallkunde in Einzeldarstellungen*, Vol.22 (Springer, Berlin, Heidelberg, New York 1971)
4.45 S. Roberts: Phys. Rev. *118*, 1509 (1960)
4.46 P.B. Johnson, R.W. Christy: Phys. Rev. B*6*, 4370 (1972); B*11*, 1315 (1975)
4.47 T.E. Furtak, D.W. Lynch: Phys. Rev. Lett. *35*, 960 (1975)
4.48 R. Kofmann, P. Cheyssac, J. Richard: Surf. Sci. *77*, 537 (1978)
4.49 G.A. Bootsma, O.L.J. Gijzeman, L.J. Hanekamp: In Ref.4.16, Chap.5
4.50 R.C. O'Handley, D.K. Burge: Surf. Sci. *48*, 214 (1975)
4.51 O. Hunderi: Surf. Sci. *57*, 451 (1976)
4.52 U. Merkt, P. Wissmann: Thin Solid Films *57*, 65 (1979)
4.53 U. Merkt, P. Wissmann: Z. Phys. Chem. Neue Folge *115*, 55 (1979)
4.54 U. Merkt, P. Wissmann: Surf. Sci. *96*, 529 (1980)
4.55 U. Merkt, P. Wissmann: Vak. Tech. *29*, 41 (1980)
4.56 J.D. Fedyk, P. Mahaffy, M.J. Dignam: Surf. Sci. *89*, 404 (1979)
4.57 P.R. Mahaffy, M.J. Dignam: Surf. Sci. *97*, 377 (1980)
4.58 F.L. Baudais, A.J. Borschke, J.D. Fedyk, M.J. Dignam: Surf. Sci. *100*, 210 (1980)
4.59 R. Kötz, B. Hayden, E. Schweizer, A.M. Bradshaw: Surf. Sci. *112*, 229 (1981)
4.60 D.A.G. Bruggeman: Ann. Phys. Leipzig *24*, 636 (1935)
4.61 D.E. Aspnes, J.B. Theeten, F. Hottier: Phys. Rev. B*20*, 3292 (1979)
4.62 D.E. Aspnes, J.B. Theeten: Phys. Rev. Lett. *43*, 1046 (1979)
4.63 D.E. Aspnes, E. Kinsbron, D.D. Bacon: Phys. Rev. B*21*, 3290 (1980)
4.64 F.H.P.M. Habraken, G.A. Bootsma: Surf. Sci. *87*, 333 (1979)
4.65 F.H.P.M. Habraken, G.A. Bootsma, P. Hofmann, S. Hachicha, A.M. Bradshaw: Surf. Sci. *88*, 285 (1979)
4.66 K. Bärwinkel, H.J. Schmidt: Thin Solid Films *59*, 373 (1979)
4.67 V.F. Drobny, D.L. Pulfrey: Thin Solid Films *61*, 89 (1979)
4.68 F.H.P.M. Habraken, E.Ph. Kieffer, G.A. Bootsma: Surf. Sci. *83*, 45 (1979)
4.69 F.H.P.M. Habraken, C.M.A.M. Mesters, G.A. Bootsma: Surf. Sci. *97*, 264 (1980)
4.70 H. Albers, W.J.J. van der Wal, O.L.J. Gijzeman, G.A. Bootsma: Surf. Sci. *77*, 1 (1978)
4.71 H. Albers, J.M.M. Droog, G.A. Bootsma: Surf. Sci. *64*, 1 (1977)
4.72 H. Albers, W.J.J. van der Wal, G.A. Bootsma: Surf. Sci. *68*, 47 (1978)
4.73 P.K. de Bokx, F. Labohm, O.L.J. Gijzeman, G.A. Bootsma, J.W. Geus: Appl. Surf. Sci. *5*, 321 (1980)
4.74 P.H. Holloway, J.B. Hudson: Surf. Sci. *43*, 123 (1974)
4.75 J.J. Vrakking, F. Meyer: Appl. Phys. Lett. *18*, 226 (1971)
4.76 T.H. Allen: J. Vac. Sci. Technol. *13*, 112 (1976)
4.77 F.H.P.M. Habraken: "Sorption and Reactivity of Oxygen on Copper Single Crystal Surfaces"; Ph. D. Thesis, University of Utrecht (1980)
4.78 F.H.P.M. Habraken, G.A. Bootsma: In Proc. IVC- 8, ICSS-4 ECOSS-3, Cannes, 1980, p.555
4.79 M. Jardinier-Offergeld, F. Bouillon: In *Reaction Kinetics in Heterogeneous Chemical Systems*, ed. by P. Barret (Elsevier, Amsterdam 1975) p.347
4.80 M. Jardinier-Offergeld, F. Bouillon: J. Vac. Sci. Technol. *9*, 770 (1972)
4.81 A.M. Bradshaw, P. Hofmann, W. Wyrobisch: Surf. Sci. *68*, 269 (1977)
4.82 P. Hofmann, W. Wyrobisch, A.M. Bradshaw: Surf. Sci. *80*, 344 (1979)
4.83 B.E. Hayden, W. Wyrobisch, W. Oppermann, S. Hachicha, P. Hofmann, A.M. Bradshaw: Surf. Sci. *109*, 207 (1981)
4.84 B.E. Hayden, E. Schweizer, R. Kötz, A.M. Bradshaw: Surf. Sci. (to be published)
4.85 J.J. Carroll, A.J. Melmed: Surf. Sci. *16*, 251 (1969)
4.86 A.J. Melmed, J.J. Carroll: J. Vac. Sci. Technol. *10*, 164 (1973)
4.87 F. Labohm: Private communication

5. The Implications for Surface Science of Doppler-Shift Laser Fluorescence Spectroscopy[1]

D.M. Gruen, A.R. Krauss, M.J. Pellin, and R.B. Wright

With 11 Figures

5.1 Introduction

Lasers are already useful in various aspects of surface science and can be expected to find wider fields of application to surface-related phenomena in the future.

Their impact has been in two chief areas: laser-induced surface reactions and laser spectroscopy of surfaces.

This paper will deal with the recent development of a particular aspect of surface-related spectroscopy: laser-induced fluorescence of sputtered atoms. Before undertaking a detailed discussion of this subject, we want at least to mention some of the other areas referred to above to indicate the scope of the applications of lasers to the study of surfaces.

In the field of laser-induced surface reactions, the ability to deposit energy in precisely controllable time and space domains is the key to success. Thus materials processing (heat treatment of alloys and forgings, cutting, welding, drilling, etc.) has already matured into commercial practice. Laser annealing of defects in ion-implanted silicon [5.1] and surface etching of LSI circuit materials by laser-generated neutral species have been demonstrated [5.2]. The production of sinterable ceramic powders by CO_2-laser-driven gas phase reactions [5.3] and "laser-plasma-spraying" processes for applying refractory coatings to materials are in the offing, as are photochemically deposited thin metallic films with better than 1-μm definition [5.4]. A potentially exciting field is laser control of reactions at or near catalytic surfaces. Some results have been reported but the detailed mechanisms are still unclear [5.5].

1 Based on work performed under the auspices of the Basic Energy Sciences, Division of Materials Sciences, U.S. Department of Energy under Contract #31-109-ENG-38.

An important discovery in the field of surface science has been the observation of enormously enhanced cross sections (up to factors of 10^6) for Raman scattering from molecules adsorbed on metal surfaces. This so-called surface enhanced Raman spectroscopy (SERS) enables the determination of detailed structural information about adsorbed molecules even in submonolayer coverages, and in both solid-gas and solid-liquid environments. Although SERS has been a subject of intense study, its interpretation is still controversial with at least six alternative mechanisms under active consideration [5.6].

Returning to the topic of this paper, recent experiments have demonstrated that Doppler-shift laser-induced fluorescence spectroscopy provides a sensitive, high-resolution and facile method for obtaining both sputtered neutral atom and ion velocity distributions and relative sputtering yields [5.7-11]. These findings are leading to the development of LFS as a quantitative technique for the study of sputtered species [5.12,13]. The exploitation of this technique has only begun but its possible impact on surface science could be dramatic.

Sputtering occurs whenever an energetic particle transfers sufficient momentum to exceed an atom's surface binding energy. Sputtering is found in vacuum tubes, ion pumps, arcs, and as a result of plasma wall interactions in fusion machines. Sputtering is used for thin film deposition, occurs with ion implantation, and is the basis of SIMS for surface analysis and depth profiling.

The sputtering phenomenon is ubiquitous and has an exceedingly complex physical basis. SIGMUND theory by and large provides a rationalization for *total* sputtering yields as a function of target mass and of projectile mass and energy [5.14]. However, the sputtered products consist not only of atoms in their ground electronic states, but to a certain and variable extent of atoms, ions, molecules, and molecular ions in ground and excited states. A comprehensive theory must enable one to calculate the *partial* yields into each of the different sputtering channels. Such calculations have to be compared to measurements made on well-characterized surfaces. Theories of secondary ion and excited-state emission have in fact been elaborated but it is difficult to test their predictions quantitatively against currently available data [5.15,16]. This is due in part to the fact that small changes in surface composition can drastically alter the distribution among sputtered products. Such changes occur readily because intended or unintended surface modification occurs either as a result of surface contamination or of the bombardment process itself. Careful control of surface conditions must therefore be combined with partial yield measurements to give ultimately a consistent data base against which the theoretical apparatus can be tested. Only in this way will knowledge of excitation, deexcitation, ionization, and neutralization processes which control the charge states and states of excitation of the sputtered species be enhanced. Although it is known in a general way that these processes are strongly dependent on surface composition, a detailed understanding is lacking concerning the basis for this extreme sensitivity to surface chemistry.

It is here that Doppler-shift laser fluorescence spectroscopy (LFS) can be expected to make a fundamental and potentially far-reaching contribution. Laser-induced fluorescence can be used, in principle, to access ground and excited states of neutral and ionized atoms and molecules. Using this technique, it becomes practical for the first time to measure, for example, velocity and angular distributions of sputtered neutral atoms and ions under the *same* experimental conditions while carefully monitoring the chemical state of the surface.

For many materials and particularly for most clean metal surfaces, the majority of sputtered species are neutral atoms and it has been a difficult experimental problem to measure their energy distributions reliably. It is now possible to make such measurements with high accuracy (±25 m/s) and great detection sensitivity (10^8 atoms/cm^3) using LFS. Similar measurements can be made on ions, and by combining neutral and ion energy distribution measurements made under identical experimental conditions the surface ionization coefficient and its energy dependence can be experimentally determined. By measuring this important quantity as a function of surface composition, for example as a function of oxygen coverage, much can be reliably learned about the effect of surface chemistry on charge transfer processes.

5.2 Charge Transfer Processes at Surfaces

In this section, a brief review will be given of certain aspects of charge transfer processes at surfaces, particularly as they relate to theories of secondary ion emission. The aim is to stress the diversity of theoretical approaches in this field and to point out that the basic theoretical question can be reduced to modeling the behavior of the ionization probability $R^+(E)$ correctly. The discussion is based on that given on this topic in [5.16].

Secondary ion sputtering yields can be approximately calculated using expressions based either on the SAHA-LANGMUIR [5.17,18] equilibrium model or the ANDERSEN-HINTHORNE [5.19,20] local thermal equilibrium (LTE) model. Contrary to experiment these expressions are independent of the kinetic energy and imply ion energy distributions identical to those of sputtered neutrals. There are other discrepancies between the predictions based on equilibrium models and the experimental results. If LTE is valid, then the electronic states of all sputtered species should have Boltzmann populations corresponding to a single temperature. However, the observed optical emission from excited states of uranium sputtered by 3-keV Kr^+ were found to be characterized by a temperature of ~4000 K [5.21], while a laser fluorescence spectroscopic study of the low-lying metastable states of uranium yielded a temperature of ~900 K [5.22]. Very similar results were obtained for the low-lying levels of Fe [5.23]. The effective "temperatures" therefore appear to be a function of the degree of excitation, with the ion "temperature" perhaps being the limiting value.

Another prediction of the equilibrium models is that the secondary ion intensities should be lower than the photon yields. Absolute measurements however indicate that the converse is true with the secondary ion yield typically exceeding the photon yield by a factor of 100 [5.24,25].

As already mentioned, a major difficulty with the equilibrium theory is its inability to predict energy distributions. Not only are the secondary ion energy distributions shifted to higher energies compared to the sputtered neutral distributions, but the energy distributions of excited atoms are quite different from both ions and neutrals in that they display *threshold* kinetic energies below which no excitation occurs. These threshold energies are far in excess of the typical energies observed in secondary ion or ground-state neutral energy distributions [5.26,27].

Secondary ion *yields* seem to be rather insensitive to the actual ionization mechanisms while the kinetic energy distributions contain potentially more information concerning these processes. One therefore wishes to focus attention on $R^+(E)$, the ionization coefficient and its dependence on energy. The ionization probability $R^+(E)$ is defined by the relation

$$N^+(E) = R^+(E)\, N(E) \quad . \tag{5.1}$$

For sputtering by random collision cascades [5.28],

$$N(E) \sim \frac{E}{(E + E_b)^3} \frac{\cos\phi_e}{\cos\phi_i} \tag{5.2}$$

where E_b is the surface binding energy, approximately equal to the heat of sublimation, and ϕ_e and ϕ_i are the angles of emission and incidence respec-

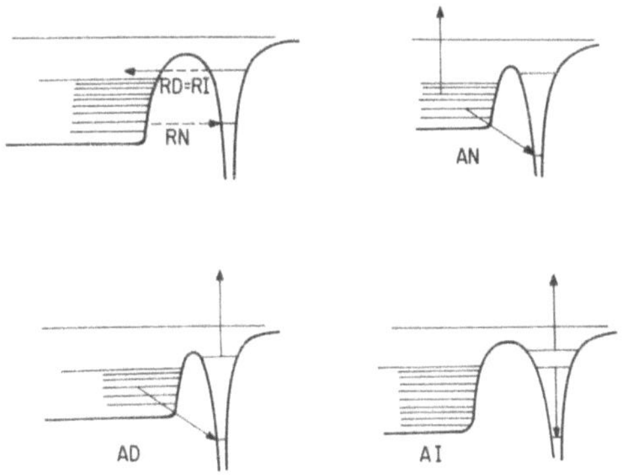

Fig.5.1. Charge exchange and excitation- deexcitation processes associated with the surface, sputtered-atom interactions. A neutral excited atom may deexcite by a resonant charge transfer back to the surface. This process may be thought of as either resonance ionization (RI) or resonance deexcitation (RD). If the sputtered atom was positively ionized the reverse process, resonance neutralization (RN) can occur. A sputtered positive ion may be neutralized by an interatomic Auger transition (AN). Interatomic Auger transitions can also cause deexcitation without changing the charge state (AD). If the sputtered atom escapes to vacuum as an excited neutral, an intraatomic Auger transition can result in deexcitation by autoionization (AI)

tively. This energy spectrum exhibits a peak at $E_b/2$ and has a long high-energy tail which falls as E^{-2}. Several charge transfer processes associated with the atom-surface interaction are shown in Fig.5.1 and have been analyzed by HAGSTRUM [5.29]. VAN DER WEG and BIERMAN [5.30] assumed that the sputtered particle leaves as an ion and is subsequently neutralized by interatomic Auger and resonance processes. They obtained the expression

$$R^+ \sim \exp(-ca/v) \tag{5.3}$$

where $a \sim 2$Å is a parameter representing the atom-surface interaction distance, $c \sim 4\times10^{14}\ s^{-1}$, and v is the velocity.

JOYES [5.31,32] on the other hand, assumed that the sputtered atom is neutral with one core electron excited to a valence shell. As long as the atom does not undergo resonance deexcitation while near the surface, it will eventually undergo autoionization by Auger deexcitation. BLAISE and SLODZIAN [5.33-36] and MACDONALD [5.37] have predicted autoionizing processes for sputtered atoms in which two valence electrons are excited as a result of the energy level rearrangement during surface crossing. Calculations were performed in the one-electron approximation and the resulting ionization probability for an autoionization mechanism was found to be

$$R^+(E) \sim 1 - \exp(-c'a/v) \quad . \tag{5.4}$$

The level-crossing argument can also be applied to compounds and leads to a mechanism for direct ionization via a "bond breaking" process [5.38]. If the bond is ionic and the sputtering event is rapid, the charge of the sputtered atoms will reflect the molecular valence. A quantitative calculation has not been performed. However, if the ionic ground state and excited covalent levels cross, the probability that the sputtered particles are ejected as ions depends on the ionic-covalent interaction potential and the separation velocity.

Another approach identifies the ionization process as the result of the evolution of the energy levels of the sputtered atom from those of the solid to those of an isolated atom. The process is illustrated in Fig.5.2, and is similar to the NEWNS-ANDERSON [5.39] approach to the chemisorption problem except that the atom-surface interaction potential V_{ak} is now time dependent. SCHROEER et al. [5.40] have calculated the ionization probability in the adiabatic approximation, assuming that the sputtered atom is initially in a neutral, unexcited state. The resulting expression is

$$R^+(E) = \left(\frac{E_b}{I - \phi}\right)^2 \left(\frac{h\nu}{a(I - \phi)}\right)^{2m} \qquad (5.5)$$

where $m \sim 1.25$ is a fitting parameter, and $a \sim 1.5$ Å. SROUBEK [5.41] has used the same basic model as SCHROEER, but employed a molecular-orbital approach to evaluate the transition probability. He obtained an expression similar to that of (5.5) but with m varying from 1/4 to 3/2 depending on the projected state density in the atomic orbital and the occurrence of level crossings as the atom leaves the surface. More recently [5.42] the calculation has been extended to consider sputtering of a multielectron atom but was only evaluated in a one-dimensional model. The total ion yield was found to vary as $\exp[-(I - \phi)e/kT]$, in accord with experiment and the predictions of the thermodynamic models. The ionization probability was calculated to vary as

$$R^+(E) \sim e^{CE} \quad . \qquad (5.6)$$

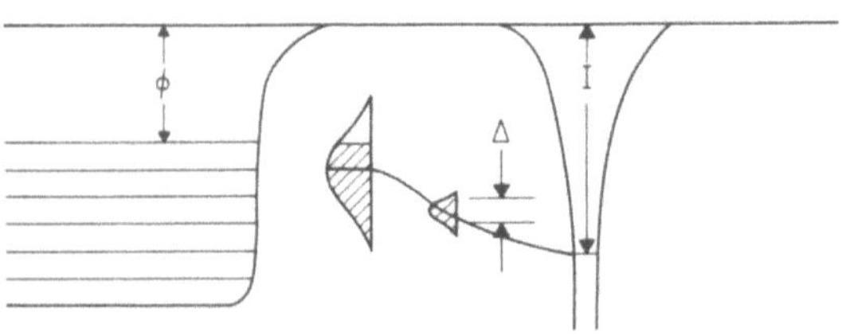

Fig.5.2. Evolution of an electronic energy level of a sputtered atom from that of the bulk band structure to that of an isolated atom. The level is described by a resonance halfwidth Δ, which decreases as the atom leaves the surface

Further refinements to this model have been made by the French group [5.43-46] utilizing a many-body theory of time-dependent perturbations to describe ionization produced by the surface-sputtered atom interaction.

As has already been pointed out, incomplete data and contradictary results make it unlikely that a fundamental theory of excited-state and secondary ion emission is in the offing. These complex processes involve different mechanisms under different surface conditions depending on rather subtle changes in those conditions. The development of the theory will surely benefit from experiments made possible by the new laser fluorescence spectroscopic technique to be described in the next section.

5.3 Laser Fluorescence Spectroscopic Measurements of Fluxes and Energy Distributions of Sputtered Particles

Resonance fluorescence excitation of sputtered particles offers a number of unique advantages for studying sputtered particles and gaining insight into the sputtering process. Using this technique it is possible to measure total sputtering yields, partial yields into excited state and ion sputtering channels, as well as energy and angular distributions. Because of the great selectivity of LFS, different atomic and molecular products can be identified and due to its intrinsic sensitivity, the technique can be employed for investigations of light ion sputtering. Since both neutral and ionic species can be accessed, measurements of $R^+(E)$ are now possible on surfaces whose composition can be simultaneous monitored.

Tunable radiation from dye lasers is available to access most of the resonance transitions of atoms in the periodic table of elements. Schematics of two types of laser spectrometers are shown in Figs.5.3,4. Figure 5.3, displays a CW tunable dye laser system used to study sputtered uranium atoms while Fig.5.4 shows a pulsed laser system employed in a study of sputtered zirconium atoms [5.12,13]. In both cases narrow-bandwidth laser radiation intersects a sputtered particle beam and excites resonance transitions of the atoms under investigation. Cross-sectional views of the target chambers used in these experiments to measure the $v_{\parallel}$ (Fig.5.5) and $v_{\perp}$ (Fig.5.6) number density velocity distributions and relative sputtering yields are shown. For details of the experimental procedures, the reader is referred to [5.11-13].

The spontaneously emitted fluorescence radiation can be proportional to the particle flux or number density depending on experimental conditions [5.13]. If the intersecting angle between laser beam and sputtered particle

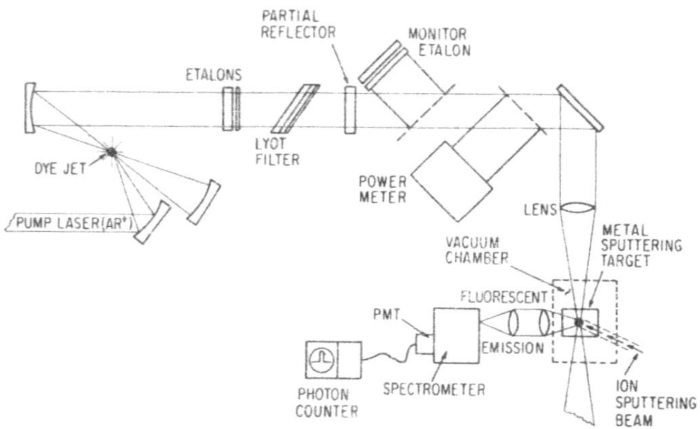

Fig.5.3 ▲

▼ Fig.5.4

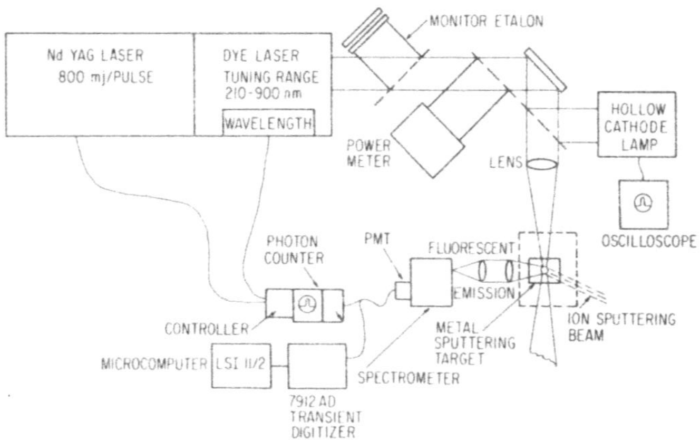

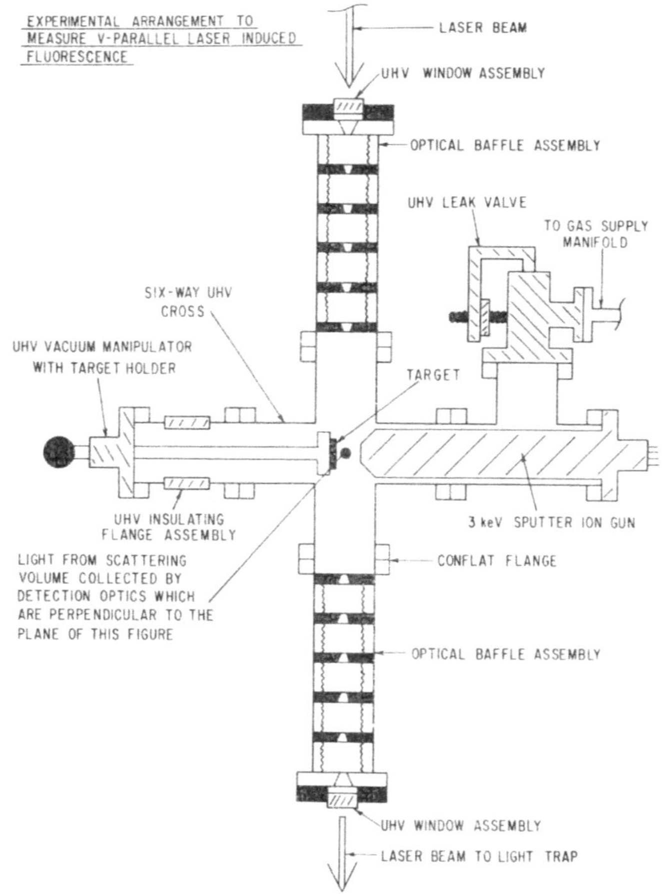

Fig.5.5 (captions for all figures see opposite page)

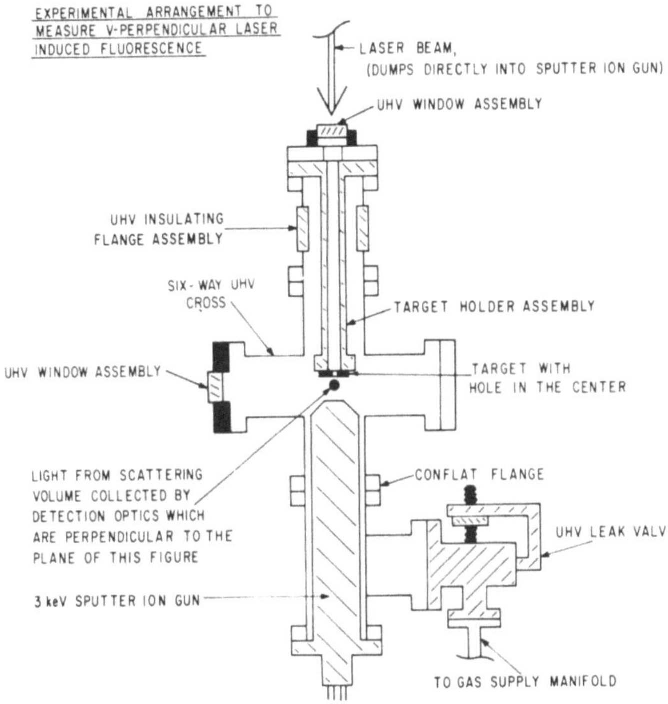

Fig.5.6. Cross-sectional view of the target chamber used to measure the $v_\perp$ sputtered atom number density velocity distribution

beam is different from 90°, the laser excitation frequency is Doppler-shifted proportional to the particle velocity. By detuning the laser from the unshifted resonance frequency, the spectrum of the velocity component of the atoms in the laser beam direction can be obtained. The velocity resolution that can be achieved with LFS making use of Doppler-shifted excitation depends on two factors. The first concerns the collimation of the beam of sputtered particles and the second has to do with the bandwidth of the laser radiation. By using etalons, the linewidth of CW lasers can be reduced to a few MHz and that of

---

Fig.5.3. Diagram of the experimental arrangement used to study the sputtering process by the LIF technique

Fig.5.4. Diagram of apparatus used in the DSLS experiments pertaining to the sputtering process

Fig.5.5. Cross-sectional view of the target chamber used to measure the $v_{\parallel}$ sputtered atom number density velocity distribution and the relative sputtering yields pulsed systems to the GHz range. Velocity resolutions of a few times 10 m/s can be achieved with CW lasers and a few times 10^3 m/s with pulsed systems. One can thus measure energy distributions of sputtered particles and therefore ionization and excitation coefficients with much higher accuracy than has been possible up to now.

Detection sensitivities achieved with LFS in laboratory experiments are in the range 10^3 - 10^8 particles/cm^3 with the higher sensitivities attainable with CW excitation.

We will now describe how LFS has been used to determine the number density velocity distributions of neutral zirconium atoms sputtered from a zirconium metal surface in the a^3F_2 ground state and in two metastable states ($a^3F_{3,4}$) which lie 570 and 1241 cm^{-1}, respectively, above the ground state [5.12]. These number density velocity distributions have been determined for both Ar^+ and Kr^+ primary ions at a variety of primary ion energies (1.5-3 keV). The sputtering yields for atoms in all three states relative to the value for ground-state zirconium sputtered with 3-keV Kr^+ were determined over an extended primary energy range (750-3 keV) for Ar^+ and Kr^+ as the primary ions.

The interpretation of LFS data has been discussed both in the context of molecular beams (e.g., [5.47]) and of sputtering (see [5.11-13] and the references listed there). Quantitative evaluation of the LFS results requires an understanding of three effects which have only recently begun to be properly taken into account: power broadening, transit-time broadening, and velocity-dependent fluorescence detection efficiencies. A detailed theoretical treatment of LFS data including these three effects has been given in the above-cited references; we will make use of the results of these considerations in the following discussion of the data.

Consider an isolated atom with the energy-level diagram of Fig.5.7 and traveling with a velocity relative to a laser beam propagating along the â axis. The Doppler-shifted transition frequency from level 1 to level 2 can be expressed as

$$\nu_0 = \nu_0\left(1 + \frac{v\ \cos\theta}{c}\right) = \nu_0\left(1 + \frac{v_\perp}{c}\right) \quad , \tag{5.7}$$

where $v_\perp$ is the perpendicular velocity component of the sputtered atom's velocity v, and θ is the emission angle of the atom defined with respect to the target surface normal (i.e., with respect to the laser direction in the present case). One can therefore convert the number density distribution n(v) into $n(v_\perp)$ resulting in

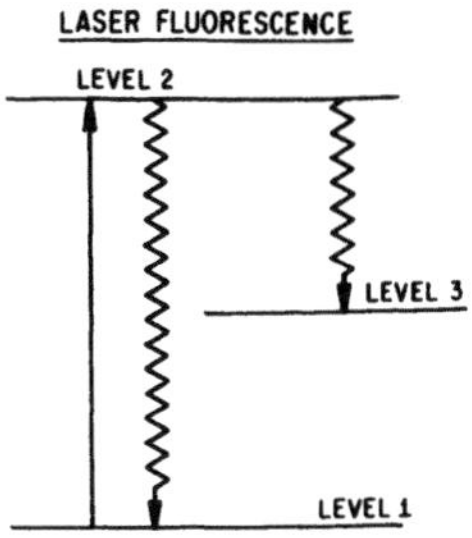

Fig.5.7. Energy-level model diagram with state 2 having an optically allowed transition to the ground state and another (3), but with no transition between states 1 and 3. Levels 1 and 2 are coupled by laser radiation

$$n(v_\perp)dv_\perp \propto \left(\frac{2^n}{M_2^{n-1}}\right)\left(\frac{2\pi}{2n+1}\right)\frac{v_\perp dv_\perp}{(v_\perp^2+v_b^2)^{n+1/2}} \quad . \tag{5.8}$$

This equation was used to compare the Thompson-Sigmund model for linear collision cascade sputtering with the measured LFS velocity distributions.

Linear collision cascade theory yields velocity distributions which can therefore be characterized by two parameters n and v_b. Here, n is a variable which is generally measured to be $1 \le n \le 2$, though theory would place it closer to 2; and v_b is the velocity corresponding to the surface binding energy E_b, i.e., $v_b = [2E_b/M_{Zr}]^{1/2}$, due to the assumed planar potential barrier encountered by the sputtered atom in the near-surface region.

The LFS data on sputtered Zr atoms can now be compared with linear collision cascade theory. A partial energy - level diagram shown in Fig.5.8 for zirconium displays the transitions used in the LFS measurements. The sputtering properties of ground-state zirconium atoms were probed using the a^3F_2 (0 cm^{-1}) $\to z^5F_1^0$ (16 787 cm^{-1}) transition. This transition, although weak ($gA = 8.8 \times 10^5\ s^{-1}$) produced sufficient excited zirconium atoms for detection. Fluorescence involving the $z^5F_1^0 \to a^5F_1$ (4871 cm^{-1}) transition was monitored. This transition with its relatively large branching ratio ($gA = 1.6 \times 10^6\ s^{-1}$) was spectrally far removed from the excitation source, thus minimizing scattered light problems. Alternate transitions from the $z^5F_1^0$ state are also significant giving $g\sum_{i\neq 3} A_{2i} = 1.08 \times 10^6\ s^{-1}$ [5.48].

The sputtering properties of the a^3F_3 metastable level which lies 570 cm^{-1} above the ground state were determined using the a^3F_3 (570 cm^{-1}) $\to z^3F_4^0$ (17 556 cm^{-1}) $\to a^3F_4$ (1241 cm^{-1}) excitation-monitor scheme. Transition strengths were $gA = 8.7 \times 10^5$ and $7.1 \times 10^6\ s^{-1}$, respectively. The total decay rate from $z^3F_4^0$ was $g\sum_i A_i = 8.24 \times 10^6\ s^{-1}$, and the energies of excitation and emission were 16 986 and 16 315 cm^{-1}, respectively.

Similarly, measurements of the a^3F_4 level which lies 1241 cm^{-1} above the ground state utilized a a^3F_4 (1241 cm^{-1}) $\to z^3D_3^0$ (18 244 cm^{-1}) $\to a^3P_2$

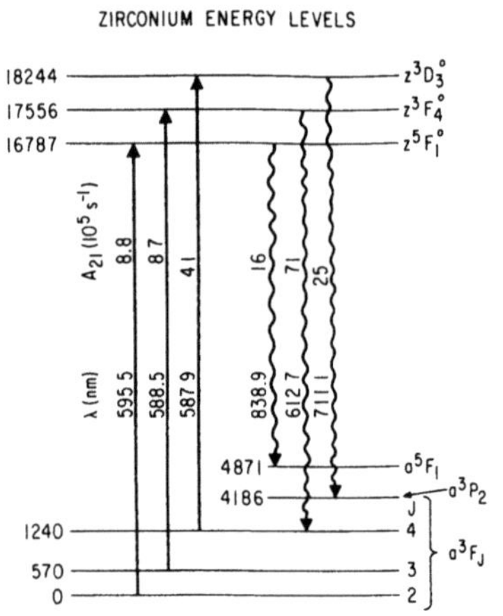

Fig.5.8. Diagram of the energy levels of neutral zirconium atoms which were investigated here. Vacuum level energies in cm-1 are displayed on the left-hand side, transition strength on the right- hand side

Table 5.1. Sputtering yields for Kr^+ ion normally incident on zirconium metal. All values are relative to the 3-keV value and the LFS data are only for ground-state neutral zirconium atoms. Binding energy was 6.305 eV

Primary ion energy [keV]	DSLS neutral atom	Theory [5.49]	Theory [5.50]
3.0	1.0	1.0	1.0
2.0	0.46 ± 0.1	0.68	0.89
1.5	0.33 ± 0.06	0.52	0.81
0.750	0.24 ± 0.06[a]	0.27	0.65

[a]Adjusted for lower beam current.

(4186 cm^{-1}) excitation-emission scheme. This had $gA = 4.1\times10^6$ and 3.5×10^6 s^{-1}, respectively, with a total decay rate from $z^3D_3^0$ of $g\sum_i A_i = 9.02\times10^6$ s^{-1}. The energies of excitation and emission were 17003 and 14058 cm^{-1}, respectively.

Relative sputtering yield measurements of ground-state (a^3F_2) zirconium for various primary ion energies are shown in Table 5.1. All sputtering yield measurements were indexed to the value obtained with a normally incident 3-keV Kr^+ primary ion beam sputtering a^3F_2 zirconium atoms using the signal function derived in [5.12]. Included for comparison are two theoretical values. The first is from a semiempirical theory by SMITH [5.49]. For primary ion energy changes this is only slightly different from the ratio of the primary energies. The second theory is by SIGMUND [5.50] and is considerably more complex.

The velocity distribution of ground-state neutral zirconium atoms sputtered by normally incident 3-keV Kr^+ ions is displayed in Fig.5.9. The points

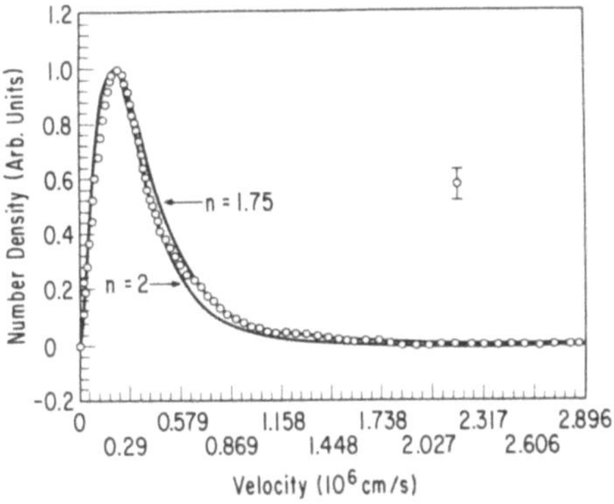

Fig.5.9. Number density velocity distribution of neutral ground-state zirconium atoms sputtered by normally incident 3-keV Kr^+. (o) represent deconvoluted data averaged over several runs. The solid curves were generated using linear collision cascade theory. The "n" are those in (5.8) with n=1.75 and 2. Average error bars are displayed on the right-hand side

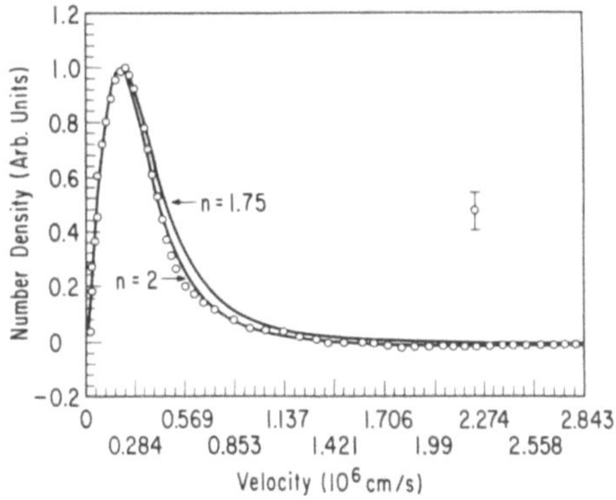

Fig.5.10. Number density velocity distribution of zirconium atoms sputtered in the a^3F_3 level, 570 cm^{-1} above the ground state. The primary ion was normally incident 3-keV Kr^+. The solid curves are identical to those of Fig. 5.9. An average error bar is displayed on the right-hand side

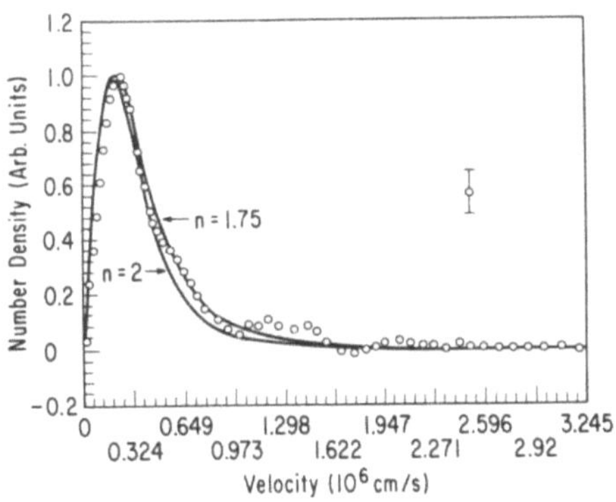

Fig.5.11. a^3F_4 zirconium atom number density velocity distribution. The primary ion was normally incident 3-keV Kr^+. The solid curves are identical to those of Fig.5.9. An average error bar is displayed on the right-hand side

are the result of averaging several measurements after each had been deconvoluted from an appropriate power and transit-time-broadened Lorentzian. It was possible additionally to deconvolute the excitation laser linewidth from the data. However, the spectral narrowness of the Nd/YAG pumped dye laser source was sufficient to make this unnecessary. All frequencies higher than 50 Hz were suppressed in the Fourier transform. Further it was necessary to account for the velocity-dependent photon yields [5.12]. The solid curve of Fig.5.9 was constructed using (5.8) with n = 2 or 1.75 and $\nu_b=3.652\times10^5$ cm/s.

Table 5.2. Relative sputtering yields of sputtered neutral zirconium atoms in the $a^3F_{3,4}$ energy levels. Primary ion is 3-keV normally incident Kr^+. All values are indexed to the 3-keV value Kr^+ value for neutral ground-state Zr given in Table 5.1

Primary ion energy [keV]	a^3F_3	a^3F_4
3.0	0.39 ± 0.08	0.24 ± 0.06
2.5	0.29 ± 0.05	0.16 ± 0.04
2.0	0.15 ± 0.06	0.09 ± 0.03
1.5	0.07 ± 0.03	0.04 ± 0.02
0.750	0.01 ± 0.01	0.01 ± 0.01

Velocity distributions taken for primary Kr^+ energies in the range 1.5-3 keV showed, within experimental error, no dependence on primary ion energy. This fundamental prediction of linear collision cascade theory was verified by the LFS data. Again n = 1.75 or 2 were used as parameters with $v_b = 3.652 \times 10^5$ cm/s.

Figures 5.10,11 represent number density velocity distributions of zirconium atoms sputtered by normally incident 3-keV Kr^+ in the a^3F_3 and a^3F_4 metastable levels; Table 5.2 shows the partial relative sputtering yields for these states. Again the velocity distributions appear to be well fit by linear collision cascade theory. This result is remarkable since SCANIIR experiments, in general, show that atoms which are sputtered in excited states have velocity distributions shifted to much higher velocities than those expected for ground-state atoms [5.27]. An indication of the magnitude of this effect in zirconium was seen by monitoring the $z^5F_1^0 \rightarrow a^3F_2$ SCANIIR transition in zirconium. The fluorescence profile of atoms sputtered in the $z^5F_1^0$ state showed a FWHM of 4.5 ± 0.6 cm^{-1}, about a factor of 10 larger than the Doppler width profiles of atoms sputtered either in the ground state or the $^3F_3$ and $^3F_4$ low-lying metastable states. Not only do the low-lying metastable levels have Boltzman distributions characteristic of temperatures that are for lower than those associated with SCANIIR levels, but their velocity distributions are much closer to those of atoms sputtered in ground states than to those sputtered either in high-energy excited states or as ions.

This important result with possibly far-reaching consequences for the development of the theory of charge transfer at surfaces would have been difficult to obtain by a technique other than LFS.

An extension of the work on neutral zirconium atoms described above clearly calls for similar measurements on zirconium ion velocity distributions. Experiments directed toward this end are in progress in our laboratory and should be useful for establishing a reliable method for the determination of $R^+(E)$. Furthermore, by simultaneously monitoring surface compositions using Auger spectroscopy or ISS, it should become possible to probe in detail the dependence of $R^+(E)$ on surface chemistry.

References

5.1 C.W. White, J. Naragan, R.T. Young: Science *204*, 461 (1979)
5.2 J.I. Steinfeld: Proc. Int. Conf. Lasers, (1979) p. 241
5.3 J.S. Haggerty, W.R. Cannon: M.I.T. Energy Laboratory Report EL-78-037 (1978)
5.4 T.F. Deutsch, D.J. Ebrlich, R.M. Osgood: Appl. Phys. Lett. *35*, 175 (1979)
5.5 C.T. Lin, T.D.Z. Atvors: J. Chem. Phys. *68*, 4233 (1978)
5.6 G.C. Schultz, R.P. Van Duyne: Surf. Sci. *101*, 425 (1980)
5.7 A. Elbern, D. Rusbüldt, E. Hintz: Proceedings of the International Symposium on Plasma Wall Interactions, Jülich, Federal Republic of Germany, October 1976
5.8 A. Elbern, E. Hintz, B. Schweer: J. Nucl. Mater. *76-77*, 143 (1978)
5.9 W. Husinsky, R. Bruckmüller, P. Blum, F. Viehböck, D. Hammer, E. Benes: J. Appl. Phys. *48*, 4734 (1977)
5.10 W. Husinsky, R. Bruckmüller: Surf. Sci. *80*, 637 (1979)
5.11 R.B. Wright, M.J. Pellin, D.M. Gruen, C.E. Young: Nucl. Instrum. Methods *170*, 295 (1980)
5.12 M.J. Pellin, R.B. Wright, D.M. Gruen: J. Chem. Phys. *74*, 6448 (1981)
5.13 R.B. Wright, M.J. Pellin, D.M. Gruen: Surf. Sci. (in press)
5.14 P. Sigmund: In "Sputtering by Ion Bombardment: Theoretical Concepts", ed. by R. Behrisch, Topics in Applied Physics, Vol.47 (Springer, Berlin, Heidelberg, New York 1981)
5.15 R.B. Wright, D.M. Gruen: J. Chem. Phys. *72*, 147 (1980)
5.16 A.R. Krauss, V.E. Krohn: *Mass Spectrometry,* Specialist Reports, Vol.6 (The Chemical Society, London 1981)
5.17 I. Langmuir, K.H. Kingdon: Proc. R. Soc. London *107*, 61 (1925)
5.18 I. Langmuir: Trans. Am. Ceram. Soc. *54*, 2798 (1932)
5.19 C.A. Andersen, J.R. Hinthorne: Science *175*, 853 (1972)
5.20 C.A. Andersen, J.R. Hinthorne: Anal. Chem. *45*, 1421 (1973)
5.21 R. Wright: Private communication
5.22 R.B. Wright, M.J. Pellin, D.M. Gruen, C.E. Young: Nucl. Instrum. Methods *170*, 295 (1980)
5.23 R. Behrisel et al.: J. Nucl. Mat. *93-94*, 645 (1980)
5.24 P. Williams, I.S.T. Tsang, S. Tsuji: Nucl. Instrum. Methods *170*, 591 (1980)
5.25 I.S.T. Tsang, N.A. Yursuf: Appl. Phys. Lett. *33*, 999 (1979)
5.26 I.S.T. Tsang, N.A. Yursuf: Nucl. Instrum. Methods *170*, 357 (1980)
5.27 S. Dzioba, O. Auciello, R. Kelly: Radiat. Eff. *45*, 235 (1980)
5.28 M.W. Thompson: Philos. Mag. *18*, 377 (1968)
5.29 H. Hagstrum: Phys. Rev. *96*, 336 (1954)
5.30 W.F. Van der Weg, D.J. Bierman: Physica Utrecht *44*, 177, 206 (1969)
5.31 P. Joyes: J. Phys. Paris *29*, 774 (1969); *30*, 243, 365 (1968)
5.32 P. Joyes: Radiat. Eff. *19*, 235 (1973)
5.33 G. Blaise: Radiat. Eff. *18*, 235 (1973)
5.34 G. Blaise, G. Slodzian: J. Phys. Paris *35*, 243 (1974)

5.35 G. Blaise, G. Slodzian: Rev. Phys. Appl. *8*, 247 (1973)
5.36 G. Blaise, G. Slodzian: J. Phys. Paris *31*, 93 (1970)
5.37 R.J. MacDonald: Surf. Sci. *43*, 653 (1974)
5.38 G. Slodzian: Surf. Sci. *48*, 161 (1975)
5.39 D.M. Newns: Phys. Rev. *178*, 1123 (1969)
5.40 J. Schroeer, T. Rhodin, R. Bradley: Surf. Sci. *34*, 571 (1973)
5.41 Z. Sroubek: Surf. Sci. *44*, 47 (1974)
5.42 Z. Sroubek, J. Zavadil, F. Kubec, K. Zdansky: Surf. Sci. *77*, 603 (1978)
5.43 G. Blaise, A. Nourtier: Surf. Sci. *90*, 495 (1979)
5.44 A. Blandin, D. Hone, A. Nourtier: J. Phys. Paris Lett. *36*, L-109 (1975)
5.45 A. Blandin, A. Nourtier, D.W. Hone: J. Phys. Paris *37*, 369 (1976)
5.46 A. Nourtier: J. Phys. Paris *38*, 479 (1977)
5.47 I.V. Hertel, H. Hoffmann, K.J. Rost: Phys. Rev. Letters: *36*, 861 (1976)
5.48 C.H. Corliss, W.R. Bozman: Natl. Bur. Stand. U.S. Monogr. *53* (1962)
5.49 D.L. Smith: J. Nucl. Mater. *75*, 20 (1978)
5.50 P. Sigmund: Phys. Rev. *184*, 383 (1969)

6. Analytical Electron Microscopy in Surface Science

J. A. Venables

With 9 Figures

6.1 Introduction

Several surface science techniques use electrons. They are used either as a probe of the surface, or as the emitted particles, as for example in photoemission, or sometimes both, such as in Auger electron spectroscopy. Electron microscopy techniques use magnetic fields (typically with axial symmetry) to focus electron beams to a fine spot and/or to preserve information as to the point of origin of the electrons.

Combinations of these two approaches are therefore of particular interest in the study of surfaces on a microscopic scale. Real surfaces are often very heterogeneous. Examples include the production of fine-scale structures for microelectronics or catalysis, the selective attack at surface defects such as in corrosion, or the changes in surface mechanical properties produced by surface segregation, interdiffusion, or ion implantation. In all these cases, as in the supposedly 'simpler' problems of surface structure, physi- and chemisorption, or crystal growth on surfaces, there is a microscopic aspect which can be approached by the techniques discussed in this paper.

Before describing such techniques, however, it should be noted that useful microscopic information might be obtained in other ways at less effort and expense. For example, the topography of a sample, on a coarser scale than ~10 nm, may be seen by standard scanning electron microscopy (SEM), even though surface-sensitivity features may be lost on transferring the sample to a relatively poor vacuum instrument. Such SEM techniques were reviewed by BROERS [6.1]. Fine details of small-particle topography, crystallography, and defect structure, down to ~1-nm scale, such as required for characterization of catalyst particles or the early stages of crystal growth can be done by conventional transmission electron microscopy (CTEM), as reviewed by HEINEMANN [6.2]. A particularly successful example of an indirect surface (CTEM) technique is the gold decoration/replica method of observing surface steps, particularly on alkali halide surfaces [6.3]. Suitably refined, this method has led to detailed studies of the vapour-solid interface during condensation and evaporation [6.4], of the liquid-solid interface during freezing and melting [6.5], and of the nucleation and (island) growth of metals on alkali halides [6.6]. These results are not reviewed here.

CTEM can also be used to observe catalytic and other dynamic processes directly using 'environmental' cells, which can support gas pressures sometimes approaching one atmosphere around the sample. One approach has been described by BAKER [6.7]; experiments of this type have been particularly successful in high-voltage electron microscopes, where the extra penetrating power of up to 1-MeV electrons, and greater space around the sample, have been used to advantage [6.8]. These microscopes, however, do not exploit surface-sensitive techniques as such; typically they do not contain any surface analysis facilities, and cannot see features which are destroyed by surface contamination or by transfer through the atmosphere.

Surface-sensitive microscopy is also possible using particles other than electrons, notably with the field ion microscope (FIM) and the associated chemical analysis technique, the 'atom-probe'. These techniques have been reviewed by van OOSTROM, MUELLER, and EHRLICH [6.9]. The related electron technique, which uses the same point geometry, is the field emission microscope (FEM), reviewed by MUELLER, GOMER, and more recently JONES [6.10]. While these techniques are extremely powerful, they are in some cases limited by the point geometry, and by the high fields needed to cause ion or electron emission. In contrast, electron microscope techniques can be applied more generally at a large range of magnifications, both to thin-film samples using TEM, and to bulk samples using SEM and other methods. The new generation of SEM, reflection (REM), and photoelectron microscopy (PEM) techniques should appeal especially to those who use the more well established surface techniques on bulk samples, since these same samples can be used for microscopic examination.

In the space available it is not possible to do full justice to the range of techniques available or to many examples. There are also several groups of potential readers with different backgrounds. This chapter is aimed at surface scientists without detailed knowledge of electron microscopy; a companion paper [6.11] covers similar material for electron microscopists interested in surface science. Another review [6.12] emphasizes these surface sensitive techniques within the framework of quantitative microanalysis of materials. Hopefully, the three papers between them give a reasonably complete account and bibliography; but the present paper is selective in the sense indicated above.

6.2 Analytical Electron Microscopy Techniques

Introductory articles aimed at a general physics audience describing modern analytical electron microscopy techniques have been given by the author [6.13] and others [6.14,15]. Here we recap the main points and problems briefly.

Electron microscope images are formed in one of two ways, either by scanning (S) or by conventional (C) imaging. In the scanning techniques, the function of the pre-specimen optics is to form as small a spot on the sample containing as much current as possible. The image is formed sequentially (as in television), and can contain information from a variety of signals. Signals which are surface sensitive are indicated schematically in Fig.6.1 [6.11]. These signals are the sample's response to the incoming electron beams; they have different signal strengths and different specificity to particular features of the sample. For example, secondary electrons form a strong signal (10^{-2} to 1 detected electrons per incoming electron) but are not very specific. However, they are widely used to produce images of surface topography which are intuitively recognizable; surface science applications are described in Sect.6.3.

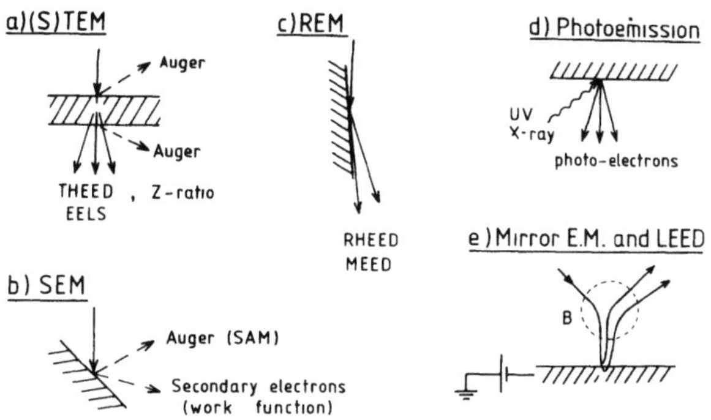

Fig.6.1a-e. Schematic illustrations of five surface-sensitive microscope configurations 6.11 : (a) (S)TEM, (Scanning) transmission electron microscopy; (b) SEM, scanning electron microscopy; (c) REM, reflection electron microscopy; (d) photoemission microscopy (PEM); (e) low energy, and mirror electron microscopy

In contrast, Auger electrons or characteristic X-ray emissions are weak signals (10^{-7} to 10^{-4} detected particles per incident electron), but are very specific spectroscopies relating to chemical composition. Scanning techniques are well adapted to computer control and analysis because of serial recording; problems include having bright-enough sources to give reasonable signal-to-noise ratios with weak signals, and radiation damage caused by high current densities in the spot. Examples of surface-sensitive spectroscopies are given in Sect.6.4.

Conventional imaging is historically older and at first sight less versatile. Electrons of a well-defined energy are used for imaging, with the immediate post-specimen lens being the most crucial optical element. Normally, electrons which have lost negligible amounts of energy are used for imaging, as in CTEM and the corresponding transmission high-energy electron diffraction (THEED) patterns. However, CTEM is unrivalled for taking high-resolution pictures quickly (seconds) since the image is exposed all at once (parallel recording) using a photographic plate, or in some cases an electronic camera which integrates the signal over time.

The equivalent conventional reflection microscopy (CREM) using elastically scattered (RHEED) patterns has been substantially developed recently, as described in Sect.6.5.2. Similarly, recently developed and projected photoemission microscopies use conventional imaging, with either post-specimen acceleration, or trapping of low-energy electrons in a magnetic field as described in Sect.6.4.3. In either case the advantage of conventional imaging is that of parallel recording; for photoemission this is particularly important, since photoelectrons yields are small and even synchrotron radiation sources do not have comparable fluxes to electron sources [6.11,16].

The pressure to find higher brightness electron sources has thus come from the scanning instruments, in an attempt to overcome the disadvantages of serial recording (typically 10^6 picture points in 50 s for an SEM picture using secondary electrons at 10-nm resolution) as against parallel recording (typically 10^8 picture points in 5 s at 0.5 - nm resolution for a good CTEM picture), assuming a similar tungsten filament source. This factor of 10^3 (considerably greater at constant resolution) has been largely retrieved by the development of the tungsten field emission source (FEG) by CREWE and co-workers [6.17], who have succeeded in imaging single atoms using scanning transmission (STEM) techniques [6.18].

The use of a FEG source has had a similar, if less dramatic, effect on bulk sample SEM techniques. Ironically, in the present context, the fact that the FEG needed UHV conditions (typically $1 \cdot 10^{-10}$ torr) for operation was as important in bringing electron microscopy and surface science together as any direct wish of the scientists concerned; once the gun used UHV technology it was a relatively straightforward matter to use UHV construction throughout. Interfacing a UHV specimen chamber to a conventionally pumped microscope has been attempted many times, especially for CTEM [6.19-21]. However, it has never become routine, nor has such a design been commercialized.

The above discussion perhaps indicates that there is now a move to use both scanning and conventional imaging, often in the same instrument, exploiting the advantages of both. Parallel recording of diffraction patterns in scanning (including scanning transmission) instruments is highly advantageous [6.22,23]; electron energy loss spectra can in some cases be recorded in parallel [6.24], and novel imaging energy analyzers are available for use with CTEMs [6.25,26] to take images in particular (chemical-specific) loss peaks. Scanning attachments and high-brightness (and high-coherence) sources have been attached to CTEMs for analysis purposes, and post-specimen optics to STEMs [6.27] to match particular conditions at the sample to specific detectors. Consequently the new generation of 'analytical' electron microscopy techniques [6.28] aims to use any suitable signal emanating from the sample, either caused by or consisting of an electron beam, to characterize that sample as fully as possible on a microscopic scale. Here we are particularly concerned with those electron microscope signals which are surface sensitive, or equivalently with surface signals which can be spatially localized. A schematic illustration of some of these signals is given in Fig. 6.1 [6.11].

6.3 Scanning Electron Microscopy of Surfaces

Scanning electron microscopy (SEM) is perhaps the easiest microscopic technique to combine with other surface studies. In a relatively crude form it is a standard attachment on most electron guns used for Auger spectroscopy, RHEED or LEED, sometimes achieving a resolution of a few μm. This is already useful, as it allows 'good' parts of the sample to be chosen for detailed analysis; pictures usually leave something to be desired. A standard commercial SEM using 3 magnetic lenses and a tungsten filament gun can achieve a resolution around, or perhaps slightly better than, 10 nm; there are some commercial examples of such machines being interfaced to UHV sample chambers, but in general either the UHV or the SEM performance is not quite optimal. Our own SEM instrument [6.29] aims to combine these two features in a more or less optimized fashion. Using a FEG, we have obtained 5-nm resolution with $1\text{-}2\cdot 10^{-10}$ torr in the sample chamber [6.30]; the design is commercially available.

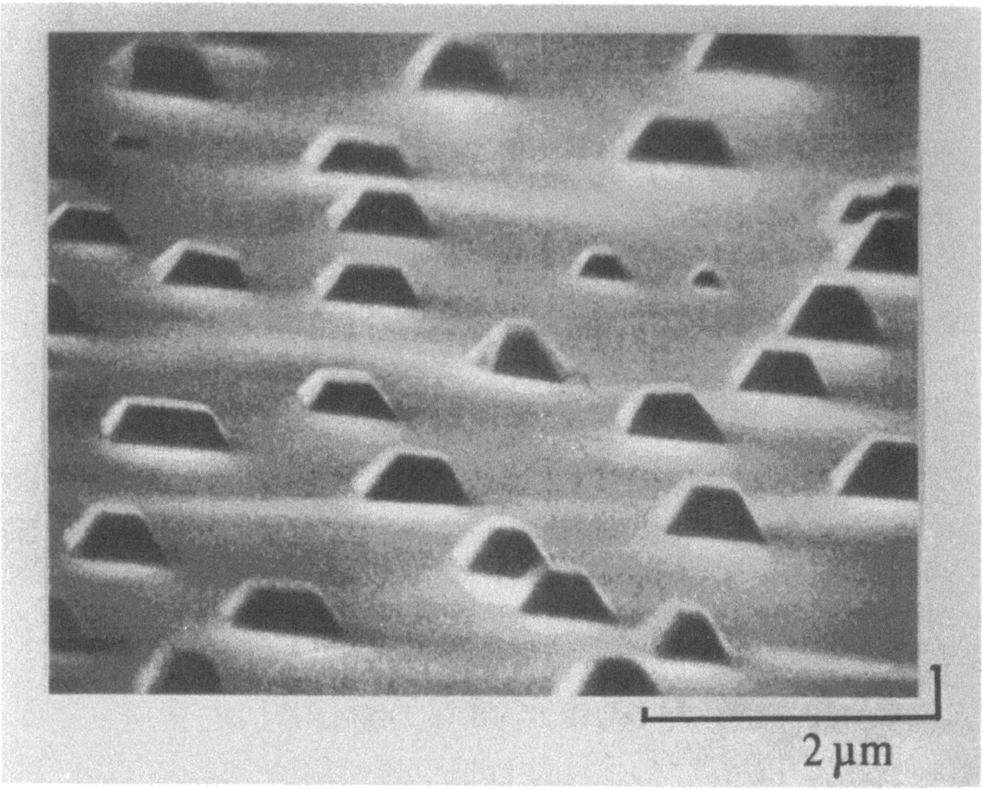

Fig.6.2. Form of Ag islands deposited on Mo{100} surface at 550°C with 20 ML deposited [6.32]. SEM picture at 10° glancing incidence, 60 kV

A fully fledged SEM is at present more than a simple 'bolt-on' addition to a UHV specimen chamber; but if it could be miniaturized, it could be just that. The most impressive development in this direction to date is by VENEKLASEN et al. [6.31]. They achieved 30-nm resolution at 4-5 kV beam energy using a thermal field-emission source in a small electrostatic gun which can fit inside a conventional cylindrical mirror analyser (CMA).

The most routinely useful feature of the SEM is its ability to produce striking pictures of the surface topography, in which the sample is viewed along the beam direction and appears to be illuminated from the position of the detector. This arises because the number of secondary electrons, both emitted and collected, rises as the sample is tilted towards the detector. The '3-d' effect this gives is particularly pronounced near glancing incidence; we have found such pictures most useful in our studies of the Stranski-Krastanov (or layer plus island) crystal growth mode [6.32-34]. An example of Ag growing on Mo{100}[6.32] is shown in Fig.6.2. The Ag crystals are grown in the {100} orientation with <001> Ag parallel to <011> Mo; {111} are the side faces exposed. The number density of the crystals and their size and shape can easily be measured.

Previous Auger electron spectroscopy (AES) work on this system had shown that one monolayer of Ag was formed before island growth commenced ([6.35], SORIA and POPPA disagree as to the coverage of the breakpoints). However, detailed analysis of Auger amplitudes versus deposition time, which show a strong breakpoint when island growth begins, is only possible if the shape of the islands is known [6.33]. Taking the shape to be that observed by SEM, the number density of islands, N, can be measured from the slope ratio at the breakpoint for $N \gtrsim 10^{10}$ cm^{-2}. Taken together, the SEM-AES results span a range of $10^5 < N < 10^{14}$ cm^{-2}, with good agreement where the data overlap. Number densities have been determined for Ag/Mo{100} [6.32], Ag/Si{111} [6.33] and Ag/W{110} [6.34]. The data for N(T) are all very similar [6.34] even though the structure of the intermediate layer and the crystallography are quite different in the three cases; this type of result is clearly crucial for more detailed understanding of this intriguing crystal growth mode.

Another invaluable feature of SEM is the ability to check the surface preparation procedure. Our experience with preparing Mo{100} surfaces by oxidation and heating [6.32] indicated that although the surface could be readily cleaned (as indicated by Auger spectroscopy), achieving surface flatness required much longer annealing times than we had been led to believe from the surface science literature. In particular, the achievement of good LEED or RHEED patterns did not indicate flatness over more than ~100 nm, whereas nucleation phenomena (e.g., Fig.6.2) may require surfaces which are flat over many microns. The evolution of this Mo{100} surface on annealing is shown in Fig.6.3. Decoration phenomema were observed at the edge of the flat regions [6.32]. Such observations lead us to expect that heterogeneous nucleation phenomena are often important, and that studies of surface reactivity need to pay particular attention to surface topography.

Another area where SEM has been particularly useful is in the study of equilibrium forms of small crystallites [6.36-40]. By measuring the relative size of low-index facets, the anisotropy of surface free energy can be measured. As the temperature is raised the entropy contribution to the free energy of higher index faces becomes more important, resulting in a more rounded equilibrium form [6.38,39]. The case of Pb/graphite is particularly favourable for in situ observation by ultrahigh-vacuum SEM, because of the negligible evaporation rate of Pb near its melting point. An example of the

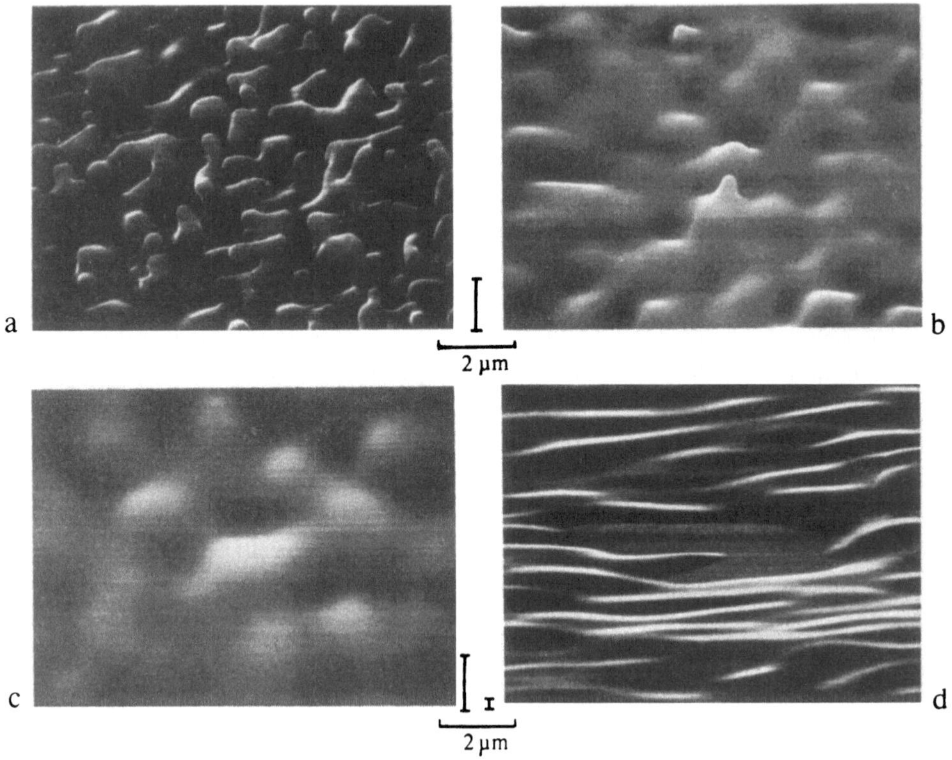

Fig.6.3a-d. SEM pictures of the preparation of the Mo{100} surface [6.32]. The crystal has been heated in $6 \cdot 10^{-6}$ torr O_2 to 1150°C for 10 hours to remove carbon. (a) Molybdenum oxide has been removed by heating to 1870°C for 3 min at $p < 10^{-8}$ torr, but the surface is very rough; (b) surface after 7 min at 1870 C; (c) after 11 min at 1870°C; (d) after a further 10 min at 2000°C. Here the {100} 'plains' are surrounded by shallow 'hills'; the plains appeared featureless at a magnification of up to $2 \cdot 10^5$ times, and seem to be essentially perfect {100} facets. [Note that (d) is viewed at ~7° incidence, whereas the others are ~45°]

appearance of small Pb crystals on graphite is shown in Fig.6.4 [6.39]. The flat Pb crystals which coexist with the equilibrium forms are kinetically determined, and have great difficulty in approaching equilibrium. Often they are unable to do so before melting intervenes, which suggests a sizeable nucleation barrier [6.38,39]. The melting of Pb on graphite has also been studied by in situ CTEM, and a sizeable superheating observed for these flat crystals [6.40]. The comparison of TEM with SEM shows that, despite the superior resolution of TEM, the observation of three-dimensional forms is much easier by SEM.

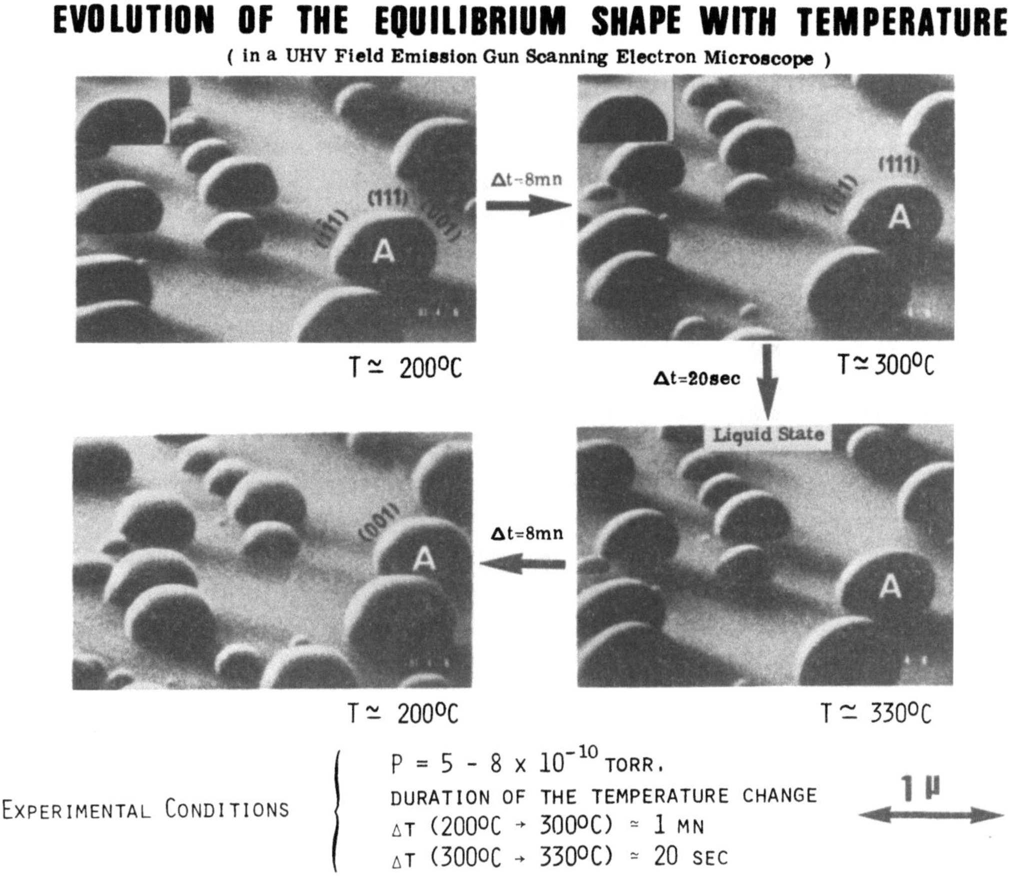

Fig.6.4a-d. Dynamic SEM pictures (taken from videotape) of the evolution of equilibrium forms of Pb on graphite with temperature T and time t [6.39]. (a) At T = 200°C the equilibrium form has {001} and {111} facets; (b) after 8 min at 300°C the {001} facets have become rounded [see insets in (a,b)]; (c) liquid drops are truncated spheres; (d) on regrowth from the liquid, the crystal A adopts the double-positioned orientation to (a)

There are many more possible examples. But the stage has now been reached where SEM examination of the surface either in situ, or before and after the experiment is almost essential to avoid misinterpretation of the results of wide-beam surface science experiments.

6.4 Surface Spectroscopies and Microscopy

Electron spectroscopy is widely used in surface science as a means of determining surface composition, electronic structure and chemical state and, at the highest energy resolution, molecular vibrations. If the point of emission can be spatially localized, the corresponding analytical microscopy can also be performed.

6.4.1 Auger Electron Spectroscopy and Microscopy

The most widely used spectroscopic technique is Auger electron spectroscopy (AES). When spatially localized it becomes micro-Auger spectroscopy, or scanning Auger microscopy (SAM), which is the combination of SEM with AES. Such machines are now commercially available[1], and several reviews have been written over the last few years, especially for electron microscopy and surface science conferences [6.41,42]. These SEM-SAM combination instruments have found many applications in research and process control; the analysis volume has been reduced over the corresponding SEM-X-ray microanalysis techniques by a factor approaching 10^5, because of the small escape depth of the Auger electrons, and the fact that most of them are produced within the incident probe, at least for low-atomic-number samples [6.41,43]. Thus the interest in SAM is not confined to the surface science community.

The information in such spectra when used on a microscopic scale is in the form of spectra I(E) from each sample point x,y, i.e., $I_{xy}(E)$. This represents rather a large amount of information if taken literally. Consequently subsets of the data are often used, such as a few point analyses I(E) from features of interest on the sample, or line analyses $I_x(E)$ taken over restricted energy ranges to analyze for a few elements. Relatively coarse pictures (I_{xy} maps or area scans) are usually taken at a fixed energy, or at two neighbouring energies [6.44] or using ratio signals [6.45] in order to make quantitative estimates of composition.

Such pictures, even if crude, can be very informative. An example of grain boundary plus surface diffusion of Ag on Au is shown in Fig.6.5 [6.46]. Despite the poor signal-to-noise ratio, there is sufficient information to tell that the Ag has diffused through the grain boundary and about 2 μm over the surface. Similar informative pictures have been produced showing the effect of surface steps on the surface diffusion on Pd on W{110} [6.47], the

1 Manufacturers offering such equipment include JEOL, Leybold-Heraeus, Perkin Elmer (Physical Electronics), Riber, VG Microscopes and Varian.

Fig.6.5. (a) SEM image of Au film $\simeq$ 100 nm thick containing grain boundaries. Silver was plated on the reverse side, and the fron surface cleaned by ion bombardment; (b) after diffusion at 160°C for 30 s, a silver Auger map delineates the grain boundaries (500 s at 3.10^{-8} A probe current'; (c) after further diffusion at 185°C for 600 s, showing further grain boundary and surface diffusion (500 s at 7.10^{-8} A, all 60 kV) [6.46]

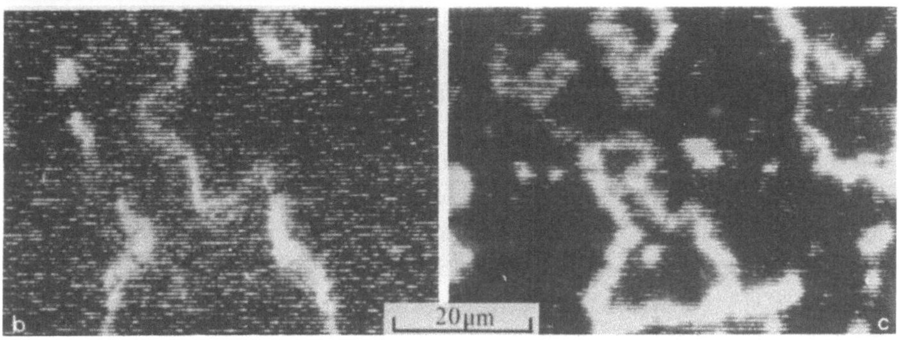

precipitation of surface carbides during methane reactions with iron [6.48], and numerous studies of semiconductor devices and the elements responsible for their failure [6.49].

Despite these successes, SAM has various problems which require further development. The Auger signals are weak, having a yield (into 4π sterad) which varies between 10^{-2} and 10^{-4} for a monolayer [6.41,50]. At a maximum the analyzer collects about 5% of this emission, typically with an energy resolution of $\simeq$10% of the Auger peak width, and the peak itself rides on a considerable background. Coupled with the need to reduce the beam current (while increasing the current density) to form a small probe, the detector Auger current in SAM is very small, typically 10^{-15}-10^{-14} A. This leads to long exposure times (typically 500 s) and to relatively noisy information, as seen in Fig.6.5. Because of the large current density in the probe and the long exposure, radiation damage is a problem for all but the hardiest samples [6.51]; however, metals and many semiconductors seem not to damage significantly at all at SEM voltages.

These problems lead to several lines of current development. Computer control of scanning, data collection, and analysis is important in that it allows the dwell time on each point of the sample to be matched to the required signal-to-noise ratio, and allows the most efficient quantification procedure to be implemented [6.52]. Several leads need to be pursued, including the minimization of topographic and backscattering effects on the desired 'chemical' signal [6.45], the quantification of such effects, and the use of the loss electrons in the tail of the Auger peak to produce depth-dependent information.

Improvement of (field-emission) SEM-SAM columns and analyzers which will increase the Auger signal is needed. Our experience to date is encouraging [6.30,52] and suggests that sub-10 nm SAM will be possible with probe currents of between 10^{-9} and 10^{-8} A. Finally, the use of thin-film samples is required for work at the highest resolution in order to remove the effect of backscattered electrons. Estimates [6.11,53] indicate that this technique, with the Auger electrons taken from the input side of the film as indicated in Fig.6.1a, should be capable of detecting 10% of the atoms in the surface layer of 2-nm diameter, i.e., 10-20 atoms.

At this level AES will become comparable in resolution with the more widely used X-ray and electron energy-loss spectroscopy (EELS) (cf. Fig. 6.1b), which are the main chemical techniques presently available in scanning transmission electron microscopes (STEM). Recent X-ray work [6.54] has analyzed $\simeq$30 atoms in 1.5-nm diameter alloy catalyst particles, and EELS may become even more sensitive, because of superior detection efficiency [6.24]. Neither of these techniques are specifically surface sensitive. However, in the limit of very thin samples (< 10 nm) such a distinction may become rather academic, especially if it is known independently (by in situ deposition, for example) that the atoms concerned are on the surface.

All electron probes are capable of damaging certain samples, and a fully focussed STEM beam is an ideal device for boring 1-nm holes in, say, an Al_2O_3 catalyst support film [6.55]. Thus there are often strong arguments for using the worst resolution necessary for the problem in hand. A good example is the use of taper sectioning to analyze planar structures, such as lasers or diffusion couples. Van OOSTROM et al. [6.56] achieved 2.5-nm depth resolution using a 5-μm diameter SAM beam on a shallow taper at the edge of an ion beam crater. They obtained a signal-to-noise ratio sufficient for 1% analyzing accuracy. However, a counter example is the case of a dilute distribution of small (catalyst) particles containing elements not present elsewhere on the sample. In this case [6.53] the signal-to-noise ratio improves markedly as the probe size is reduced towards the particle size.

6.4.2 Secondary Electron Spectroscopy and Work Function Imaging

Secondary electrons also contain surface-sensitive information, especially because the yield and onset energy is influenced by the surface potential, which for a uniform sample is the same as the work function. The energy distribution of the secondaries can be visualized using the same analyzer as for Auger spectroscopy, especially if the sample is negatively biased by 10-20 V to bring the onset energy into the range of the analyzer. Scanning techniques can then produce spatially resolved information as for other spectroscopies.

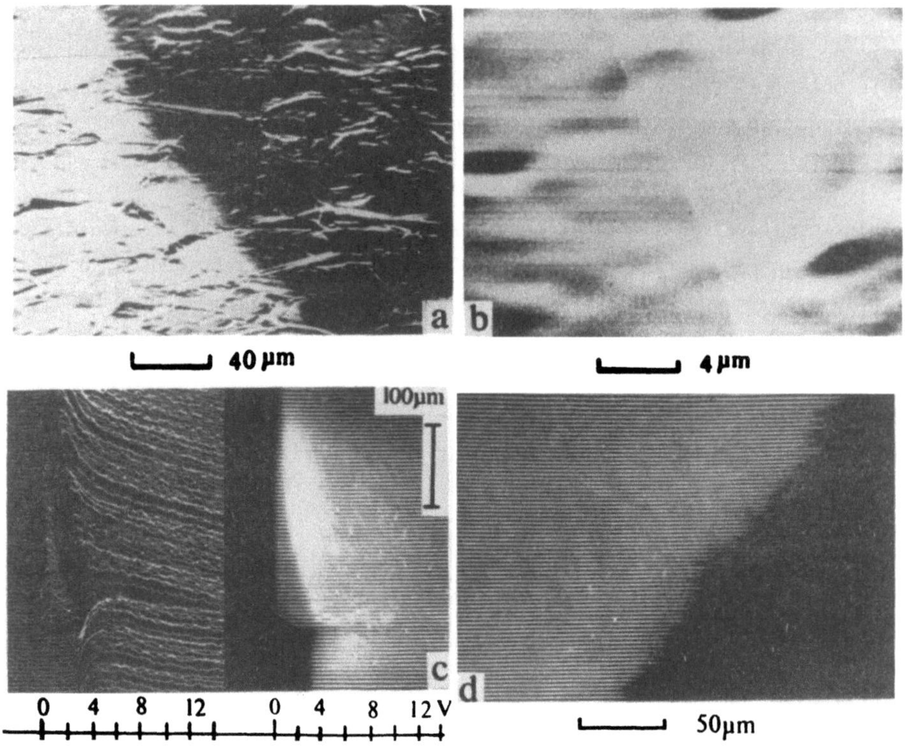

Fig.6.6a-d. Low-energy secondary electron observations of a Cs monolayer boundary on polycrystalline W [6.57,58]. (a,b) Secondary electron contrast from the edge at low (a) and higher (b) resolution, Cs layer to the left; (c) line analyses across this edge, with low-energy spectra (horizontal) with line scan (vertical) in y- and z-modulated display; (d) work function map across the edge

These techniques have recently been developed and applied to visualizing Cs monolayers on tungsten [6.57,58]. Cs was evaporated past a shutter. In normal SEM, the Cs half of the crystal (Fig.6.6a) appears bright, corresponding to more secondary electrons, and at higher resolution (Fig.6.6b) this monolayer edge is seen to be sharp, due to second-layer diffusion over the edge [6.58]. Low-energy spectra can be taken during a line scan across the edge, and the resulting multiple graphs [line analyses $I_x(E)$] are shown in Fig.6.6c. This shows the work function drop of $\simeq 3$ V on going from clean W to Cs/W. On the Cs side of the boundary the repulsive patch fields influence the trajectories of electrons emitted in the range 0-3 eV. Nonetheless, some low-energy electrons escape, probably by multiple reflection at the surface, and carry the information about the work function [6.57]. A feedback loop on the sample potential can be made to track this onset energy; with such a device a work function map of this same Cs monolayer edge can be made. So far an energy resolution of < 40 mV and spatial resolution < 0.4 µm has been achieved [6.57].

This work function signal is very strong, and can be applied to dynamic phenomena such as surface diffusion of Cs/W [6.58]. The techniques also work for much smaller work function differences, such as Ag/W [6.34] or O/W, as shown earlier at lower spatial resolution [6.59].

6.4.3 Photoemission and Energy-Loss Spectroscopy

These two surface spectroscopies have not yet been fully developed into the corresponding microscopies, though it is clear from the preceeding discussion that they can be developed in future.

Interest in photoemission microscopy has a long history [6.60], but although the high contrast of the instrument has been noted, specific surface science experiments do not seem to have been undertaken. The conventional photoelectron microscope (CPEM) uses a negatively biased ($\sim$ -50 kV) sample to accelerate the emitted electrons into an optical column similar to a CTEM. The problem is one of adequate photon intensity, and a channel plate intensifier has been used to detect the electrons [6.61]. Because of chromatic abberation and intensity problems the resolution has typically been 0.1 µm, although < 10 nm should be obtainable with such a system. With the new sychrotron radiation sources such a microscope should be particularly valuable, and it may be estimated that, with energy filtering of the emergent electrons as well as the incident photons, chemical- (and depth-) specific images should be obtainable at around 1- to 10-nm spatial resolution. The depth specifity would arise from choosing the energy of the emerging electrons and consequently their escape depth via a varying mean free path [6.11].

Although such a microscope has not yet been built, two other approaches show promise at the 1-μm resolution level. The first involves bombarding a thin target in situ to produce characteristic X-rays which then emit photoelectrons into a conventional CMA [6.62]. Scanning the target beam then produces a picture, for which 1-μm resolution is hoped; 15 μm has been obtained. The second relies on the spiralling of the emergent low-energy electrons in a high-magnetic field and the divergence (i.e., the magnification) of the field lines with decreasing field strength [6.63]. With an initial B_1 of $\sim 10^5$ gauss and a final B_2 of 10 gauss, the magnification $m = (B_1/B_2)^{\frac{1}{2}}$ is 100. Essentially all emergent electrons can be collected and energy filtered to a few meV energy resolution; the spatial resolution is about 1 μm. This second approach, and the CPEM, are good examples of overcoming an intensity limitation by conventional imaging (parallel recording) of the outgoing electrons.

Electron energy-loss spectroscopy has become widely used as an analytical tool in the (S)TEM geometry of Fig.6.1a. It is also widely used in surface science studies of bulk samples. The development of the corresponding microscopy seems straightforward though it has not yet been done. The geometries which seem promising are either I) reflection from bulk samples (Fig.6.1c) using a narrow-angle cone of forward-scattered loss electrons into a magnetic sector spectrometer as in STEM, or II) the SEM geometry, using a relatively high-energy electrostatic analyzer (e.g. CMA) which can approach the beam energy. In either case signal levels should be relatively high, and localized electronic information on surfaces should be obtainable. The prospects for microscopy using low-energy electron beams, which are used for surface vibrational studies [6.64] (Chap.14), however, look rather poor.

Energy-filtered images have been obtained in both TEM and SEM geometries. While in transmission these have been used largely to observe chemical-specific images, in SEM they have been used to reduce the unwanted effects of the electron cascade in the sample. Energy-filtered backscattered electrons have been used to see subsurface dislocations [6.65] in a way which should provide a very useful complement to surface studies, especially of surface reactivity.

6.5 Diffraction Techniques and Microscopy

6.5.1 Transmission Electron Diffraction and Microscopy

The relation between transmission high-energy electron diffraction (THEED) and transmission electron microscopy (both CTEM and STEM) is very close. All high-resolution studies of crystals and of defects in crystals rely on observing both the micrograph and the corresponding diffraction pattern, and on interpreting the observed contrast in terms of a suitable dynamical diffraction theory.

Over the last decade it has been realized that such techniques are in principle surface sensitive, despite the fact that they observe the whole thickness of the thin-foil sample which is typically 2-200 nm thick. In visualizing surface features, one either uses the thinnest possible samples, or exploits particular combinations of thickness and diffraction conditions to enhance surface features.

At the highest resolution, single atoms can be seen both by CTEM [6.66] and STEM [6.18]. Recently such studies have begun to make contact with surface science, as surface diffusion has been observed of heavy atoms supported on very thin graphite surfaces [6.67]. In principle, such studies offer a complement to FIM studies (e.g., [6.9]), but so far only diffusion around surface steps has been observed in the vicinity of room temperature. Direct observation of atomic diffusion across the basal plane of graphite would require atomic resolution at low (< 77 K) temperatures, which has not yet been achieved. At somewhat lower resolution small catalyst particles, down to $\sim$ 1-2 nm in size can be observed by similar CTEM [6.68] or STEM [6.54,55,69] techniques. While not all such studies have exploited surface-sensitive microscopy techniques as such, the possibility for useful future work in this area is considerable.

Atomic height surface steps and surface reconstructions have been observed by CTEM in a variety of materials by several groups. Observations have been made on MgO [6.70], graphite [6.71], silicon [6.72], and metals [6.73-78]. Under certain conditions, up to 10% contrast from a single step can be achieved. Understanding this contrast, however, requires detailed knowledge of the exact diffraction conditions.

In a few laboratories ultrahigh-vacuum chambers have been constructed for TEM with in situ deposition or other reaction facilities [6.19-21]. In such cases the progress of monolayer deposition can be followed directly. An example is shown in Fig.6.7 [6.74]. Here {111} Au crystals have been condensed on a

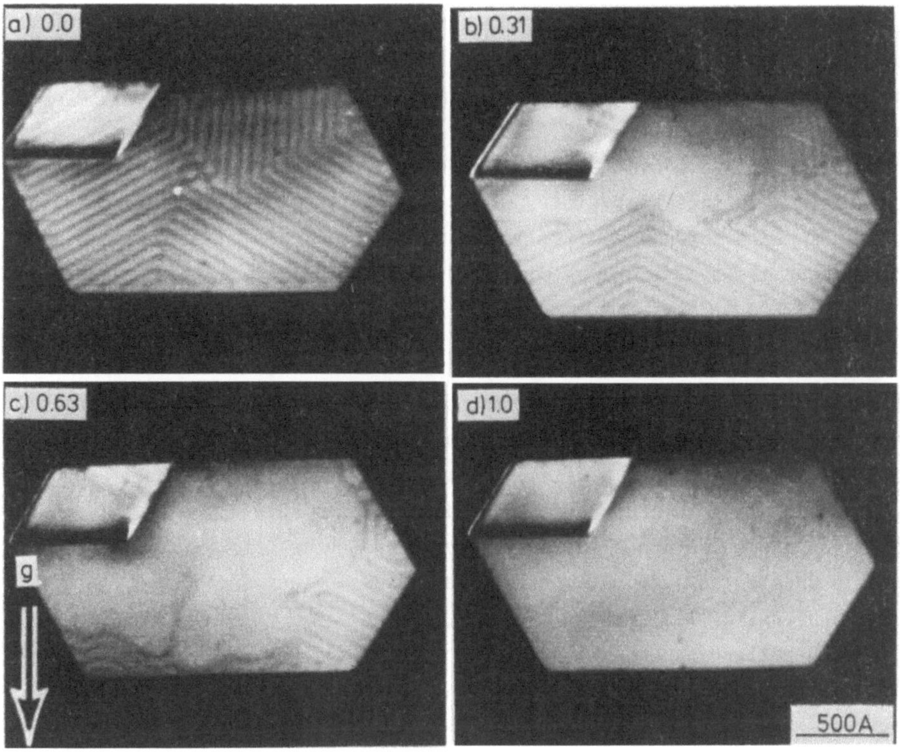

Fig.6.7a-d. CTEM images of Au (23 × 1) surface structure and its removal by submonolayer Ag deposits [6.74]. (a) A {111} gold crystal is grown in situ onto MoS_2. The parallelogram in the top left-hand corner is a twin crystal, and 3 domains of the (23 × 1) surface reconstruction, with 6.3-nm period are seen; (b-d) successive depositions of Ag at 190°C form a monolayer island which removes the reconstruction (parameter: Ag coverage in monolayers). All pictures are {111} dark field images at 100 kV

MoS_2 substrate and a crystal oriented to see the three domains of the (23 × 1) superlattice on the Au surface [6.75]. In situ deposition of submonolayer amounts of Ag removes the superlattice as shown in Figs.6.7b-d. Detailed experiments of this type have been made for several metal-metal combinations, including Ag/Au {111} [6.74], Pd/Ag {111} [6.73], Pb/Ag {111} [6.77], and Ag/Ag{111} [6.78]. In all cases 2-d monolayer growth was observed. This work follows on from a large amount of work on metals deposited in situ onto various insulators, especially MoS_2, MgO, and graphite, which has provided much of our knowledge about island growth systems [6.19-21].

Transmission diffraction has also been shown to be surface sensitive by the observeration of adsorbed rare-gas monolayers (Xe and Kr) on graphite

[6.79,80]. By using an electron microscope with which to perform the diffraction experiments, small areas (≃ 2-μm diameter) can be chosen and the substrate examined for defects prior to the experiment. Quantitative scattering measurements can be done [6.80].

Cleanliness in a converted TEM is however always a problem, and most such machines lack surface-analytical facilities. The ingenuity required even to approach surface science conditions can be appreciated from the descriptions of the apparatus by the groups concerned [6.19-21,81].

6.5.2 RHEED and Reflection Microscopy

Reflection high-energy electron diffraction (RHEED) is very surface sensitive, like LEED, and less than a monolayer deposited is sufficient to obliterate the substrate signal in certain cases. The angular resolution is very fine, usually much finer than in LEED, and recent developments have shown that the LEED projection can be obtained [6.82], but with enough sensitivity to detect higherto undetected submonolayer surface phases in such systems as Ag/Si {111} [6.83] and In/Si {111} [6.84].

A RHEED pattern has a similar geometry to THEED, but uses a bulk sample at glancing incidence (Fig.6.1c). Consequently the corresponding reflection electron microscopy (REM) can be done using a transmission microscope, but with the sample stood on end. This has recently been done by the Tokyo [6.85-88] and Arizona [6.89-92] groups. In the first case, Si {111} surfaces have been cleaned in situ by heating to 1200°C for some 40 min inside a liquid helium cryopumped sample chamber. The surface partially evaporated and they were able to obtain images of monatomic surface steps, whose movement was captured on videotape [6.85]. The most spectacular observation was of surface steps terminating in a dislocation as shown in Fig.6.8. The contrast observed in sensitive to deviation from the exact Bragg reflecting position, as in TEM, and the sense of the steps and of the screw dislocations which terminate them can be determined from pictures such as Fig.6.8. The contrast is thought to arise from small rotations of the surface planes (≃ 10^{-4} rad [6.90]) in the neighbourhood of the steps. Additional Fresnel diffraction effects occur away from the exact line focus which is obtained from the strongly tilted sample [6.86,90].

The {111} surface has the (7×7) superlattice structure at room temperature, and undergoes a transition to the unreconstructed (1×1) surface at 830°C. This transformation has been visualized, as seen in Fig.6.9 [6.87]. Again this sequence of events has been seen on videotape [6.85]. The nucle-

Fig.6.8a,b. Reflection microscope images of surface steps on Si{111}, with one of them ending at a screw dislocation (arrowed). Note that the step contrast depends on the sense of the step and on the sign of the deviation [+ve for (a), -ve for (b)] from the exact 444 Bragg reflection used. The dislocation image also reverses, and is about 0.5 μm wide in (a) and (b) [6.87]

ation and growth of the (7×7) phase on cooling could be seen to nucleate at the upper side of the steps, and to proceed across the terraces in a striplike manner. These strips correspond to separate nucleation events, which are in general not in registry with each other; consequently 'out-of-phase' boundaries are formed running across the terraces. This is the first time a surface phase transition has been observed in such detail.

Other work has started to observe submonolayer deposits of Ag and Au on Si {111} and {100} [6.88], and the oxidation of Cu surfaces [6.92]. It is clear that rather flat surfaces are required to use reflection imaging effectively. The examples shown in Fig.6.8,9 correspond to an angle of incidence $\alpha = 1.2^{\circ}$ (at 100 kV), and the step separation of 0.2 μm corresponds to the surface being within 0.1° of the exact {111} orientation. The steps observed have to be rather parallel to the beam direction, since the fore-

Fig.6.9a,b. Reflection images of the (1 × 1) to (7 × 7) phase transition on Si{111} at 830°C, taken with the 444 reflection. The dark regions are the (7 × 7) phase, and expand across the terraces on cooling, i.e., from (a) to (b) 'out-of-phase' boundaries can also just be discerned [small arrows in (a)] in the (7 × 7) structure [6.87]

shortening in these pictures is of order to 50 to 1. In all this work, as in the SEM work described in Sects.6.3,4, surface structure and adsorbed phases turn out to be richer in microscopic detail and surface defects than expected on the basis of conventional 'wide-beam' surface science experiments.

6.5.3 LEED and Low-Energy Microscopy

If diffraction experiments are conducted at lower energy than RHEED (20-200 kV), the angle of incidence α increases, until at LEED energies (50-200 V) normal incidence is used to obtain comparable surface sensitivity. However, for microscopic studies, comparable electron optical performance cannot be obtained at low energies, largely because of larger chromatic aberrations and sensitivity to stray fields. Consequently the geometry of Fig.6.1e has been used, in which a high-energy electron beam is decelerated just at the sample surface (by biasing the sample negatively inside a electrostatic lens); a magnetic field is often used to separate the incoming and outgoing beams. With such a system, LEED, mirror, and 'almost-mirror' microscopy can be performed.

Such microscopes have been reviewed elsewhere [6.11]. In general the spatial resolution has not been better than 1 μm, though a good design goal is 0.1 μm; the sensitivity to surface potential or work function can be as good as 10 mV [6.93]. Low-voltage SEM is also possible using the same principles and a resolution of 0.3 μm has been demonstrated at 300 V; the increase in surface sensitivity below 1 kV is very noticeable [6.94]. More work needs to be done on the optimum mode of operation of THEED, THEED, and LEED microscopes, i.e., on the choice of energies and angles of incidence, and of conventional or scanning imaging. However, the standard has been set by pictures such as those shown in Figs.6.7-9; it seems difficult to believe that the low-voltage electron optical performance will be adequate to the challenge, except for some special applications.

References

6.1 A.N. Broers: Crit. Rev. Solid State Sci. *4*, 333 (1974)

6.2 K.W. Heinemann: In *Chemistry and Physics of Solid Surfaces, III*, ed. by R. Vanselow (CRC, Boca Raton, FL 1981)

6.3 G.A. Bassett: Philos. Mag. *3*, 1042 (1958); H. Bethge: In *Molecular Processes on Solid Surfaces*, ed. by E. Drauglis, R.D. Gretz, R.I. Jaffee (McGraw-Hill, New York 1969) p.569; M. Krohn, H. Bethge: In *Crystal Growth and Materials*, ed. by E. Kaldis, H.J. Scheel (North-Holland, Amsterdam 1977) p.142

6.4 H. Höche, H. Bethge: J. Cryst. Growth *33*, 246, 255 (1976); *42*, 110 (1977); H. Meyer, H. Dabringhaus, A. Maas, B.J. Stein: J. Cryst. Growth *30*, 225 (1975); H. Meyer, B.J. Stein: J. Cryst. Growth *49*, 707 (1980)

6.5 G. Grange, B. Mutaftschiev: Surf. Sci. *47*, 723 (1975); *54*, 445 (1976); J. Cryst. Growth *49*, 343 (1980); B. Mutaftschiev: In *Chemistry and Physics of Solid Surfaces, III*, ed. by R. Vanselow (CRC, Boca Raton, FL 1981)

6.6 J.L. Robins: Thin Solid Films *32*, 151 (1976); M. Harsdorff: Thin Solid Films *32*, 103 (1976); R. Kern, G. LeLay, J.J. Metois: In *Current Topics in Materials Science*, Vol.3, ed. by E. Kaldis (North-Holland, Amsterdam 1979) p.131

6.7 R.T.K. Baker: In *Chemistry and Physics of Solid Surfaces I*, ed. by R. Vanselow (CRC, Boca Raton, FL 1976) p.375; Catal. Rev. Sci. Eng. *19*, 161 (1979); R.T.K. Baker, R.D. Sherwod: J. Catal. *61*, 378 (1980); *66*, 56 (1980)

6.8 E.P. Butler: Rep. Prog. Phys. *42*, 833 (1979); W. Thoni, P.L. Gai, P.B. Hirsch, Philos. Mag. *35*, 781 (1977)

6.9 A. van Oostrom: Crit. Rev. Solid State Sci. *4*, 353 (1974); E.W. Müller: In *Chemistry and Physics of Solid Surfaces I*, ed. by R. Vanselow (CRC, Boca Raton, FL 1976) p.85;

G. Ehrlich: In *Chemistry and Physics of Solid Surfaces III*, ed. by R. Vanselow (CRC, Boca Raton, FL 1981); J. Vac. Sci. Technol. *17*, 9 (1980); Phys. Today *34*, 44 (1981)
6.10 R. Gomer: In *Chemistry and Physics of Solid Surfaces II*, ed. by R. Vanselow (CRC, Boca Raton, FL 1979) p.119;
J.P. Jones: In *Chemical Physics of Solids and Surfaces*, Vol.8, ed. by M.W. Roberts, J.M. Thomas (Chemical Society, London 1980)
6.11 J.A. Venables: Ultramicroscopy *781* (1981)
6.12 J.A. Venables: In *Quantitative Microanalysis with High Spatial Resolution* ed. by G. Lorimer, M.H. Jacobs, P. Doig (The Metals Society, London 1981) p. 205
6.13 J.A. Venables: Phys. Bull. *26*, 359 (1975);
J.A. Venables, A.P. Janssen, C.J. Harland: Phys. Bull. *29*, 307 (1978)
6.14 R.W. Carpenter: Phys. Today *34*, 34 (1981)
6.15 L.M. Brown: J. Phys. F*11*, 1 (1981)
6.16 M. Isaacson, M. Utlaut: Optik Stuttgart *50*, 213 (1978);
D. Sayre, J. Kirz, R. Feder, D.M. Kim, E. Spiller: Ultramicroscopy *2*, 337 (1977)
6.17 A.V. Crewe: In *Progress in Optics*, Vol.11, ed. by E. Wolf (North-Holland, Amsterdam 1973) p.225;
A.V. Crewe, M. Isaacson, D. Johnson: Rev. Sci. Instrum. *40*, 241 (1969)
6.18 A.V. Crewe, J. Wall, J.P. Langmore: Science *168*, 1138 (1970);
J.P. Langmore, J. Wall, M. Isaacson: Optik Stuttgart *38*, 385 (1973);
M. Isaacson, M. Ohtsuki, M. Utlaut: In *Analytical Electron Microscopy*, ed. by J.J. Hren, J.I. Goldstein, D.C. Joy (Plenum, New York 1979) Chap.13
6.19 U. Valdrè, E.A. Robinson, D.W. Pashley, M.J. Stowell, T.J. Law: J. Phys. E*3*, 501 (1970);
D.W. Pashley: Recent Progress Surf. Sci. *3*, 23 (1970); in *Epitaxial Growth*, ed. by J.W. Matthews (Academic, New York 1975) A, p.2
6.20 H. Poppa, R.D. Moorhead, K. Heinemann: Nucl. Instrum. Methods *102*, 521 (1972);
H. Poppa: In *Epitaxial Growth*, ed. by J.W. Matthews (Academic, New York 1975) A, p.215
6.21 K. Takayanagi, K. Yagi, K. Kobayashi, G. Honjo: J. Phys. E*11*, 441 (1978);
G. Honjo, K. Yagi: In *Current Topics in Material Science*, Vol.6, ed. by E. Kaldis (North-Holland, Amsterdam 1980) p.195
6.22 J.A. Venables, C.J. Harland, A.P. Janssen: Inst. Phys. Conf. Ser. *41*, 313 (1978)
6.23 J.M. Cowley: Inst. Phys. Conf. Ser. *52*, 239 (1980) ; Proc. Electron Microsc. Soc. Am. *37*, 372 (1979); Ultramicroscopy *3*, 433 (1979); Chem. Scr. *14*, 33 (1978-1979)
6.24 H. Shuman: Ultramicroscopy *6*, 163 (1981)
6.25 R.F. Egerton, J.G. Philip, P.S. Turner, M.J. Whelan: J. Phys. E*8*, 1033 (1975);
C. Colliex, V.E. Cosslett, R.D. Leapman, P. Trebbia: Ultramicroscopy *1*, 301 (1976)
6.26 G. Zanchi, J.Ph. Perez, J. Sevely: Optik Stuttgart *43*, 495 (1975);
B. Jouffrey, Y. Kihn, J.Ph. Perez, J. Sevely, G. Zanchi: Proc. 9th Int. Cong. for Electron Microscopy, Toronto, 1978, Vol.3, p.292
6.27 A.J. Craven and T.W. Buggy: Ultramicroscopy *7*, 27 (1981)
6.28 J.J. Hren, J.I. Goldstein, D.C. Joy (eds.): *Analytical Electron Microscopy* (Plenum, New York 1979)
6.29 J.A. Venables, A.P. Janssen, P. Akhter, J. Derrien, C.J. Harland: J. Microsc. Oxford *118*, 357 (1980)
6.30 J.A. Venables, A.P. Janssen: Ultramicroscopy *5*, 297 (1980);
J.A. Venables, G.D. Archer: Proc. 7th European Cong. on Electron Microscopy, The Hague, 1980, Vol.1, p.54

6.31 L.H. Veneklasen, N. Yew, J. Wiesner: Proc. 8th Int. Conf. on Electron and Ion Beam Science and Technology (ed. R. Bakish, The Electrochemical Society, 1978) 11; L.H. V
L.H. Veneklasen, G. Todd, H. Poppa: Proc. 9th Int. Conf. on Electron Micrsocopy, Toronto, 1978, Vol.1, p.12;
G. Todd, H. Poppa, L.H. Veneklasen: Thin Solid Films *57*, 213 (1979)
6.32 K. Hartig, A.P. Janssen, J.A. Venables: Surf. Sci. *74*, 69 (1978)
6.33 J.A. Venables, J. Derrien, A.P. Janssen: Surf. Sci. *95*, 411 (1980)
6.34 P. Akhter, J. Derrien, A.P. Janssen, J.A. Venables: Inst. Phys. Conf. Ser. *52*, 383 (1980);
P. Akhter: Thesis, University of Sussex (1980);
J.A. Venables, P. Akhter, J. Derrien, G.D.T. Spiller: Proc. ICSS-4, ECOSS-3, Cannes, 1980, Vol.1, p.42;
G.D.T. Spiller, P. Akhter, J.A. Venables: To be published
6.35 K. Hartig: Thesis, Ruhr University Bochum (1976);
F. Soria, H. Poppa: J. Vac. Sci. Technol. *17*, 449 (1980)
6.36 B.E. Sundqvist: Acta Metall. *12*, 67 (1964)
6.37 J.C. Heyraud, J.J. Mêtois: Acta Metall. *28*, 1789 (1980)
6.38 J.J. Metois, J.C. Heyraud: J. Cryst. Growth (in press)
6.39 J.J. Metois, G.D.T. Spiller, J.A. Venables: To be published
6.40 G.D.T. Spiller: Thesis, Imperial College, 1979 (to be published)
6.41 J.A. Venables, A.P. Janssen: Proc. 9th Int. Conf. on Electron Microscopy, Toronto, 1978, Vol.3, p.280;
A.P. Janssen, J.A. Venables: In SEM '79', ed. by O. Johari (SEM Inc., AMF O'Hare 1979) Vol.2, 259; Inst. Phys. Conf. Ser. *52*, 357 (1980)
6.42 A. van Oostrom: Surf. Sci. *89*, 615 (1979);
R. Browning, M. Prutton: Phys. Technol. *10*, 259 (1979)
6.43 A.P. Janssen, J.A. Venables: Surf. Sci. *77*, 351 (1978);
M.M. El-Gomati, A.P. Janssen, M. Prutton, J.A. Venables: Surf. Sci. *85*, 309 (1979);
J. Kirschner: Appl. Phys. *14*, 351 (1977)
6.44 R. Browning, P.J. Bassett, M.M. El-Gomati, M. Prutton: Proc. R. Soc. London A*357*, 213 (1977); Surf. Sci. *68*, 328 (1977)
6.45 A.P. Janssen, C.J. Harland, J.A. Venables: Surf. Sci. *62*, 277 (1977)
6.46 A.P. Janssen, J.A. Venables, J.C.M. Hwang, R.W. Balluffi: Philos. Mag. *36*, 1537 (1977)
6.47 R. Butz, H. Wagner: Surf. Sci. *87*, 69, 85 (1979); Proc. 7th Int. Vac. Cong. and 3rd Int. Conf. on Solid Surfaces, Vienna, 1977, p.1289
6.48 A.M. Brown, M.P. Hill: Inst. Phys. Conf. Ser. *52*, 379 (1980); Carbon *19*, 51 (1981)
6.49 R.L. Gerlach, N.C. McDonald: In SEM '76 (ed. O. Johari, SEM Inc. AMF O'Hare, 1976) Vol.1, p.199;
N.C. McDonald, C.T. Hovland, R.L. Gerlach: SEM '77, Vol.1, p.201 (1977); SEM '79, Vol.1, p.213 (1979)
6.50 P. Staib, G. Staudenmaier: Proc. 7th Int. Vac. Cong. and 3rd Int. Conf. on Solid Surfaces, Vienna, 1977, p.2355
6.51 H.E. Bauer, H. Seiler: Proc. 7th European Cong. on Electron Microscopy, The Hague, 1980, Vol.3, p.214
6.52 D.J. Fathers, C.J. Harland, G.D. Archer, J.A. Venables: Work in progress
6.53 D.B. Wittry: Proc. 7th European Cong. on Electron Microscopy, The Hague, 1980, Vol.3, p.14
6.54 H. Dexpert, E. Freund, J. Lynch: In Ref. 6.12, p.101
6.55 C.E. Lyman, personal communication.
6.56 A. van Oostrom, L. Augustus, W. Nijman, L. Leswin: J. Vac. Sci. Technol. *17*, 40 (1980)
6.57 A.P. Janssen, P. Akhter, C.J. Harland, J.A. Venables: Surf. Sci. *93*, 453 (1980)

6.58 P. Akhter, J.A. Venables: Surf. Sci. *103*, 301 (1981)
6.59 M.G. Wells, D.A. King: J. Phys. C*7*, 4053 (1974)
6.60 L. Wegmann: J. Microsc. *96*, 1 (1972);
G.F. Rempfer, K.K. Nadakavukaren, O.H. Griffith: Ultramicroscopy *5*, 437 (1980)
6.61 W.J. Baxter, S.R. Rouze: Rev. Sci. Instrum. *44*, 1628 (1973); J. Appl. Phys. *46*, 2429 (1975)
6.62 C.T. Hovland: Appl. Phys. Lett. *30*, 274 (1977); Proc. 7th Int. Vac. Cong. and 3rd Int. Conf. on Solid Surfaces, Vienna, 1977, p.2363;
J. Cazaux: Mater. Sci. Eng. *42*, 45 (1980); J. Phys. Paris *38*, L473 (1977);
J. Perrin, D. Mouze, J. Cazaux: Proc. ICSS-4, ECOSS-3, Cannes, 1980, Vol.1, p.291
6.63 G. Beamson, H.Q. Porter, D.W. Turner: J. Phys. E*13*, 64 (1980); Nature London *290*, 556 (1981)
6.64 H. Ibach, H. Hopster, B. Sexton: Appl. Surf. Sci. *1*, 1 (1977);
S. Andersson: In *Chemistry and Physics of Solid Surfaces III*, ed. by R. Vanselow (CRC, Boca Raton, FL 1981)
6.65 P. Morin, M. Pitaval, D. Besnard, G. Fontaine: Philos. Mag. *40*, 511 (1979); Scanning *2*, 217 (1979); In *Microbeam Analysis*, 1980, ed. D.B. Wittry, p.22
6.66 H. Hashimoto, A. Kumao, K. Hino, H. Yotsumoto, A. Ono: Jpn. J. Appl. Phys. *10*, 1115 (1971);
R.M. Henkelman, F.P. Ottensmeyer: Proc. Natl. Acad. Sci. U.S.A. *68*, 3000 (1971)
6.67 M. Utlaut: Phys. Rev. B*22*, 4650 (1980)
6.68 R.D. Moorhead, H. Poppa, K. Heineman: J. Vac. Sci. Technol. *17*, 248 (1980);
G.R. Millward: J. Catal. *64*, 381 (1980)
6.69 M.M.J. Treacy, A. Howie, C.J. Wilson: Philos. Mag. A*38*, 569 (1978);
J. Catal. *53*, 265 (1980)
6.70 K. Kambo, G. Lehmpfuhl: Optik Stuttgart *42*, 187 (1975);
G. Lehmpfuhl, Y. Uchida: Ultramicroscopy *4*, 275 (1979)
6.71 S. Iijima: Optik Stuttgart *48*, 437 (1977); *48*, 193 (1978); Proc. 9th Int. Cong. on Electron Microscopy, Toronto, 1978, Vol.1, p.252; Proc. Electron Microsc. Soc. Am. *38*, 376 (1980)
6.72 S. Iijima: Ultramicroscopy *6*, 41 (1981)
6.73 D. Cherns: Philos. Mag. *30*, 549 (1974);
R.L. Hines: Thin Solid Films *35*, 229 (1976);
J.P.F. Levitt, A. Howie: J. Microsc. *116*, 89 (1979)
6.74 Y. Tanashiro, H. Kanamori, K. Takayanagi, K. Kobayashi, K. Yagi, G. Honjo: Proc. ICSS-4, ECOSS-3, Cannes, 1980, Vol.1, p.683
6.75 J.C. Heyraud, J.J. Mêtois: Surf. Sci. *100*, 519 (1980)
6.76 K. Yagi, K. Takayanagi, K. Kobayashi, N. Osakabe, Y. Tanashiro, G. Honjo: Surf. Sci. *86*, 174 (1979)
6.77 K. Takayanagi, G. Honjo: Proc. ICSS-4, ECOSS-3, Cannes, 1980, Vol.1, p.267;
K. Takayanagi: Surf. Sci. *104*, 527 (1981)
6.78 G. Lehmpfuhl, K. Takayanagi: Ultramicroscopy *6*, 195 (1981)
6.79 J.A. Venables, H.M. Kramer, G.L. Price: Surf. Sci. *55*, 373 (1976); *57*, 782 (1976)
6.80 P.S. Schabes-Retckiman, J.A. Venables: Surf. Sci. *105*, 536 (1981)
6.81 G.L. Price, J.A. Venables: Surf. Sci. *49*, 264 (1975)
6.82 S. Ino: Jpn. J. Appl. Phys. *16*, 891 (1977)
6.83 Y. Gotoh, S. Ino: Jpn. J. Appl. Phys. *17*, 2097 (1978);
T. Ichikawa, S. Ino: Surf. Sci. *97*, 489 (1980)
6.84 S. Baba, M. Kawaji, A. Kinbara: Surf. Sci. *85*, 29 (1979);
T. Aiyama, S. Ino: Surf. Sci. *82*, L585 (1979)

6.85 K. Yagi: Proc. Electron Microsc. Soc. Am. *38*, 290 (1980) [visual presentation of video-recording at the meeting]
6.86 N. Osakabe, Y. Tanashiro, K. Yagi, G. Honjo: Surf. Sci. *97*, 393 (1980); *102*, 424 (1981)
6.87 N. Osakabe, K. Yagi, G. Honjo: Jpn. J. Appl. Phys. *19*, L309 (1980)
6.88 K. Yagi, N. Osakabe, Y. Tanashiro, G. Honjo: Proc. ICSS-4, ECOSS-3, Cannes, 1980, Vol.2, p.1007;
K. Yagi: Private communication
6.89 P.E. Højlund-Nielsen, J.M. Cowley: Surf. Sci. *54*, 340 (1976); Ultramicroscopy *1*, 145 (1975)
6.90 P.S. Turner, J.M. Cowley: Ultramicroscopy *6*, 125 (1981)
6.91 J.M. Cowley, J.L. Albain, G.G. Hembree, P.E. Højlund-Neilsen, F.A. Koch, J.D. Landry, H. Schuman: Rev. Sci. Instrum. *46*, 826 (1975)
6.92 G.G. Hembree, J.M. Cowley: In SEM '79 (ed. O. Johari, SEM Inc., AMF O'Hare, 1979)Vol.1, p.145; Oxid. Met. *13*, 331 (1979)
6.93 R. Pantel, M. Bujor, J. Bardolle: Surf. Sci. *62*, 589 (1977)
6.94 H. Soezima: Surf. Sci. *85*, 610 (1979)

7. He Diffraction as a Probe of Semiconductor Surface Structures

M. J. Cardillo

With 15 Figures

7.1 Introduction

The outermost layers of many single-crystal materials undergo reconstruction, i.e., the geometry at the surface differs from that expected for an ideal termination of the bulk. Most of these reconstructions result in a change in the size of the repeating surface unit cell and therefore can be identified by the simple observation of the diffraction spot pattern in a low-energy electron diffraction (LEED) apparatus. In particular, the common semiconductors, which are primarily covalent materials with strong directional bonding, all undergo surface reconstruction with some faces exhibiting one or more metastable surface structures. These may be considered to be driven by the high potential energy of the half-filled orbitals generated when the bulk bonds are broken to make surface atoms. The resulting periodicities of these surfaces have in many cases been known from LEED for over twenty years. Despite numerous experimental and theoretical studies the geometrical configurations of almost all these surfaces are still to be resolved. In Table 7.1 are listed a selection of the reconstructions observed for some low-index faces of common semiconductors. At the present time the geometries of GaAs{110} and Si{111} 2×1 are the only ones generally accepted as well determined. In both cases the structures are close to the ideal bulk termination geometry.

There are a variety of new and exciting probes of surface structure which are at early stages of development. These include Surface Extended X-Ray Absorption Fine Structure (SEXAFS), glancing incidence X-ray diffraction, ion scattering, and atom diffraction which is the topic of this chapter. Each of these probes may play a key role in the determination of surface structures. However, the principal and most productive tool for surface structural studies has been LEED; it therefore serves as a standard by which to compare atom diffraction.

Table 7.1.

Si{111}	- (7 × 7),(2 × 1), (1 × 1)?
Si{100}	- (2 × 1) ↔ c(2 × 4) disordered?
Si{110}	- (5 × 1), c(1 × 2) (5 × 4),...
Ge{111}	- (2 × 8), (1 × 1)?
Ge{100}	- c(2 × 4)
Ge{110}	- (2 × 2),...
GaAs{111}	- (2 × 2), ($\sqrt{19}$ $\sqrt{19}$)R 23.9°, 3 × 3
GaAs{100}	- (2 × 8), (2 × 4), (1 × 6)
GaAs{110}	- (1 × 1) "relaxed"

The information contained in the array of LEED diffraction beams observed generally, and in their variation with incident beam voltage, is truly comprehensive in terms of structural details. However, a theoretical computation of these beam intensities if fraught with difficulties, particularly in the case of semiconductors. One of the basic problems in LEED structural analysis is the penetration and subsequent backscattering of the electron beam from depths approaching 10 Å or 5 layers into the bulk. Strong multiple scattering occurs between and within each layer. In addition the semiconductor crystals may not become periodic in the direction perpendicular to the surface until 5 layers into the bulk. Thus a dynamic calculation of the LEED intensities must include the complexities of multiple scattering, possibly different geometries for 3-5 layers, and a sufficiently accurate description for the electron-solid potential.

The analysis of the diffraction of thermal energy light neutral particles, in particular He atoms, appears to be an attractive complement. At the low energies employed, typically 10-100 meV, the atoms strictly do not penetrate with the result that the observed diffraction pattern is due to the periodic potential of the outermost *exposed* layer. Only a single geometry is in question. However, in contrast to 100-eV electrons, for which the scattering is dominated by the atomic cores, thermal He atoms reflect from the surface potential at distances in the range of van der Waals diameters. Although experiments and theory have shown that the diffraction pattern is extremely sensitive to the details of this weak interaction potential, the information about nuclear positions is quite indirect. Thus the analytical aspects of the development of He diffraction into a quantitative surface structural tool may be divided into two substantial problems. The first is the derivation of a satisfactory description of the surface scattering potential from the diffraction beams. The second is the relation of the potential to the nuclear positions.

The semiconductor surfaces offer an exciting prospect for this development. There is a continuing challenge to resolve the geometries of these surfaces. He diffraction patterns have shown the semiconductor potentials to be strongly corrugated at the van der Waals distances of closest approach so that there are large phase-shift variations over the unit mesh. This implies sufficient structural sensitivity that a solution of the He scattering problem will give rise to useful geometric information. Strongly corrugated surfaces, however, present somewhat of a theoretical challenge. As will be discussed below, important progress has been made in both aspects of this problem.

In this review progress in the study of He diffraction from semiconductor surfaces will be surveyed and is organized as follows. Experimental results for the disordered Si{100} (2×1) surface will be discussed first along with a simple but informative analysis of the scattering. Subsequently, the study of a known structure, the GaAs{110} surface, will be described. There we have treated the theoretical aspects of atom diffraction from strongly corrugated surfaces as unknowns. Finally recent results from the Si{111} 7×7 surface will be described, from which a new type of reconstruction model has been proposed.

7.2 Si{100}: Disordered Dimer Array?

7.2.1 Si{100} Periodicity

The Si{100} (2×1) reconstruction was first observed with LEED by SCHLIER and FARNSWORTH [7.1] in 1959. They proposed the dimerization of adjacent rows of Si atoms in order to saturate the adjacent half-filled orbitals on the surface. For an accurately cut and annealed crystal ($\pm 1/2^{\circ}$ in orientation) there are generally monoatomic steps spaced 100-300 Å so that a surface diffraction pattern is the superposition of the scattering from each single terrace region. For Si{100} alternate layers are rotated 90° so that the two-domain Si{100} 2×1 presents a square array of diffraction beams in reciprocal space. Many LEED studies of the Si{100} surface have been carried out, and with two exceptions all have observed just the two-domain (2×1) pattern. LANDER and MORRISON [7.2] observed a $c(2 \times 4)$ periodicity with intensity streaks extending from the 1/4 order spots. POPPENDIEK et al. [7.3] reported a sharp $c(2 \times 4)$ after careful annealing by reducing the surface temperature from 1300 C at $1\text{-}2^{\circ}/s$. The 1/4 order beams they observed amount to $\sim 1\%$ of the integer beams, i. e., they are sharp but weak.

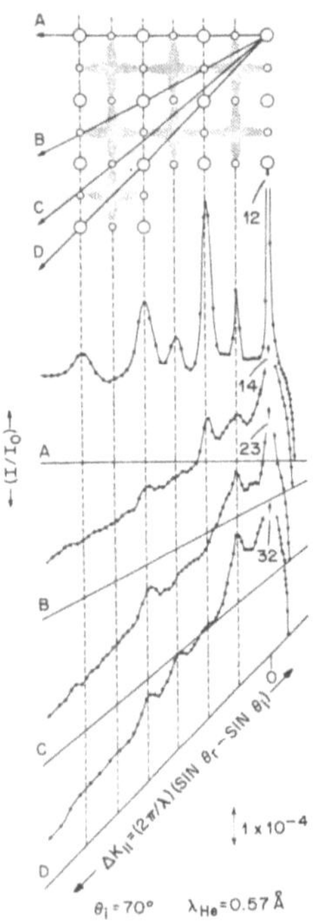

Fig.7.1

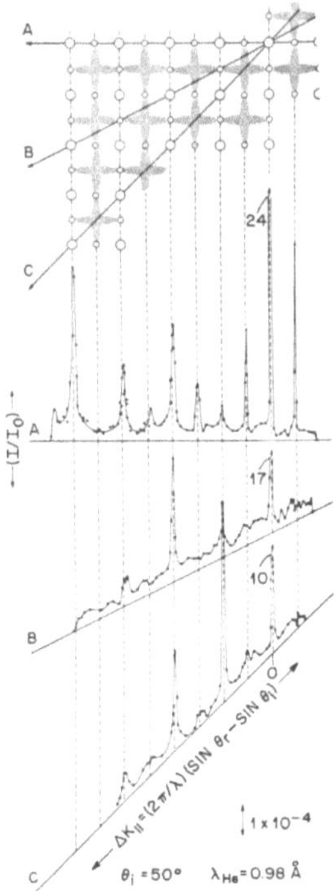

Fig.7.2

Fig.7.1. Diffraction scans plotted vs $\Delta K_{||}$, the reciprocal net coordinate. Each scan is plotted at the azimuth angle ϕ so that a vertical correspondence exists between the diffraction data and the reciprocal net above. The wavelength λ = 0.57 Å and the detector acceptance angle $\Delta\theta_r = 1.5^{\circ}$

Fig.7.2. Diffraction scans plotted as in Fig.7.1 but for a wavelength λ = 0.98 Å and $\Delta\theta_r = 0.55^{\circ}$

In the atom diffraction study [7.4], particular attention was paid to different ways of annealing the crystal, and consistantly a sharp two-domain (2×1) reciprocal net was observed with additional disordered intensity suggesting a secondary reconstruction. The additional intensity forms a diffuse cross (Figs.7.1,2) which would appear to include patches of $c(2\times 4)$ and $p(2\times 2)$ or $c(2\times 2)$ periodicity. In Figs.7.1,2 the azimuth dependence of the diffraction patterns for different wavelengths are plotted versus $\Delta K_{||}$, the parallel momentum transfer $[\Delta K_{||} = (2\pi/\lambda)(\sin\theta_r - \sin\theta_i)]$, from which the reciprocal net illustrated above is derived. Note in Fig.7.2 that the detector acceptance angle has been substantially decreased so that both the background and the disordered diffraction intensities are considerably reduced. However, the "extra" intensity is clearly observable and in

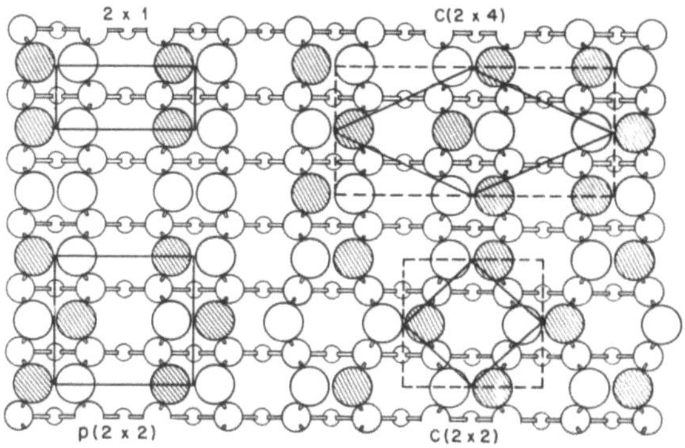

Fig.7.3. Illustration of the asymmetric dimer model in different periodic arrays according to the registry of the buckling rows

addition it appears structured, i.e., an intensity minimum is actually observed along the axis of the diffuse cross. This is a striking example of the sensitivity of atom diffraction. The structure observed here lies within a diffusely scattered intensity which is generally unobservable in display LEED systems. The observation of directional disorder based on He diffraction has been discussed [7.4] in terms of the distorted dimer model based on recent total energy calculations of CHADI [7.5]. In Fig.7.3 the buckled dimer model and the possible periodicities involved are illustrated. CHADI's calculations clearly favor the basic dimerization model over several others tested for the principal (2×1) reconstruction and, in particular, indicate that a slight tilting of the dimer bond with respect to the surface plane lowers the surface energy on the order of 0.3 eV per pair and removes a metallic dispersion in the valence band. An alternation of the tilting so that every other dimer in a row is different reduces the energy somewhat more (0.1 eV/pair). The total energy calculations cannot distinguish between the $p(2 \times 2)$, $c(2 \times 4)$, and $c(2 \times 2)$ unit meshes all of which were considered possible based on the atom diffraction results. We have interpreted the streaks (crosses in two domains) observed with atom diffraction as a second-order reconstruction perpendicular to the dimer direction which involves much less energy compared to the dimerization, consistent with the proposed buckling or alternatively a twist. However, because of the weak forces involved in the buckling, adjacent rows are not strongly coupled and their registry is not well ordered at room temperature. Thus it has been proposed that the equilibrium surface at 300 K has domains of locally registered rows of buckled (or twisted) dimers which may be ~25 Å in length and give rise to the splitting in the diffuse intensity. The domain edges can actually be small regions of $p(2\times 2)$ which may be easily converted to $c(2\times 4)$ if the surface temperature were lower. Alternatively there may be regions of $c(2\times 2)$ which would be favored at step edges if the steps formed a significant angle with respect to the [10] direction. These regions would *not* be easily converted to $c(2\times 4)$ since it would require breaking a row of bonds. In summary, the disorder in the registry of adjacent dimer rows is interpreted in terms of the weak coupling involved in the perturbation of the symmetric dimer. This gives rise to the diffusely structured cross observed in the He diffraction experiment.

7.2.2 Diffraction Scans and Qualitative Features of the Si{100} Surface

Diffraction scans for a wavelength $\lambda \sim 1$ Å are plotted in Fig.7.4 for the [10] azimuth. Several interesting features can be observed. A sharp alternation in the intensity of adjacent beams can be observed in some regions ($\theta_i \sim 20^\circ$) and in other data in [7.4]. This arises from the two domains – one of which contributes only to integer peaks while the other contributes to both the integer and half-order peaks in this direction. In the spectra at $\theta_i = 70^\circ$ and 60°, a broad maximum in the intensity envelope of both half and integer-order beams can be observed displaced 70° from the specular beam. We can associate this maximum with scattering from one domain, i.e., in the direction of the strong second-order reconstruction. In Fig.7.3 this corresponds to scattering across the dimer rows. This type of envelope maximum is termed a rainbow maximum; from classical scattering considerations it can be directly related to the slope of the potential at the points of inflection. For hard wall scattering from an oscillatory potential shape function $\xi(x)$, the cross section varies as the inverse of the second derivative of the potential, i.e., $\xi''(x)^{-1}$. This gives rise to an integrable singularity displaced an angle $\Delta\theta_{rb}$ from specular which is the rainbow maximum,

$$\Delta\theta_{rb} = -2\tan^{-1}\xi''(x) \quad . \tag{7.1}$$

If the first Fourier component of the potential dominates the shape of the corrugation function such that $\xi(x) \sim \frac{1}{2} d_\perp \cos(2\pi x/L)$, then

$$d_\perp \sim (L/\pi)\tan(\Delta\theta_{rb}/2) \quad . \tag{7.2}$$

A qualitative insight to the Si{100} surface can be made using this relation. Referring to Fig.7.4, the value $\Delta\theta_{rb} \sim 70^\circ$ is obtained. However,

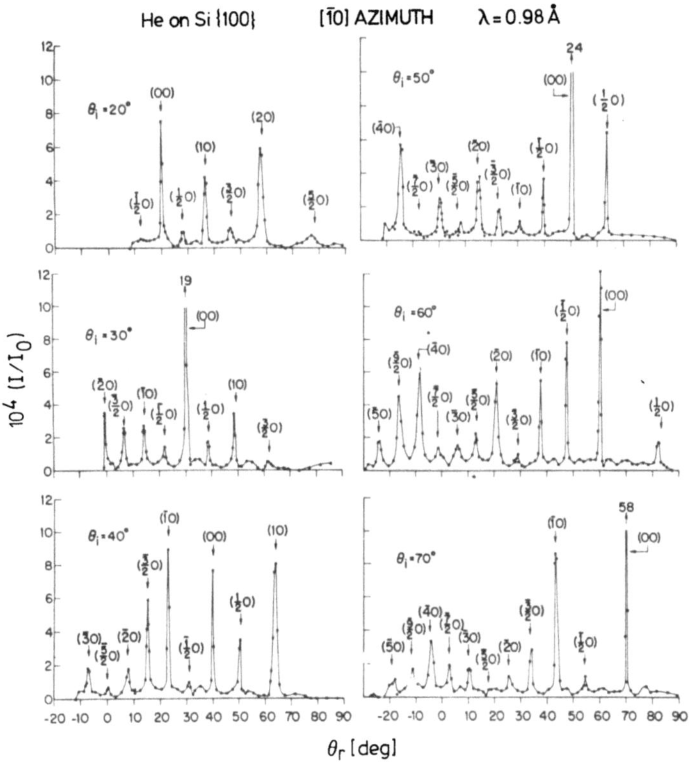

Fig.7.4. Diffraction scans for Si{100} in the 10 direction for λ = 0.98 Å and $\Delta\theta_r$ = 0.55°

this angular displacement includes the effect of refraction due to the attractive portion of the He surface potential. For the low-energy beam, E = 21 meV, a weak attractive well of depth D ~ 5 meV can increase $\Delta\theta_{rb}$ by ~10°. In [7.4] diffraction scans for λ = 0.57 Å (E = 63 meV) are plotted for the [01] direction and a strong rainbow maximum can be observed at $\Delta\theta_{rb} \sim 60°$. At this higher energy the attractive potential will have a small refractive effect. Therefore, we use $\Delta\theta_{rb}$ = 60° and L = 2 × 3.84 Å = 7.68 Å, to estimate an effective corrugation across the dimerized rows, i.e., $d_\perp$ ~ 1.4 Å. This corrugation is also consistent with the rainbow positions observed in Fig. 7.4 for θ_i = 60° and 70° if an attractive well of depth D ~ 5 meV is assumed.

This simple analysis illustrates the relation between strongly corrugated surfaces and angular distributions of diffracted beams. Simple approximate parameters, such as a principal corrugation of $d_\perp$ ~ 1.4 Å, are important in comparing models of the surface which have substantially different contours.

Other features in the data of Fig.7.4 and in [7.4] can be similarly analyzed to give an approximate picture of the He-Si{100} surface potential [7.6]. A reliable analysis to derive a full description of the potential is a substantial computational undertaking. It is not clear at this point that such an effort is warranted until the surface can be well ordered.

7.2.3 Specular Intensities

Structure in specular intensity scans is often associated with surface resonances. However, in the case of Si{100} these resonance features are generally weak. This is likely the result of a lack of detailed long-range order. There are, however, broad structural features in the specular intensity scans which are due to interference in $\Delta k_\perp$. These features may be observed in the results plotted in Fig.7.5. Semiclassical scattering analysis indicates that the specular intensities are dominated by the flat regions of of the potential [7.6], so that the envelope structure observed is principally related to the peak-to peak corrugation amplitudes. The interference oscillations for simple oscillatory potentials vary approximately as the Bessel function $J_0^4(c)$, where $c = dk_i \cos\theta$ [7.7]. This expression is very close to Bragg's law, i.e.,

$$d_\perp = n\lambda/(2\cos\theta_i) \quad . \tag{7.3}$$

As in Sect.7.2.2, if we consider only the 63-meV data ($\lambda = 0.57$ Å), to reduce the effects of refraction by the attractive potential, the separation of the two principal maxima in Fig.7.5 yields $d_\perp \sim 1.4$ Å. The scan at 21 meV appears shifted but similar in features to the 63-meV results and is consistent with the interpretation in terms of $\Delta k_\perp$ interference. It is improbable that the agreement between $d_\perp$ obtained in this simple analysis and that obtained from the classical rainbow analysis in fortuitous. Similar agreement has been obtained for GaAs{110} and subsequently confirmed by rigorous diffraction calculations. In general, the diffraction of He from semiconductor surfaces has shown that interference analysis of specular intensities is a useful and nearly independent probe of corrugated surfaces which complements the full diffraction analysis. Further illustration of its utility will be shown below.

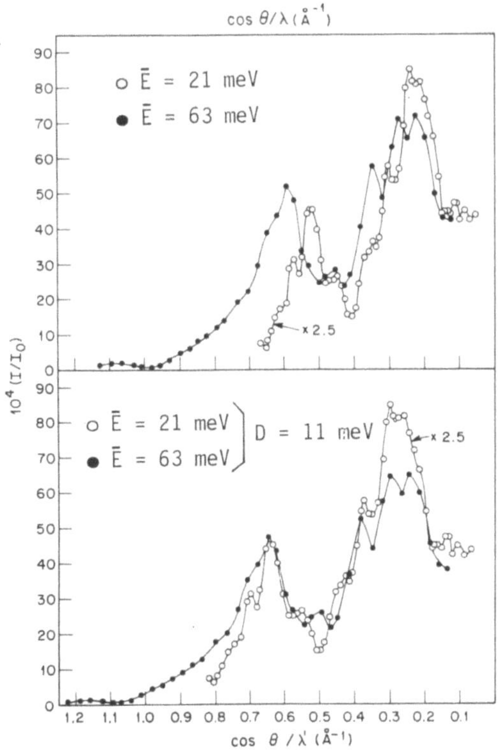

Fig.7.5. Specular intensity variation in [11] direction plotted vs cosθ/λ for two energies. In the lower curve the abscissa has been empirically adjusted for a wavelength shift but not for refraction [7.4]

7.2.4 Structural Models for Si{100}

Although a theoretical fit of the Si{100} diffraction data is not yet available, some comments about structural models can be made. All the data for the Si{100} surface are consistent with the distorted dimer model of the surface. The observation of a stable sharp (2 × 1) reconstruction with a weak second-order perturbation in the perpendicular direction, which does not produce long-range registry of adjacent rows, is in accord with the energies of alternating buckled dimers proposed by CHADI (or alternating twisted dimers). In addition a simple classical diffraction analysis suggests a principal corrugation in the dimer direction around 1.4 Å, i.e., a slope at the point of inflection of ~30°. A second corrugation distance in the perpendicular direction of ~0.4 Å is also indicated by the data. Using a hard ball model in which the He-Si distance of closest approach is taken as d_{CA} = 3.3 Å (justified below), a one-dimensional diagram of the tilted dimer is schematically drawn in Fig.7.6 in which the charge contour

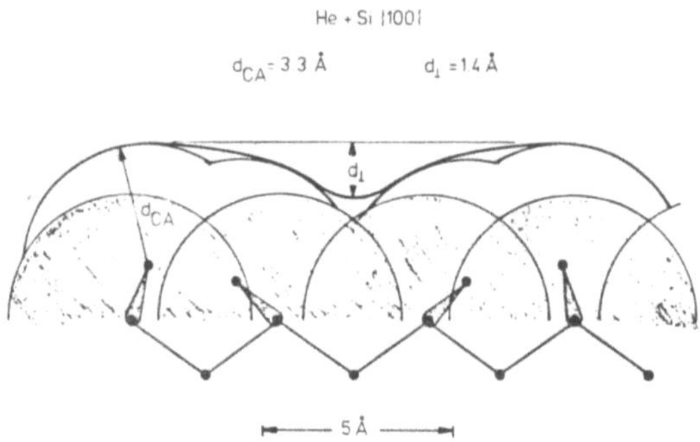

Fig.7.6. Schematic illustration more dimension of a charge contour appropriate for thermal He scattering based on the tilted dimer model using a distance of closest approach d_{CA} = 3.3 Å

which would account for the distance $d_\perp$ = 1.4 Å is sketched. An alternative model [7.3], consisting of missing alternative rows of Si atoms, can be shown in such a diagram to have a principal corrugation much larger than that derived here and can therefore be ruled out.

7.3 GaAs{110}

The study of the Si{100} surface has proven to give an interesting start to the development of He diffraction into a quantitative structural tool for semiconductor surfaces. Several aspects of the problem have become better defined. For corrugated surfaces, simple scattering theories such as the eikonal or sudden approximation are not adequate, in particular in their treatment of multiple scattering. The diffraction probabilities are strongly coupled. However, they do allow qualitative features of the potential to be described. An accurate computation of the diffraction intensities for realistic potentials is a substantial undertaking. The attractive potential must be well understood since the phase-shift variation over the unit mesh is large and the potential energy may also vary significantly over the unit mesh. Furthermore, a relation between the potential and nuclear positions is still required to connect the diffraction data to geometric information. A problem specific to the Si{100} is the lack of long-range order in the secondary reconstruction which makes an accurate comparison with theory difficult.

In order to progress further with strongly corrugated surfaces, we have studied He diffraction from a known semiconductor structure, GaAs{110} [7.8]. By treating the diffraction calculation and the theory of the potential as unknowns, we wish to develop a suitable description of atom diffrac-

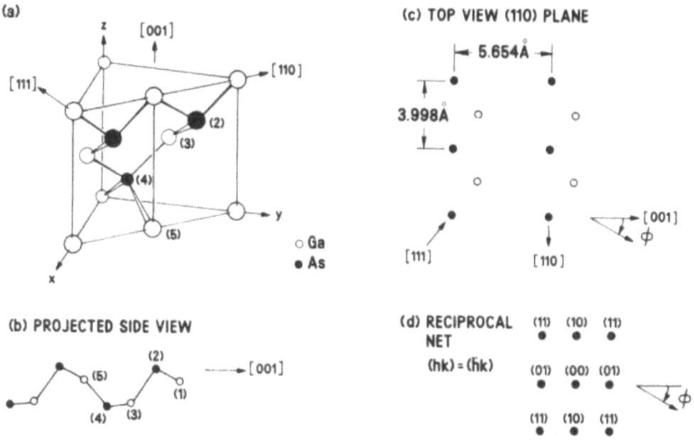

Fig.7.7a-d. GaAs crystal structure showing (a) {110} planar face, (b) projected side view of the {110} plane, (c) top view of outermost layer, (d) reciprocal net indexing and definition of ϕ

tion to permit quantitative structural studies of the other corrugated surfaces. The surface of GaAs{110} is illustrated in Fig.7.7 along with conventions for angles and diffraction beams used in this paper. The surface may be thought of as ridges and troughs of gallium-arsenic bonded chains. The surface is "rough" in the [01] direction, across the troughs, and relatively "smooth" in the [10] direction, along the troughs. The surface reconstructs by tilting the surface Ga-As bond by $\sim 27^{\circ}$ but does not change its periodicity from what is termed (1×1) [7.9]. The surface bears somewhat of a resemblance to the tilted dimer model of the Si{100} surface.

7.3.1 Diffraction Scans

In Fig.7.8 are shown a selected set of diffraction scans with $\lambda = 0.98$ Å ($\bar{E} \sim 21$ meV) for a series of θ_i with $\phi = 0$ and 180°. Note that the two sets of scans have gross similarities in their envelopes but that there is a clear asymmetry in the diffraction pattern which can be ascribed to the tilting of the surface Ga-As bond (Ga and As are sufficiently large and close in atomic number that this difference can be ruled out as the cause of the asymmetry). These and many other scans for different ϕ and λ have been analyzed at various levels of theory. Simple one-dimensional classical scattering analysis of rainbow maxima, as seen in the data for example at $\theta_i = 55^{\circ}$, for which the surface is assumed to have glide plane symmetry, indicates a gross corrugation across the troughs of $\xi_x \sim 1.0$ Å and a corrugation along the troughs of $\xi_y \sim 0.3$ Å. The hard-wall eikonal wave ap-

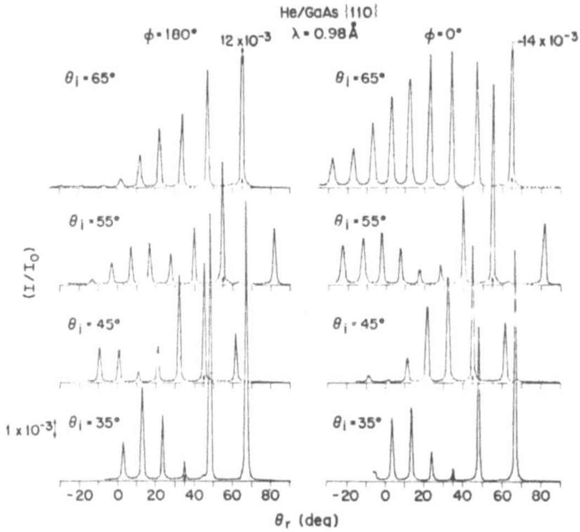

Fig.7.8. Diffraction intensities for $\phi = 0^\circ$ and 180° at $\lambda = 0.98$ Å. θ_i is measured from the surface normal

proximation, although not expected to be quantitatively accurate for this surface, has also been used for a qualitative comparison to the classical scattering analysis. It yields values of $\xi_x \sim 1.2$ Å and $\xi_y \sim 0.36$ Å, similar to the classical analysis. In these simple analyses the role of the attractive potential has been ignored. However, the angular displacements of the rainbow maxima for high-energy data ($E = 63$ meV, $\lambda = 0.57$ Å) are similar to those of Fig.7.8. This indicates that the attractive potential plays a small role, even at $E = 21$ meV, $\lambda = 0.98$ Å, i.e., the laterally averaged well depth is likely to be less than 10 meV.

7.3.2 Specular Intensity Scans

A number of specular intensity versus polar angle (θ_i) scans were recorded for various azimuths (ϕ) and wavelengths. In contrast to the disordered Si{100}, these scans show sharp resonance oscillations. In Fig.7.9 one set of scans in the vicinity of $\phi = 0$ at $\lambda = 0.98$ Å is shown. The sharp features are associated with bound-state resonances. These resonances arise as follows. For specific incident kinematic conditions the incoming particle can enter a scattering state associated with the attractive potential so that it is effectively bound in the direction normal to the surface. If the surface is not strongly corrugated, a useful assumption is that the particle is moving nearly freely so that the conservation condition for parallel momentum, K, and energy, E_i, can be written

$$(K + G)^2 = (2m/\hbar^2)(E_i + |E_j|) \qquad (7.4)$$

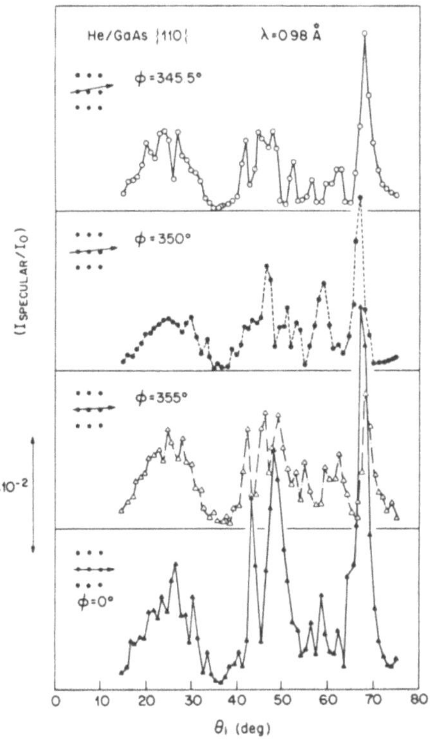

Fig.7.9. Specular intensity versus θ_i for a series of azimuths near $\phi = 0°$

where G is a reciprocal net vector and E_j is a bound-state energy level of the laterally averaged potential for motion normal to the surface. Note that (7.4) is the equation of circles in the K plane centered at various G. A plot of the positions in the K plane at which the resonances occur should allow the appropriate G for each circle to be indexed. The radii of the circles then determine the various values of E_j. This procedure has been successfully employed to obtain the He surface potential for several weakly corrugated surfaces [7.10].

For moderately corrugated surfaces, many more beams can strongly couple to resonances with the result that the density of transitions is high. There will then be many intersections of these kinematic circles, which are associated with free particle state degeneracies. These degeneracies are lifted by splitting these states to form gaps in the bands of allowed energies. These splittings cause deviations of the resonance positions from the kinematic circles. Where these splittings can be identified, the Fourier components of the surface potential can be derived.

For the case of GaAs{110}, which is strongly corrugated, a plot of the resonance positions in the K plane yields a high density of points but no readily identifiable circles. The coupling between beams is sufficiently strong that there is substantial splitting all over the K plane and no kinematic analysis was successful. Thus no information about the attractive potential has been derived from these resonance features to date. The analysis of resonance scattering from strongly corrugated surfaces remains an area requiring serious study. However, information about the potential has been obtained from a comprehensive analysis of the diffraction patterns [7.11].

In addition to sharp oscillations in the data of Fig.7.9, there are broad envelopes similar to those observed for Si{100} specular scans. These envelope features again represent Bragg-like interference in the specular beam and can be analyzed for structural information about vertical displacements as in the case of Si{100}. The most consistent feature apparent in all the $\lambda = 0.98$ Å data is the maximum observed near $\theta_i = 25^\circ$. If the effect of the attractive potential is again ignored, the relation $2d_\perp \cos\theta = n\lambda$ yields $d_\perp/n = 0.54$ Å. For $n = 2$, $d_\perp \sim 1.1$ Å in agreement with the diffraction analysis. The $n = 1$ maxima should then appear around $\theta_i = 64^\circ$. Adding the effect of a uniform well depth of $D \leqq 5$ meV brings the assignment into agreement with the features of the data at $\theta_i = 26$ and 68°. For the recurrent maxima at $\theta_i = 45^\circ$, the value $d_\perp/n = 0.75$ Å is obtained. This distance corresponds to the difference between the two corrugations obtained in the diffraction analysis, i.e., $d_x \sim 1.0 - 1.1$ Å, $d_y \sim 0.3$ Å. Similar analyses with the eikonal approximation have been done for a variety of other specular intensity scans with comparable parameters for the corrugation being deduced. In general, the results of either Bragg's law or the eikonal analysis of the specular intensity are consistent with diffraction analysis, i.e., the principal corrugations across the troughs is $\zeta_x \sim 1.0 - 1.1$ Å and along the troughs $\zeta_y \sim 0.3$ Å.

7.3.3 Rigorous Calculation of Diffraction Intensities

The diffraction intensities for GaAs{110} have been extensively analyzed theoretically by LAUGHLIN [7.11]. A Green's function technique for carrying out dynamic close-coupled calculations for realistic soft-walled but strongly corrugated potentials was developed. The calculation is accomplished through a recursion relation starting well inside the repulsive potential and integration out to the vacuum. It can routinely handle potentials of any shape or corrugation with good numerical stability and convergence properties. Unitarity is identically satisfied. Using this analytical procedure the diffraction intensities were found to be generated by a soft-wall potential which can be roughly described at the He atom classical distance of closest approach as having an effective peak-trough sinusoidal corrugation of $\sim$1.0 Å, confirming the qualitative results of the approximate analysis described above. In addition the more rigorous theory demonstrated that the requirement that the potential be soft is necessary in order to accurately reproduce the way the rainbow maxima collapse into the specular beam as θ_i approaches grazing incidence (Fig.7.8). The average attractive potential well depth is also sensitive to this rainbow collapse and a good fit of the complete set of data at E = 21 meV indicates D $\sim$ 7 meV.

7.3.4 The Original of the He/GaAs Potential

Having obtained a reasonable description of the effective He/GaAs{110} scattering potential, LAUGHLIN [7.11] then considered its origin and developed an empirical connection between He surface potentials and the positions of the atomic cores which is based on the charge density of the surface. Noting the general similarity in form of the energy terms for interacting neutral gases in the Gordon-Kim theory [7.12], an expression for the repulsive He atom-target interaction was derived for which the leading term is

$$E = T\rho_t \tag{7.5}$$

where ρ_t is the charge density of the target and $T = 9\times10^4$ meV Å^3. Correcting empirically for the improper accounting of exchange and correlation and adding the r^{-6} attractive term the following expression is obtained

$$E = T\rho_t - X\rho_t^{1/3} - \sum C_6/r^6 \tag{7.6}$$

where X = 40 meV Å^3.

The first term is equivalent to a result independently derived by ESBJERG and NORSKOV [7.13] for the He surface repulsion based on a local density calculation for embedding He in a uniform electron gas. Both results quantitatively associate the leading term in the scattering potential of He (E = 10 - 100 meV) with the surface charge density at an extremely low value (10^{-5} a.u). Thus for the He surface potential to be amenable to direct calculation based on these results, charge densities must become available at an unprecedented level of accuracy.

7.3.5 Computation of Rarified Charge Densities

After the specific form of the relation between the He surface potential and charge densities was proposed, HAMANN tested the result [7.14] against the GaAs{110} potential obtained from the diffraction study using a surface linear augmented plane wave (SLAPW) method recently developed. The exponentially decaying charge densities can be treated accurately and selfconsistently with this technique with the principal approximation being the local exchange and correlation potential.

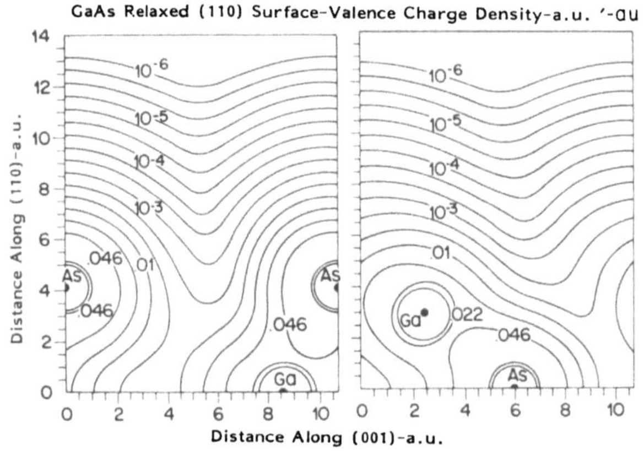

Fig.7.10. Charge densities for the relaxed (1 × 1) surface of GaAs{110} calculated in [7.14]. The key contour at 2×10^{-5} is appropriate for the estimation of the corrugations observed in He diffraction

The results for the Hamann calculation for GaAs{110} are shown in Fig. 7.10. Charge densities are plotted for two cuts across the troughs. The first cut is through the As nuclei and the second through the Ga nuclei. The principal corrugation across the troughs is found to be $\zeta_x \sim 0.9$ Å and along the troughs $\zeta_y \sim 0.3$ Å, in remarkable agreement with the analysis of the diffraction results. As a test of the structural sensitivity, the surface was "unrelaxed" to the ideal bulk termination geometry in which the Ga-As bond is in the horizontal plane. The principal corrugation parameters dropped to $\zeta_x = 0.6$ Å and $\zeta_y \sim 0.1$ Å so that the relaxed and unrelaxed surfaces are clearly distinguished. In addition to GaAs{110}, HAMANN has also shown these calculations to be appropriate for the He potential derived for H chemisorbed on Ni{110} based on the results of RIEDER and ENGEL [7.15]. There the fit to the experimental corrugation function enabled a quantitative specification of the binding geometry of the hydrogen atom on the Ni{110} surface.

7.3.6 Summary

The calibration experiment of He diffraction from a known structure, GaAs {110}, must be termed successful in that it has stimulated important progress. It has provided a basis for significant theoretical steps forward in the quantitative development of He diffraction into a useful structural probe. A detailed theoretical investigation of scattering from GaAs{110}, a strongly corrugated surface, has been carried out [7.11]. In this study a new computational technique was developed, a failure of the hard-wall approximation was documented, and a description of the He surface potential was discussed. This has led to an empirical expression for the repulsive potential in terms of the target charge densities which quantitatively confirms a similar proposal of ESBJERG and NORSKOV. Based on this relationship HAMANN has applied SLAPW calculations out to the appropriate charge densities. His results show that the charge density profiles and He diffraction results are consistent with and sensitive to the geometry of the GaAs{110} surface and yield the bonding geometry for H on Ni{110}.

7.4 Si{111} (7 × 7)

The reconstruction of the Si{111} surface to form a (7 × 7) unit mesh has been known for 20 years [7.16]. Despite extensive study, no convincing evidence to support any structural model has emerged. In part this is due to the size of the unit mesh which contains 98 top and nearly co-planar second-layer atoms. This makes even a kinematic analysis of a core-scattering probe, such as LEED, a substantial task and hardly amenable to testing a range of structural models. In addition dynamic effects associated with the penetration of the electron beam to several layers and the electron-solid potential render kinematic analyses of dubious accuracy.

The diffraction pattern of He atoms reflecting only from the outermost exposed layers may prove a useful alternative. Although it is unlikely that a complete diffraction calculation for a specific structural model will prove much less of a task, it is possible that key features of the diffraction pattern from the outermost exposed layer will yield to the simplifying aspects of approximate analyses in terms of characteristic shapes and distances.

7.4.1 Diffraction Scans

In Fig.7.11 a series of diffraction scans [7.17] is plotted in the [01] direction for a wavelength $\lambda = 0.98$ Å ($\bar{E} = 21$ meV) and an acceptance angle $\Delta\theta_r = 0.56^\circ$ for a sequence of incident angles θ_i. For the (7×7) surface, high angular resolution is required in order to discriminate against the nearest out-of-plane beams. The overall intensity distribution in these scans may be contrasted to the GaAs{110} patterns in that there are no smooth features in the envelope of the diffraction pattern indicative of rainbow scattering. Yet there is substantial structure at large angles characteristic of rough surfaces. We may estimate the expected corrugation for the unreconstructed Si{111} for comparison. In Sect.7.3 the value of the charge density appropriate for the He distance of closest approach was discussed [$\rho \sim \bar{E}(\text{meV})/10^5$ Å^3 meV]. For silicon one may estimate that the effective distance of closest approach for a charge density of 2×10^{-4} Å^{-3} is approximately 3.5 Å. Using this distance to draw overlapping spheres it becomes clear that a relatively smooth surface will result which has a maximum vertical corrugation of ~0.5 Å. In this dilute region of charge, there will be some smoothing of the cusped regions, so that one may expect a local corrugation within the (7×7) unit mesh in the vicinity of 0.2 - 0.3 Å. Thus the absence of smooth envelopes in Fig.7.11 is not associated with the local potential of the undisturbed Si{111} surface. If the (7×7) reconstruction were of the "smooth" type, such as long wavelength undulations or buckling, this same argument would require the observation of intense rainbow maxima characteristic of extended regions of the slowly varying slope of the surface potential.

7.4.2 Specular Intensity Interference

An important indication as to the nature of the (7×7) reconstruction was inferred recently [7.18] from a He specular intensity scan taken at a wavelength of $\lambda = 0.57$ Å and a detector acceptance angle $\Delta\theta_r \sim 1.1^\circ$. The data are reproduced in Fig.7.12. There a series of rapid broad oscillations may be observed which can be associated with interference in the normal momentum, $\Delta K_\perp$, similar to that for Si{100} and GaAs{110}.

The first four oscillations from $\theta_i = 25^\circ$ to $\theta_i = 52^\circ$ can be immediately fit to a vertical distance by the simple Bragg relation $2d_\perp\cos\theta = n\lambda$ and a value $d_\perp \sim 3.2$ Å is obtained. Using this value a series of maxima can be generated assuming that two large nearly flat regions separated by a vertical spacing of 3.2 Å exist within the (7×7) unit mesh. They are indicated

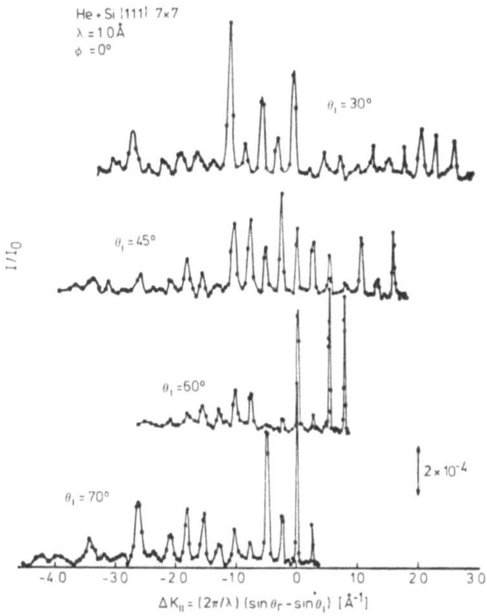

Fig.7.11. Normalized scattered He intensity vs $\Delta K_{\parallel}$ for scans in the $[01]^*(\phi=0^{\circ})$ direction. The grating relation allows diffraction peaks at multiples of 0.27 $Å^{-1}$

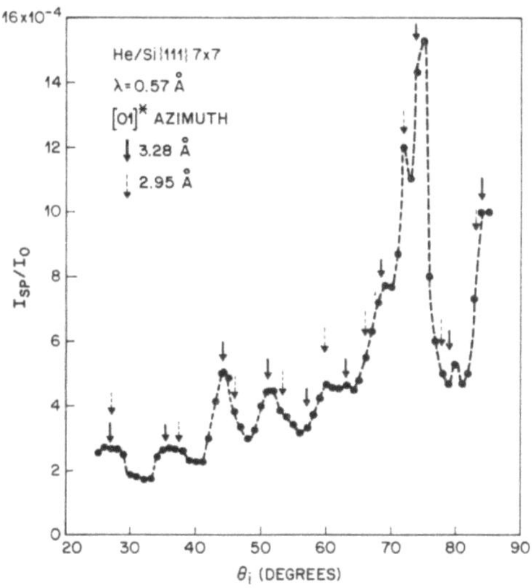

Fig.7.12. Specular intensity scan for E_i=0.63 eV He beam. The dark arrows show expected maxima for a vertical spacing of 3.2 Å. The dashed arrows correspond to a secondary distance of 2.9 Å. Note the difference of these distances is 0.3 Å

on Fig.7.12 by solid arrows. One can pursue this analysis and postulate a second vertical distance based on the shape of these peaks, i.e., at 44° the oscillation is sharp with a slight shoulder, neighboring peaks appear broader, and the peak at 60° appears split as if two similar distances are beating. This leads to a secondary distance of 2.9 Å which along with a Debye-Waller factor accounts reasonably well for the shape of the entire scan. There is some disagreement as to whether two distances with values this close can show independent oscillations, or if a single oscillation for the average value should be observed. This point is not relevant to the rest of the discussion concerning the model of the (7 × 7). However, it is worth pointing out that calculations of specular intensities, even in the rigorous close coupling form, do not fit the data for corrugated surfaces well. The calculations generally show a more "coherent" result for the scattering than the data indicate. This may be related to the fact that the ratio of the He surface collision time to surface vibrational periods is not sufficiently small. At this point it seems reasonable to simply in-

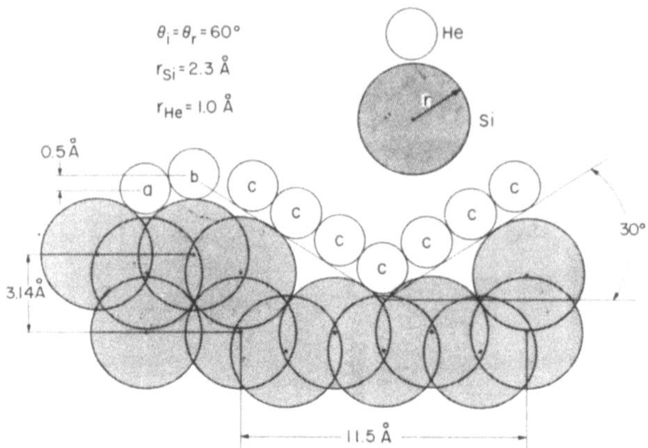

Fig.7.13. Specular trajectory illustration for $\theta_i = 60^\circ$ showing an estimate of the extent of the exposed third layer required for the He to specularly scatter out

dicate that the data suggest the occurrence of two distances and resolve this point in future studies. Both of these distances are close to, and their average equals, the double layer spacing of the Si{111} crystal. Since the He atom is a completely nonpenetrating probe, these distances are appropriate to the outermost exposed layer. The angular resolution is $\Delta\theta_r > 1^\circ$, so that the sensitivity to coherent structural effects is limited to distances < 100 Å on the surface. It is well known that the Si{111} 7×7 surface anneals to terraces $\sim 10^3$ Å in extent, even at the cost of making multilayer steps [7.19]. Therefore, these spacings must lie within the (7×7) unit mesh and are not due to random steps. Thus, it can be concluded that the third layer (second double layer) of the Si{111} surface is exposed to the He atom due to the (7×7) reconstruction.

A simple trajectory analysis at $\theta_i = 60^\circ$, in which the distance of closest approach to the Si atoms is taken to be 3.5 Å, is illustrated in Fig.7.13 to show that the extent of the exposed region is large—of the order of half the unit mesh dimension. Furthermore, the correspondence of the predicted Bragg maxima with the oscillations of Fig.7.12 is obtained without the inclusion of the attractive potential. Unless the potential well depth is particularly small ($D \lesssim 3$ meV), this implies that the areas of the interfering regions are large compared to the range of the attractive potential. Adding 2 Å to the repulsive distance of closest approach (3.5 Å) yields a radius of 5.5 Å or an area of ~ 90 Å^2 as the minimum exposed region of the third layer. This confirms the trajectory argument for large open areas.

7.4.3 A Model of the Si{111} (7 × 7)

In Fig.7.14 the implications of these results in terms of a structural model are illustrated. The Lander model [7.20] of the (7 × 7) is drawn. The ordered array of vacancies were postulated to result in a strong local relaxation (benzene type of electronic delocalization) which returns more energy than the cost of the vacancies. The analysis presented here, however, requires third layer exposure at least to the extent drawn in Fig.7.14b. This shallow etched region of exposed third layer is the minimum area required to satisfy the data. For this more extended region to form as an exothermic unit, a more complex argument for the surface relaxation is required. It would seem that a region of the size drawn in Fig.7.14b would generate a (5 × 5) reconstruction. Since the Si{111} periodicity is (7 × 7), the open area has been increased in the model in Fig.7.14c to occupy one-half the unit mesh. At this point the (7 × 7) units can be thought of as strong interacting.

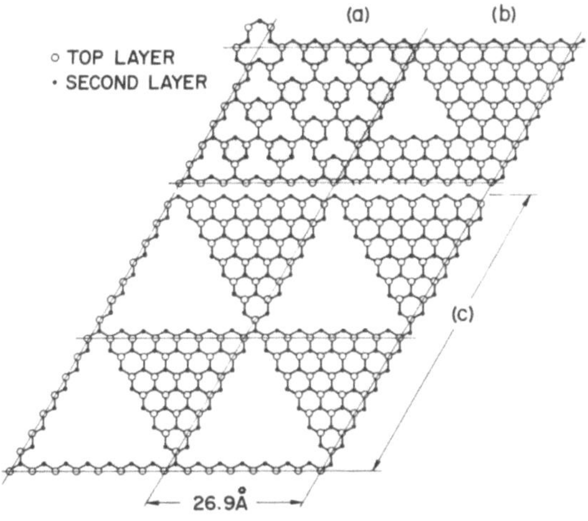

Fig.7.14. Models for the Si{111} (7 × 7) reconstruction: a) Lander vacancy model; b) minimum open area to satisfy the He specular data; c) proposed model based on the extent of the exposed third layer region

It is clear that the data of Fig.7.13 do not indicate any of the details of the model in Fig.7.14c – just the extent of the exposed third layer. However, a model of this form does generate interesting possibilities as to the nature of the driving forces for the reconstruction. Some arguments in this regard are presented here since this model is significantly different from previous proposals for the nature of the (7 × 7) reconstruction. (The author acknowledges RICE [7.21] for key ideas and clarifications in the development of these arguments. See [7.22] for two concurrent analyses of the Si{111} (7 × 7) hydrogenated surface with LEED and independent derivations of a nearly triangular island model with different orientation). One approach for estim-

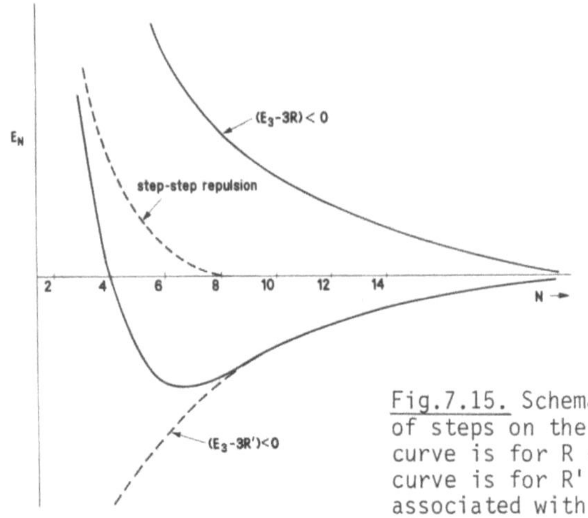

Fig.7.15. Schematic energy diagram for packing of steps on the Si{111} surface. The repulsive curve is for $R < E_3/3$ whereas the attractive curve is for $R' > E_3/3 \sim 0.8$ eV. The minimum is associated with the generalized step-step repulsion

ating the energetics of this surface is to consider it a maximum packing density of steps. The implication is that the formation of a step may actually be exothermic. The surface consists of 3-bonded atoms at the top, along the edge, and at the bottom of each $N \times N$ triangle. The number of each is $3 \times 2(N-1)$, $1/2(N-3)(N-2)$, and $1/2(N-2)(N-1)$ respectively. Labelling the energy to create a 3-bonded atom E_3, and allowing the edge atoms a relaxation energy R, the total energy of the surface unit can be written $E_T(N) = 2(N-1)(E_3-3R)$ and the energy per unit area $E_A = E_T(N)/N^2$. The issue of whether the step formation is exothermic depends on the term (E_3-3R). If $R > E_3/3$ then the model may be considered as the optimization of the closest packing of steps and the short-range step-step repulsion. Note that the E_3 should be close to 2.3 eV, half the cohesive energy of silicon. A schematic energy diagram (Fig.7.15) illustrates this point. It is simple to show that (E_3-3R) also determines the formation energy of a single linear step so that one would expect Si{111} cleaves to show spontaneous step development if it is negative. The cleaving experiments of HENZLER [7.23] do not show this development. Total energy calculations [7.24] also indicate that $E_3 > 3R$ for the three-bonded step, i.e., that the formation energy is somewhat positive.

There is however, a significant difference between an infinite linear step, with a positive energy of formation, and the array shown in the model in Fig.7.14, which should strongly effect the relaxation energy of the edge atoms. Each triangular unit has an effective perimeter which allows for relaxation of the inner top layer region. This top double layer can flatten in the electronic direction of an sp^2 bonding configuration to relieve strain. In relieving this strain it also pushes out and rolls over the edge atoms so as to have a strong parallel overlap of the adjacent half-filled orbitals. Note that a small additional term may arise from the long "one-dimensional" metallic chains which may have a square-root dispersion and lower the energy further. Thus for the postulated triangular units there are two additional energy terms compared to the linear step which can result in an exothermic reconstruction; strain relief and strong coupling of adjacent edge orbitals. The extent of atomic motion which can occur in order to relieve strain is limited and this may be a significant factor in restricting the size of the triangle. In the simplest terms, however, the near-island nature of the top double-layer units can increase R through strain relief and by significantly increasing the coupling of the edge orbitals. If $R \gtrsim 0.8$ eV then the attractive curve of Fig.7.15 is applicable.

7.4.4 Summary

In this chapter progress in the study of semiconductor surface structures using He diffraction has been reviewed. The diffraction for the Si{100} surface has revealed a lack of ordering in a weak secondary reconstruction perpendicular to the strong (2×1). The periodicity and qualitative corrugation analysis of both the diffraction scans and specular intensity scans are consistent with an asymmetric dimer model of the surface. In addition, Si{100} diffraction has led to some understanding of the analytical problems associated with strongly corrugated surfaces.

A calibration experiment for the theory of He diffraction from corrugated surfaces was undertaken using the GaAs{110} surface for which the structure is well established. This study has stimulated major theoretical advances in the computation of diffraction probabilities from rough surfaces, understanding the nature of the He surface potential, and has served as an important and successful test for the direct calculation of that potential from the very low density charge profiles involved.

Finally, experiments and some incite from results for the Si{111} 7×7 have been described. Although this surface has yet to receive detailed theoretical attention, a new model has been proposed based on a specular intensity scan [7.21]. This has led to interesting speculation about the nature of the dominant forces involved on the (7×7) surface and will stimulate much interest in future experiments and theory.

Acknowledgement. The author acknowledges the collaboration and helpful comments of G.E. Becker, E.K. Grimmelman, D.R. Hamann, R. Laughlin, D.R. Miller, T.M. Rice, S.S. Sibener, and J.C. Tully.

References

7.1 R.E. Schlier, H.E. Farnsworth: J. Chem. Phys. *30*, 4 (1959)
7.2 J.J. Lander, J. Morrison: J. Chem. Phys. *37*, 720 (1962)
7.3 T.D. Poppendieck, T.C. Ngoc, M.B. Webb: Surf. Sci. *75*, 287 (1978)
7.4 M.J. Cardillo, G.E. Becker: Phys. Rev. B*21*, 1497 (1980)
7.5 D.J. Chadi: Phys. Rev. Lett. *43*, 43 (1979)
7.6 J.C. Tully, E.K. Grimmelman: Unpublished results
7.7 V. Garibalid, A.C. Levi, R. Spadacini, G.E. Tommei: Surf. Sci. *48*, 649 (1978)
7.8 M.J. Cardillo, G.E. Becker, S.J. Sibener, D.R. Miller: Surf. Sci. *107*, 469 (1981)
7.9 C.B. Duke, et al.: J. Vac. Sci. Technol. *16*, 1252 (1979)
7.10 H. Hoinkes: Rev. Mod. Phys. *52*, 933 (1980)
7.11 R.B. Laughlin: Phys. Rev. B (in press)
7.12 R.G. Gordon, Y.S. Kim: J. Chem. Phys. *56*, 3122 (1972)
7.13 N. Esbjerg, J.K. Norskov: Phys. Rev. Lett. *45*, 807 (1980)
7.14 D.R. Hamann: Phys. Rev. Lett. *46*, 1227 (1981)
7.15 K.H. Rieder, T. Engel: Phys. Rev. Lett. *45*, 824 (1980)
7.16 R.E. Schlier, H.E. Farnsworth: J. Chem. Phys. *30*, 917 (1959); J.J. Lander, J. Morrison: J. Chem. Phys. *37*, 729 (1962)
7.17 M.J. Cardillo, G.E. Becker: Phys. Rev. Lett. *42*, 508 (1979)
7.18 M.J. Cardillo: Phys. Rev. BR*23*, 4279 (1981)
7.19 N. Osakabe, K. Yagi, G. Honjo: Jpn. J. Appl. Phys. *19*, L309 (1980)
7.20 J.J. Lander: In *Progress in Solid State Chemistry*, ed. by H. Reiss (Pergamon, Oxford 1965)
7.21 T.M. Rice: Personal communication
7.22 E.G. McRae, C.W. Caldwell: Phys. Rev. Lett. *46*, 1032 (1981)
7.23 M. Henzler: Surf. Sci. *36*, 109 (1973)
7.24 D.J. Chadi, J.R. Chelikowsky: To be published

8. Studies of Adsorption at Well-Ordered Electrode Surfaces Using Low-Energy Electron Diffraction

P.N. Ross, Jr.

With 19 Figures

8.1 Introduction

As in gas-solid heterogeneous catalysis, the adsorbed intermediate play a central role in the development of a quantiative theory for electrocatalysis. The geometric structure and the electronic and vibrational states of the adsorbate must be known in order to derive quantum-mechanical descriptions of electron transfer occurring during adsorption of an electroactive species from solution, as in the electrodeposition of metals. There are electrochemical reactions where the rate-determining step is not an electron transfer step, but a purely chemical step, such as chemisorption. This is the case, for example, in the classic hydrogen electrode reaction, $H_2 \rightarrow 2H^+ + 2e^-$, where for the Pt-group metals the rate is determined by the rate of dissociative chemisorption of hydrogen. In reactions of this type, one is seeking virtually identical information about the adsorbed state as in heterogeneous catalysis, with the additional complication of the solution phase. Thus, studies of adsorbed states have the same important relationship to physical understanding of reaction paths and kinetics in both gas-solid and electrochemical catalysis.

The use of in situ spectroscopic techniques to probe adsorbed intermediates is at the forefront of gas-solid catalytic research. Analogously, there are in situ methods which have been used with electrochemical systems to probe the geometric and electronic structure of adsorbates on electrode surfaces. Examples of these techniques are reflectance spectroscopy, ellipsometry, and surface-enhanced Raman spectroscopy (SERS). SERS seem to have the most promise of any of the in situ techniques and was a lecture topic at the 1979 ISISS [8.1]. The in situ techniques suffer from a variety of shortcomings. All lack a general applicability and none provide, even in principle, a definitive determination of adsorbate structure. A number of research groups that study the electrode-electrolyte interface have developed ex situ methods for this study, and it is these that we shall concern ourselves with here. X-ray photoelectron spectroscopy (XPS) has been used extensively for ex situ analysis of electrode surfaces [8.2] but *not* for the study of adsorbed or submonolayer electrodeposited species. In this review, we shall describe in some detail studies of submonolayer structures on electrode surfaces using the combination of classical electrochemical methods with low-energy electron diffraction (LEED). It is shown that by use of this combination of techniques the enthalpy and entropy of adsorption onto well-ordered surfaces can be determined directly. The effect of imperfections in the surface on adsorption enthalpy can be readily determined using stepped surfaces (i.e., ordered periodic imperfections). The variation of enthalpy and entropy with coverage on structurally homogeneous surfaces can provide insight into substrate-adsorbate and adsorbate-adsorbate interactions.

8.2 Thermodynamics of Electrodeposition

The deposition of a ionized species from solution onto a metal electrode surface may, without loss of generality, be considered to proceed via an intermediate state, usually termed the "underpotential" state. We shall consider only underpotential deposition reactions which may be written as simple charge transfer steps,

$$A^{Z+} + Ze^- \rightarrow A^* \tag{8.1}$$

where A^* is the adsorbed (adatom) state on the electrode surface, and A^{Z+} is the solvated cation species. Implicit in (8.1) is the fact that there is a change in the solvation shell around A. The change in total entropy for the adsorption reaction will reflect this change in solvation and the magnitude of the $T\Delta S$ term to the total change in free energy is not insignificant. As the adatom concentration increases, association occurs leading to the formation of the bulk phase, as

$$A^* + A^* \rightarrow A_b \quad . \tag{8.2}$$

The energetics of this association reaction may be used to define the thermodynamics of the underpotential state. We define the enthalpy change in (8.2) by ΔH^*,

$$\Delta H^* = -(2D_{AM} - D_{AA}) \tag{8.3}$$

where D_{AM} is the dissociation energy of A^* from the metal substrate and D_{AA} is the dissociation energy of A from A. For an energetically homogeneous surface, and assuming no interactions between adatoms, we can define an isotherm for A^* based on the magnitude of $\Delta G^* \equiv \Delta H^* - T\Delta S^*$,

$$\frac{\theta}{1-\theta} = \alpha_A^{\frac{1}{2}} \exp(-\Delta G^*/2RT) \tag{8.4}$$

where α_A is the thermodynamic activity of the bulk phase A. In an electrochemical cell at equilibrium, the thermodynamic activity of the bulk phase is related to the electrode potential via the Nernst relation

$$E = E_0 - \frac{RT}{2ZF} \ln(\alpha_A/\alpha_A Z+) \tag{8.5}$$

where E is the potential of the electrode (at which deposition of A is occurring) relative to any reference electrode and E_0 is the potential of the electrode when A is at unity activity (relative to the same reference electrode). It is convenient to define the electrode potentials only in terms of $U \equiv E - E_0$, i.e., against an A/A^{Z+} reference electrode in the same solution, so that

$$U = -\frac{RT}{2ZF} \ln(\alpha_A) \tag{8.6}$$

or

$$\alpha_A^{\frac{1}{2}} = \exp(-ZFU/RT) \quad .$$

Finally we define the ideal relation between the coverage of the substrate surface by adatoms and the electrochemical potential in terms of the standard free energy of adsorption,

$$\frac{\theta}{1-\theta} = \exp(-\Delta G_0^*/2RT) \exp(-ZFU/RT) \quad . \tag{8.7}$$

The standard free energy of absorption, ΔG_0^*, is consequently $2ZFU_{\frac{1}{2}}$, where $U_{\frac{1}{2}}$ is the potential when $\theta = 1/2$.

Isotherms recorded at variable temperature may be used to calculate isosteric heats of adsorption (W) by use of the relation

$$\left[\frac{d \ln}{dT} (\alpha_A)\right]_\theta = \frac{W}{RT^2} \quad . \tag{8.8}$$

The entropy of adsorption can be obtained as a function of coverage from the relation between $RT \ln(\alpha_A)$ and W, i.e.,

$$\Delta S^* = -(\frac{W}{T}) - R \ln(\alpha_A) \equiv -(\frac{W}{T}) + \frac{2ZFU}{T} \quad . \tag{8.9}$$

The great virtue of electrochemistry is that the relation between coverage and chemical potential can be measured directly and with great accuracy

(10^{-3} ML). Experimentally, it is more convenient to measure the derivative of θ directly, then integrate to obtain θ. This is done by the method of linear sweep voltammetry. When the potential of an electrode is changed, current flows in the external circuit which is the sum of two contributions, one representing the (dis) charging of the electrical double layer, another the (dis) charging of the adsorbate layer, e.g.,

$$i = dq/dt \tag{8.10}$$

$$= \left(\frac{dq^{ads}}{dU}\right)\left(\frac{dU}{dt}\right) + C_{d\ell}\left(\frac{dU}{dt}\right) \quad . \tag{8.11}$$

Then

$$aQ\left(\frac{d\theta}{dU}\right) = (i - aC_{d\ell}) \quad ,$$

$$\left(\frac{d\theta}{dU}\right) = (i/aQ - C_{d\ell}/Q) \tag{8.12}$$

where $a \equiv$ sweep rate and Q is the charge required to form a monolayer of adsorbed species. The double-layer capacitance $C_{d\ell}$ of the high work function metals that one normally uses as electrode substrates (e.g., Pt, Au, Hg) are typically 10 - 40 $\mu F/cm^2$, depending on the electrolyte. Since $Q \equiv 200 - 300 \mu C/cm^2$, the double-layer (dis) charging contribution to (8.12) is therefore typically 0.05 - 0.1 V^{-1} independent of sweep rate. The relationship between the experiment, which measures i versus U during a linear sweep, and the derived isotherm is shown graphically in Fig.8.1. Isotherms can be obtained from linear sweep voltammetry in the manner described only if the system is perturbed under reversible conditions, i.e., the coverage at a given potential is independent of the sweep rate. The criterion of reversibility is easily verified experimentally, and in some cases it cannot be met at any (practical) sweep rate, as is the case for the underpotential deposition of oxygen on Pt-group metals. The shape of the i-U curve contains information about the nature of the adsorbate interactions. This is illustrated in Fig.8.1b using the simplest model for interactions, one in which the isoteric heat of adsorption changes linearly with coverage. For repulsive interactions ($g > 0$), the peak maximum moves to lower potential, and the peak shape is much broader than ideal. For attractive interactions ($g < 0$), the peak maximum moves to higher potential, and the peak shape is narrower than ideal. More complex and physically more realistic models of adsorbate interaction, such as dipole-dipole or image dipole interactions, have been calculated by CONWAY and co-workers [8.3] and show qualitatively the same behavior as in Fig.8.1b but with asymmetric peak

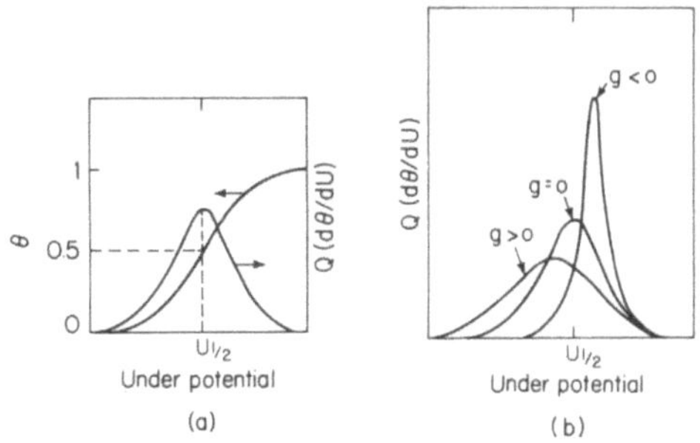

Fig.8.1. (a) The shape of the current-voltage curve ($i \equiv Qd\theta/dU$) for an adsorption process and the corresponding isotherm; (b) the change in shape of the current-voltage curve when the adsorption is not ideal ($g = 0$); $g > 0$ repulsive interaction; $g < 0$ attractive interaction

shapes. Therefore, i-U curves from structurally homogeneous surfaces (single crystals) provide at least a qualitative indication of the nature of the nonideality of the absorption reaction and the relative importance of adsorbate interactions. Further detail on these interactions can best be derived by analysis of the variation in the isoteric heat of adsorption and differential entropy of adsorption with coverage.

Some confusion frequently occurs when thermodynamic quantities derived from electrochemical measurements are compared to what may be the comparable thermodynamic quantity derived in a nonelectrochemical (i.e., no electrolyte) measurement. As defined here, the thermodynamic functions ΔG^*, ΔH^*, and ΔS^* all refer to the same reference state, *the bulk phase at unit activity*. What we measure electrochemically is the change in state variable between that reference state and the underpotential state (submonolayer adatoms on a substrate in solution). The nonelectrochemical experiment, e.g., vacuum deposition, has the same reference state and has a similar, but not identical, final state. The difference in thermodynamic quantities measured by the two methods reflects a difference in the chemical state of the ad-atoms in the two environments and not a difference in the manner of measurement per se. We shall be discussing, in a later section, detailed comparisons of thermodynamic functions determined with and without a solution phase.

8.3 Experimental Methods

The basic principle of ex situ analytical methods is to couple electrochemical cells to UHV chambers where the plethora of techniques for the study of the vacuum interface can be used to examine the electrode surface. The directness of this coupling varies from laboratory to laboratory, from very crude (pull the electrode from the cell, carry it two buildings away, and

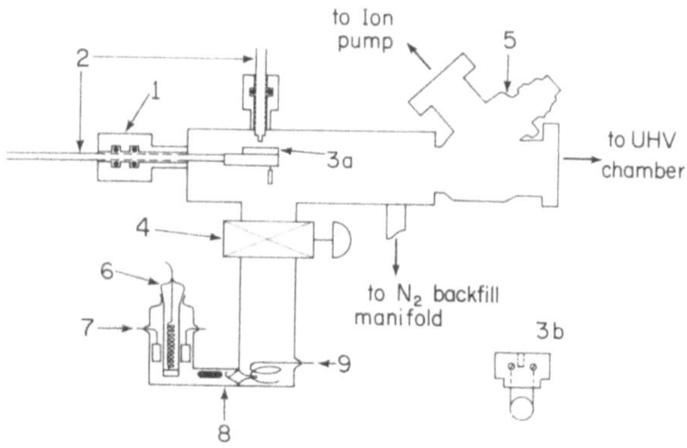

Fig.8.2. Schematic of the electrochemical-surface analysis apparatus: (*1*) O-ring feed-thrus; (*2*) transfer rods; (*3a*) and (*3b*) transfer block with crystal suspended from wires; (*4*) gate valve; (*5*) straight-thru valve; (*6*) reference electrode; (*7*) counter electrodes; (*8*) glass-break and Luggin capillary [8.5]

insert into a UHV system) to very sophisticated (direct transfer with differential pumping). Depending on the particular system under study and the type of analysis being done ex situ, crudely coupled systems suffice, e.g., RHEED analysis of metal electrodeposits [8.4]. Glove-box transfer appears to be adequate for the study of the "near-surface" region, e.g., RHEED- and XPS-type analysis. However, for definitive work on submonolayer species, the highest standards of UHV practice must be applied to the transfer of the electrode surface to and from the analytical chamber.

The experimental apparatus as used in our laboratory for the study of the underpotential states of hydrogen and oxygen on Pt is shown in Fig.8.2. The single crystals were spot welded to gold wires and suspended from a stainless steel block. The block was transferred from the manipulator in the LEED-AES bell jar to the electrochemical cell by means of two polished stainless steeel rods; vacuum sealing of the rods was accomplished using Viton O rings radially compressed. The working electrode compartment and the high-vacuum parts of the system were isolated from the electrolyte reservoir and reference electrode via a glass break. The transfer system and working electrode compartment were differentially pumped to about 1.3×10^{-6} Pa(1×10^{-8} torr) after a bakeout at 423 K. In the typical experiment, a new single crystal was mounted on the transfer block and attached to the vertical motion transfer rod with the electrochemical cell detached. The transfer block was transferred to the horizontal motion rod, the electrochemical cell attached to the system, and the entire transfer system pumped out to 2×10^{-6} Pa. The single crystal was then transferred to the ultrahigh-vacuum system, where it was cleaned by simultaneous argon ion bombardment and electron beam heating to 973 K. The ultrahigh-vacuum system was fitted with a rotary-motion feed-thru with a threaded tip with which the transfer block could be picked up and rotated, permitting ion bombardment and LEED/AES analysis to be carried out on both sides of the crystal. LEED analysis was accomplished using conventional 4-grid optics (Varian) and Auger analysis used a single pass CMA with 3 keV integral electron gun (Varian). The Auger spectrometer system also had sample imaging capability so that the entire surface of the crystal was analyzed for impurities. The structure of the surface was then determined using LEED. After structure analysis was performed, the crystal was transferred into the working electrode compartment and the electrochemical system was isolated from the LEED-AES system by closure of the straight-thru valve. The working electrode compartment was then backfilled with Research Grade argon (Airco, 5 N purity) to near ambient pressure, and the electrolyte introduced by breaking the glass seal.

Since the cell configuration employed here required complete submersion of the crystal, regions of the crystal that are not well ordered are contacted by electrolyte and thus contribute to the voltammogram, e.g., the edge. Typically the geometric area of edges on these crystals was 20% of the area of the polished faces, and the effective area even greater because of roughness. The contribution of the crystal edges to the voltammogram was measured directly by a gold masking technique. After obtaining the voltammogram for the complete crystal, gold was evaporated onto the polished faces so that just the edges and near-edge region of the Pt crystal were exposed to electrolyte. The voltammogram for these disordered regions was obtained following the same procedure as used for the crystal without the gold mask. To facilitate data manipulation the voltammograms were digitized and recorded in a multichannel analyzer. At least two other electrochemical research groups have employed systems of the type shown in Fig.8.2 to study underpotential states on single-crystal electrode surfaces [8.6,7]. However, the apparatus at our laboratory is the only system of its type in which variable temperature in the electrochemical cell is easily accomplished. The problem with this system is that the volume of electrolyte is quite large and it is difficult to purify the solutions to the extent necessary. The thin-layer cell employed by the YEAGER group [8.7] appears to solve this electrolyte purity problem by increasing the surface-to-volume ratio by three orders of magnitude.

8.4 Underpotential States of Hydrogen on Pt

If two Pt foils are suspended in water (containing sufficient salt to provide conductivity) and maintained at a potential difference > 1.23 V, hydrogen will be evolved from the negative electrode and oxygen from the positive. If we measure the current passed during a linear sweep of the voltage (relative to a third reference electrode) of either electrode, we find that current is passed at potentials well below the thermodynamic potential difference. These currents correspond to the adsorption of water dissociation products that are intermediates in the formation of hydrogen and oxygen. Formally, these adsorbed species are referred to as the underpotential states of hydrogen and oxygen. Figure 8.3 shows linear sweep voltammetry curves for a clean, annealed, polycrystalline Pt foil in dilute sulfuric acid electrolyte obtained in the apparatus of Fig.8.2. The reference electrode was a reversible hydrogen electrode in the same electrolyte with hydrogen at unit activity (101.3 k Pa pressure). Therefore, the electrode potential scale of Fig.8.3 is exactly U of (8.6), and sweeping the potential corresponds to sweeping the equivalent partial pressure of hydrogen, given by the relation

$$P_{H_2}^{1/2} = \exp(-FU/RT) \quad . \tag{8.13}$$

The first sweep was cathodic, starting at 0.4 V, and the cathodic current measured between 0.4 and 0 V corresponds to the adsorption (underpotential deposition) of hydrogen adatoms from solvated protons in solution,

$$[H_9O_4]^+ + M(e^-) \rightarrow MH + 4H_2O \quad . \tag{8.14}$$

Because of the excellent kinetics of the recombination/dissociation reaction on Pt, the equilibrium concentration of molecular hydrogen is established near the electrode surface (the diffusion boundary layer),

$$2\,MH \rightleftharpoons H_2 \quad . \tag{8.15}$$

The mirror-image anodic currents observed on reversal of the linear sweep corresponds to the reaction (8.14,15) driven in the opposite direction. The sharply increasing cathodic current below 0.05 V corresponds to the evolution of sufficient molecular hydrogen to achieve equilibrium partial pressures that approach 101.3 k Pa.

The anodic current measured at 0.8 - 1.3 V in Fig.8.3 corresponds to the adsorption of intermediates in the formation of molecular oxygen from water, i.e., the underpotential states of oxygen. We shall discuss the underpotential states of oxygen separately in a later section.

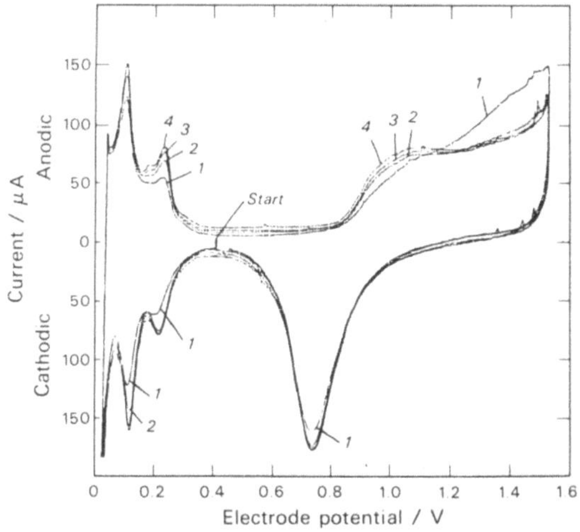

Fig.8.3. Cyclic voltammetry of a clean, annealed polycrystalline Pt foil in 0.5-M H_2SO_4 following transfer from the LEED/Auger chamber 0.50 mV/s [8.5]

There are, clearly, multiple states of hydrogen adsorbed on Pt electrode surfaces, and the challenge to electrochemists in recent years has been to produce a definitive explanation of the multiplicity of these states. The use of ex situ methods of surface analysis and systems like that in Fig.8.2 have proven to be extremely useful. LEED has been used to provide definitive determination of the surface structure of Pt electrodes before *and* after contact with electrolyte; Auger electron spectroscopy used to monitor surface contamination and/or oxidation; UHV practice used to assure no contamination of the surface occurred before *or* after contact with electrolyte. It has now been clearly demonstrated [8.5,8-10] that the multiplicity of states represents adsorption of hydrogen at sites of different geometric configuration, i.e., local ordered domains of different geometry. We shall review here some selected results from these studies.

8.4.1 Isotherms for Hydrogen on {111} and {100} Pt

Voltammetry curves for hydrogen adsorption-disorption on {111} and {100} Pt single-crystal electrodes are shown in Figs.8.4,5. In both cases, posttest LEED analysis indicated the surfaces were the (1×1) structures during the experiment. In the case of the {100} crystal, the results in Fig.8.5 were obtained with an oxygen-treated crystal in which the oxygen was left on the surface, so that the starting surface was (1×1)-O. YEAGER's group [8.11] has reported that a different result is obtained if one uses the clean (5×20) reconstructed surface of the {100} crystal, and that transformation to the (1×1) surface occurs commensurate with hydrogen chemisorption. When

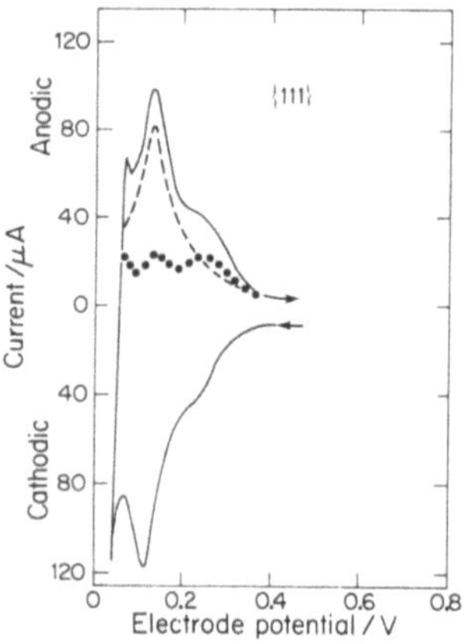

Fig.8.4. Voltammetry curve for hydrogen adsorption-desorption on Pt{111} single crystal (——); specific contribution due to crystal edge (···); curve for adsorption on the well-ordered {111} face after correction for the edge effect. 0.05-M $HClO_4$ 0.50 mV/s [8.5]

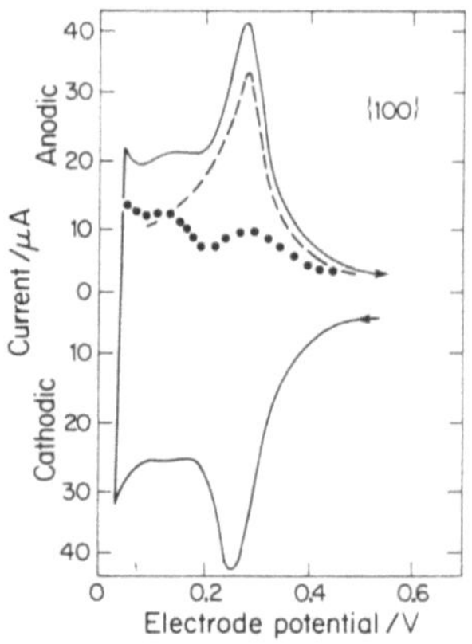

Fig.8.5. Voltammetry curves for hydrogen adsorption-desorption on Pt{100} single crystal (——); specific contribution due to crystal edge (···); curve for adsorption on the well-ordered {100} face after correction for the edge effect. 0.05-M $HClO_4$ 0.50 mV/s [8.5]

the curves in Figs.8.4,5 were corrected for edge effects using the gold masking technique described, the result was a single asymmetric peak for each surface. Isotherms for those single states, derived by integration of the corrected i-U curve, are shown in Fig.8.6. The isotherms are clearly non-Langmuirian. The isosteric heat of adsorption on both surfaces falls monotonically with coverage (Fig.8.7). It seems unlikely that progressive filling of sites of different energy would produce a monotonic function for the isosteric heat of adsorption. It seems more likely that the surface is energetically homogeneous and that adsorbate-induced interactions caused the monotonic decrease in the heat of adsorption. The classical models of ROBERTS [8.12] for dissociative adsorption on identical sites with nearest-neighbor interaction can be applied directly to compute the differential (isoteric) heat of adsorption. For a surface of uniform sites, the adsorption entropy for a diatomic molecule varies slowly with coverage for $\theta < 0.8$ [8.13], and for our purposes here we assume ΔS^* is essentially constant. Then

$$\Delta G^*(\theta) = -W(\theta) - T\Delta S_i^* \tag{8.16}$$

where $W(\theta) \equiv -\Delta H^*$ is given by one of the models of adsorbate-adsorbate interaction. If the hydrogen adatoms are mobile at the temperature of adsorption, then [8.12]

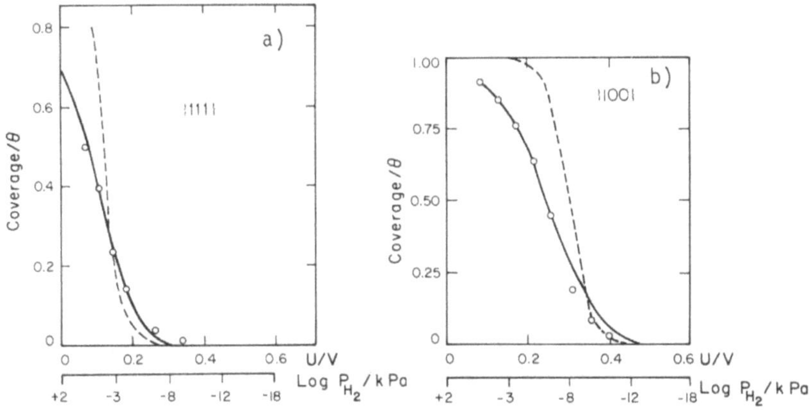

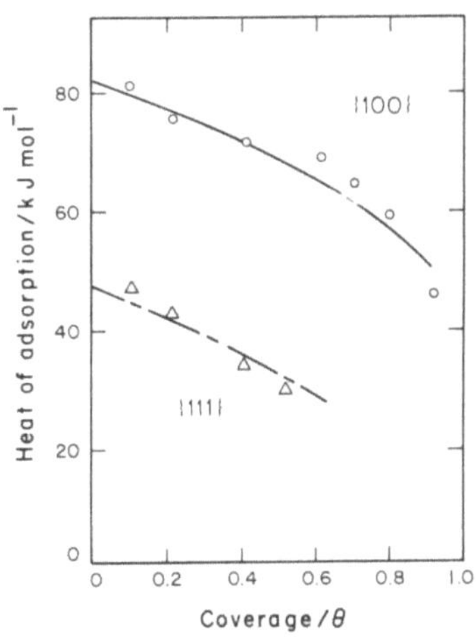

▲ Fig.8.6a,b. Isotherm for hydrogen on the well-ordered (a) {111} and (b) {100} Pt surface; measured (o); isotherms calculated from Langmuir model (---) and Robert's model for immobile adsorption with repulsive interaction (——). 300 K

◄ Fig.8.7. Isosteric heats of adsorption of hydrogen on {111} and {100} surfaces of Pt in dilute $HClO_4$ solution. Curves shown were calculated from Robert's model for immobile adsorption with repulsive interaction [8.5]

$$W(\theta) = W_0 - n\gamma\left(1 - \frac{1 - 2}{[1 - 4(1 - \rho)\theta(1 - \theta)]^{\frac{1}{2}}}\right) \qquad (8.17)$$

where ρ is the Boltzmann factor $\exp(-\gamma/RT)$, W_0 is the initial (zero coverage) heat of adsorption, and γ is the pair-wise interaction energy. For repulsive interaction, $\gamma > 0$ and the heat of adsorption decreases with coverage; for $|\gamma| < RT$, Taylor series of expansion of (8.7) yields a linear change in the heat of adsorption $W(\theta) = W_0 - n\gamma\theta$. Adsorption isotherms with a linear variation in adsorption free energy have been used extensively for the modeling of underpotential deposition (e.g., [8.14]) and it is well known that the linear function does not produce an asymmetric voltammetry curve. However, if the hydrogen adatoms are immobile, the resulting isotherm produces an asymmetric curve and a heat of adsorption function which fits the observed variation extremely well. For immobile adsorption [8.12],

$$W(\theta) - W_0 - n\gamma\left[\frac{(n-1)^2}{n}\,\theta\,\frac{(2n-\theta)}{(n-\theta)}\right] \quad . \tag{8.18}$$

The best values of W_0, the heat of adsorption at zero coverage, and γ were derived by fitting (8.7,16,18) to the isotherms for the {111}, n = 6, and the {100}, n = 4, surfaces simultaneously, then checking the resulting $W(\theta)$ function against the experimentally observed variation. The best-fit values of ΔH_i^*, ΔS_i^*, and γ are given in Table 8.1; the calculated isotherms and the calculated $W(\theta)$ functions appear as the solid curves in Figs.8.6,7, respectivley. While the theoretical isotherms are not a perfect fit to the experimental curves, the asymmetry of the voltammetry curve for the {100} surface is predicted as well as the general form of the functional relation between isosteric heat of adsorption and coverage.

Table 8.1. Thermodynamic functions[a] for the adsorption of hydrogen on Pt single-crystal surfaces

Surface	ΔH_i^* [kJ/mol]	ΔS_i^* [J/mol K]	γ [kJ/mol pair]
{111}	-46.8	-55.8	5.0
{100}	-81.1	-53.1	5.0

[a]Zero coverage at 300 K

In principle, it should be possible to calculate the ΔS^* term for mobile and immobile dissociative adsorption and determine whether the values in Table 8.1 are consistent with the assumption of an immobile state. However, in the case of adsorption at surfaces in solution, the problem of solvent orientation makes this calculation less than straightforward. Figure 8.8 shows the classical electrochemical view (the BDM theory) of the water-hydronium ion double layer [8.15] and a hypothetical view of the manner in which hydrogen adatoms are chemisorbed. The BDM double-layer model was developed specifically for the mercury electrode, but should apply to any metal electrode where there is no specific (chemical) interaction between water and the metal surface. This model should, therefore, apply to a Pt electrode at potentials positive to that for hydrogen adsorption and negative to that for OH formation. The layer of water molecules, strongly oriented with the positive end of the water dipole towards the metal surface, must be displaced by the hydrogen adatoms. There has been no model developed to

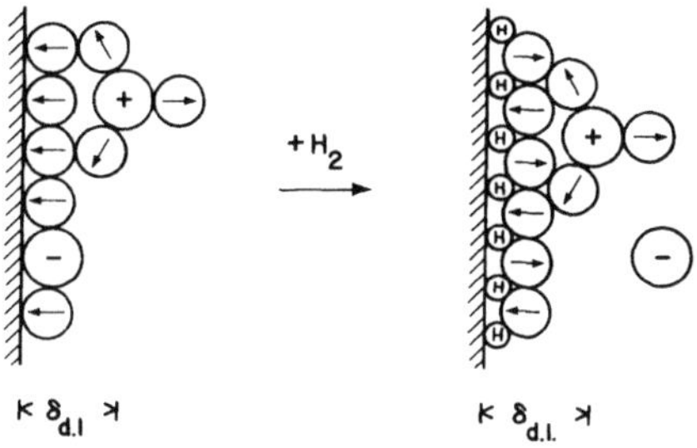

Fig.8.8. (a) Classical electrochemical view of the water-hydronium ion double layer with contact adsorption of the anion; dipole arrow points to hydrogen end of H_2O molecule; (b) hypothetical displacement of H_2O from the innermost layer upon chemisorption

predict the relaxation of the double layer following hydrogen chemisorption, so there is no firm physical basis for calculating the total change in entropy. Until a great deal more is known about the structure of the double layer when adsorbed hydrogen is present on the surface, it is not possible to calculate the "expected" values for the entropy of adsorption.

8.4.2 Hydrogen at Stepped Surfaces

As SOMORJAI and co-workers have shown [8.16], certain high Miller index faces of fcc metals form terraced surface structures with monoatomic steps. Cutting a {111} Pt rod at a progressively increasing angle along the zone line between the <111> and <110> pole produces a series of {111} terrace structures, denoted [n{111} × {111}] in the notation of LANG and SOMORJAI [8.17]. The clean, annealed surface of {110} Pt in vacuo has a (2 × 1) reconstructed surface. DUCROS and MERRILL [8.18] proposed a "sawtooth" structure consisting of tilted {111} microfacets for the {110} - (2 × 1) surface. Ball models of the sawtooth structure (Fig.8.9) indicate the "trough" in this structure has the same site coordination as along the steps on [n{111} × {111}] surfaces, and can really be regarded as a vicinal of the [n{111} × {111}] group. It would be expected, therefore, that comparison of underpotential hydrogen on {111}, [7{111} × {111}], [4{111} × {111}], and {110} would reveal the effect of adsorption at the fourfold sites (designated C^4_{110}) that exist at the steps in the [n{111} × {111}] surfaces and in the "trough" of the {110} - (2 × 1) surface. The ratio of fourfold step sites (C^4_{110}) to threefold hollow sites (C^3_{111}) on the terraces varied in this series from 0, to 1:6, to 1:3, and to 1:1, respectively. The volammetry curves shown in Fig.8.10 clearly delineate a second peak (at 0.225 V) whose peak area ratio to the primary peak was about th 1:3 and 1:1 expected if the secondary peak does represent adsorption at C^4_{110} sites and the primary peak adsorption at C^3_{111} sites.

Just as crystals produced by cutting a {111} rod at sequentially increasing angles toward the <110> pole revealed adsorption at sites having the {110} surface geometry, cutting a {111} rod at progressively increasing angles towards the <110> pole revealed adsorption at sites characteristic of {100} geometries. Figure 8.11 shows a series of voltammetry curves for a [n{111} × {100}] family of surface sturctures produced by cutting a {111} rod at selected angles towards the <100> pole. If we assume that adsorption at the steps occurs at the fourfold sites of {100} geometry (C^4_{100}) and at the terraces in the threefold sites of {111} geometry (C^3_{111}), then the ratio of step sites to terrace sites on the crystals {111} would be 0, 1:6, 1:3, and 1:2, respectively. States associated with the steps are clearly delineated in the voltammograms of Fig.8.11, and the total charge associated with the steps was about that expected if only half the C^3_{111} sites are filled, as was the case for the ordered {111} crystal. However, unlike the {111} steps, two states of hydrogen are associated with adsorption at {100} steps, one state being more strongly bound at the {100} step than at the macroscopic {100} surface. It is not clear at this time what these two states correspond to, i.e., what site geometries are associated with each state.

Voltammetry curves taken at variable temperature have indicated that the heat of adsorption for the hydrogen adsorbed at steps is larger than for hydrogen adsorbed at *either* low-index face {111} or {100}. This observation is particularly interesting for the C^4_{110} site as the standard free energy of adsorption at these sites is intermediate between that for the low-index faces, meaning that the entropy of adsorption of C^4_{110} sites is significantly more negative than the value reported in Table 8.1. Qualitatively, more negative adsorption entropy implies more restricted motion of the adsorbate, i.e., less mobility. Again, however, the problem of coupled reorganization of the double layer makes rigorous association of the total entropy change to the adsorbate alone impossible. Nonetheless, it can be said that the sum of

→

Fig.8.9a,b. Ball models of (a) the {100} - (1 × 1) unreconstructed surface (*right*) and the reconstructed {110} - (2 × 1) surface (*left*); (b) the [4{111} × {111}] stepped surface. Small clear balls represent the fourfold adsorption sites denoted C^4_{110} in the text

Fig.8.10. Voltammetry curves for hydrogen adsorption-desorption on [n{111} × {111}] stepped surfaces. 0.05 M $HClO_4$ 0.50 mV/s

Fig.8.11. Voltammetry curves for hydrogen adsorption-desorption on [n{111} × {100}] stepped surfaces. 0.05 M $HClO_4$ 0.50 mV/s [8.5]

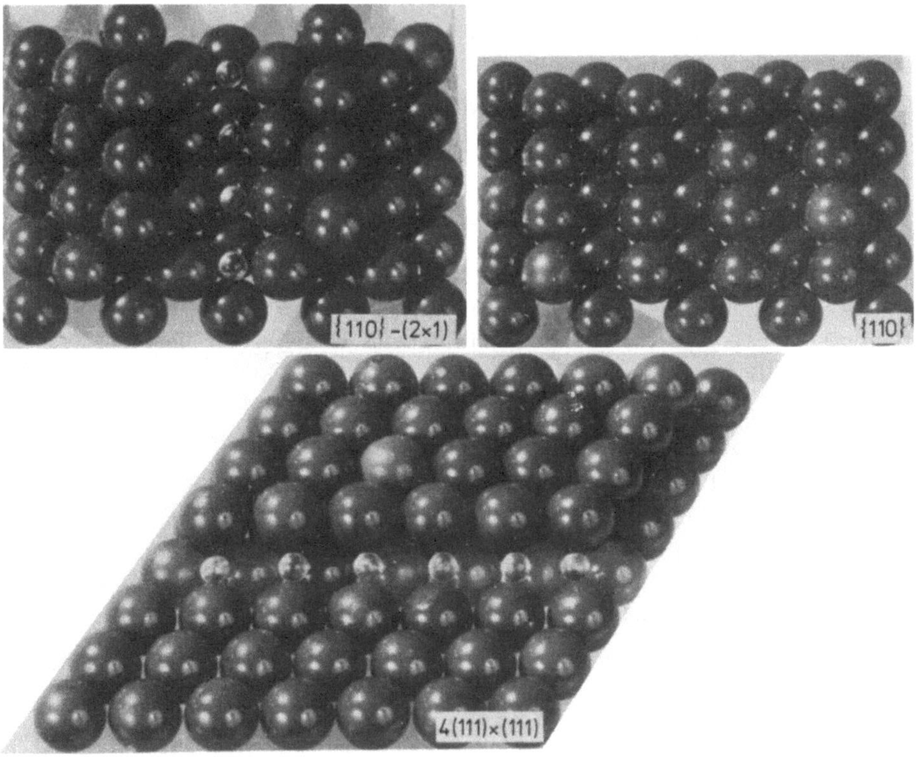

Fig.8.9

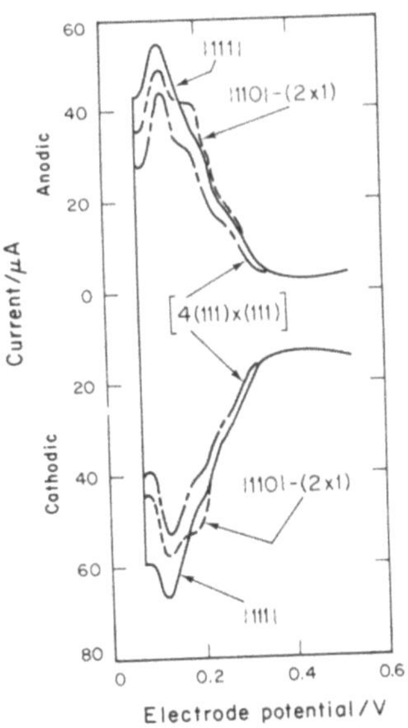

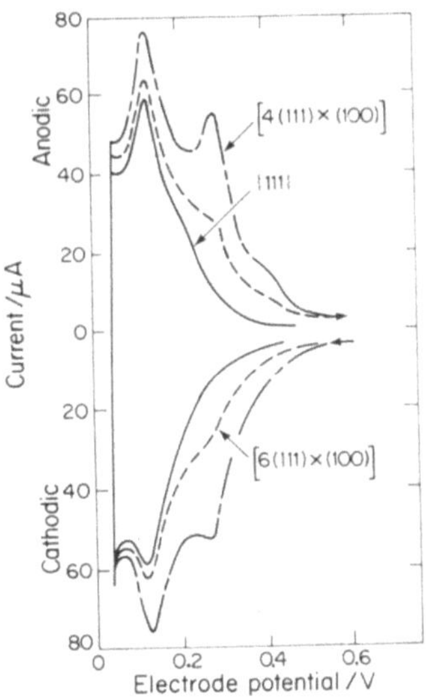

Fig.8.11 ►

◄ Fig.8.10

the adsorbate contribution and the double-layer contribution to the total entropy change indicates that the final state (hydrogen adatoms plus water dipole layer) is more ordered at monoatomic steps than at low-index surfaces.

8.5 Underpotential States of Oxygen on Pt

In the linear sweep voltammetry curve of Fig.8.3, the anodic currents observed in the potential region >0.7 V correspond to what may be termed thermodynamically as the underpotential states of oxygen. Figure 8.12 shows this region in greater detail, with the anodic limit of the potential scan increasing. The criterion for reversibility of a surface faradaic process is that the anodic and cathodic waveforms not be displaced from each other in potential [8.19]. Reversibility is required if we are to extract the thermodynamic functions via the underpotential. For underpotential oxygen on Pt, the potential region of reversibility is quite restricted, as the cathodic peaks in Fig.8.12 start to lag the anodic peaks at potentials above 0.82 V. Above 0.82 V, the anodic surface species becomes progressively more irreversible. There have been a very substantial number of studies conducted over the years to identify the molecular species corresponding to the "reversible" and "irreversible" states, but definitive identification has been elusive. The present interpretation has changed little since the 1973 paper of CONWAY and co-workers [8.20]. We shall use their interpretation of this discussion, and further restrict the discussion to the behavior of the surface in acid solution only. The initial portion of the anodic wave corresponds to dissociation of H_2O to form chemisorbed OH,

$$H_2O + M \rightarrow MOH + H_9O_4^+ + 1e^- \quad . \qquad (8.19)$$

The reversible OH species is the only species present up to ca. 25% of a monlayer (1 ML is 210 $\mu C/cm^2$ on a polycrystalline electrode). As the coverage is increased beyond 0.25 ML, the surface is reconstructed via the place exchange mechanism shown in Fig.8.13. LEED studies in our laboratory tend to support this reconstruction hypothesis [8.20]. Diffraction patterns were observed after cycling a Pt {111} surface to varying anodic potentials. As shown in Fig.8.14, the angular width of the (01) spots in the LEED increased when the anodic potential was increased above 0.8 V, which is just the potential where reconstruction starts to occur. The reconstruction process may also be viewed as an "oxide" underpotential process. Pt atoms which have undergone place exchange have largely OH nearest neighbors, as is the case for Pt

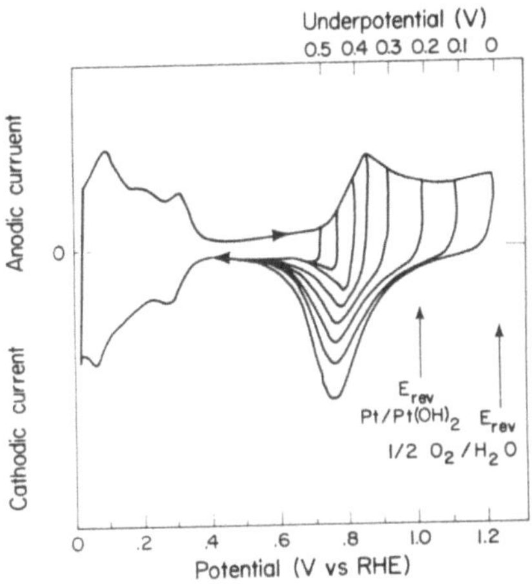

Fig.8.12. Cyclic voltammetry of polycrystalline Pt in 0.1 M HF solution; underpotential states of oxygen are observed between 0.8 - 1.23 V

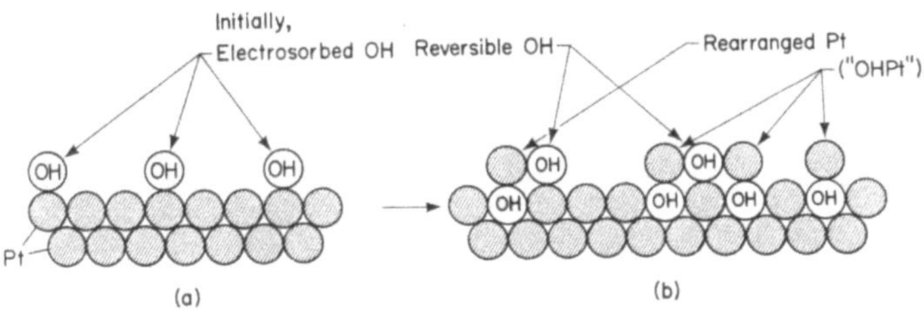

▲ Fig.8.13a,b. Place exchange mechanism for reconstruction of the Pt surface during anodic film formation [8.20]

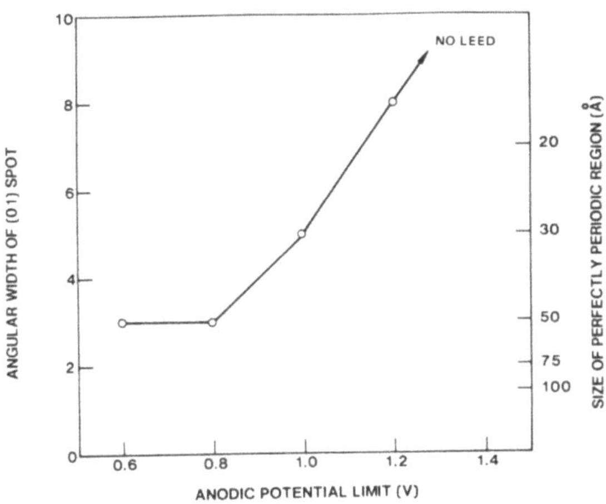

Fig.8.14. Loss of order in the {111} Pt surface during anodic film formation as observed by LEED

atoms in the hydrated oxide, $Pt(OH)_2$. The standard potential for formation of the bulk hydrated oxide $Pt(OH)_2$, via

$$Pt + 2H_2O \rightarrow Pt(OH)_2 + 2\,H_9O_4^+ + 2e^- \quad , \tag{8.20}$$

is 0.98 V [8.21]. Figure 8.12 shows clearly that the irreversible OH species forms at a potential "under" the standard potential for the hydrated oxide.

The irreversibility of the 0.5 O_2/H_2O reaction means we cannot define a thermodynamic underpotential scale equivalent to RT ln(P), as we did in the case of the H_2/H_2O reaction. What the voltammetry curve of Fig.8.12 tells us is that reaction (8.19) is kinetically dominant in the potential region below 0.98 V, and the reversibility of the reaction makes it possible to define an underpotential scale equivalent to RT $\ln(\alpha_{OH})$. Then the state variables ΔH^* and ΔS^* for the formation of surface OH can be measured in the same manner as used for hydrogen. The results are summarized in Table 8.2. The reference state for these functions corresponds to the dissociated state of OH at unit activity, i.e., the equilibrium

$$H_2 + O_2 \rightleftharpoons 2(OH)_{ads} \quad . \tag{8.21}$$

As with hydrogen, the adsorbed state of OH refers to a surface submerged in aqueous solution, so that the electrochemical adsorption process would include a term for the reorganization of the double layer. In this case, however, the enthalpy is so large this reorganization term would contribute little to the total energy. We might expect therefore that the desorption energy of OH on Pt to the gas phase as measured in a UHV apparatus and the adsorption enthalpy measured electrochemically would be reasonably close. This point will be discussed further in a later section.

Table 8.2. Thermodynamic functions[a] for OH formation on Pt single-crystal surfaces

Surface	H_i^* [kJ/mol]	S_i^* [J/mol K]
{111}	-209.9	-110.8
{110}	-226.7	-108.5

[a]Zero coverage at 300 K

The transition in the structure of the OH layer on Pt at potentials well below the potential for oxygen evolution has a profound consequence on the kinetics. At potentials close to the theoretical potential for O_2 evolution, i.e., RT $\ln(P_{O_2}) \rightarrow 0$, the coverage of the surface by OH species exceeds 1 ML. If unoccupied Pt sites are required to catalyze the dissociation of H_2O to O_2, then the number of active sites available for O_2 evolution becomes exceedingly small.

The self-poisoning nature of this reaction illustrates one of the oldest concepts in heterogeneous catalysis, Sabatier's principle, that the intermediate must be adsorbed neither too weakly nor too strongly. So-called volcano relations, which are an expression of Sabatier's principle, between the enthalpy of formation of $M(OH)_n$ and the kinetics of oxygen evolution have been demonstrated by APPLEBY [8.22]. All of the transition metals bond OH species even more strongly than Pt ($-\Delta H^* > 200$ kJ/mol OH), but the noble metals Ag and Au bond OH very weakly ($-\Delta H^* < 30$ kJ/mol OH). There is, therefore, a very large gap in adsorption energy between Pt and Ag or Au where one might expect to find the "optimum" catalytic surface, the surface more active for O_2 evolution (or oxygen reduction to water) than pure O_2. Phenomenologically, it is clear what properties the "optimal" electrode surface should have, i.e., weaker interaction with H_2O than Pt, and stronger interaction with O_2 than Pt. There is, of course, no material known which has these properties. One approach towards developing such materials is to look at modifications to the intrinsic properties of Pt by various means, e.g., doping, alloying, etc. There is some indication in preliminary work from our laboratory [8.23] that doping Pt with Group IVb and Vb elements produces a catalytically favorable modification to the intrinsic properties of pure Pt. More rigorous studies on the modified state of Pt in these alloys is in progress using UPS and XPS.

8.6 Underpotential States of Metals on Metals

Metal adsorbates on foreign metal surfaces are both fundamentally interesting and technologically important species. Technologically, they represent the initial stage for metal deposition and their structure should influence the further growth behavior and final morphology of the deposit. Also, metal adatoms are believed to have electronic properties which are substantially different from those of the bulk metal and appear to have attractive catalytic properties [8.24]. Fundamentally, metal adatoms are ideal systems for the study of lateral interactions between adsorbates and of phase formation on surfaces.

The underpotential deposition of metals on foreign metal electrode substrates is a widely studied phenomenon in electrochemistry, and the standard electrochemical methods of coulometry and linear sweep voltammetry provide means of *directly* measuring heat of adsorption, entropies of adsorption and, in most cases, absolute coverage. Absolute coverage can be accurately measured in any system if an independent method exists for determining the electrosorption valence. Ex situ determination of the valence by XPS appears possible. The marriage of classical electrochemical methods with the ex situ UHV methods of electron spectroscopy and electron diffraction should enable these metal-on-metal adsorbate systems to be studied in extraordinary detail. What makes ex situ UHV analysis particularly attractive in this case is the generally high stability of the metal adatom layer which enables it to survive the vacuum transfer intact [8.25]. The stability of these systems arises principally from the nonvolatile and relatively immobile state of the adatom on the substrate surface. In spite of the particular attractiveness of ex situ UHV analysis in these systems, there are no *completed* studies of underpotential metal deposition to report at this time, although a number are in progress. The research group of KOLB and co-workers at the Fritz-Haber Institute has, however, made detailed studies of metal adsorbates at both the metal-vacuum and metal-solution interface and concluded that the presence of the solution had little or no effect on the structure of the adsorbate, i.e., the studies yielded complementary information. It is of interest then to review the results for a system that appeared to behave "ideally" in that sense, Pb deposition on Ag{111}. The electrolytic deposition of Pb was studied [8.25] in a conventional electrochemical cell in perchlorate solution using linear sweep voltammetry. Vacuum deposition was by electron-beam evaporation, and LEED-AES-work function analyses were conducted on the deposited layer in a conventional (Varian) UHV chamber. Auger signal intensities of Pb and Ag indicated a sharp break at exactly the point where 1 ML of closely packed Pb atoms would be expected, indicating that deposition starts in a layer-by-layer growth mechanism. In the submonolayer region, LEED analysis indicated the formation of two ordered structures (Fig.8.15), a $(\sqrt{3} \times \sqrt{3})R\ 30^\circ$ structure forming at $0.2 < \theta < 0.5$, and an hcp twist structure which coexists with the $(\sqrt{3} \times \sqrt{3})R\ 30^\circ$ at $0.5 < \theta < 1.0$ ML and forms the overlayer structure at higher coverages. The LEED patterns correlated precisely with breaks in the $\Delta\phi$ versus θ curves (Fig.8.16). Vacuum deposition of Pb on Ag{111}, therefore, starts with the formation of a uniform monolayer in three steps: random adsorption followed

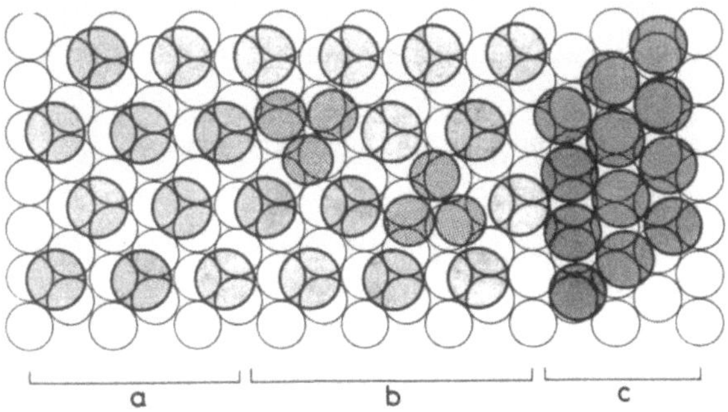

Fig.8.15

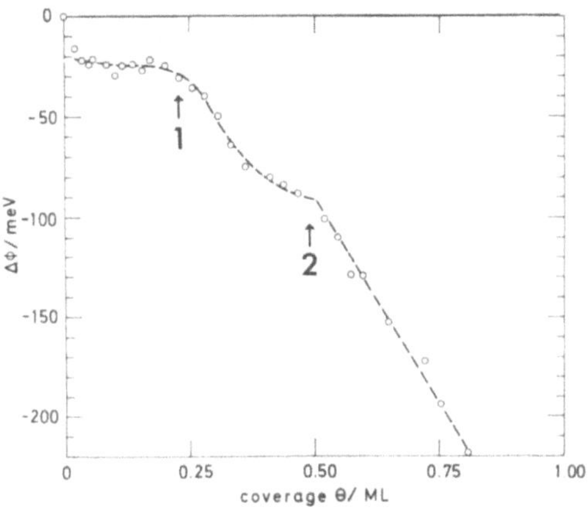

Fig.8.16

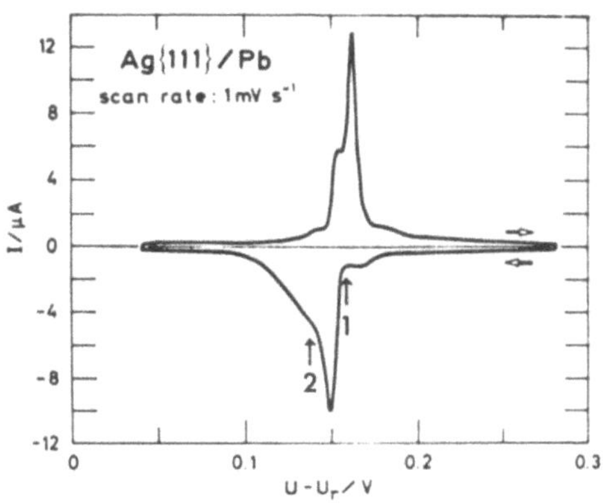

Fig.8.17

Fig.8.15. Submonolayer structures of Pb on {111} Ag surfaces deduced from LEED patterns: (a) ($\sqrt{3} \times \sqrt{3}$)R 30° structure; (b) coexisting ($\sqrt{3} \times \sqrt{3}$)R 30° and hcp twist structures; (c) the hcp twist structure [8.25]

Fig.8.16. Work function changes due to deposition of Pb on {111} Ag surfaces. 1 and 2 denote coverages where changes in LEED patterns were observed

Fig.8.17. Cyclic voltammetry of Pb deposition on {111} Ag. The potential scale is the underpotential scale, i.e., V vs Pb/Pb^{2+} electrode in the same electrolyte. 1 and 2 denote potentials where the coverages were 0.2 and 0.5 ML, respectively [8.25]

by a ($\sqrt{3} \times \sqrt{3}$) superstructure and lastly an hcp layer. Deposition of Pb from solution appears to follow *precisely* the same sequence, as indicated in the linear voltammetry curve of Fig.8.17. There are two distinct transitions indicative of three deposition processes; the transitions occurred at 0.2 and 0.5 ML, as was the case for vacuum deposition. The sharpness of the peak at 0.15 V underpotential indicates the adatoms have an attractive interaction while the broad tail to the peak representing the last Pb adatoms to be deposited is indicative of repulsive interaction. These interaction effects are consistent with the LEED structures in Fig.8.15, since second-nearest-neibhor interactions represent attractive forces, and nearest-neighbor interactions repulsive. Can we then say that electrodeposition and vacuum deposition of Pb on Ag{111} result in the formation of identical structures? The picture seems tantalizingly consistent and complementary, yet the paralles could be coincidental. However, truly definitive experiments utilizing direct coupling of the electrochemical cell to the UHV chamber appear possible. If the electrodeposition process is stopped at ca 0.5 ML, the electrolyte evaporated from the surface, the crystal transferred to the UHV chamber, and the LEED-AES work function analysis gives the expected result, then the evidence would indeed be conclusive.

8.7 Relation of the Underpotential State to the Chemisorbed State in Vacuum

At least two underpotential states of atoms studied in considerable detail appear to form bonding configurations with the substrate surface very like those formed in the absence of electrolyte, i.e., at the vacuum interface. These are hydrogen atoms on Pt surfaces in dilute perchloric or hydrofluoric acid solutions, and Pb atoms on Ag surfaces in dilute sodium perchlorate solutions. Considering the imposing experimental problems associated with definitive tests of this proposition, it seems reasonable to examine carefully the rational basis behind it. Since we are, in the present context, restricting our attention to just the thermodynamic functions ΔH^* and ΔS^* relating adatoms to the bulk phase [reaction (8.2)] the thermodynamic cycle presented in Fig.8.18 provides a convenient and direct way of relating adsorption at the metal-vacuum interface to adsorption at the metal-solution interface. In the case of the total enthalpy change ΔH^* the enthalpy changes for dehydration (L_M), adsorption (W_{AM}), and solution (S_{AM}) are all experimentally measurable quantities, particularly for the case where the solution is pure H_2O (infinite dilution). Then the enthalpy of formation of the underpotential state (ΔH^*) is related to the gas-phase heat of adsorption (W_{AM}) by

$$-\Delta H^* = W_{AM} - (L_M - S_{AM}) \quad . \tag{8.22}$$

If it takes more energy to remove H_2O from the bare metallic surface than is released when the adsorbate covered surface is immersed, then $-\Delta H^*$ is

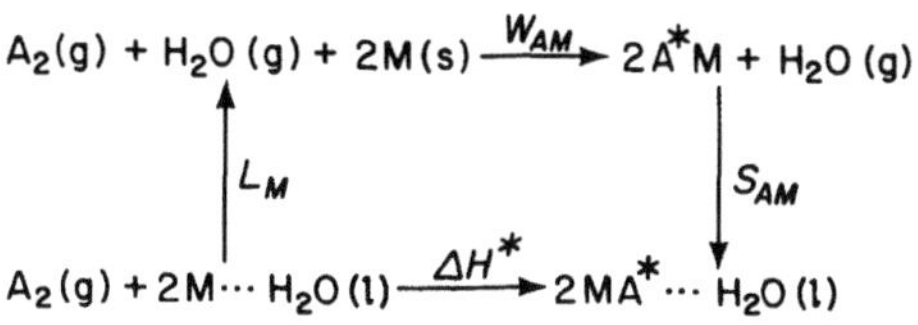

Fig.8.18. Thermodynamic cycle relating the gas-phase heat of adsorption (W_{AM}) to the solution-phase enthalpy of adsorption (ΔH^*) measured electrochemically

less than the gas-phase heat of adsorption, and vice versa. There are two sets of circumstances where it would be expected that $-\Delta H^* \simeq W_{AM}$. If the enthalpies of dehydration and immersion are essentially equal ($L_M \simeq S_{AM}$), or if the heat of adsorption is very large relative to either ($W_{AM} \gg L_M, S_{AM}$), then the gas-phase heat and the solution-phase enthalpy should be equal. The thermodynamic cycle of Fig.8.18 suggests a whole new avenue of experimental research for electrochemists who want to relate in a rigorous way the underpotential state to the comparable state in vacuum in order to understand electrode processes more fully. The key element in this new, nontraditional approach to the study of electrode processes is the study of water on all metallic surfaces of interest (as electrodes), with and without adatoms or adsorbed species of interest, using entirely UHV methods. In a way, one would be "doing electrochemistry" entirely in UHV. Strictly speaking, the surface processes observed would apply only to electrode processes at pH $\simeq$ 7 in an infinitely dilute electrolyte. Additional and experimentally more difficult work will have to be done to determine the role of cations and anions in order to extend the results to real electrolytes.

Fortunately, there is already a significant and rapidly growing body of literature on the interaction between water and clean metal surfaces. The heat of desorption of H_2O from metallic Pt surfaces (L_M) has been measured [8.26] and is only slightly higher than the heat of sublimation (60 kJ/mol vs 42 kJ/mol) indicating the specific interaction between H_2O and Pt is weak. Unfortunately the immersion heat of a hydrogen-covered Pt surface has not been measured, nor has the heat of adsorption of H_2O on a hydrogen-covered Pt surface been measured. The author attempted the latter experiment in the apparatus of Fig.8.2, but the heating configuration of the sample holder resulted in a heating rate too low for thermal desorption experiments. The weak specific interaction of H_2O with the Pt surface implied in the results of FISHER and GLAND [8.26] indicates that the presence of hydrogen on the Pt surface should have little effect on the energy released on hydration of the surface, i.e., $L_M \simeq S_{AM}$.

In the particular case of Pt electrodes in very dilute aqueous solutions, there is, therefore, a rational basis for the expectation that the adsorption energies in solution and in vacuum will be the same. For hydrogen chemisorption on Pt, a comparison of the electrochemical results from our laboratory with the vacuum experiments reported in the literature indicates that the two methods give comparable adsorption energies, although there is considerable variation in the energies for hydrogen on {111} Pt between investigators. The binding states of hydrogen on {111} Pt in vacuum has been studied in detail by ERTL and co-workers [8.27], McCABE and SCHMIDT [8.28], and LU and RYE [8.29] using the thermal desorption method. However, the latter two studies did not have LEED capability in their systems to determine in a rigorous way the structure of their single-crystal surfaces. McCABE and SCHMIDT and LU and LYE both reported an adsorption energy (zero coverage) of 73.2 kJ/mol for {111} Pt, while ERTL and co-workers found a much lower value, 41.2 kJ/mol. The Ertl group reported that the adsorption energy of hydrogen on {111} Pt was very sensitive to the annealing of the ion-bombarded surface, and that adsorption energies as high as 70-80 kJ/mol were observed on "imperfect" (nonannealed) surfaces. However, surfaces that produced "perfect" {111} - (1×1) LEED patterns produced the lower adsorption energy of 41.2 kJ/mol. Since neither of the other investigator's systems utilized LEED analysis of their surfaces, it seems likely the higher adsorption energies reported in those studies are not representative of "perfect" {111} - (1×1) surfaces. The value of 41.2 kJ/mol from the Ertl group is very close to the electrochemically determined value of 46.8 kJ/mol. Less work has been reported for hydrogen on {100} Pt. Qualitatively, hydrogen bonding is stronger on the {100} plane, as reported by both LU and RYE [8.29] and McCABE and SCHMIDT [8.28]. Quantitatively, the latter reported initial heats of adsorption of 102.5 and 115 kJ/mol for hydrogen on {100} Pt above room temperature, significantly higher than the electrochemically determined value of 81.1 kJ/mol. As before, the absence of LEED structure analysis in the work of McCABE and SCHMIDT raises questions as to the nature of the surface in their experiments. Nonetheless, their results clearly indicate that the *relative* adsorption energy of hydrogen on {100} Pt is higher than on {111} Pt by ca. 40 kJ/mol, which is what the electrochemically measured values indicated as well. Further qualitative agreement between electrochemically measured bond energies and the thermal desorption bond energy is evident when the results for stepped Pt surfaces are compared. This done in Fig.8.19 using the thermal desorption work of COLLINS and SPICER [8.30] and the electrochemical results of ROSS [8.5]. COLLINS and SPICER deconvoluted their thermal desorption spectra into a "terrace" and "step" contribution, the step contribution always occurring at higher binding energy than the terrace

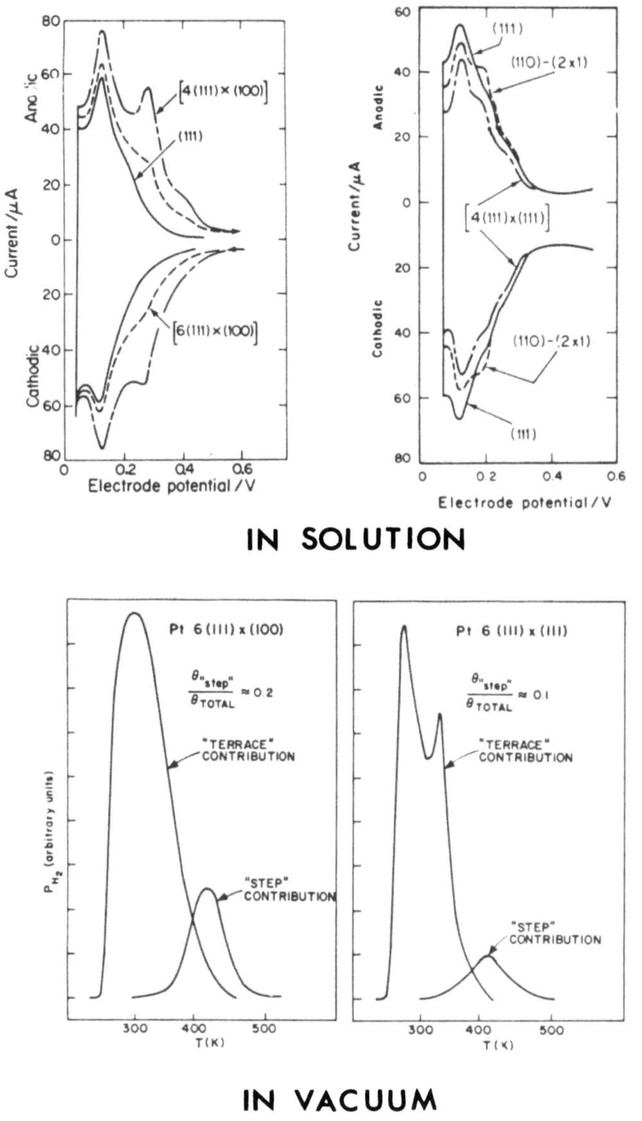

Fig.8.19. Comparison of spectra for hydrogen chemisorbed on {111} × {100} and {111} × {111} stepped Pt surfaces: by thermal desorption in vacuum [8.30]; by potential sweep in solution [8.5]

contribution. At the step of {111} × {100} geometry, the coverage by hydrogen is just twice that at the step of {111} × {111} geometry. Both these observations are *precisely* what was observed electrochemically (refer to the detailed discussion given in Sect.8.4.2). Unfortunately, COLLINS and SPICER did not report absolute binding energies for hydrogen on their surfaces, so a quantitative comparison is not possible. It would appear that at present it is not possible to make quantitative comparisons of adsorption energies of hydrogen on Pt determined by electrochemical methods with those determined using thermal desorption methods in UHV. There are, however, a precise set of experiments that would make rigorous quantitative comparison possible:

I) electrochemical measurements of adsorption energies at infinite dilution (pH 7);

II) definitive measurement of the heat of dehydration of the metallic surface (L_M) and the heat of solution of the adsorbate-covered surface;

III) definitive experiments to determine the thermodynamic heat of adsorption of hydrogen on well-ordered single-crystal surfaces in UHV.

The thermodynamic cycle of Fig.8.18 is also useful in interpreting the entropy of adsorption in solution, but adsorption entropy is not a state function easily measured in UHV systems. By their very nature, UHV methods rely on the irreversibility of the adsorption in order to measure coverage via a spectroscopic method, e.g., AES. *Equilibrium* isotherms are therefore rarely obtained in contemporary surface studies, and consequently adsorption entropies are infrequently reported. There are notable exceptions, represented by Chaps.9,16. The use of the Kelvin probe method for dynamic measurement of the change in work functions upon adsorption appears particularly promising, since one can then obtain coverages of the surface under *reversible* (equilibrium) conditions. MENZEL (Chap.16) reported equilibrium isobars for hydrogen and carbon monoxide on Ru {0001}, from which the heat of adsorption and the differential entropy of adsorption were obtained as a function of coverage. More widespread use of this technique, when applicable, should lead to more complete determinations of the thermodynamic properties of surfaces. ERTL and co-workers have also made extensive use of the Kelvin probe method, particularly for the study of hydrogen on metals, but have reported [8.13] isobars and differential entropies only for hydrogen on Pd {100}. For that system, at 370 K, the partial molar entropy of the adsorbed phase is perfectly described by the model for two-dimensional translation of a system of non-interacting particles (Volmer gas). Below 370 K, the adsorbed layer can be described by a localized model, consistent with LEED indications of ordered states below 270 K. This order-disorder phenomena at ca. 300 K is typical of hydrogen on the transition metals, and it probably occurs on Pt as well. The electrochemically measured value for the differential entropy of adsorption at zero coverage (ΔS_i), given in Table 8.1, was - 55.8 J/mol K. The reference state is the gas at 100.3 k Pa and 298 K. If we completely ignore the possible effect of water (actually the double layer including solvated protons and anions), we can calculate the partial molar entropy of the adsorbed phase ($\overline{S}_i^*$) at low coverage from the relation

$$\overline{S}_i^* = \Delta S_i^* + \overline{S}_g \quad .$$

At 100.3 k Pa and 298 K, $\overline{S}_g$ = +130.5 J/mol K [8.31], so that $\overline{S}_i^*$ = +75.2 J/mol K. This value agrees well with the value for a two-dimensional gas at $\theta \simeq 0.1$ and 298 K, +94.5 J/mol K, relative to the much lower (factor of two) value for the configurational entropy of a localized adsorbate +36.6 J/mol K. The electrochemically measured value for the differential entropy of adsorption therefore appears to be consistent with the gas-phase results indicative of a mobile hydrogen adatom state.

The observant reader will have noted that the conclusion that the hydrogen adatom layer on Pt is mobile at room temperature conflicts with the immobile adsorbate model used to interpret the isotherms [8.18]. The reason for this conflict is not clear. The analysis of adsorption entropy above ignored the possible effect of reorganization of the double layer, so the derived partial molar entropy of the adsorbed phase could be misleading. However, the general mobility of hydrogen on transition metals in vacuum suggests the mobile interpretation may be the correct one. In that case, the Roberts pair-wise repulsive interaction model does not explain the observed dependence of isosteric heat of adsorption on coverage, or the asymmetry of the isotherm.

Pt and Au are unique electrode materials in that the specific interaction of these surfaces with aqueous electrolytes is very weak and the adsorption of neutral molecules (or atoms) at the metal-solution interface is relatively undisturbed by the presence of electrolyte. The heat of adsorption of hydrogen on Pt electrodes is higher than on any other electrode surface, yet, in vacuum, there are many metal surfaces at which hydrogen is bound much more strongly than at Pt, e.g., W. Other metals adsorbing hydrogen have much stronger specific interaction with water, resulting in lower net heats of adsorption of hydrogen on these metals in aqueous solution. Although Au does not chemisorb many neutral molecules, e.g., H_2, CO, C_2H_4, it is the ideal surface for studying the adatom states of metals with a minimum of interference from the electrolyte. However, aside from these two metals, there appear to be few other electrode surfaces at which adsorption occurs with a minimum interaction between the solvent and the adatom. The underpotential states of hydrogen on Pt are, therefore, atypical states for electrode processes, and, in general, thermodynamic functions derived from underpotential measurements will *not* be the same as those measured in vacuum.

Acknowledgments. This work was supported by the Assistant Secretary for Fossil Energy, Office of Coal Utilization and Extraction, Advanced Concepts Division of the U.S. DOE under contract number W-7405-ENG-48 through the Fuel Cell Projects Office, NASA Lewis Research Center, Cleveland, OH.

References

8.1 R.P. Van Duyne: "Surface Enhanced Raman Spectroscopy", in *Chemistry and Physics of Solid Surface III*, ed. by R. Vanselow (CRC, Boca Raton, FL 1981)
8.2 R.O. Ansell, T. Dickinson, A.F. Povey, P.M.A. Sherwood: J. Electroanal. Chem. *98*, 69, 79 (1979) and references therein
8.3 B.E. Conway, H. Augerstein-Kozlowska: J. Electroanal. Chem. *113*, 63 (1980)
8.4 H.O. Beckman, H. Gerischer, D.M. Kolb, G. Lempfuhl: Symp. Faraday Soc. *12*, 51 (1978)
8.5 P.N. Ross: Surf. Sci. *102*, 463 (1981)
8.6 A.T. Hubbard, R. Ishikawa, J. Katekaru: J. Electroanal. Chem. *86*, 271 (1978);
T.E. Felter, A.T. Hubbard: J. Electroanal. Chem. *100*, 473 (1979)
8.7 P. Hagans: Ph. D. Thesis, Chemistry Dept., Case Western Reserve University, Cleveland, OH (1980);
E. Yeager: Surf. Sci. *101*, 1 (1980)
8.8 P.N. Ross: J. Electrochem. Soc. *126*, 67 (1979)
8.9 E. Yeager, W. O'Grady, M. Woo, P. Hagans: J. Electrochem. Soc. *102*, 348 (1978)
8.10 K. Yamamoto, D. Kolb, R. Koetz, G. Lempfuhl: J. Electroanal. Chem. *96*, 233 (1975)
8.11 E. Yeager: J. Electrochem. Soc. *128*, 168C (1981)
8.12 J.K. Roberts: *Some Problems in Adsorption* (Cambridge Univ. Press, Cambridge 1939) pp.21-41
8.13 R.J. Behm, K. Christmann, G. Ertl: Surf. Sci. *99*, 320 (1980)
8.14 B.E. Conway, H. Augerstein-Kozlowska, F.C. Ho: J. Vac. Sci. Technol. *14*, 351 (1977)
8.15 J. O'M. Bockris, M.A. Devanathan, K. Müller: Proc. Soc. London A*297* (1963)
8.16 D.W. Blakely, G.A. Somorjai: Surf. Sci. ??? 413 (1977)
M.A. Van Hove, G.A. Somorjai: Surf. Sci. *92*, 489 (1980)
8.17 B. Lang, R. Joyner, G.A. Somorjai: Surf. Sci. *30*, 440, 454 (1972)
8.18 R. Ducros, R. Merrill: Surf. Sci. *55*, 227 (1977)
8.19 P. Stonehart, H. Augerstein-Kozlowska, B.E. Conway: Proc. R. Soc. London A*310*, 541 (1969)
8.20 H. Augerstein-Kozlowska, B.E. Conway, W.B. Sharp: J. Electroanal. Chem. *43*, 9 (1973)
8.21 M. Pourbaix: *Atlas of Electrochemical Equilibria in Aqueous Solutions* (Pergamon, New York 1966)
8.22 A.J. Appleby: Catal. Rev. Sci. Eng. *4*, 221 (1970)
8.23 P.N. Ross: EPRI Final Report EM-1553 (1980)
8.24 D.M. Kolb: In *Advances in Electrochemistry and Electrochemical Engineering*, Vol.11, ed. by H. Gerischer, C. Tobias (Wiley, New York 1978) p.125
8.25 K. Takayanagi, D. Kolb, K. Kambe, G. Lempfuhl: Surf. Sci. *100*, 407 (1980)

8.26 G.B. Fisher, J.L. Gland: Surf. Sci. *94*, 446 (1980)
8.27 K. Christmann, G. Ertl: Surf. Sci. *60*, 365 (1976);
K. Christmann, G. Ertl, T. Pignet: Surf. Sci. *54*, 365 (1976)
8.28 R. McCabe, L. Schmidt: In Proc. 7th Int. Vacuum Congr. and 3rd Int. Conf on Solid Surfaces (Vienna, 1977) p.1201
8.29 K. Lu, R. Rye: Surf. Sci. *45*, 677 (1974)
8.30 D. Collins, W. Spicer: Surf. Sci. *69*, 85 (1977)
8.31 J.A. Dean (ed.): *Lange's Handbook of Chemistry*, 11th ed. (McGraw-Hill, New York 1973)

9. Low-Energy Electron Diffraction Studies of Physically Adsorbed Films*

S. C. Fain, Jr.

With 7 Figures

9.1 Introduction

Physically adsorbed monolayers on crystal surfaces provide experimental realizations of two-dimensional model systems of condensed matter [9.1-4]. This paper will concentrate on the structure, thermodynamics, and phase transitions of krypton, argon, and nitrogen monolayers adsorbed on the cleavage surface of graphite crystals. The method of investigation used by CHINN, SHAW, DIEHL, TONEY and the author has been low-energy electron diffraction (LEED), and results so obtained will be emphasized [9.5-12]. Results from other laboratories, often using otner techniques, and for other gases and other substrates will be referred to where they are directly related to the systems of primary interest.

The widespread interest in the field of physical adsorption is evident from recent conference proceedings [9.2-4], and no attempt will be made here to review the entire field [9.13]. Reference will be made to several papers from the first three International Summer Institutes in Surface Science (ISISS 1973, 1975, 1977) to maintain some continuity with earlier proceedings [9.14-20].

9.2 Background

The initial impetus for LEED investigations of physically adsorbed films on graphite was provided by the excitement generated in Seattle by heat capacity studies of phase transitions in helium films on a large-surface-area graphite product Grafoil [9.1,17,21]. BRETZ and DASH were able to deduce from such studies that they had observed an order-disorder transition of a

---

* This material is based upon work supported by the National Science Foundation under Grant No. DMR 80-06334

$(\sqrt{3} \times \sqrt{3})R\ 30^{\circ}$ structure of helium on graphite [9.22]. DASH raised the possibility of structural studies of adsorbed films by electron, X-ray, or neutron diffraction. Initially use of LEED for the lighter inert gases seemed doubtful due to the electron-beam desorption effects seen at Brookhaven for multilayer Kr, Ar, and Ne films on Nb{100} [9.23]. However, the image intensifier described by van OOSTRUM for field ion microscopy [9.15] suggested the possiblity of LEED studies at much lower currents than in the Brookhaven studies. Construction of a LEED instrument equipped with a channel plate image intensifier [9.24] was begun in 1974.

At this time, the first results on neutron diffraction from an adsorbed film were published. DASH had spent some time with Lawrence Passell's group at Brookhaven and they chose N_2 on graphite for their first study [9.25]. Before the low-current LEED instrument was completed in July, 1976, neutron diffraction had been used to study N_2, Ar, Kr, O_2, and He films on graphite. This neutron work was reviewed by McTAGUE at ISISS 1977 [9.20]. KJEMS discussed some advantages of neutron studies over LEED studies [9.26]. This chapter emphasizes examples where LEED studies have been very useful.

In Europe, the principal impetus for diffraction studies of physically adsorbed films came from the volumetric isotherms on exfoliated graphite by THOMY and DUVAL which HOBSON had discussed at ISISS 1973 [9.14]. THOMY and DUVAL hypthesized that xenon and krypton monolayers on graphite had solid, liquid, and gas phases in analogy with ordinary bulk solids [9.27]. A group in Marseilles headed by BIENFAIT set out to study the heavier rare gases using a conventional LEED apparatus. By the time our instrument was working this group had published a number of papers on Xe on graphite [9.28] and had in press a paper on Kr on graphite [9.29]. VENABLES in Sussex reported high-resolution lattice parameter measurements of Xe on graphite using transmission high-energy electron microscopy (THEED) [9.30]. The question arose as to how a diffraction pattern could look so much like the $(\sqrt{3} \times \sqrt{3})R\ 30^{\circ}$ pattern and not be it. If the layer were not in registry, why should it have the correct orientation? This will be discussed in Sect.9.5.1. After studying the LEED work of KRAMER and SUZANNE [9.29], the THEED work of VENABLES et al. [9.30] and the volumetric isotherm work of THOMY et al. [9.31], it was decided to search for the out-of-registry Kr structure hypothesized by THOMY in his 1968 thesis [9.32].

9.3 LEED Instrument

The main features of the instrument developed in Seattle, and shown schematically in Fig.9.1, are described in a paper written in Summer 1976 as we were completing construction [9.5]. One key feature has been the simple electron gun[1] recommended by M. Barney Webb and Max Lagally of University of Wisconsin-Madison. When operated at low currents, the beam size at the sample some 20 cm from the end of the gun is about 0.3 mm in diameter (FWHM). The transfer width of this gun has been measured and is of the order of 100 to 200 Å [9.33,34]. The more useful concept of resolving power has been recently discussed [9.35]. As this gun was not designed especially for LEED, it may not be optimal. However, it works sufficiently well that we have not yet made a search for a better design which uses the full potential of the low-current requirement of our LEED system.

For four years a simple helium-gas-cooled sample holder based on a design used for field ion microscopy was used [9.5]; in 1980, we installed a commercial sample holder which operates more successfully below 30 K.[2] Accurate and precise temperature measurement using sensors compatible with the ultrahigh-vacuum bakeout presents a problem recognized in the earlier LEED work at Brookhaven [9.23]. Initially, we used an iron-doped-gold versus Chromel-P thermocouple [9.5], but in SHAW's work with Ar it was discovered that the heat being conducted down the gold-iron lead was causing a temperature error of about 2 K [9.9,36]. The Kr results were later corrected for this error [9.10]. Subsequently, we switched to commercial silicon diodes which are reasonably sensitive and stable [9.12][3]. However, great care has to be taken to be certain that the graphite surface and the diode are at the same temperature.

Four commercial LEED grids are used to reject most of the inelastically scattered electrons, which do not contribute to the structural information. For some preliminary work we used a simple phosphor screen after the grids; after adding the one-stage channel electron multiplier plate, experiments were much easier to carry out [9.5]. With this multiplier plate the electron gun could be operated at 3×10^{-9} A and still give plenty of inten-

---

1 Cliftronics, Clifton N.J. CE-406S with 1mm G2 aperture and tungsten filament.

2 Air Products and Chemicals, Allentown, PA, Model LT-3-110.

3 Lake Shore Cryotronics, Westerville, Ohio, Model DT-500K-T05.

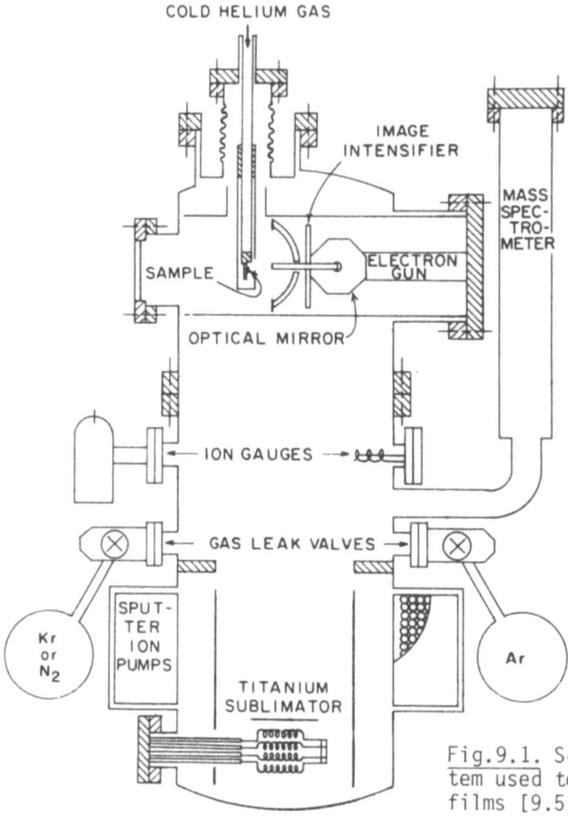

Fig.9.1. Schematic diagram of LEED system used to study physically adsorbed films [9.5]

sity at the phosphor screen. At this current the beam was small enough to fit into one facet of a good graphite crystal, permitting high-resolution LEED studies.

With Kr and Ar monolayers, no evidence was found for electron-beam effects at currents as high as 10^{-7} A (spot size 0.8 mm). At similar currents UNGURIS et al. saw striking effects for Kr and Ar monolayers on Ag{111} [9.37]. Thus, it turns out that the electron-beam effects that we designed our instrument to minimize are not so serious on graphite. The real gain was being able to fit the beam onto one crystal and having sharp enough spots to permit measurements of small splittings observed for Kr and Ar. The resolution obtained in this way is not significantly greater than that of some other LEED systems [9.18,33,34] but it is much better than that of the original Marseilles apparatus [9.28,29]. Very similar LEED optics have now been built at Marseilles [9.38] and some results obtained with this optics will be mentioned in Sect.9.5.1 [9.39].

In the work done so far there is no way of isolating the sample from the sputter-ion and titanium sublimation pumps. Usually, the sputter-ion pumps are turned off and the entire chamber is backfilled through gas leak values to pressures up to 10^{-3} torr. Because the gas hitting the sample is at room temperature, a thermomolecular correction is applied to the apparent pressure in order to be able to compare the results with those taken under more closely isothermal conditions [9.9-12]. So far this correction has been used by Webb's group as well [9.40], but not by VENABLES [9.41] or SUZANNE [9.39]. Our glass ion gauge was calibrated by Boeing Technology Services for Kr, Ar, and N_2 on a dynamic standard utilizing a capillary flow meter and a fixed orifice. The uncertainty in that calibration is less than ±20% in the range from 10^{-7} to 10^{-4} torr. For work with Ar and N_2, a small 60-K cold finger not shown in Fig.9.1 was used to trap CH_4 when the ion pumps were turned off.

9.4 Krypton on Graphite

As mentioned in Sect.9.2, by the time the Seattle instrument was ready, the LEED study of KRAMER and SUZANNE for Kr on graphite was already in press [9.29]. The only feature left to study seemed to be to search for the incommensurate Kr phase.[4] THOMY had proposed in 1968 that such a phase could exist if the Kr monolayer took up its spacing in the bulk crystal of d = 4.06 Å (at 77 K) [9.32,42]. He also noted that the commensurate $(\sqrt{3}\times\sqrt{3})30^\circ$ structure proposed by LANDER and MORRISON for Xe on graphite [9.43] might also exist for Kr on graphite ($d = d_0 = 4.26$ Å). These two possibilities are shown in Fig.9.2, along with a schematic representation of THOMY's 77-K Kr adsorption isotherm [9.42]. The main features shown by the isotherm are large steps which were interpreted to be due to layer-by-layer condensation; three large steps are shown in Fig.9.2, and five or six large steps could be distinguished before the pressure came too close to the bulk condensation pressure P_0 (77 K) = 1.75 torr to control [9.27]. Sandwiched between the first- and second-layer steps was a small event which THOMY believed represented the transition from a commensurate $(\sqrt{3}\times\sqrt{3})30^\circ$ monolayer to an incommensurate monolayer with lattice constant closer to its

4 An incommensurate phase is defined as one in which the ratio of the number of adsorbate molecules to the number of substrate adsorption sites is not a simple fraction such as the 1/3 value for the $(\sqrt{3}\times\sqrt{3})$ 30° phase.

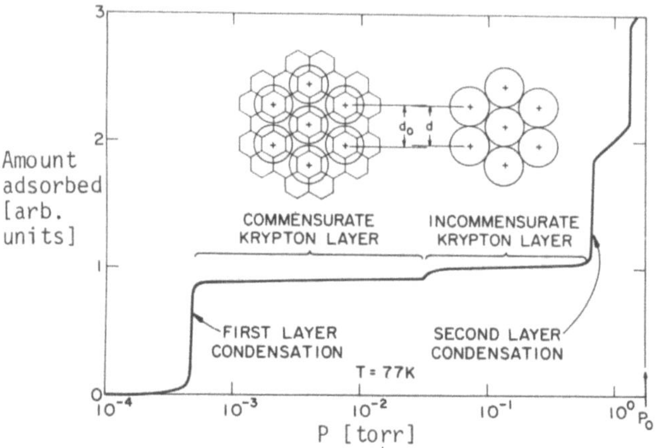

Fig.9.2. Schematic representation of 77-K volumetric adsorption isotherm for Kr on exfoliated graphite [9.42]. First-, second- , and third-layer condensation steps are shown, as well as the commensurate and incommensurate structures proposed for the monolayer

value in the bulk crystal. In 1972, THOMY et al. studied this transition as a function of temperature [9.31] and thus provided the information necessary to extrapolate the pressure for this transition down to temperatures where the pressure would be low enough ($<10^{-4}$ torr) for LEED experiments [9.7].

In the first day of operation with the channel plate we obtained the $(\sqrt{3} \times \sqrt{3})30^{\circ}$ phase at monolayer condensation as well as a new pattern at a lower temperature where the pressure was about 10^2 greater than monolayer condensation. This was a multiple scattering pattern from a compressed incommensurate structure such as shown in Fig.9.2. These patterns were published in the first report on the apparatus [9.5], followed by three detailed isotherms [9.6,7]. The main interest in the isotherms was to study the transition from the commensurate structure to the incommensurate structure. Within the accuracy of the LEED data, the results were consistent with a second-order transition similar to that obtained in the one-dimensional model of FRANK and van der MERWE [9.44] when the "natural misfit" (to be discussed in Sect.9.5.2) was varied. BAK showed that mean-field theory predicts a first-order transition to the hexagonal incommensurate structure [9.45]. A number of further experimental studies have investigated this phase transition. Recently, NIELSEN and co-workers have seen evidence for a small first-order transition at temperatures close to those of our work, but a bit lower [9.46]. On the other hand, BIRGENEAU and colleagues favor the second-order-phase-transition idea at the higher temperatures they have studied

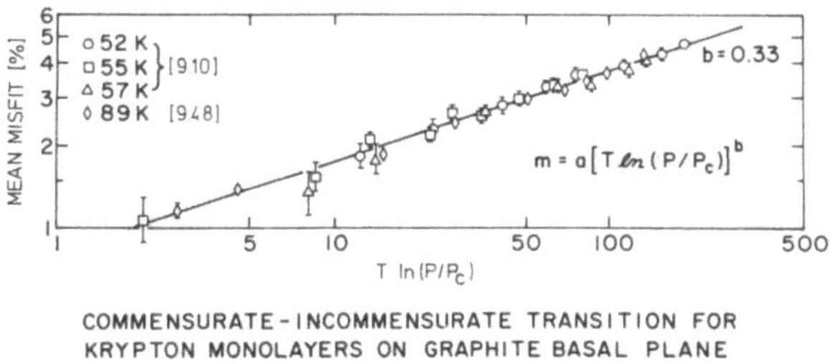

Fig.9.3. Mean misfit, $(d_0 - d)/(d_0 = 4.26$ Å), vs chemical potential for LEED data compared with X-ray data of [9.48]. The critical pressures P_c for the LEED data were adjusted to give the best agreement with the 89 K X-ray data [9.10]

[9.47]. Both of these recent studies utilize high-intensity X-ray sources and a special graphite product "ZYX" with less surface area than Grafoil, but much higher perfection.

STEPHENS et al. obtained data accurate enough to rule out the one-dimensional Frank-van der Merwe model and found instead a fit to a power law in chemical potential [9.48]. By adjusting the critical pressure for the onset of the transition, we could fit our data to a power law with the same coefficients as those found by STEPHENS et al. at 89 K, as shown in Fig.9.3 [9.10]. However, there is no theoretical model which gives an exponent close to the experimental value [9.49], and VENABLES has suggested that the exponent is an accident of the interparticle potential [9.41]. Further results published near 97 K follow the same power law and show a significant loss in long-range order right at the transition [9.50].

Recently, BIENFAIT's group has found a physically adsorbed system, Xe on Cu{110}, which undergoes a uniaxial commensurate-incommensurate transition which is more closely related to solvable theoretical models [9.51]. They found an exponent close to that predicted by several theoretical models, but HALDANE and VILLAIN have pointed out that the observed behavior may be influenced by the strong induced dipoles produced by Xe adsorption on metal surfaces [9.52].

9.5 Argon on Graphite

9.5.1 Rotational Epitaxy of an Incommensurate Monolayer

The problem of rotational epitaxy of multilayer metal films of different bulk lattice constants has a long history and is usually explained in terms of misfit dislocation networks which extend through several layers [9.19,53]. In the process of making calculations to explain neutron studies of Ar monolayers, NOVACO and McTAGUE discovered that the energy of the monolayer was lower for some orientations with respect to the substrate than for others [9.54]. The striking difference from the typical metal-on-metal film case was that the angle was not along a symmetry direction and varied with the misfit between the film and the substrate. This can be explained as due to the competition between static strains of transverse and longitudinal nature [9.54] and has recently been discussed in detail by several authors [9.49,55,56].

In our LEED experiments with Ar on graphite, we were able to directly observe the rotational epitaxy; this is a feature that cannot be measured in neutron experiments [9.20,57]. The cold trap mentioned in Sect.9.3 was necessary to prevent CH_4 contamination of the Ar films. By varying the substrate temperature or Ar gas pressure the misfit between the substrate and monolayer could be varied; the variation in orientation angle with misfit [9.8] was almost exactly that predicted by NOVACO and McTAGUE [9.54] *in advance* of the experiment! This was indeed an unusual case for surface physics and was due to the simplicity or Ar on graphite and to the perfection of our crystals. In all of the work of WEBB et al. for Xe, Kr, and Ar on Ag{111}, the monolayers are oriented along a substrate symmetry direction instead of the Novaco-McTague direction [9.37,40]. This does not prove that graphite is more perfect than Ag, for the potential variation along the graphite substrate is much stronger than for Ag{111}, making the energy gain larger on graphite [9.40]. The energy gain in going from the $(\sqrt{3} \times \sqrt{3})30^\circ$ orientation to the orientation of lowest energy is calculated to be only 0.1 K per atom [9.54]. For the effect to be observed at all, there must be a large number of atoms in a given coherent domain of the substrate.

In reviews of the subject VILLAIN remarked that the theory should be tested for Ne, because there the assumptions made are more suspect [9.49]. CALISTE and SUZANNE were able to do experiments for Ne; once again the agreement was very good [9.39]. In both cases the perturbations of the adsorbed layers due to the substrate lateral forces are small. A more difficult problem occurs when the misfit becomes smaller as in the case of Kr, because the perturbations become more significant. Several authors have made theoretical calculations for this problem based on elastic continuum approximations [9.49,55,56]. SHIBA calculated the dependence of rotation angle on mean misfit for several ratios of longitudinal c_L to transverse c_T sound velocities. His curve for the ratio of an unconstrained monolayer with central forces ($c_L = \sqrt{3}\, c_T$) is drawn in Fig.9.4 for the value of his parameter ℓ_0 which causes the commensurate-incommensurate transition to occur at the 4% natural misfit we originally proposed [9.6].

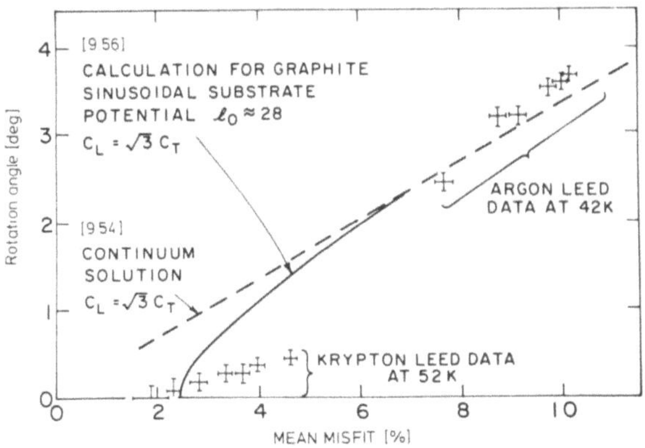

Fig.9.4. Theory and experiment for rotation angle vs mean misfit [9.8,10,56]

SHIBA predicted a continuous transition in angle; also shown in Fig.9.4 are angles deduced by DIEHL from our original Kr data assuming a continuous transition in which the size of the basic LEED spots does not vary with misfit [9.10]. These data are consistent with a continuous transition, but produce a smaller rotation than SHIBA predicted. Because the intrinsic substrate perturbations on the adsorbed monolayer are larger for small misfits, we first argued that the experimental orientation results for Kr and Xe were not likely to be affected by steps and defects since we observed intrinsic results for Ar [9.8]. However, there are many possible subtle influences of defects for small misfits [9.49], and the results for Kr and Xe rotation may not be free from the influences of defects. All of the Kr work was on one crystal, and although the phase transition results are similar to those of SCHABES-RETCHKIMAN and VENABLES, the angles are half as large [9.41].

9.5.2 Thermodynamics of an Incommensurate Monolayer

In interpreting the Kr results, it was necessary to understand how the lattice constant b of a monolayer would vary with the gas vapor pressure above the monolayer for the hypothetical case of a perfectly smooth surface with no lateral periodicity. {The quantity $(d_0-b)/(d_0=4.26$ Å) is the natural misfit used in theoretical models [9.6,44,49,56]}. We were able to reproduce the change in lattice constant with change in vapor pressure obtained in the detailed calculations of PRICE and VENABLES [9.58]; we took appropriate derivatives of the bare interparticle potential and ignored the entropy terms and substrate corrections to the potential. In the case of Kr on graphite, the substrate periodicity produces strong effects on the actual mean misfit observed [9.6], but for Ar the calculations of NOVACO and McTAGUE showed that these effects were much smaller [9.54]. An Ar isotherm was obtained, and following discussions with HALSEY [9.59] the compressibility derived from LEED data was compared to some simple calculations [9.9a]. The Ar data are shown in Fig.9.5 along with one of the Kr isotherms. Clearly the Kr is strongly influenced by the substrate to stay at d = 4.26 Å as long as possible. Interparticle potentials [9.60] for Kr and Ar are shown in Fig.9.6, and the ranges of d values observed by LEED are shown [9.7,9a]. The procedures for obtaining the compressibility from the data and from the potentials are described elsewhere [9.9a]. In Table 9.1, values of the compressibility at two lattice constant values are given along with values calculated from two different Ar pair potentials [9.9]. BRUCH et al. have compared the experimental values with more sophisticated calculations [9.61] and with similar data from Xe, Kr, and Ar on Ag{111} [9.37]. They concluded that the compressibilities for Ag{111} scale roughly as expected from the law of corresponding states, and obtained good agreement between theory and experiment for Ag{111} [9.37] and for Ar on graphite [9.61].

Measurements of the intensities of the Ar and graphite LEED spots as a function of vapor pressure were made at a number of temperatures between 43 K and 55 K. These always showed the appearance of ordered Ar adsorption when the coverage increased as judged by the attenuation of the graphite LEED spots [9.9a,36]. This indicates that the Ar monolayer phase transition near 50 K inferred from neutron scattering [9.57] and heat capacity [9.62] does not involve much loss of rotational order in the dense phase [9.9a]. X-ray measurements are currently being made with much higher resolution than the previous work [9.63], but it is difficult to conclude anything about the orientational ordering from such measurements.

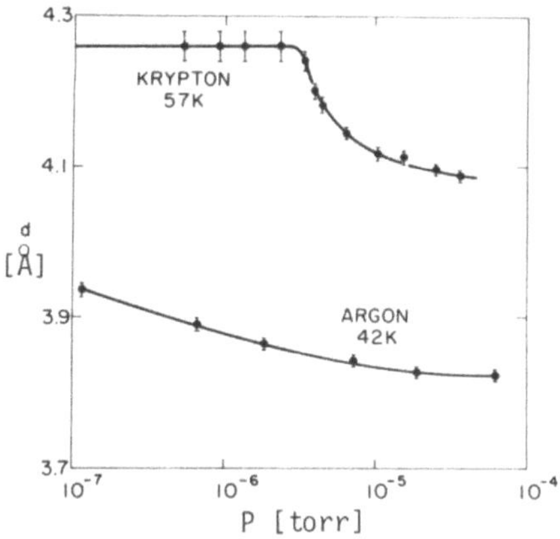

◄ Fig.9.5. Argon and krypton isotherms with lattice constants measured from LEED patterns [9.9,10]. In the krypton case, a second-order transition was assumed for data analysis

Fig.9.6. Pair potentials for Ar and Kr [9.60] with range of distances observed by LEED [9.7,9] for graphite substrate shown. The energy divided by Boltzmann's constant is given in degrees Kelvin ▼

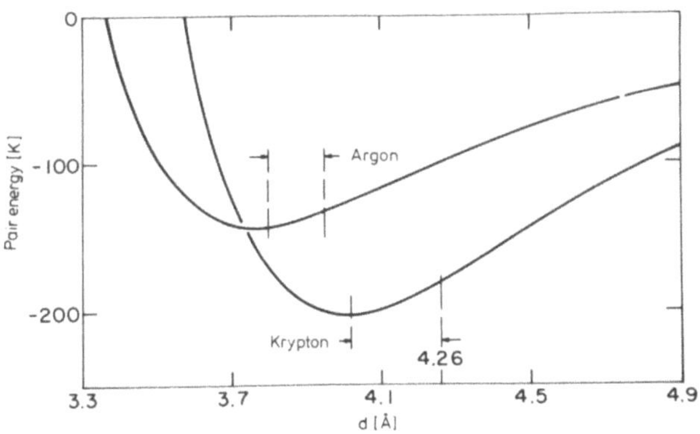

Table 9.1. Two-dimensional isothermal compressibility in units of 10^{-3} Å^2/K

	d = 3.86 Å	d = 3.905 Å
Data for Ar on graphite		
Isosteres [9.9b] (very imprecise)	3.3 ± 2.0	3.9 ± 2.0
Isotherms [9.9a]	3.2 ± 0.2	4.2 ± 0.4
Isotherms [9.9b,36] (improved analysis)	2.7 ± 0.2	3.8 ± 0.2
Simple calculations for Ar		
Barker potential [9.9a]	2.6	3.8
6-12 potential [9.9a]	2.5	3.2
Calculations for Ar on graphite		
Without registry correction [9.61]	2.4	3.1
With registry correction [9.37,61]	2.6	3.8
Data for Ar on Ag{111} [9.37]	- - -	2.7 ± 0.4
Calculations for Ar on Ag{111} [9.37]	- - -	2.7

The intensity measurements were also used to determine the dependence of Ar monolayer condensation pressure on temperature [9.9]. A comparison of this result with volumentric measurements [9.64] showed that the thermomolecular correction mentioned in Sect.9.3 is necessary [9.9a].

9.5.3 Overlayer-Substrate Spacing for an Incommensurate Monolayer

At ISISS 1975, WEBB and COHEN described a method used to obtain the overlayer-substrate spacing for Xe on Ag{111} [9.18]. WEBB later suggested that this method be applied to Ar on graphite. Although detailed intensity averaging was used for Xe on Ag{111} [9.18], the clean graphite specular-beam intensity is quite kinematic [9.65], and it was hoped that data for only one set of angles would suffice. The ratio of the specular-beam intensity from the argon monolayer on graphite to that from bare graphite did show definite oscillations related to the overlayer-substrate spacing [9.11]. The Fourier transform of these oscillations as a function of momentum transfer was compared with the Fourier transform of kinetic model calculations to determine that the Ar-graphite mean distance was 3.2 ± 0.1 Å [9.11]. Subsequently, a "background" term was removed as in EXAFS [9.66]. The transforms of the data and one model calculation are compared in Fig.9.7 after subtraction of simple linear corrections to the ratios [9.67]. It can be seen that both transforms exhibit a large peak near 3.2 A which is identical to the distance used for the model calculation because the energy dependence of the Ar phase shifts is so weak [9.67].

9.6 Nitrogen on Graphite

Diatomic nitrogen is a relatively inert gas which is widely used to estimate the surface area of powders. Very careful measurements are necessary to see structure in volumetric adsorption isotherms on graphite [9.68] because the N_2 vapor pressure is so low at temperatures ($T < 48$ K) where there is a large density difference between coexisting fluid and solid phases. This seemed to offer the possibility of using LEED to study the phase transition near 48 K, which has been variously interpreted as a solid-liquid-gas triple point [9.69], an incipient triple point [9.70], and a Potts lattice gas order-disorder transition [9.71]. After making extensive measurements of the graphite LEED intensity versus pressure for a number of different temperatures, we were unable to distinguish between these models [9.12].

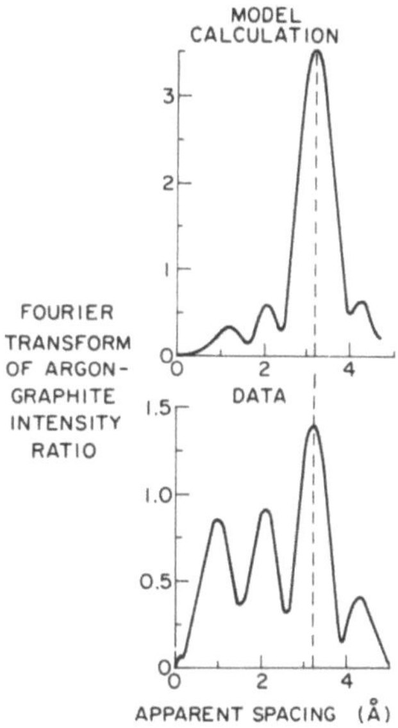

Fig.9.7. Fourier transforms of model calculations and experimental data compared for Ar on graphite [9.11,67]

In attempting to understand the N_2 commensurate - incommensurate transition better [9.12] DIEHL noted subtle differences near 30 K which led her to find a LEED pattern which indicated ordering of the N_2 molecular axes previously observed by neutrons [9.72]. However, a number of LEED spots expected from naive arguments were absent. ENGEL told us of an example where a glide plane had produced such systematic absences [9.73]. Indeed one of the hypothesized structures had the required glide planes [9.74]. All was well until the sample was tilted, for then most of the previously absent spots appeared. ENGEL then told us that such behavior was expected due to dynamical effects in LEED [9.75]. We were able to determine a low-temperature structure from our LEED patterns and results for the structure and the phase transition have been published [9.76].

9.7 Conclusions

The focus of the work discussed here has been on graphite substrates, in part due to the continuing interest in this substrate for heat capacity studies and in part due to the wide variety of other techniques applied to graphite. There are still several multiatomic inert gases which could be studied on graphite for which high-resolution LEED studies might provide useful new information. We have done preliminary work on O_2, which is an interesting case where there is not complete agreement between measurements made by neutrons and by X-rays [9.77]. Suzanne's group in Marseilles is continuing studies on graphite and already has some results for CF_4 in addition to the results for Ne mentioned here [9.38].

As amply illustrated by the studies of thermodynamics and structure of Xe, Kr, and Ar on Ag{111} [9.37,40], and of the commensurate-incommensurate transition of Xe on Cu{110} [9.51], metal substrates also offer a wide variety of interesting possibilities. Some layer compounds such as $NbSe_2$ are also suitable for LEED studies and the list goes on. The field will continue to be an interesting one. X-ray studies can be used where higher resolution is appropriate [9.46,50,63]. LEED instruments can be improved and the intrinsic surface specificity of electrons will continue to keep LEED in a unique position for the study of adsorbed films.

Acknowledgements. I wish to especially thank my colleagues M.D. Chinn, C.G. Shaw, R.D. Diehl, and M.F. Toney, who have made these experiments possible. I have had room in the text to mention only a few of the many others who contributed ideas and challenged me to clarify my own ideas. I dedicate this chapter to the memory of my wife Carolyn whose support and encouragement was important for my work described here.

References

9.1 J.G. Dash: *Films on Solid Surfaces* (Academic, New York 1975)
9.2 "Phases Bidimensionnelles Adsorbées", J. Phys. Paris *38*, Colloque C-4 (1977)
9.3 J.G. Dash, J. Ruvalds (eds.): *Phase Transitions in Surface Films* (Plenum, New York 1980)
9.4 S.K. Sinha (ed.): *Ordering in Two Dimensions* (Elsevier, North Holland, New York 1980)
9.5 M.D. Chinn, S.C. Fain, Jr.: J. Vac. Sci. Technol. *14*, 314 (1977)
9.6 M.D. Chinn, S.C. Fain, Jr.: Phys. Rev. Lett. *39* , 146 (1977)
9.7 S.C. Fain, Jr., M.D. Chinn: J. Phys. Paris Colloq. *38*, C4-99 (1977)
9.8 C.G. Shaw, S.C. Fain, Jr., M.D. Chinn: Phys. Rev. Lett. *41*, 955 (1978)
9.9a C.G. Shaw, S.C. Fain, Jr.: Surf. Sci. *83*, 1 (1979)
9.9b C.G. Shaw, S.C. Fain, Jr.: Surf. Sci. *91*, L1 (1980)
9.10 S.C. Fain, Jr., M.D. Chinn, R.D. Diehl: Phys. Rev. B*21*, 4170 (1980)
9.11 C.G. Shaw, S.C. Fain, Jr., M.D. Chinn, M.F. Toney: Surf. Sci. *97*, 128 (1980)
9.12 R.D. Diehl, C.G. Shaw, S.C. Fain, Jr., M.F. Toney: In *Ordering in Two Dimensions*, ed. by S.K. Sinha (Elsevier North Holland, New York 1980) p.199
9.13 O.E. Vilches: Annu. Rev. Phys. Chem. *31*, 463 (1980)
9.14 J.P. Hobson: Crit. Rev. Solid State Sci. *4*, 22 (1974)
9.15 A. van Oostrom: Crit. Rev. Solid State Sci. *4*, 353 (1974)
9.16 J.C. Tracy, J.M. Burkstrand: Crit. Rev. Solid State Sci. *4*, 380 (1974)

9.17 J.G. Dash: Crit. Rev. Solid State Sci. *6*, 206 (1976)
9.18 M.B. Webb, P.I. Cohen: Crit. Rev. Solid State Sci. *6*, 253 (1976)
9.19 J.H. van der Merwe: Crit. Rev. Solid State Mater. Sci. *7*, 109 (1978)
9.20 J.P. McTague, M. Nielsen, L. Passell: Crit. Rev. Solid State Mat. Sci. *8*, 135 (1978)
9.21 M. Schick: In *Phase Transitions in Surface Films*, ed. by J.G. Dash, J. Ruvalds (Plenum, New York 1980) p.65
9.22 M. Bretz, J.G. Dash: Phys. Rev. Lett. *27*, 647 (1971)
9.23 H.H. Farrell, M. Strongin, J.M. Dickey: Phys. Rev. B*6*, 4703 (1972)
9.24 Acta Electron. *14*, no.1 (January 1971); *14*, no.2 (April 1971); *16*, no.1 (January 1973)
9.25 J.K. Kjems, L. Passell, H. Taub, J.G. Dash: Phys. Rev. Lett. *32*, 724 (1974)
9.26 J.K. Kjems, L. Passell, H. Taub, J.G. Dash, A.D. Novaco: Phys. Rev. B*13*, 1446 (1976)
9.27 A. Thomy, X. Duval: J. Chim. Phys. Chim. Biol. *66*, 1966 (1969); *67*, 286, 1101 (1970)
9.28 J. Suzanne, J.P. Coulomb, M. Bienfait: Surf. Sci. *40*, 414 (1973); *44*, 141 (1974); *47*, 204 (1975); J. Cryst. Growth *31*, 87 (1975)
9.29 H.M. Kramer, J. Suzanne: Surf. Sci. *54*, 659 (1976)
9.30 J.A. Venables, H.M. Kramer, G.L. Price: Surf. Sci. *55*, 373 (1976); *57*, 782 (1976)
9.31 A. Thomy, J. Regnier, X. Duval: Colloq. Int. CNRS *201*, 511 (1972)
9.32 A. Thomy: "Changements de phase bidimensionnels dans l'adsorption de molécules simples sur le graphite"; Ph. D. Thesis, Université of Nancy, France (1968)
9.33 G.-C. Wang, M.G. Lagally: Surf. Sci. *81*, 69 (1979)
9.34 D.G. Welkie, M.G. Lagally: Appl. Surf. Sci. *3*, 272 (1979)
9.35 T.-M. Lu, M.G. Lagally: Surf. Sci. *99*, 695 (1980)
9.36 C.G. Shaw: Ph. D. Dissertation, University of Washington (1979)
9.37 J. Unguris, L.W. Bruch, E.R. Moog, M.B. Webb: Surf. Sci. *109*, 522 (1981)
9.38 S. Calisti: Ph. D. Thesis, University of Aix-Marseille II, Marseille, France (1981)
9.39 S. Calisti, J. Suzanne: Surf. Sci. *105*, L255 (1981)
9.40 J. Unguris, L.W. Bruch, E.R. Moog, M.B. Webb: Surf. Sci. *87*, 415 (1979)
9.41 P.S. Schabes-Retchkiman, J.A. Venables: Surf. Sci. *105*, 536 (1981)
9.42 J. Regnier, A. Thomy, X. Duval: J. Chim. Phys. Phys. Chim. Biol. *74*, 926 (1977)
9.43 J.J. Lander, J. Morrison: Surf. Sci. *6*, 1 (1967)
9.44 F.C. Frank, J.H. van der Merwe: Proc. R. Soc. London A*198*, 205, 216 (1949); A*200*, 125 (1949)
9.45 P. Bak: In *Solitons and Condensed Matter Physics*, Springer Series in Solid-State Science, Vol.8, ed. by A.R. Bishop and T. Schneider (Springer, Berlin, Heidelberg, New York 1978) p.216
9.46 J. Bohr, M. Nielsen, J. Als-Nielsen: Proceedings of the Third European Conference on Surface Science, Paris, 1980, ed. by D.A. Degras, M. Costa (Societe Francaise duVide, Paris 1980) Vol.1, p.112; M. Nielsen, J. Als-Nielsen, J. Bohr, J.P. McTaque: Phys. Rev. Lett. *47*, 582 (1981)
9.47 R.J. Birgeneau, E.M. Hammonds, P. Heiney, P.W. Stephens, P.M. Horn: In *Ordering in Two Dimensions*, ed. by S.K. Sinha (Elsevier North Holland, New York 1980) p.29
9.48 P.W. Stephens, P. Heiney, R.J. Birgeneau, P.M. Horn: Phys. Rev. Lett. *43*, 47 (1979)
9.49 J. Villain: In *Ordering in Strongly Fluctuating Systems*, ed. by T. Riste (Plenum, New York 1980) p.214; Surf. Sci. *97*, 219 (1980); in *Ordering in Two Dimensions*, ed. by S.K. Sinha (Elsevier North Holland) p.123

9.50 D.E. Moncton, P.W. Stephens, R.J. Birgeneau, P.M. Horn, G.S. Brown: Phys. Rev. Lett. *46*, 1533 (1981)
9.51 M. Jaubert, A. Glachant, M. Bienfait, G. Boato: Phys. Rev. Lett. *46*, 1679 (1981)
9.52 F.D.M. Haldane, J. Villain: J. Phys. Paris *42*, 1673 (1981)
9.53 J.W. Matthews: In *Epitaxial Growth, Part B*, ed. by J.W. Matthews (Academic, New York 1975) p.559
9.54 A.D. Novaco, J.P. McTague: Phys. Rev. Lett. *38*, 1286 (1977); J. Phys. Paris Colloq. *38*, C4-116 (1977)
9.55 J.P. McTague, A.D. Novaco: Phys. Rev. B*19*, 5299 (1979)
9.56 H. Shiba: J. Phys. Soc. Jpn. *46*, 1852 (1979); *48*, 211 (1980)
9.57 H. Taub, K. Carneiro, J.K. Kjems, L. Passell, J.P. McTague: Phys. Rev. B*16*, 4551 (1977)
9.58 G.L. Price, J.A. Venables: Surf. Sci. *59*, 509 (1976)
9.59 G.D. Halsey: Surf. Sci. *72*, 1 (1978)
9.60 J.A. Barker: In *Rare Gas Solids*, ed. by M.L. Klein, J.A. Venables (Academic, New York 1976) p.212
9.61 M.S. Wei, L.W. Bruch: J. Chem. Phys. *75*, 4130 (1981)
9.62 T.T. Chung: Surf. Sci. *87*, 348 (1979)
9.63 J.P. McTague, J. Als-Nielsen, J. Bohr, M. Nielsen: Unpublished manuscript
9.64 F. Millot: J. Phys. Paris Lett. *40*, L-9 (1979)
9.65 J.J. Lander, J. Morrison: J. Appl. Phys. *35*, 3593 (1964)
9.66 E.A. Stern: Contemp. Phys. *19*, 289 (1978);
C. Bouldin, E.A. Stern: Phys. Rev. (in press)
9.67 M.F. Toney: Unpublished
9.68 Y. Larher: J. Chem. Phys. *68*, 2257 (1978)
9.69 T.T. Chung, J.G. Dash: Surf. Sci. *66*, 559 (1977)
9.70 D.M. Butler, J.A. Litzinger, G.A. Stewart, R.B. Griffiths: Phys. Rev. Lett. *42*, 1289 (1979)
9.71 S. Ostlund, A.N. Berker: Phys. Rev. Lett. *42*, 843 (1979);
A.N. Berker: In *Ordering in Two Dimensions*, ed. by S.K. Sinha (Elsevier North Holland, New York 1980) p.9
9.72 J. Eckert, W.D. Ellenson, J.B. Hastings, L. Passell: Phys. Rev. Lett. *43*, 1329 (1979)
9.73 K.H. Rieder, T. Engel: Phys. Rev. Lett. *45*, 824 (1980)
9.74 C.R. Fuselier, N.S. Gillis, J.C. Raich: Solid State Commun. *25*, 747 (1978)
9.75 B.W. Holland, D.P. Woodruff: Surf. Sci. *36*, 488 (1973)
9.76 R.D. Diehl, M.F. Toney, S.C. Fain, Jr.: Phys. Rev. Lett. *48*, 177 (1982)
9.77 P.W. Stephens, P.A. Heiney, R.J. Birgeneau, P.M. Horn, J. Stoltenberg, O.E. Vilches: Phys. Rev. Lett. *45*, 1959 (1980)

10. Monte Carlo Simulations of Chemisorbed Overlayers

L. D. Roelofs

With 15 Figures

10.1 Introduction

Monte Carlo (MC) simulation is an increasingly popular method for treating complex many-body systems in several branches of physics. The simulation of lattice gas models of adlayer systems in surface physics [10.1-13] is one facet of this development and is the subject of this review.

The increased interest in MC studies applied to surface physics has occurred together with recent experimental concentration on phase diagrams of surface systems. There now exist temperature-coverage phase diagrams for H/Pd{100} [10.14],H/Ni{111} [10.15],H/W{100} [10.16], H/Mo{100} [10.17], O/W{110} [10.18], O/Ni{111} [10.19], S/Pt{111} [10.20], and others [10.21, 22]. Similar information is also available for several physisorption systems (see Chap.9). MC simulation in general must be based on considerable experimental knowledge, including not only the phase diagram itself, but also anything that is known regarding binding sites, order of transitions, etc. However a simulation in turn provides tests of the interpretations of the experimental results or of theoretical calculations.

The motivation for simulation of surface systems is usually to reproduce the experimentally observed cooperative behavior of a surface or overlayer in order to confirm or test microscopic models of the overlayer. Principally, one is interested in the interactions between adatoms or between adatoms and substrate atoms, however, sometimes thermodynamic and ordering properties are of interest. We will discuss this further in Sect.10.2, where we will also compare MC to other methods of treating cooperative systems.

In Sect.10.3 we will discuss the specific application of MC to lattice gases. We will consider briefly the usefulness of the lattice gas approximation in chemisorption systems and then introduce two common simulation modes, fixed coverage and fixed chemical potential, for studying them.

In Sect.10.4 we will discuss in more detail some of the results that have been obtained using MC for lattice gas systems. We will consider square, rectangular, triangular, and hexagonal systems.

We will conclude with a brief summary and discussion of the outlook for future work.

10.2 Motivation for Monte Carlo Simulation of Surface Systems

In many surface systems one sees cooperative, i.e., essentially many-body, behavior within the surface or overlayer. This results from the importance of the adatom interactions to the physics of the surface, and provides a nearly unique opportunity to see purely two-dimensional cooperative behavior.

It is convenient to divide surface systems into two classes, that in which the binding is strongly site specific and that in which the lateral adatom-substrate potential is only weakly corrugated. Strong chemisorption systems usually fall into the first class and will be our chief concern in this review, although one can also treat physisorption systems (Chap.9) [10.23,24] which generally fall into the second class. For the case in which the adatoms are strongly bound to specific sites the lattice gas approximation is appropriate. In this approximation one treats the overlayer as a set of binding sites with possibly varying binding energies, each of which can be either empty or singly occupied. The Hamiltonian contains interactions between adatoms that depend only on the relative locations of the sites they occupy.

To illustrate we imagine a very simple adsorption system, that of a c(2 × 2) overlayer on a square lattice [10.1,5,8-10,12]. Such an overlayer can be produced by a simple pairwise lattice gas Hamiltonian involving only pairwise interactions,

$$\mathcal{H} = E_1 \sum_{<ij>_1} n_i n_j + E_2 \sum_{<ij>_2} n_i n_j + ,,, \tag{10.1}$$

where $<ij>_k$ refers to a kth-neighbor pair of sites i and j, and n_i is the ith site occupation variable which is allowed to take on the values 0,1. To get the c(2 × 2) overlayer E_1 obviously must be repulsive ($E_1 > 0$) and E_2 attractive or not too repulsive. Some cases are given in [10.8,9]. Lattice gas Hamiltonians appropriate for other systems (see Sect.10.4) often involve a larger number of significant interactions, but even when the Hamiltonian is as simple as (10.1), one is left with a very difficult statistical mechanics problem. The exact results [10.25] available for (10.1) are limited to the case E_2 (and all longer range interactions) = 0 at coverage $\theta = 1/2$.[1,2] In other symmetry lattices (e.g., rectangular, triangular, and hexagonal) the situation is about the same. To go beyond what is known exactly, we can use MC, mean field, various renormalization group methods, series expansions or Monte Carlo renormalization group (MCRG). In this section we will give a brief introduction to the MC method for lattice gases (a more general point of view has been successfully developed in [10.11]), describe the results which can be obtained via MC, and then compare the method to those listed above.

10.2.1 Introduction to the Monte Carlo Method

The name Monte Carlo derives from the element of chance in a MC calculation. Consider the partition function of a system

$$Z = \sum_{\{n_i\}} \exp[-\beta\mathscr{H}(\{n_i\})] \quad , \tag{10.2}$$

$\{n_i\}$ denotes the set of occupation number of the sites, $i \in \{1,2,\ldots,N\}$, $\beta \equiv 1/k_BT$, and $\mathscr{H}$ is the Hamiltonian of the system expressed in terms of the occupation variables. Other observables which can also be expressed in terms of $\{n_i\}$ can be calculated using

$$\langle A\rangle = \frac{1}{Z} \sum_{\{n_i\}} A(\{n_i\})\exp[-\beta\mathscr{H}(\{n_i\})] \quad . \tag{10.3}$$

Again the sum is taken over all possible configurations of the system. The goal of Monte Carlo calculations is to obtain expectation values like (10.3) through a (partially) random selection of states. The simplest approach is to generate configurations of the system randomly, calculate the values of the observables needed and form the sums in (10.2,3). The disadvantage to this naive approach is cost, since at a given temperature an exceedingly small fraction of the possible configurations of a statistical system occur with significant weight, and thus such a calculation must be extended to a usually impractical number of configurations. To reduce computing costs it

---

1 In this example the coverage is defined as the ratio of the number of occupied sites to the total number of sites. When we discuss specific experimental systems we will use the surface science coverage convention, i.e., coverage is the ratio of the number of adatoms to the number of first-layer substrate atoms. This differs from a convention based on the number of binding sites when there is more than one binding site per surface atom.

2 There is some recent progress on the $\theta \neq 1/2$ problem which may be exact [10.26], also [10.9]

is necessary to concentrate on the more probable configurations. In practice this can be accomplished with the following algorithm. With the system initially in some state a change in the system is generated *randomly* within some space of allowed changes. The resulting change in energy, ΔE, is calculated using the Hamiltonian of the system. This change is always accepted if the energy is thereby lowered and is accepted with probability $p = \exp(-\Delta E/k_BT)$ if the energy is increased by the change[3]. Whether or not a change is made, values of the observables under consideration are calculated and accumulated in *unweighted* averages. The new configuration (or the previous configuration if the change was not accepted) is used as the initial state for the next step. This algorithm seeks out the most probable states and allows excursions from the most probable states into others, weighting them with the probability of their occurrence. The MC estimate for a given observable can be written

$$\langle A \rangle_{MC} = \frac{1}{M} \sum_{i=1} m_i A_i \qquad (10.4)$$

where the sum is now taken over all the states of the system with m_i being the number of times the ith state appears in the simulation (for most i's $m_i = 0$). A_i is the value of the observable in the ith state, and M is the total number of Monte Carlo steps taken. Because the proposed algorithm causes states to occur with relative probability $\exp(-E/k_BT)$, we clearly have in the limit of large M

$$\frac{m_i}{M} = \frac{\exp(-E_i/k_BT)}{\sum_i \exp(-E_i/k_BT)} \quad , \qquad (10.5)$$

and hence the average obtained by the Monte Carlo method (10.4) will approach the true statistical mechanical result (10.3) for large M.

It can be seen that while the calculation of any observable expressable as a function of $\{n_i\}$ is straightforward, the calculation of the free energy

$$F = -kT \ln(Z) \qquad (10.6)$$

or Z itself is not. The free energy is occasionally useful since it provides a criterion for determining first-order transition points. Therefore consider-

---

3 This algorithm actually represents one limit in a continuum of possible MC schemes. The factor $\exp(-\Delta E/kT)$ may be divided up in any fashion between the propagation between states and the average to determine the measurables. The particular choice of division is refered to as the "importance sampling" of the algorithm; see [10.11]

able effort (see references in [10.11]) has been devoted to developing accurate calculations of F. Since for most overlayer simulations, standard MC methods seem adequate for determining transition points we shall not discuss these methods further.

10.2.2 Results Obtainable via Monte Carlo

Equation (10.4) shows that any observable expressible in terms of the local variables can be calculated using MC. For example, the order parameter appropriate to a given ordered phase can be obtained. That of the $c(2 \times 2)$ overlayer produced by (10.1) is

$$\psi = \frac{1}{N}(N_A - N_B) \tag{10.7}$$

where N_A is the number of atoms in one $c(2 \times 2)$ sublattice and N_B the number in the other. We can write ψ as

$$\psi = \frac{1}{N}\sum_j p_j n_j \tag{10.8}$$

where p_j is the phase factor, $\exp[i(\pi n + \pi m)]$, associated with the jth site located at $\underline{r}_j = (na, ma)$. Other phases also have similar simply defined order parameters. By tracking all the order parameters as a function of temperature and coverage one can obtain a phase diagram for the system being modeled. For adlayer studies this is the most important result obtainable via MC. Specific examples and a discussion on the determination of transition order will be given in Sects.10.3,5.

When one is able to duplicate an experimental phase diagram, the interaction energies in the model Hamiltonian can be expected to be close to the corresponding parameters in the real system. This conclusion is subject to the *caveat* that a uniqueness proof would be quite difficult and would in any case have to be considered for each individual system. In any case it is by this means that one attempts to determine adatom interactions by MC simulation. Since at least for chemisorption systems these interactions are very difficult to calculate (Chap.11) [10.27], it is useful to have results to which comparisons can be made. In systems in which several interactions are significant, simulation is usually the only method of "measuring" interactions {for a few substrate-adsorbate combinations field ion microscopy can be used to determine adatom interactions [10.13] (Sect.10.5)}.

It is also rather easy to calculate correlation functions during MC simulations. It is useful to do so for two reasons. First, the determination of the order of a given transition can be made by examining the behavior of the correlations as a function of temperature near the transition (see Sect. 10.3). Also since ultimately the comparison is usually to low-energy electron diffraction (LEED) which measures the correlation function of the surface, the most direct comparison between simulation and experiment is the correlation function,

$$\Gamma(r) = \langle\langle n_{\underline{0}} n_{\underline{r}} \rangle\rangle = \left\langle \sum_i n_{\underline{r}_i} n_{\underline{r}_i + \underline{r}} \right\rangle \quad , \tag{10.9}$$

where the outer angular brackets denote a thermal average as in (10.3). In (10.9) $\underline{r}_i$ is the position vector of the ith site and is used interchangably with i itself to subscript local variables. In keeping with the lattice gas approximation we will restrict $\underline{r}$ in (10.9) to being one of the lattice vectors $\underline{r} \in \{r_i\}$.

A kinematic LEED experiment measures the Fourier transform of (10.9) [10.28]

$$I(\underline{q}) = \sum_{\underline{r}_i} \langle\langle n_{\underline{0}} n_{\underline{r}_i} \rangle\rangle \exp(i\underline{q} \cdot \underline{r}_i) \quad , \tag{10.10}$$

and so the most convenient (and also the most economical) way of looking at the pair correlation function is to calculate the average of the complex square of the Fourier transform of the lattice configuration (the calculation can be reduced to integer multiplies and complex addition [10.29]) and so model the expected LEED pattern directly.

An additional use to which the capability to look at correlation functions of their transforms may be put is to examine partially ordered phases, for which a structure model must be verified. The case treated by ROELOFS et al. [10.6] for O/Ni{111} is an example of this. An antiphase domain model structure proposed to explain a LEED pattern with some braodened beams and some split beams was reproduced by the simulation and found to give the correct LEED pattern (Sect.10.4).

A third use for MC simulation is to model the effect of instrumental limitations or substrate defects. The key limitation in LEED experiments is the broadening of the beams by the instrument response function [10.30], which is equivalent [10.5] to a real-space "window" through which the LEED instrument views the surface. This effect can be crudely modeled just by choosing a lattice size which roughly coincides with the LEED window [10.1]. However this procedure intertwines statistical finite size effects [10.31] with the effect of the instrument. A more defensible procedure is to simulate a larger lattice or scale the results up to a larger lattice by the use of finite size scaling [10.31] and at the end convolute with an appropriate instrument response function. The effect of imperfect temperature control can also be modeled in a straightforward way.

The case of substrate defects is also conveniently handled by MC simulation. The work of FLAGG and KLEBAN [10.32] on the effect of steps in the O/W{110} phase diagram is instructive.

Finally, we also mention that by applying finite size scaling MC has been used [10.9] to calculate critical exponents (Chap.11) [10.33].

10.2.3 Comparison to Other Methods for Treating Statistical Systems

MC would seen to be well suited for the problems connected with chemisorption systems, however, there exist other methods for treating statistical systems to which we can make comparisons. There are essentially four other such techniques which we will consider in historical order: mean field, series expansions, renormalization group (RG) methods, and Monte Carlo RG (MCRG).

The mean-field technique is well known from standard statistical mechanics texts [10.34] and is frequently used as a first treatment of a complex system since it has the advantages of simplicity and economy. The relaxation method developed by JAMES [10.35] and used extensively to treat the absorption of molecules on surfaces [10.35,36] is particularly useful for complicated systems [10.29,37]. Unfortunately, as the dimension of space decreases the importance of fluctuations increases and the usefulness of mean field decreases. The dimensionality of interest for adlayer critical phenomena is $d = 2$, rather far from the regime where mean field gives accurate results[4]. This is reflected in the extremely poor results given by mean field for overlayer phase diagrams [10.9,10,29,37]. Transition temperatures may be in error by more than a factor of 2, and there are qualitative errors in the order of transitions and presence or absence of various phases. Recent simulation work [10.39] has also shown that mean field greatly errs in predicting the effect of trio interaction in producing asymmetries in phase diagrams, thus demonstrating that the reliance on a mean-field decoupling in [10.10] for studying that effect is unjustified.

The method of series expansions [10.40] is based on a direct perturbative calculation of the partition function or various measurables. This method, extended through the use of Padé approximants, was at one time the only re-

4 Mean field gives exact results for critical behavior above a certain dimensionality of space, d_c (4 for most universality classes) [10.38]

liable way to calculate critical exponents. However, the method is and was severely restricted to simple Hamiltonians since each additional bond introduces another perturbative parameter. The method has achieved many successes, particularly in the calculation of exponents [10.40] and has been used [10.41] to calculate the effect of including E_2 (along with E_1) in (10.1). The difficulties in moving away from $\theta = 1/2$ are prohibitive. The chief drawbacks of this method are the rapidly accumulating number of contributions in higher perturbation orders and the restriction to the simplest models.

It is no exaggeration to say that the application of the RG method [10.42] is the most significant advance of the last decade in statistical mechanics. It is most powerfully (but perhaps not so intuitively) applied as an expansion in the dimension of space about the dimension at which mean field works [10.38]. (The expansion parameter is typically $\varepsilon = 4 - d$ and so there is difficulty reaching $d = 2$). Alternatively one can imagine renormalizing [10.43] a group of spins into a "block spin". The cell renormalization methods of NIEMEIJER and Van LEEUWEN [10.43,44] have been extremely successful in studying critical behavior, especially of the Ising [10.44] and Potts [10.45,46] models. This method is less well suited to the calculation of phase diagrams [10.9,44]. The Migdal-Kadanoff[5] (M-K) RG approach is however thought to produce more accurate phase diagrams and has been used extensively by BERKER and co-workers to calculate lattice gas-phase diagrams [10.23,47]. This method has the advantages of simplicity, economy, and to some extent glamor. In known systems with duality it is impressively accurate, although we have not made a detailed comparison in more complex systems. There are two problems with its use. First, on some occasions the errors M-K make in exponents are crucial to the occurrence of a particular phase or other qualitative feature in the phase diagram [10.48]. Secondly, the method is really applicable only to systems with nearest-neighbor interactions only [these interactions, however, may be much more complicated than the simple $n_i n_j$ of (10.1)]. Thus one must map the lattice gas system onto a corresponding nearest-neighbor model by means of a "prefacing transformation" [10.23,49]. This prefacing transformation is the weak point in a M-K phase diagram calculation since it cannot be accomplished without severe approxi-

5 The key difference between N-vL and M-K RG's is that the former ignores higher order bonds which increase the complexity of the calculation, while the latter never allows them to occur at all by moving other bonds. Thus in the M-K approach no bonds are "lost" and transition temperatures tend to be more accurate, although critical exponents are less accurate.

mations. Since however M-K RG is so economical and has demonstrated some success, an optimal approach to lattice gas-phase diagrams might be to do an initial M-K RG study and then adjust transition temperatures and check dubious points using MC simulation.

The final method we wish to mention is the Monte Carlo renormalization group (MCRG) recent developed by SWENDSEN [10.50]. This combination of MC methods with RG transformations has so far been applied to the calculation of critical exponents [10.50-53] and to limited[6] phase diagram calculations [10.53] for the model specified by (10.1). The method seems to allow more direct calculations of exponents if T_c is known than standard MC methods based on finite size scaling, but no detailed comparison of relative economy is available. For calculating phase diagrams the MCRG seems to be more useful for locating tricritical points [10.53] than standard MC, but in situations where no unusual finite size effects occur MC is probably more economical.

We turn now to a discussion of the implementation of MC for lattice gases.

10.3 Monte Carlo Methods for Lattice Gases

We will now consider the application of MC simulation to lattice gas models. The applicability of a lattice gas description to a chemisorption system must be tested experimentally. One must know, for example, that the adatoms are either confined to the surface or in equilibrium with some reservoir whose chemical potential is under experimental control. (The former situation is common but not universal in chemisorption systems over restricted coverage and temperature intervals [10.7]; the latter is achievable in physisorption systems (Chap.9). Both desorption and bulk dissolution must be considered. It is also important to consider the possiblity of reconstructions of the substrate. In some cases these merely renormalize the interaction constants of the lattice gas [10.6], while in others [10.16,17,54] new degrees of freedom corresponding to substrate position variables must be added to the Hamiltonian. Incommensurate phases do not necessarily rule out the use of the lattice gas approximation since it is the wave vector $\underline{q}_0$ of the order

6 One point on the phase diagram, the tricritical point marking the change from second-order transitions to first-order transitions as θ is decreased (or by symmetry increased) from $\theta = 1/2$, is obtained.

which is incommensurate with the reciprocal lattice. The real-space spacings can still all be commensurate with the real-space lattice with antiphase domains providing the "incommensuration." If the antiphase domain case applies, then the lattice gas approximation survives. If not, one must account for the motion of adatoms within their binding sites.

Some surface reactions also rule out a simple lattice gas description. For example, above an oxygen coverage of about 0.4 monolayers, the Ni{111} surface begins to oxidize [10.55]. At that coverage therefore a lattice gas description appropriate to oxygen chemisorbed on Ni{111} breaks down.

Assuming all these questions have been answered for a particular system one can program a computer to simulate adatoms on a lattice of the correct symmetry, interacting with one another via a parametrized lattice gas Hamiltonian. There are two popular simulation modes: one in which a fixed number of simulated adatoms are confined to the lattice (θ mode), and the other in which the simulated adatoms are in equilibrium with a (fictitious) bath of adatoms whose chemical potential is specified (μ mode).

10.3.1 Simulation Mode

The θ mode constitutes the most direct application of MC techniques to most chemisorption systems, and it has been used in several simulations of specific overlayer systems [10.1-7]. With it one choses a particular coverage θ and temperature T, and then the simulation determines the phase of the system and the thermodynamic and order parameters at that point. A phase diagram is produced by running at many (θ,T)'s. In thermodynamic language, the various averages are calculated using the canonical partition function. This mode resembles the experimental situation in a chemisorption system experiment since there desorption is precluded by the strong binding energy and adsorption is kinetically limited by the low gas pressure. Note that the surface is not in equilibrium with the surrounding gas in such an experiment.

The other mode arises from the simulation of model magnetic systems (e.g., Ising) and the transcription to the lattice gas system using the standard map, $n_i = (s_i + 1)/2$, where s_i is a spin variable at site i which can assume the values ±1 (the Yang-Lee analogy solves some lattice gas systems which transform to the Ising problem [10.1,37,56]). In the simulation of a model magnetic system these spins, localized on the site of the lattice, flip in response to their interactions with one another and with an external magnetic field H. One then obtains a phase diagram in (H,T) space; the magnetization M is one of the parameters calculated.

The spins correspond to occupied or vacant sites, the magnetization becomes the coverage, and the magnetic field becomes a chemical potential. Flipping a spin in the magnetic simulation corresponds to adding or removing (rather than moving) an adatom in the lattice gas simulation. Therefore in this case the chemical potential rather than the coverage is held fixed. This is the grand canonical partition function approach since now both the energy and the number of particles are allowed to fluctuate. Since the adatoms actually move on and off the lattice to and from a fictitious reservoir, this mode corresponds experimentally to a situation in which the overlayer is actually in equilibration with the surrounding gas, i.e., some physisorption experiments. In the μ mode the coverage is one of the averages calculated so that the (μ,T) phase diagram produced by the simulation can be converted to a (θ,T) phase diagram for comparison with experiment. The μ mode has been used in several lattice gas studies [10.7-10,12].

Usually only one or the other of these two models is used ([10.6] being the exception), and one can find strong expressions of preference [10.8,11]. Actually the modes are to some extent complementary, and for optimum efficiency it is necessary to use both [10.6,29].

10.3.2 Microscopic Dynamics

The coverage constraint in a θ-mode simulation is implemented by allowing only changes in system configuration which conserve the number of adatoms, e.g., hops between different sites. This brings us to the subject of the microscopic dynamics of an MC simulation.

First it should be noted that if one is interested in static properties, then the microscopic dynamics are irrelevant so long as the lattice is able to achieve equilibrium. However, when dynamic processes such as diffusion are the point of interest, then the MC dynamics should correspond closely to those in the real system. We will concentrate on the former case.

As noted above, the configuration changes allowed in a θ-mode simulation are (possibly long-range) hops between sites. In a θ-mode simulation one attempts to add or subtract an atom from the lattice including in the change in energy ΔE the cost or benefit of the applied chemical potential.

These are not the only possibilities, but they are the most convenient to program. In cases of severe metastability near first-order transitions it may be worthwhile to allow for more complex changes that more closely correspond to transitions between the two phases involved, thus allowing the system to find a more economical route between the two free-energy minima.

To avoid biasing the result, time reversal symmetry must be maintained, i.e., the reverse processes must also be considered. The programming complications are such that to our knowledge this approach has not yet been tested.

10.3.3 Order of Transitions

Assuming a particular transition point (T_c, μ_c) or (T_c, θ_c) has been located; the next question to answer is the order of the transition. To be precise we wish to differentiate between critical or continuous transitions, i.e., those in which the scale of the fluctuations (the correlation length) is divergent, and first-order transitions, i.e., those in which the free energy has a discontinuous slope at the transition point. The latter usually implies discontinuities in the order parameter and other measurables. When the particular transition is strongly first order, i.e., fluctuations are not significant, then a criterion based on order parameter or coverage discontinuity (in μ mode) succeeds. However, to differentiate a weakly first-order transition from a continuous transition is more difficult. Three approaches are popular.

As mentioned above, the widths of diffraction beam profiles are determined by the correlation length of the overlayer. A divergence of this quantity indicates that the transition is by definition critical. Below the transition region the long-range order of the Bragg part of the intensity dominates, and the width of the beams is determined by the instrument response function (or the lattice size in a simulation). Above T_c only the critical intensity remains. The correlation length is thus easily accessible, and its rapid decrease for $T > T_c$ causes a sharp and rapid increase in diffraction beam width. KORTAN's data [10.57] for the O/Ni{111} p(2 × 2) order-disorder transition shows precisely this effect, which can also be seen in simulations [10.5].

An alternative approach is based on watching for the hysteresis which occurs at a first-order transition due to the discontinuity in the energy (latent heat). This can be masked by the slow rate of transitions between phases, i.e., "time hysteresis" obscures the real hysteresis.

A third technique for determining the order of a transition is based on the character of the free energy as a function of the order parameter near the transition point [10.58]. If the transition is first-order there are (at least) two different values of the order parameter, say ψ_1 and ψ_2, (for which the system has different symmetry) at which the free energy has relative minima. By contrast, in the case of a continuous transition there is a single "mushy" well in the free energy.

In practice one implements this method by forming a histogram of the values of ψ from the states considered by the simulations. The two essential considerations for this method are: the calculation must be done near the transition point so that both wells (if there are more than one) can be explored by the simulation; and the run must be extended to long enough time so that the system can make the transition between the wells. If the barrier between the wells is very high, it can be reduced relative to the thermal fluctuations by reducing the size of the simulated lattice. Such reduction makes the transition look more nearly second order. The greatest advantage of this method over the other two is that it requires consideration of only one phase diagram point (after the transition point has been accurately determined).

There are two situations in which these methods for determining orders of transition can be frustrated. The first is that of tricriticality. A tricritical point is a point on a phase boundary which divides a line of first-order transitions from a line of second-order transitions [10.8,53, 59]. Tricritical phenomena are of general interest [10.46,60] since the critical behavior usually differs from that along the second-order line.

When one approaches a tricritical point $\mathscr{T}$ from the first-order side (along the phase boundary) the coverage jump diminishes and *vanishes* while the correlations become more and more important and actually diverge weakly at $\mathscr{T}$. If $\mathscr{T}$ is approached from the second-order side the extent of the region over which critical fluctuations occur decreases to 0. Thus all methods for distinguishing the transition order break down as one approaches $\mathscr{T}$. Moreover the finite size dependence is anomalous at a tricritical point [10.53]. By failing for these reasons to determine transition order in the vicinity of the tricritical point, MC can err in locating the position of the tricritical point.

A second problematic case arises when the number of degenerate ground states of the ordered phase is larger than the dimension of space. In this event a kinetic effect first noticed by LIFSHITZ [10.61] and recently developed quantitatively by SAFRAN [10.62] can result in the "trapping" of the system in a disordered state. Therefore, near but below a critical point the system may fluctuate into a metastable disordered state and remain for an inappropriate length of time. Clearly such an occurrence can lead to an *physically* misleading double-peaked order-parameter distribution. A rather dramatic example of this effect was encountered by NOVOTNY and LANDAU [10.63] in a simulation of the 2-d Baxter - Wu model (four degenerate ground states) which is known by exact solution to exhibit a continuous transition [10.64].

However, the time dependence of the simulation (cf. [Ref.10.3, Fig.3]) strongly suggests a metastable disordered state. Our own studies of the p(2 × 2) order-disorder transition on triangular or hexagonal lattices (again four degenerate ground states) are plagued by analogous difficulties. The use of several shorter runs to reduce the importance of trapping has been suggested [10.63]. We are exploring the use of finite size scaling to surmount the problem.

It may be worth noting that the two difficult cases – tricriticality and degeneracy of ground state greater than spatial dimensionality – may be related since an alternate definition of a tricritical point is the confluence of *three* simultaneously critical phases [10.8]. The reader is referred to other reviews of MC techniques [10.11,65] for further discussions of other technical aspects.

10.4 Monte Carlo Simulation Results

In this section we will discuss the adlayer simulation studies that have appeared. We will begin with the square lattice results [10.1,8,9,12] which have via the inclusion of trio interactions recently been applied [10.10] to the system H/Pd{100} [10.14]. We will then consider the simulations [10.3,4, 13] of the rectangular symmetry systems O/W{110} [10.18] and Si/W{110} [10.13] Then we will consider triangular symmetry [10.59] which occurs for atop binding on fcc{111} surfaces. Finally we will discuss hexagonal symmetry simulations [10.6,7] which apply to fcc{111} surfaces with binding in the threefold hollows [10.15,19].

10.4.1 Square Lattice Simulations

For the case of square symmetry, several simulations were performed before a suitable experimental system was found for comparison. DOYEN et al. [10.1] used a map from the solved [10.25] Ising problem to the lattice gas given by (10.1) to determine the order-disorder transition temperature of the c(2 × 2) state at $\theta = 1/2$, and extended the treatment to some other coverages by MC simulation. BINDER and LANDAU [10.1,8-10,12] have done more extensive calculations on (10.1), considering the cases $R = E_2/E_1 = 0.0, -0.5, -1.0, +0.25$ ([10.9] corrects [10.8]), +0.5 and 1.0. The cases with $R \leq 0$ are of more interest in overlayer systems, and the $(T - \theta)$ phase diagrams are given in Fig.10.1. One sees that increasing the magnitude of R increases the temperature of the tricritical point and broadens the single-phase region

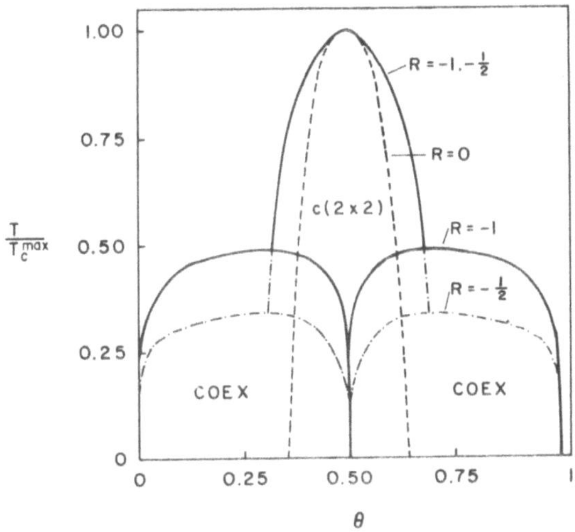

Fig.10.1. Phase diagrams for the square lattice with E_1 repulsive and three different values of E_2. $R \equiv E_2/E_1$. The region labeled c(2 × 2) is a pure phase and those labeled COEX are a coexistence of c(2 × 2) and disordered phases. These do not occur for R = 0. Tricritical points occur at the intersection of the second-order phase boundary with the boundary of the coexistence regions [10.10]

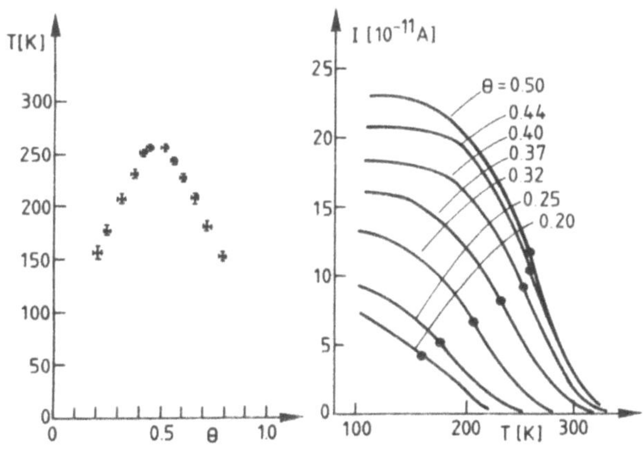

Fig.10.2. Experimental phase diagram for H/Pd{100} and raw data, LEED intensity vs temperature for several coverages [10.10]

although the broadening appears to be saturated already at R = -1/2. The phase diagram observed by BEHM et al. [10.14] for H/Pd{100} has two remarkable features. First there is the asymmetry about θ = 1/2 and second, the width of the c(2 × 2) region is larger than that in most calculated phase diagrams.

Figure 10.2 shows the experimental phase diagram and the measured $I_{1/2}$ versus T curves used to obtain it. An asymmetric phase diagram of this sort implies the loss of particle-hole degeneracy which can be achieved via the introduction of three adatom forces [10.66]. Recent work by BARTELT and EINSTEIN [10.39] and the approximate decoupling treatment in [10.10], however, show that small values for the short-range trio (T_1 in Fig.10.3) do not significantly alter the shape of the phase diagram near $\theta = 1/2$. However, BARTELT [10.39] has recently shown that if, together with T_1, another trio interaction (T_2 in Fig.10.3) is included, one can obtain a first- order contribution to the anisotropy of the $c(2\times 2)$ phase boundary near $\theta = 1/2$. T_2 was not included in the discussions of [10.10,14]. The trio interaction T_1 has by itself a profound effect on the occurrence and stability of both the low- and high-coverage coexistence regions [10.10].

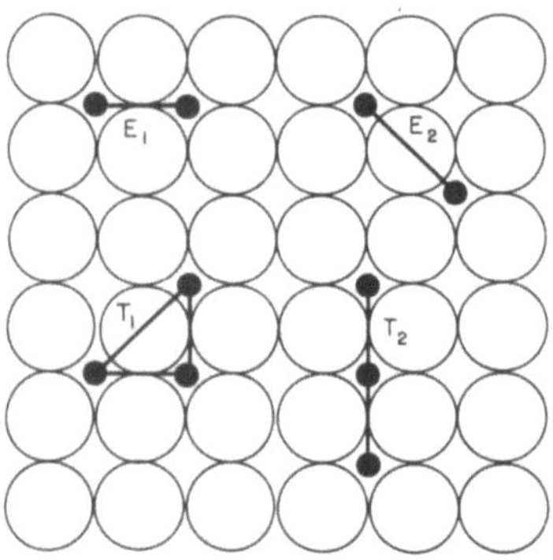

Fig.10.3. Adatom interactions for H/Pd{100}. T_2 has not yet been included in simulation studies although it together with T_1 is necessary to produce asymmetry near $\theta = 1/2$

Since detailed simulations including T_2 have not yet been accomplished, the adatom interaction estimates for the system H/Pd{100} given in [10.10,14] are unlikely to be correct.

The width of the $c(2\times 2)$ single-phase region in the experimental phase diagram has a revealing "explanation" given by BINDER and LANDAU [10.10]. They pointed out that the use of the inflection point of a plot of $I_{1/2\ 1/2}$ versus T[14] leads to a broader $c(2\times 2)$ region than obtained by (μ-mode) simulations. The lesson one learns here is that it is important to tailor the MC to the experiments so that the same conventions can be used for determining phase boundaries. It should be noted that [10.10] is still not completely consistent in this regard since their transition temperature definition is the 1/2-intensity point rather than the inflection point used in the experiment.

10.4.2 Rectangular Lattice Simulations

In this section we will actually discuss case where the substrate lattice has *centered* rectangular symmetry since the experimental substrate to be considered is W{110}. We will discuss mostly the case of oxygen adsorption for which a phase diagram was obtained by WANG et al. ([10.18], of the three plausible phase diagram interpretations of the data, the authors seem to prefer choice b; see [10.67]). Figure 10.4 shows this phase diagram. The phase boundaries without data points are schematic and very difficult to locate using LEED. Figure 10.5 shows the W{110} surface, the oxygen binding sites as determined by IV-LEED analysis [10.68][7], and the p(2 × 1) structure which is observed.

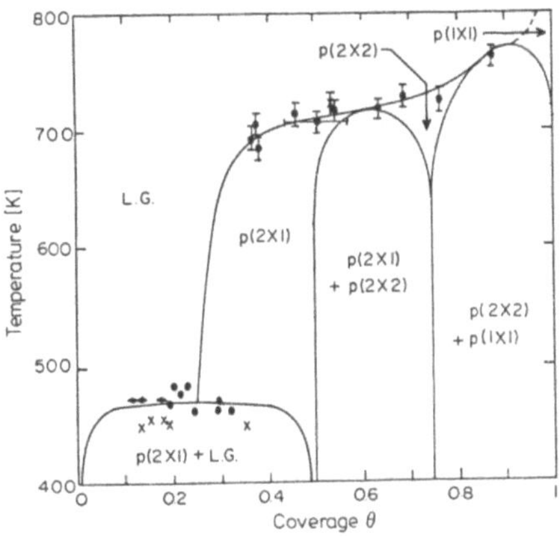

Fig.10.4 ▲

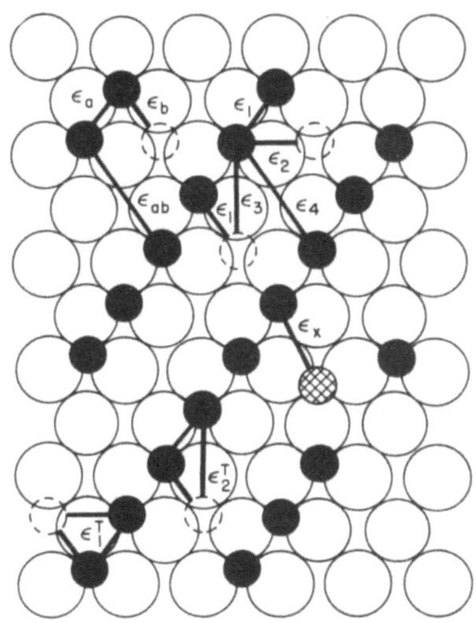

Fig.10.5 ►

Fig.10.4. Experimental phase diagram for O/W{110}. L.G. refers to a disordered phase. [10.67]

Fig.10.5. W{110} surface with oxygen atoms (···) occupying the binding sites suggested by IV-LEED analysis (see however [10.65]). The adatoms are shown in a p(2 × 1) arrangement which is seen experimentally. Interaction sets used in various simulation studies are also shown. The cross-hatched adatom occupies an equivalent binding site, giving addition bonds such as ε_X, which have not been included in the simulations

7 This binding site assignment is not universally accepted. Ref.3 specifically adopts it while Ref.4 makes no choice. Chemical arguments seem to suggest instead the twofold symmetric bridge position as shown in Fig. 10.6. See [10.69].

Since simulations of this system have been carried out by three groups [10.2-4] it is interesting as a comparison between independent results. The first group [10.2] used the set of interactions ε_a, ε_b, and ε_{2b} with ε_a and ε_{2b} attractive and ε_b repulsive, and obtained a fit to the low coverage part of the phase diagram with $\varepsilon_a = \varepsilon_{2b} = -0.048$ eV and $\varepsilon_b = 0.052$ eV. WILLIAMS et al. [10.3,70] and CHING et al. [10.4,71] pointed out that by symmetry in the notation of ERTL and SCHILLINGER [10.2] the bonds ε_a and ε_b are identical and that hence the results of that work were internally contradictory. Both groups did simulations using interactions ε_1, ε_2, ε_3, and ε_4 as shown in Fig.10.5. To reduce the number of parameters WILLIAMS et al. [10.3,70] chose to constrain $\varepsilon_2 = \varepsilon_3$, and found best fits to the disordering temperature at two coverages ($\theta = 0.25$ and $\theta = 0.53$) with the values $\varepsilon_1 = 0.09$ eV, $\varepsilon_2 = \varepsilon_3 = +.0.075$ eV, and $\varepsilon_4 = -0.03$ eV. CHING et al. [10.71] worked initially with a model which did not include ε_3; they could fit the same data well with $\varepsilon_1 = -0.072$ eV, $\varepsilon_2 = 0.080$ eV, and $\varepsilon_4 = -0.049$ eV. They also tested William's model and found agreement with the two transition temperatures, thus making the point that too little experimental information was available or being used to uniquely determine the parameters.

The observed phase diagram is grossly asymmetric about $\theta = 1/2$ with a $p(2\times 2)$ phase occuring at higher coverage. CHING et al. [10.4] therefore extended their treatment to include some short-range trio interactions, ε_1^t and ε_2^t in Fig.10.5. They found that trio interations with roughly the strength of 1/4 ε_1 could produce the observed asymmetry. EINSTEIN [10.66] pointed out that there are three additional trios of comparable strength which CHING et al. had not considered. Including these allows the strength of the trios to be reduced.

One additional comment should be made on these simulations. There is a fully equivalent (by symmetry) set of threefold sites [10.3] which seen not to be allowed in any of the simulations (see however [10.68]). The cross-hatched atom in Fig.10.5 is in a site in this equivalent set. Including this new set of sites has several effects. The symmetry of the binding site lattice is altered from a "primitive" centered rectangle to a centered rectangle with a two-site basis. Secondly, the inclusion of these sites would lead to additional bonds such as ε_x in Fig.10.5, which need to be determined.

The second MC simulation with centered rectangle symmetry concerns the system Si/W{110} studied by TSONG and CASANOVA [10.13] using field ion microscopy (FIM). In this case the binding sites (visualized directly) are the "long bridge" sites shown in Fig.10.6, and so there is a single set of sites with centered rectangular symmetry. This study is interesting because it

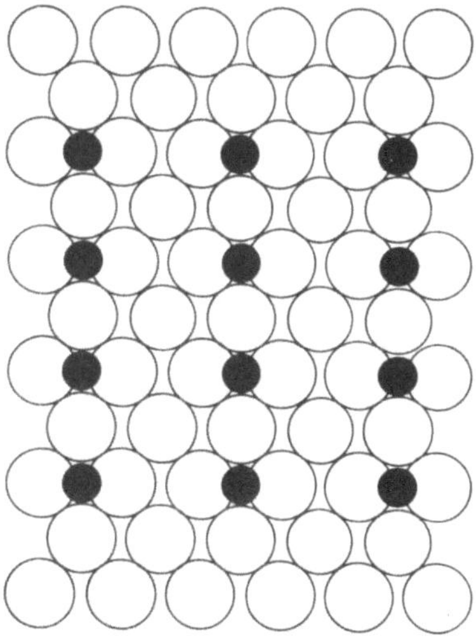

Fig.10.6. Proposed structure for Si/W {110} [10.13]

illustrates the one other method of determining interaction energies experimentally. Using FIM one (patiently) records configurations assumed by pairs of adatoms and does statistics to determine the relative energies of the various bonds. TSONG and CASANOVA made such a determination for five pairwise bonds, and then directly observed the superstructure of the saturated adlayer (Fig.10.6) using FIM. Finally via a MC simulation using those interactions, they produced the observed superstructure. Thus for this phase on a FIM tip, MC and FIM statistics give agreeable results. Unfortunately, the number of substrates and adsorbates which can be studied by FIM is very limited.

10.4.3 Triangular Lattice Simulations

Lattice gases with triangular symmetry occur when one has atop binding on fcc{111} surfaces. They are especially interesting because two common superstructures, the $(\sqrt{3} \times \sqrt{3})R30^{\circ}$ and $p(2 \times 2)$ phases, should disorder via transitions in the universality classes of the interesting 3- and 4-state Potts models (Chap.11) [10.72]. Several phase diagrams have been stimulated, but there has not been sufficiently detailed experimental study to which comparisons can be made. ROELOFS et al. [10.6] concentrated on the $p(2 \times 2)$ phase (see Fig.10.7a) simulating chemical potential-temperature (θ-T) and coverage-temperature (θ-T) phase diagrams (Fig.10.8) for the interaction values $E_2/E_1 = 0.1$ and $E_3/E_1 = -0.05$ (see Fig.10.10 for a definition of the bonds). Only low coverages were studied, but this set of interactions should also

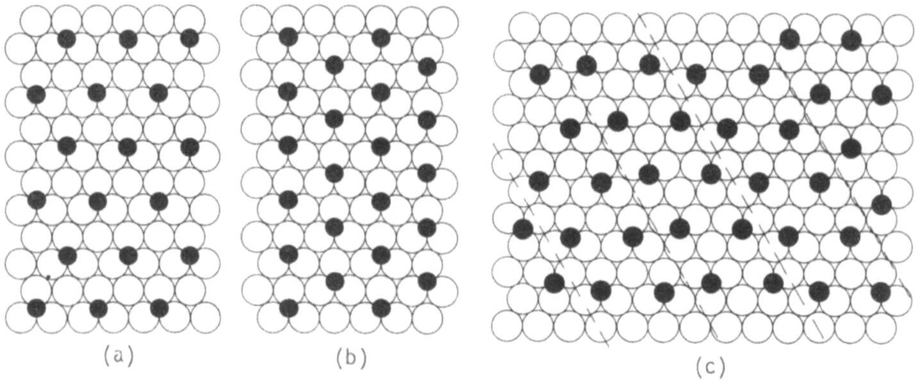

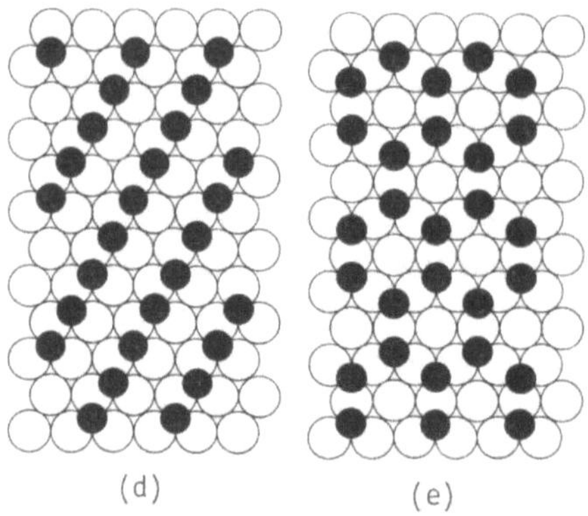

Fig.10.7a-e. Phases on triangular and hexagonal lattices (a) p(2 × 2); (b) ($\sqrt{3} \times \sqrt{3}$)R30°; (c) antiphase domain phase of O/Ni{111} (d) p(2 × 1); (e) p(2 × 2) honeycomb

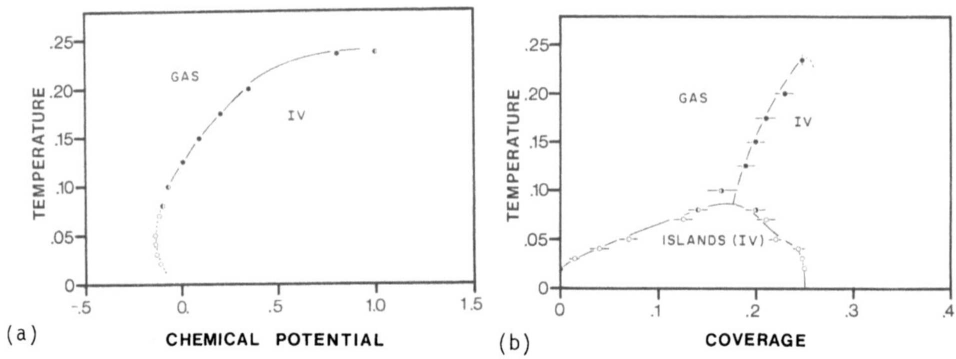

Fig.10.8a,b. Low-coverage phase diagram for triangular lattice with E_2/E_1 = 0.1 and E_3/E_1 = -0.05. Solid points denote second-order transitions and open points first-order transitions. Points for which the order is uncertain are half-filled. "IV" denotes the p(2 × 2) phase of Fig.10.7a. (a) (T-μ) phase diagram; (b) (T-θ) phase diagram [10.6]

produce a $(\sqrt{3}\times\sqrt{3})R30^{\circ}$ phase (see Fig.10.7b) at coverages near 1/3, and a $p(2\times1)$ phase near $\theta=0.5$. The phase diagram contains a tricritical point at $T = 0.08$ dividing the line of second-order transitions from first-order transitions. In the light of the anomalous size scaling behavior recently seen by LANDAU and SWENDSON [10.53] near a tricritical point, the tricritical temperature may have been overestimated.

LANDAU [10.59] has carried out a more extensive study of the $(\sqrt{3}\times\sqrt{3})R30^{\circ}$ phase for $E_3 = 0$ and $E_2/E_1 = -1.0$ (E_1 repulsive). Second-order transitions from both low- and high-coverage $(\sqrt{3}\times\sqrt{3})R30^{\circ}$ phases to a disordered phase are seen, and because of the next-nearest- neighbor attraction, there are tricritical points terminating the lines of continuous transitions. Below the tricritical points there are again coexistence regions. Near $\theta = 1/2$ the situation is somewhat complex.[8]

10.4.4 Hexagonal Lattice Simulations

The last lattice symmetry we will consider is hexagonal, i.e., triangular with a two-site basis. This case occurs when there is binding in both types of threefold hollows on fcc{111} surfaces or basal planes of hcp metals. (The two types of threefold hollows are differentiated by the presence or absence of a second-layer substrate atom beneath it and will be refered to as the hcp and fcc sites, respectively). Some ordered superstructures on hexagonal lattices are shown in Fig.10.7.

Two experimental systems with this symmetry have been extensively studied, O/Ni{111} [10.19,57] and H/Ni{111} [10.15,74,75]. We consider the second case first.

The phase diagram of H/Ni{111} is shown in Fig.10.9. The coverage scale gives the ratio of the number of H atoms to surface-layer Ni atoms. Originally the ordered state was thought to have $p(2\times1)$ symmetry (Fig.10.7d) [10.15], and the binding sites were thought to form a triangular lattice either through atop binding or binding in just one set of threefold hollows. In this case the observed asymmetry of the phase boundary about 1/2 monolayer would require very strong trio interactions. Later however DOMANY et al. [10.74] pointed out that if the structure were actually a $p(2\times2)$ honeycomb

8 According to LANDAU [10.73] the phase diagram in [10.59] is incorrect in the region near the bicritical point at $\mu = 0$ or $\theta = 1/2$. The correct phase diagram has a fixed line (see [10.48]) of continuous transitions intervening between the first-order line at $\mu = 0$ and the bicritical point.

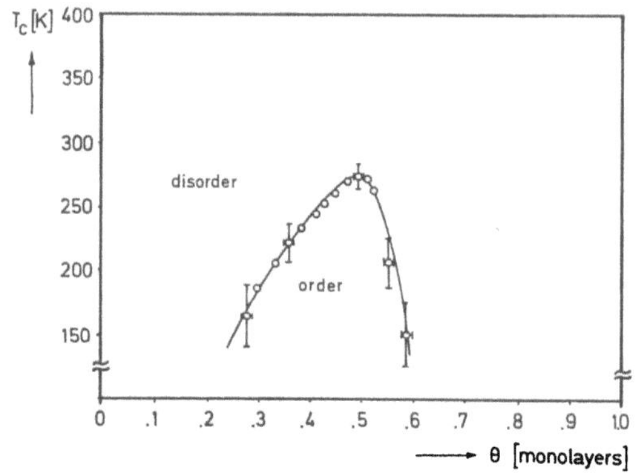

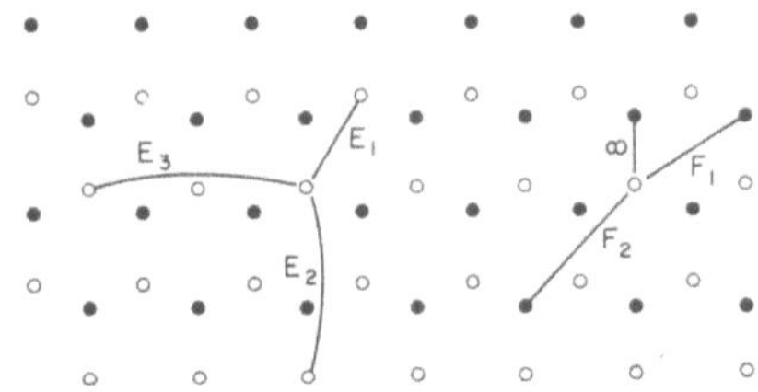

Fig.10.10 ▲

◀ Fig.10.9

Fig.10.9. Experimental phase diagram H/Ni{111}. Ordered state is now thought to be the p(2 × 2) honeycomb as in Fig.10.74 [10.15]

Fig.10.10. Interactions for hexagonal lattice. Solid circles denote fcc sites and open circles hcp sites. Interactions E couple same types of site, and F different types of sites. There may also be a binding energy difference E_0 between the two types of sites

(Fig.10.7e) and both sets of threefold troughs acted as binding sites giving a hexagonal lattice, then one would expect particle-hole symmetry only about $\theta = 1$, since that corresponds to filling 1/2 the adsorption sites. IV-LEED analysis [10.75] verified this binding site proposal.

Diffraction intensities are quite weak because H has a small scattering cross section, and thus the phase diagram has not been extended to lower coverages to look for the existence of an island region. Furthermore no detailed study of the beam profiles, again because of the weak scattering, has been done to check the order of the transition, but there is apparently no hysteresis observed, suggesting that the transitions are continuous.

ROELOFS et al. [10.6] have done an extensive set of simulations on a hexagonal lattice for application to O/Ni{111} to be discussed below, and in one of their phase diagrams studied the honeycomb structure for $\theta \leq 0.5$. The set of interactions parametrizing the lattice gas Hamiltonian is shown in Fig.10.10. For the case of H/Ni{111} only the shorter range interactions, viz., F_0 ($\to \infty$), E_0, E_1, F_2 are important as these determine whether the (2 × 1) or the honeycomb form is obtained. Figure 10.11 shows the phase diagram obtained with the following interactions: $E_1 = 1.0$, $F_1 = 0.20$, $E_2 = 0.20$, $F_2 = 0.20$, $E_3 = -0.20$, and $E_0 = 0.05$. E_0 is actually a binding energy difference between the two types of sites. This simulation clearly reproduces the general shape of the observed phase boundary, especially the asymmetry about $\theta = 0.5$, but also the correct width of the ordered region is obtained.

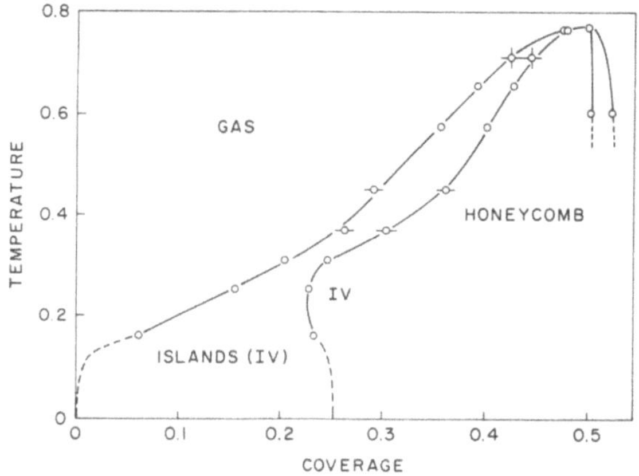

Fig.10.11. Simulated hexagonal lattice phase diagram showing existence of honeycomb phase and strong asymmetry about θ = 0.5 reminiscent of Fig.10.9. IV refers to a pure p(2 × 2) phase

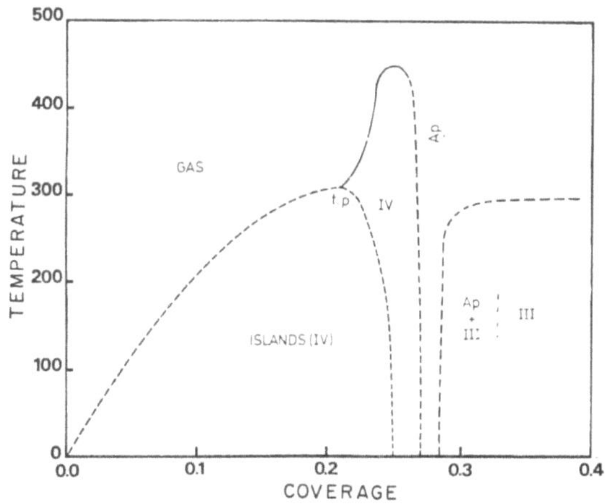

Fig.10.12. Experimental phase diagram for O/Ni{111} [10.19]. See text for description of phases. "t.p." denotes a (deduced) tricritical point

Although the phase diagram of Fig.10.10 depicts a first-order transition, this is a case of four-fold degenerate ground states and we have not been able to establish with certainty the order of the transition. One should not take the above set of parameters too seriously except to note that E_1 must be considerably more repulsive than F_1 to get the honeycomb rather than the p(2 × 1) ordered state, and to observe that the honeycomb structure can form near θ = 0.5 even with some fairly significant binding energy difference between the two sublattices.

The second system which has this symmetry is O/Ni{111}, whose experimental phase diagram [10.19] is shown in Fig.10.12. This system was also originally thought to be triangular with only one set of threefold hollows acting as binding sites [10.57], and indeed IV-LEED analysis [10.76] in this case could not determine whether the p(2 × 2) state formed in the fcc or the hcp sites. However, this phase diagram contains an (at least partially) ordered region (Ap in Fig.10.12) whose diffraction pattern is not consistent with the Ni{111} reciprocal lattice and has been analyzed [10.19,29] and found to require occupation of both sets of threefold hollows. The structure contains domains of $p(2\times 2)$ in one set of sites separated by antiphase boundaries from domains of $p(2\times 2)$ in the other set of sites as shown in Fig.10.7c. The validity of the lattice gas approximation was verified for this system with $\theta \leq 0.4$ and $T \leq 500$ K [10.19]. Recently NARUSAWA et al. [10.77] have performed an ion scattering study of O/Ni{111} and have suggested that oxygen induces vertical reconstructions (~ 0.15 Å) of the first-layer nickel atoms. Simulation of course determines the interactions as they *are*, and so, although this reconstruction introduces no new degrees of freedom into the Hamiltonian, to compare to the simulation result, eventual direct calculations will also need to account for the reconstruction. The $p(2\times 2)$ order-disorder transition is second-order near $\theta = 0.25$ and has been analyzed to determine critical exponents (Chap.11) [10.78]. The $(\sqrt{3}\times\sqrt{3})R30^{\circ}$ transition is first order and displays considerable hysteresis.

The interactions necessary to describe this phase diagram were defined in Fig.10.10. Note that the interactions must extend to at least the range of E_3 in order to produce an island $p(2\times 2)$ region (attractive E_3). The adjacent site, opposite sublattice bond, F_0, was set to ∞ by not allowing such configurations to occur in the simulations, because that configuration implies overlap of the oxygen atoms.

Clearly it is useful to obtain some constraints on the possible values of the adatom interactions thus reducing the parameter space to be searched. Some qualitative features of the phase diagram and the (extrapolated) $T = 0$ information can be used to some advantage. For example, the occurrence of the coexistence between $p(2\times 2)$ and gas requires that E_3 be attractive, i.e., $E_3 < 0$. Secondly, by assuming that the Ap phase persists all the way to $T = 0$ temperature (measurements below ~ 150 K were precluded by kinetic effects [10.19]) we can obtain constraints relating E_3, E_2, and F_2. These guided the initial parameter choices.

Initially five low-coverage hexagonal-lattice phase diagrams were simulated [10.6]. Altogether eight sets of interactions were considered and are listed in Table 10.1. We will discuss in detail the phase diagram obtained using the first set listed.

Set 1 was based on the constraints described in the preceeding section. Figure 10.13 shows the full phase diagram generated for this set. In overall appearance it is rather similar to the experimental result (Fig.10.12). The

Table 10.1. Adatom interaction sets considered in [10.6]. (N.D. - not determined)

Set #	E_1	F_1	E_2	F_2	E_3	E_0	T^{IV}_{max}	AP
1	1.0	0.667	0.05	0.167	-0.06	0.05	0.164	Yes
2	1.0	0.667	0.05	0.167	-0.06	0.0	N.D.	Yes
3	1.0	0.667	0.10	0.167	-0.06	0.10	0.19	N.D.
4	1.0	0.667	0.10	0.167	-0.06	0.20	0.19	N.D.
5	1.0	0.667	0.05	0.167	-0.06	0.30	0.196	No
6	1.0	0.667	0.05	0.167	-0.10	0.30	0.248	No
7	1.0	0.667	0.0	0.167	-0.10	0.05	0.195	Yes
8	1.0	0.20	0.20	0.20	-0.20	0.05	0.34	No

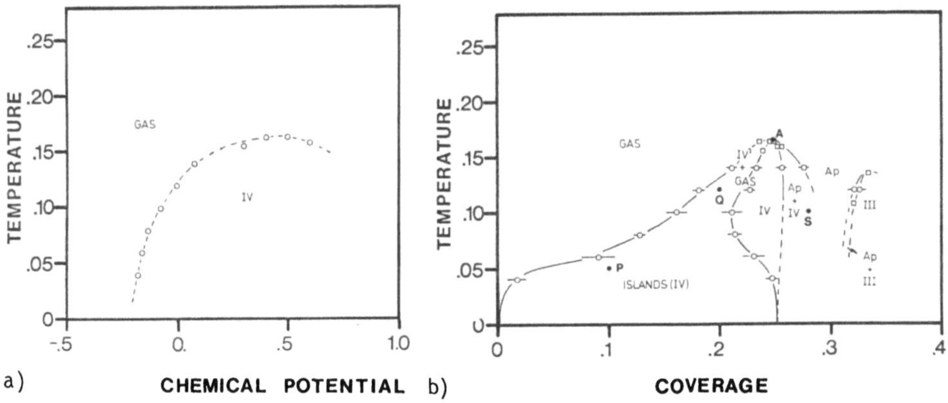

Fig.10.13a,b. Simulated phase diagram with interaction set 1 in Table 10.1. (a) (T-μ) phase diagram; (b) (T-θ) phase diagram

ordered phase observed experimentally are reproduced in the simulation. The region labelled "IV" is a long-range p(2 × 2) phase, and "III" represents a ($\sqrt{3}$ × $\sqrt{3}$)R30° phase (an inverse coverage convention in Roman numerals is used to label the phases). There are also coexistence regions and disordered phases.

As mentioned previously, there is difficulty in determining the order of the transition from p(2 × 2) to gas. At low temperatures and coverages the transition is unambiguously first order and, therefore, in the (T,θ) phase diagram (Fig.10.13b) there is a low-temperature coexistence ["Islands (IV)"]. However, at higher temperatures we are unsure as to the order of the transition. In Fig.10.13 we display the result assuming a first-order transition. If in fact the transition is actually continuous, the coexistence region labelled "IV + Gas" in Fig.10.13b would be replaced by a single second-order boundary.

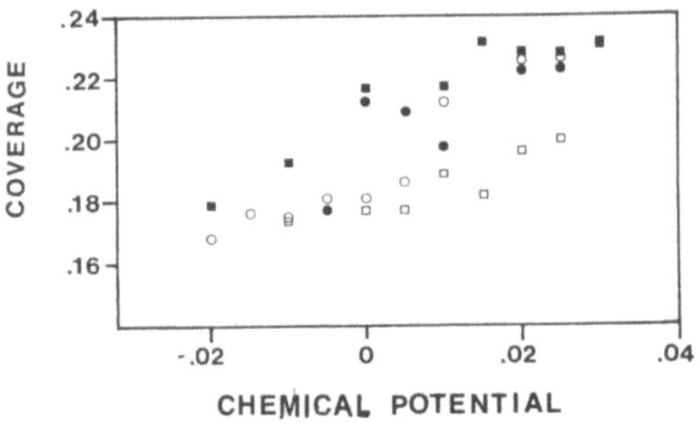

Fig.10.14. Coverage versus chemical potential from (μ-mode) simulations near the p(2×2)-gas order-disorder transition at T = 0.12 for set 1. Two full sweeps from the ordered state at μ = 0.025 to the gas at μ = -0.02 and back are shown. Filled points are from the first sweep and open points from the second. Circles denote points calculated during the decrementation of μ, and squares those calculated as μ was increased to its original value. The coverage gap suggests a first-order transition

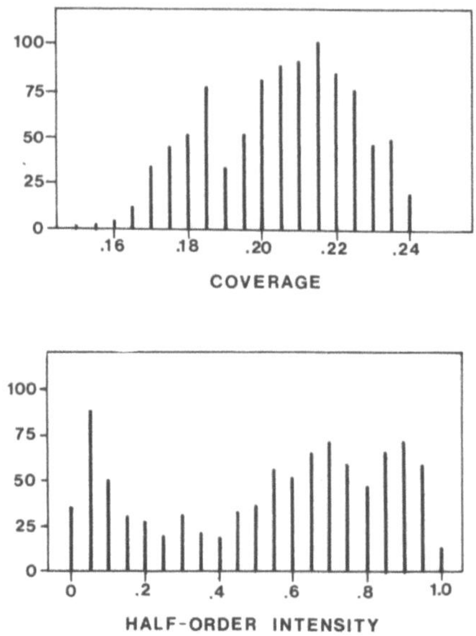

Fig.10.15a,b. Histograms of coverage (a) and order-parameter squared (b) from a long run near the transition studied in Fig.10.14

Our investigation of this question, if inconclusive, is extensive. Consider the transition with T fixed at 0.12. A plot of coverages versus chemical potential is shown in Fig.10.14. The circles represent points obtained while the chemical potential was decremented. The plot contains two runs, both of which began with a fully ordered p(2×2) overlayer and μ = 0.025. The quantity μ was then decremented (between calculations) to μ = -0.02. In neither case was there a single calculation during which the transition took place. After reaching -0.02, μ was sequentially incremented back up to 0.03, all at T = 0.12. Although the hysteresis is not reproducible, there is apparently a coverage gap, suggesting a first-order transition and providing the two points to delineate the coexistence region at T = 0.12. Figure 10.15 shows histograms of both the coverage and half-order LEED intensity [i.e., square of p(2×2) order parameter] for a long run (~ 1000 hops per adatom) at the point T = 0.12, μ = 0.005, very close to the transition. The values sorted into the histograms were extracted from the simulated lattice at intervals of one hop per adatom. The coverage histogram is bimodal with peak values at 0.185 and 0.215. These values correspond well with the limits of the coverage gap in Fig.10.14. The order-parameter distribution is also bimodal. Here, however, we note that the possibility of trapping in the disordered state cannot be ruled out. Of particular significance is the fact that the distribution of values in the upper peak is extremely broad, suggesting critical behavior. There are two possible interpretations of this evidence. Either a second-order transition is being preempted by a weakly first-order transition, or the distributions reflect trapping in a metastable disordered state.

The region to the right of the $p(2\times 2)$ region is the "Ap" phase. In this region we hope to reproduce the correct LEED pattern [10.19] which contains some split extra spots and some broadened spots. Direct simulation of LEED patterns showed nearly the correct behavior, but where splitting was expected there was only a slight dip in the center of the beam. This deficiency was related to the fact that even with E_0 so small $(0.05\times E_1)$ the $p(2\times 2)$ domains in the favored sites were larger than those in the repressed sites. By decreasing E_0 the splitting could be made complete. This establishes that E_0 is very small for this system.

By usual criteria the transition from the $p(2\times 2)$ to the "Ap" phase seems to be first order, but again we cannot yet rule out the possibility that the trapping mechanism has produced this appearance.

Finally we turn to the $(\sqrt{3}\times\sqrt{3})R30^\circ$ regions, (III) and (III + Ap). The $(\sqrt{3}\times\sqrt{3})R30^\circ$ also occurs in the (T,μ) phase diagram, but is not shown in Fig.10.13a since that region was investigated primarily with the θ-mode simulation, which does not calculate μ; however, some limited μ-mode runs showed that the transition temperature maximum of the $(\sqrt{3}\times\sqrt{3})R30^\circ$ phase occurs in the neighborhood of $\mu = 1.3$. The transition out of the $(\sqrt{3}\times\sqrt{3})$ $R30^\circ$ phase is unambiguously first order everywhere along the boundary. There is no $(\sqrt{3}\times\sqrt{3})$ coordination anywhere on the disorder side of the boundary. The first-order nature of the $(\sqrt{3}\times\sqrt{3})R30^\circ$ transition is consistent with the predictions of DOMANY and SCHICK [10.79]. They found that, although for large E_0, this transition may be continuous and in the 3-state Potts universality class, for E_0 zero or small, the transition must be first order.

Comparison of Figs.10.12,13b shows that the agreement between simulation and experiment is fair. The various long-range ordered phases are reproduced, as is the short-range ordered "Ap" phase. There are some discrepancies. Relative to the peak disorder temperature of the $(\sqrt{3}\times\sqrt{3})R30^\circ$ phase, the $p(2\times 2)$ at $\theta = 0.25$ disorders at a higher temperature in the experiment than in the simulation. Also we must still resolve the question of the order of the $p(2\times 2)$ order-disorder transition. Finally there may be differences in the shape of the "Islands (IV)" coexistence region in the simulation and experiment.

10.5 Summary and Discussion

This review has attempted to give an introduction to some of the technical aspects of MC simulations, their motivation, some of the subtleties and pitfalls, and a sample of the experiments upon which simulations are based.

MC techniques perhaps augmented by renormalization group methods remain among the most useful for developing whole phase diagrams for comparison with experiment and are likely to become even more important as computer technology continues to develop. Especially promising are simulations on dedicated minicomputers, since large memory and word size are not needed, while the marginal difference in speed can be compensated by greater accessibility.

However, one cannot yet point to a real physical system for which a simulation including all of the important parameters has reproduced the experimental phase diagram. This is not so much a consequence of technique, as it is of the difficulty of the problem or the insufficiency of experimental input. The root of the problem is often the large number of interaction parameters to be determined. For example in O/Ni{111} there are six interactions, not including any trios. Also there are pitfalls to be avoided such as curious behavior recently found in the triangular lattice [10.73] and the difficulty of determining transition order in cases of ground-state degeneracy larger than spatial dimension.

The current state of the art is that success for the system O/Ni{111} seems close at hand. N/Pd{100} and H/Ni{111} are also within reach, although in both cases additional experimental work at lower temperatures would be useful. No sufficiently powerful calculations of adatom interactions yet exist for comparison, although there is also significant recent progress in this area which is beyond the scope of this review.

For the future one expects that other systems in several symmetries, especially square and triangular, will be carefully studied. Hopefully more experimental phase diagrams will be obtained. The nucleus of simulation results available now will speed future MC studies.

Acknowledgments. It is a pleasure to acknowledge the contribution of T.L. Einstein, R.L. Park, and A.R. Kortan both to some of the work described in this paper and to the paper itself. In addition thanks are due to D.P. Landau, M.G. Lagally, G. Ehrlich, S.C. Ying, P.J. Estrup, T.T. Tsong, and K. Hayek for useful discussions and/or allowing the use of their results prior to publication. Also the author is very grateful to N.C. Bartelt for extending the hexagonal lattice simulations and thereby revealing the ambiguity of the order of the p(2 × 2) order-disorder transition. It is also necessary to apologize to those authors whose work should be, but is not, because of limitations of space, included in this review.

References

10.1 G. Doyen, G. Ertl, M. Plancher: J. Chem. Phys. *62*, 2957 (1975)
10.2 G. Ertl, D. Schillinger: J. Chem. Phys. *66*, 2569 (1977)
10.3 E.D. Williams, S.L. Cunningham, W.H. Weinberg: J. Chem. Phys. *68*, 5 (1978)
10.4 W.Y. Ching, D.L. Huber, M.G. Lagally, G.-C. Wang: Surf. Sci. *77*, 550 (1978)
10.5 L.D. Roelofs, Robert L. Park, T.L. Einstein: J. Vac. Sci. Technol. *16*, 478 (1979)
10.6 L.D. Roelofs, T.L. Einstein, N.C. Bartelt: Phys. Rev. B (to be published)
10.7 L.D. Roelofs, A.R. Kortan, T.L. Einstein, R.L. Park: J. Vac. Sci. Technol. *18*, 492 (1981);
Robert L. Park, A.R. Kortan, T.L. Einstein, L.D. Roelofs: In *Ordering in Two Dimensions*, ed. by S.K. Sinha (North Holland, New York 1980)
10.8 K. Binder, D.P. Landau: Surf. Sci. *61*, 577 (1976)
10.9 K. Binder, D.P. Landau: Phys. Rev. B*21*, 1941 (1980)
10.10 K. Binder, D.P. Landau: Surf. Sci. *108*, 503 (1981)
10.11 K. Binder (ed.): *Monte Carlo Methods in Statistical Physics*, Topics in Current Physics, Vol.7 (Springer, Berlin, Heidelberg, New York 1979) p.337
10.12 D.P. Landau: In Ref.10.11
10.13 T.T. Tsong, R. Casanova: Phys. Rev. Lett. *47*, 113 (1981);
R. Casanova, T.T. Tsong: Surf. Sci. *109*, L497 (1981)
10.14 R.J. Behm, K. Christmann, G. Ertl: Surf. Sci. *99*, 320 (1980)
10.15 R.J. Behm, K. Christmann, G. Ertl: Solid State Commun. *25*, 763 (1978)
10.16 R.A. Barker, P.J. Estrup: J. Chem. Phys. *74*, 1442 (1981)
10.17 P.J. Estrup: J. Vac. Sci. Technol. *16*, 635 (1979);
C.H. Huang: Ph. D. Thesis, Brown University (1979)
10.18 G.-C. Wang, T.-M. Lu, M.G. Lagally: J. Chem. Phys. *69*, 479 (1978)
10.19 A.R. Kortan, R.L. Park: Phys. Rev. B*23*, 6340 (1981)
10.20 K. Hayek: Private communication
10.21 E. Bauer: In *Phase Transition in Surface Films*, ed. by J.G. Dash, J. Ruvalds (Plenum, New York 1980) p.267
10.22 L.D. Roelofs: Appl. Surf. Sci. (in press)
10.23 A.N. Berker, S. Ostlund, F.A. Putnam: Phys. Rev. B*17*, 3650 (1978)
10.24 L.W. Bruch, J.M. Phillips: Surf. Sci. *91*, 1 (1980)
10.25 L. Onsager: Phys. Rev. *65*, 117 (1944)
10.26 E. Muller-Hartmann, J. Zittartz: Z. Phys. B*27*, 261 (1977)
10.27 T.L. Einstein: CRC Crit. Rev. Solid State Mat. Sci. *7*, 261 (1978) [and references therein]
10.28 M.B. Webb, M.G. Lagally: Solid State Phys. *28*, 306 (1973)
10.29 L.D. Roelofs: Ph. D. Thesis, University of Maryland (1980)

10.30 Robert L. Park, J.E. Houston, D.G. Schreiner: Rev. Sci. Instrum. *42*, 60 (1971)
10.31 M.E. Fisher, A.E. Ferdinand: Phys. Rev. Lett. *19*, 169 (1967); A.E. Ferdinand, M.E. Fisher: Phys. Rev. *185*, 832 (1969); M.E. Fisher, M.N. Barber: Phys. Rev. Lett. *28*, 1516 (1972); M.E. Fisher: In *Proc. Intern. Summer School Enrico Fermi 1970,* Course 51, Varenna, Italy (Academic, New York 1971)
10.32 P. Kleban, R. Flagg: Surf. Sci. *103*, 552 (1981)
10.33 H.E. Stanley: *Introduction to Phase Transitions and Critical Phenomena* (Oxford University Press, New York 1971)
10.34 F. Reif: *Fundamentals of Statistical and Thermal Physics* (McGraw-Hill, New York 1965) p.430
10.35 H.M. James, J.C. Raich: Phys. Rev. *162*, 649 (1968)
10.36 A.B. Harris, A.J. Berlinsky: Can. J. Phys. *57*, 1852 (1979) [and references therein]
10.37 R.M. Roberts, L.D. Roelofs: Unpublished
10.38 A.Z. Patashinkii, V.L. Pokrovskii: *Fluctuation Theory of Phase Transitions* (Pergamon, Oxford 1979) p.25
10.39 N.C. Bartelt, T.L. Einstein: To be published
10.40 C. Domb, M.S. Green (eds.): *Phase Transitions and Critical Phenomena*, Vol.3 (Academic, New York 1974)
10.41 N.W. Dalton, D.W. Domb: J. Math. Phys. *10*, 1271 (1969)
10.42 M.E. Fisher: Rev. Mod. Phys. *46*, 597 (1974)
10.43 Th. Niemeijer, J.M.J. Van Leeuwen: In *Phase Transitions and Critical Phenomena*, Vol.6, ed. by C. Domb, M.S. Green (Academic, New York 1977) p.425
10.44 K.R. Subbaswamy, G.D. Mahan: Phys. Rev. Lett. *37*, 642 (1976)
10.45 R.B. Potts: Proc. Cambridge Philos. Soc. *48*, 106 (1952)
10.46 B. Nienhuis, A.N. Berker, E.K. Riedel, M. Schick: Phys. Rev. Lett. *43*, 737 (1979)
10.47 A.N. Berker: In *Ordering in Two Dimensions,* ed. by S.K. Sinha (North Holland, New York 1980) p.9
10.48 J.V. Josê, S. Kirkpatrick, L.P. Kadanoff, D.R. Nelson: Phys. Rev. B*16*, 1217 (1977)
10.49 L.D. Roelofs, T.L. Einstein, N.C. Bartelt: Phys. Rev. Lett. *47*, 1348 (1981) [response to comment by M. Schick]
10.50 R.H. Swendsen: Phys. Rev. Lett. *42*, 859 (1979)
10.51 R.H. Swendsen: Phys. Rev. B*20*, 2080 (1979)
10.52 C. Rebbi, R.H. Swendsen: Phys. Rev. B*21*, 4094 (1980)
10.53 D.P. Landau, R.H. Swendsen: Phys. Rev. Lett. *46*, 1437 (1981)
10.54 K.H. Lau, S.C. Ying: Phys. Rev. Lett. *44*, 1222 (1980); in *Ordering in Two Dimensions,* ed. by S.K. Sinha (North Holland, New York 1980) p.193
10.55 P.H. Holloway, J.B. Hudson: Surf. Sci. *43*, 123, 141 (1974)
10.56 T.D. Lee, C.N. Yang: Phys. Rev. *87*, 404, 410 (1952); M.E. Fisher: In *Lectures in Theoretical Physics* (University of Colorado Press, Boulder, CO 1964) Vol.VIIc
10.57 A.R. Kortan, P.I. Cohen, Robert L. Park: J. Vac. Sci. Technol. *16*, 541 (1979)
10.58 S.J. Knak Jensen, O.G. Mouritsen, E.K. Hansen, Per Bak: Phys. Rev. B*19*, 5886 (1979)
10.59 D.P. Landau: Phys. Rev. Lett. *28*, 449 (1972); B. Mihura, D.P. Landau: Phys. Rev. Lett. *38*, 977 (1977)
10.60 M.E. Fisher, S. Sarbach: Phys. Rev. Lett. *41*, 1127 (1978)
10.61 I.M. Lifshitz: Zh. Eksp. Teor. Fiz. *42*, 1354 (1962) [Sov. Phys. JETP *15*, 939 (1962)]
10.62 S.A. Safran: Phys. Rev. Lett. *46*, 1581 (1981)

10.63 M.A. Novotny, D.P. Landau: Phys. Rev. B*24*, 1468 (1981)
10.64 R.J. Baxter, F.Y. Wu: Phys. Rev. Lett. *31*, 1294 (1973); Aust. J. Phys. *27*, 357 (1974);
R.J. Baxter: Aust. J. Phys. *27*, 368 (1974)
10.65 K. Binder: Adv. Colloid Interface Sci. *7*, 279 (1977)
10.66 T.L. Einstein: Surf. Sci. *84*, L497 (1979)
10.67 M.G. Lagally, T.-M. Lu, G.-C. Wang: In *Ordering in Two Dimensions*, ed. by S.K. Sinha (North Holland, New York 1980) p.113
10.68 M.A. Van Hove, S.Y. Tong: Phys. Rev. Lett. *35*, 1092 (1975)
10.69 G. Theodorou: Surf. Sci. *81*, 379 (1979)
10.70 E.D. Williams, S.L. Cunningham, W.H. Weinberg: J. Vac. Sci. Technol. *15*, 417 (1978)
10.71 W.Y. Ching, D.L. Huber, M. Fishkis, M.G. Lagally: J. Vac. Sci. Technol. *15*, 653 (1978)
10.72 E. Domany, M. Schick, J.S. Walker, R.B. Griffiths: Phys. Rev. B*18*, 2209 (1978)
10.73 D.P. Landau: Conf. on Phase Transitions on Surfaces, Univ. of Maine, Orono, August 1981
10.74 E. Domany, M. Schick, J.S. Walker: Solid State Commun. *30*, 331 (1979)
10.75 K. Christmann, R.J. Behm, G. Ertl, M.A. Van Hove, W.H. Weinberg: J. Chem. Phys. *70*, 4168 (1979)
10.76 P.M. Marcus, J.E. Demuth, D.W. Jepsen: Surf. Sci. *53*, 501 (1975)
10.77 T. Narusawa, W.M. Gibson, E. Törnquist: Phys. Rev. Lett. *47*, 417 (1981)
10.78 L.D. Roelofs, A.R. Kortan, T.L. Einstein, R.L. Park: Phys. Rev. Lett. *46*, 1465 (1981);
M. Schick: Phys. Rev. Lett. *47*, 1347 (1981)
10.79 E. Domany, M. Schick: Phys. Rev. B*20*, 3828 (1979)

11. Critical Phenomena of Chemisorbed Overlayers

T. L. Einstein

With 8 Figures

11.1 Introduction

Critical phenomena refer to the singular behavior of various thermodynamic properties, particularly the divergence of the size of fluctuations near second-order phase transitions. This behavior has surprisingly little to do with the microscopic Hamiltonian describing the system or the species undergoing the phase change; instead, it is determined by the dimensionality of the system and the symmetry of the ordered state. Based on these attributes, transitions can be classified into a small number of universality classes. All members of such a class have common critical exponents which describe the singularities near the transition. The dependence on spatial dimensionality d is quite strong. For d = 1, fluctuations are so strong that long-range order is impossible (except when long-range forces are present). In fact, for a pure 2-d system, long-range order as conventionally defined is, in principal, also impossible (two recent reviews are [11.1]), although in realistic experimental systems these limitations enter only weakly. On the other hand, similar considerations lead to predictions of two-stage melting in 2-d, with a distinctive "hexatic" intermediate phase [11.2].

For a 2-d lattice it is well known that long-range order can exist in a situation with only short-range interactions: witness the celebrated ONSAGER solution to the Ising model [11.3]. Other less familiar models have also recently been solved in 2-d (e.g., hard hexagon or three-state Potts class [11.4a], triple spin or Baxter-Wu or four-state Potts [11.4b], and eight vertex [11.4c]), while none have been solved exactly in 3-d (save the spherical model [11.5a]; the infinite-d spin "Ising" model has a nearly identical solution [11.5b,c]). A number of theoretical techniques, described in Sect.10.2.3 (this chapter introduces many of the lattice gas concepts relied on here), have been developed to deal generally with the features of a 2-d lattice. A particularly favorable trait of 2-d second-order transitions is that the critical region is orders of magnitude larger than in 3-d, making experimental observation considerably easier.

To test the theories associated with 2-d critical phenomena, one needs genuinely 2-d systems, rarities in nature. Atoms adsorbed on essentially passive surfaces offer the desired 2-d properties. Several recent conferences have explored this exciting area [11.6,7]. Fortunately, their proceedings, in most cases, have been published [11.6], making this progress available to a wide audience. Two excellent reviews [11.8,9] will appear soon, too. The current presentation will only be able to focus on a limited selection of topics. The hope is to whet the reader's appetite by presenting some highlights of special interest to those studying adsorption. We assume some familiarity with surface science but very little with progress in static critical phenomena. (An outstanding series of review articles is in [11.10].)

In Sect.11.2, a general introduction to critical properties is given. The ideas of critical exponents, corrections to scaling, and crossover phenomena are presented in terms appropriate to surface science. Section 11.3 discusses order parameters and universality classes. "Universal" features do not depend on any of the explicit parameters of a system, just on symmetries and dimensionality. This section presents a general scheme for determining what sorts of universality classes can be found in commensurate adsorption systems. Section 11.4 discusses how LEED measurements can probe critical phenomena and explores some of the problems that arise in actual experiments, notably substrate disorder. Section 11.5 describes the few experiments undertaken to measure critical exponents for adsorbed overlayers. The only chemisorption case so studied, O/Ni{111}, is examined carefully. Section 11.6 mentions some implications of this work.

11.2 Important Concepts

11.2.1 Lattice Gas Model

An important feature of all chemisorption and most physisorption systems is that, at low coverages, the adatoms sit at specific positions relative to the substrate atoms. Thus, it is appropriate to describe these systems with lattice gas Hamiltonians [see (10.1)]. For present purposes, all but the first few pairwise interactions E_i can be neglected. We remind the reader that these E_i fall off rapidly in magnitude with increasing separation and, in the case of chemisorption on transition metals, are anisotropic and can be attractive or repulsive. These pairwise interactions, as well as higher order interactions, were reviewed in ISISS 1977 [11.11]; there has since been progress in calculating signs and rough magnitudes for H on narrow-band transition metals [11.12], as well as in general understanding [11.13]. We emphasize the distinction from physisorption, where the E_i's originate essentially from van der Waals forces and so are invariably strongly repulsive at small spacings R_i, then (decreasingly) attractive for larger R_i.

In chemisorption, as opposed to physisorption, the binding energy is so much larger than the E_i's that the fractional coverage $\theta \equiv \mathcal{N}/N$ [N being the number of lattice sites, $\mathcal{N} = \sum n_i$ the number of adatoms] is essentially fixed at temperatures of order to E_i/k_B, at which 2-d phase transitions occur (in contrast to physisorption). The phase transitions that concern us here are order-disorder processes. These are the only kind that occur in simple systems with only a few repulsive pair interactions and no attractions. The intuitive picture is that adatoms try to avoid each other but are "forced" closer together by increased coverage. For example, in the case of a square lattice with only nearest neighbor repulsions, the adatoms occupy sites randomly at low θ, but (at low temperatures) for some θ between 1/4 and 1/2 form an ordered checkboard pattern with only one "color" of site occupied (to avoid the penalty energy E_1). As temperature increases, the entropy gained by occupying both sites outweighs the energy considerations in the free energy, and the $c(2\times 2)$ pattern disorders (cf. R = 0 curve [11.14] in Fig.10.1). The important feature in determining critical behavior is the number and symmetry of equivalent sites available for disordering. A second kind of transition occurs at low coverages when attractive interactions are present. Then, ordered islands form, e.g., in the square lattice gas case, an attractive E_2 leads to ordered checkboard islands at coverages well below 1/4 (in fact down to 0). With increasing temperature, adatoms "evaporate" into the gas background, with associated latent heat; the transition is first order. At higher coverages, one again encounters order-disorder type transition (cf. R = -1, -1/2 plots [11.14] in Fig.10.1).

We recall the expression for expectation values of observables given in (10.2). At high temperatures, all sites are occupied with equal probability, formally $\langle n_i \rangle = \theta$ and more importantly $\langle n_i\, n_j \rangle = \theta^2$, for all i,j. Below the transition temperature T_c, the lattice sites can be imagined as dividing into a [small] number of sets of equivalent sites having some lower symmetry than the lattice as a whole. Each set has the same number of sites.

As we shall see more clearly below, the idea is that below T_c one of these sets is preferentially occupied.

11.2.2 Critical Exponents and Scaling Laws

The measurable thermodynamic properties are usually defined in terms of the analogous magnetic Hamiltonian for spins on a lattice. Thus the independent variables are magnetic field h and temperature T [or often more conveniently, reduced temperature $t \equiv (T - T_c)/T_c$]. In chemisorption experiments one measures the divergences of the correlation length $\xi\{\langle s_i\, s_j\rangle \sim \ln(R_{ij}) \times \exp(-R_{ij}/\xi)\}$ in 2-d and the susceptibility $\chi[|-(\partial^2F/\partial h^2)|_T]$ with decreasing $|t|$,

$$\xi \propto |t|^{-\nu} \quad ; \quad \chi \propto |t|^{-\gamma} \quad . \tag{11.1a}$$

The "magnetization", i.e., the order parameter (to be defined shortly) vanishes as

$$m \propto (-t)^{\beta} \quad [t < 0 \quad \text{only}] \quad . \tag{11.1b}$$

Expressions for these measurables in magnetic language are given in most general reviews [11.15,16]; we shall encounter their adsorption appearance in Sect.11.4. In addition, we mention the specific heat

$$C = (kT^2)^{-1}(\langle\mathscr{H}^2\rangle - \langle\mathscr{H}\rangle^2) \propto |t|^{-\alpha} + \text{const.} \tag{11.1c}$$

While this quantity cannot be measured in chemisorption, it can sometimes in physisorption and is useful in Monte Carlo simulation. Other exponents are not readily accessible.

This Greek alphabet soup is dramatically simplified by the *scaling hypothesis*. This experimentally corroborated idea states that the divergence of ξ (i.e., the long-range correlations of fluctuations near T_c) is responsible for all singular behavior; ξ is the only relevant length in determining the singular behavior at criticality [11.16]. This hypothesis is consistent with the independence of critical exponents from details of the microscopic Hamiltonian and the lattice constant. One consequence of the hypothesis was used implicitly in writing (11.1a,c), namely, that α, γ, and ν are the same above and below T_c (note that the proportionality constants are *not* identical). More generally, only two exponents are independent; this statement is expressed in "[hyper]scaling laws,"

$$\alpha = 2(1 - \nu) - \nu(d - 2) \quad , \tag{11.2a}$$

$$\gamma = 2(\nu - \beta) + \nu(d - 2) \quad . \tag{11.2b}$$

The laws are written so as to emphasize their particularly simple form for $d = 2$.

11.2.3 Corrections to Scaling

When confronting experimental curves, (11.1) indicates that fits of observables $G(T)$ require the nonlinear form,

$$G(T) = A_{\pm}|T - T_c|^{\lambda} \quad . \tag{11.3}$$

Note that in general T_c must be determined simultaneously with the exponent λ and the [nonuniversal] amplitude A. To do the fit, we usually want $G(T_c) = 0$, not ∞; for divergent quantities we need merely fit the inverse. Further complications do exist, however. The simple form of (11.1,3) is just the first term of a power series, $t^{\lambda}[1 + Dt + \ldots]$. Often there are other weaker ["confluent"] singularities at T_c leading to corrections of the form

$$|t|^{\lambda}[1 + D_{1,\pm}|t|^{\Delta 1} + D_{2,\pm}|t|^{\Delta 2} + \ldots] \quad , \tag{11.4}$$

where Δ_1 and Δ_2 are not integers [11.17]. Since all the D's are nonuniversal and the Δ's are model dependent, general statements are difficult, if not impossible. For the particular case of the 2-d Ising model, for which a detailed solution is available (but for which corrections of the form (11.4) do not occur [11.18]), ROELOFS [11.19] investigated the dependence of the effective value of ν on the range of t analyzed. For $t_{min} = 0.005$ and 0.02, he found the deviation (underestimation) to be less than 5% for $t_{max} \leq 0.09$ and 0.06, respectively (cf. Fig.11.1). There is some evidence

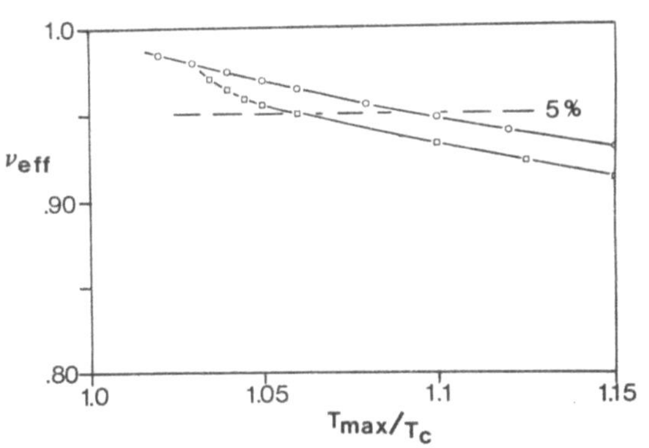

Fig.11.1. Variation of the effective value of ν vs the upper limit of the fitting range, T_{max}, for the Ising model solution for inverse correlation length [11.20]. The circles are for T_{min}=1.005 and the squares for T_{min}=1.02 [11.19]

[11.18] that such effective exponents, deduced for some common range, obey the scaling laws of (11.2), although it is not presently certain how generally this is true. Under certain circumstances, associated with a parameter being "marginal" (not changing under scale increases at the "fixed point" representing the transition), the correction is logarithmic [11.21],

$$|t|^{\lambda}(1 - u_{\pm}\ln|t|)^{\tau} \quad . \tag{11.5}$$

In this case the correction term becomes worse [larger] as t_{min} decreases. The size of the correction is partially determined by the nonuniversal [i.e., dependent on specifics of the system] coefficient u. We shall encounter such a case below, the four-state Potts model [11.21][1]. Since conclusive experimental demonstration of logarithmic corrections requires decades of t range, it is particularly favorable to investigate lower dimensionalities, where critical regions are relatively large.

11.2.4 Crossover Phenomena [11.22]

Often real systems do not correspond perfectly to simple models but contain small admixtures of other models having different-symmetry ground states. Some examples for magnetic systems are weakly anisotropic systems or layered systems with primarily 2-d coupling and weak 3-d coupling between layers. We shall encounter, in Sect.11.3.4, 5 the case of {111} fcc [or {0001} hcp] surfaces, when a honeycomb lattice of threefold sites is split by a weak crystal field into two triangular lattices. Such systems have the form

$$H = H_0 + gH_1 \tag{11.6}$$

where $\langle H_1 \rangle_{g=0} = 0$.

Far from T_c, the system appears to be in the universality class associated with H_0, as though g were zero. Very close to T_c, the system senses the symmetry of the full H and responds according to its [different] universality class. Note that this new behavior implies not only a different set of critical exponents but also a different, g-dependent critical temperature. At intermediate temperatures the system crosses over from one class to the other. By the scaling hypothesis, the intermediate behavior can be written as, e.g., for susceptibility,

---

1 If u vanishes, so does the logarithmic correction. Such may be the case for the Baxter-Wu model [11.4b], which is in the same universality class as the four-state Potts model but exhibits pure power law behavior [11.21].

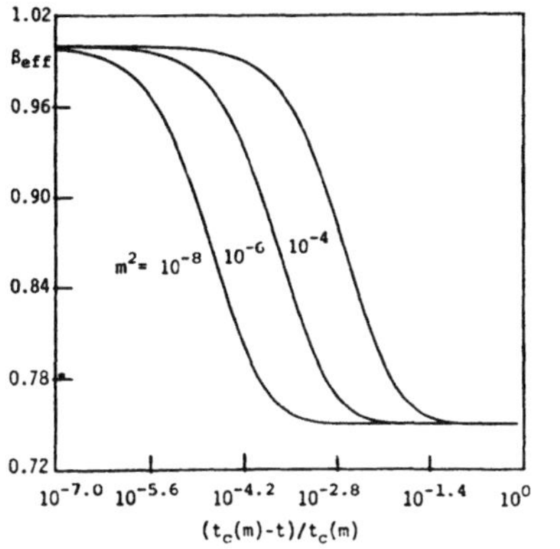

Fig.11.2. Illustration of effective exponents in a crossover regime for what amounts to three different values of g [11.23]

$$\chi(T,g) \propto t^{-\gamma}X(g/t^{\phi}) \tag{11.7}$$

where t is reduced with respect to $T_c(g = 0)$ and the crossover scaling function $X(z)$ is normalized by $X(0) = 1$. The crossover scaling exponent ϕ has the form $\phi = \zeta\nu$, where ζ is an exponent associated with scaling the field of H_1 and ν is familiar by now. The crossover region thus occurs for $t \sim g^{1/\phi}$; its width is determined by the explicit, system-dependent form of $X(z)$.

In a graph of ln(G) vs ln|t|, the curve is nearly linear for large and small |t|, and bends smoothly between these limits. The effective exponent, defined by $\partial \ln(G)/\partial \ln|t|$, has the value associated with H_0 for large |t| and with H for small |t|. In between, it varies continuously from one to the other, and cannot be neatly defined as some weighted average. Figure 11.2 illustrates this idea [11.23]. Note that *if* we knew we were in a well-defined critical region, a way to fine-tune a fit would be to minimize the curvature of ln(G) vs ln|t| with respect to T_c.

The crossover also applies near tricritical points, described in Sect. 10.3.3. Tricritical points have their own set of exponents, often with smaller β than that of the adjoining second-order line [11.24]. In this vicinity one could see (depending on some details) the tricritical exponents until one gets close to that line, then crossover to those of the line (or to a weak first-order transition). To illustrate the complexities inherent to this region, we cite a very recent impressive study of the disordering of $(\sqrt{3} \times \sqrt{3})$ Kr physisorbed on (ZYX exfoliated) graphite [11.25]. With X-rays from a synchrotron, resolution (HWHM = 3.5×10^{-4} Å^{-1}) two orders of magnitude better than LEED studies of Sect.11.5 were obtained. The measured value of β, 0.065 ± 0.015, was somewhat larger than the predicted β_M of the tricritical point but much less than the predicted second-order line β [which is (nearly) double β_M].

11.2.5 Fisher Renormalization

Another more general caveat concerns Fisher renormalization [11.26]. Although ideal exponents are calculated with the external field (i.e., the chemical potential for a lattice gas) as the independent variable, for chemisorption it is the conjugate variable, the coverage, that one can control. When a quantity, in this case θ, that would vary critically at a second-order transition is constrained, the measured exponents will change. Figure 11.3 illustrates schematically the essence of this situation, showing lines of constant μ in a second-order phase boundary in a θ-T plot. The singular variation of these lines at the critical point has the effect of exponentially renormalizing the reduced temperature by $1/(1 - \alpha)$ to lowest order. Accordingly, divergences in $|t|$ are renormalized; $\lambda \to \lambda/(1 - \alpha)$, where $\lambda = \beta, \gamma$, and ν, but $\alpha \to -\alpha/(1 - \alpha)$. Note that the scaling laws (11.2) still hold for these renormalized exponents. Note also that, since $\alpha = 0$ for the Ising model, fixing θ does not renormalize exponents (although there are subtle logarithmic effects). Inspection of Fig.11.3 suggests that one expects to measure the ideal, unrenormalized exponents until one gets close to the boundary, where there is crossover to the renormalized values. The crossover region happens at smaller and smaller reduced temperatures as one approaches the symmetry line interacting with the apex of the boundary. To date, no conclusive experimental observation of Fisher renormalization has been reported; with their large critical regions and intrinsically constrained θ, chemisorption systems present a propitious hunting ground.

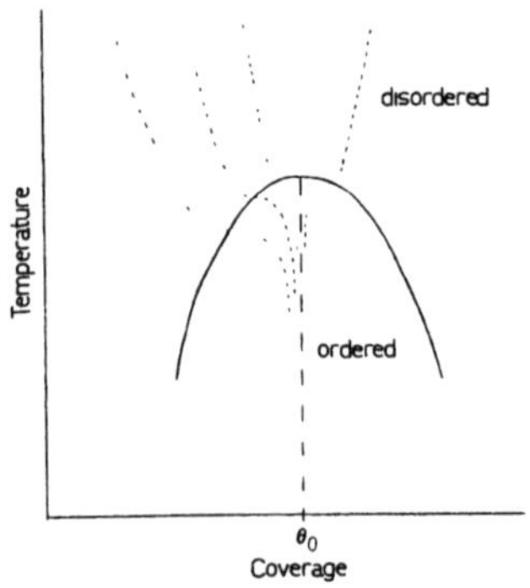

Fig.11.3. Illustration of the source of Fisher renormalization. The solid line denotes second-order transitions. The (dashed) lines of constant chemical potential vary singularly at the transition, except at the maximum, $T_c(\theta_0)$. Constraining θ rescales temperature near $T_c(\theta)$ over a range which vanishes near $T_c(\theta_0)$ [11.19]

The implications of Sect.11.2.4,5 for experimentalists are twofold. When it is feasible, measure critical exponents away from a crossover regime before tackling that complicated domain. When near such a regime, plot ln(G) vs $\ln|t|$ to see if well-defined critical behavior is monitored.

11.3 Universality Classes for Atoms on a 2-d Lattice

11.3.1 Order Parameters

The details of phase transitions are best discussed in terms of an order parameter ψ which can be a scalar, a vector, or a tensor of higher rank. By definition it vanishes above T_c, when there is no long-range order, and it is usually normalized to unit magnitude for a fully ordered state. In a second-order transition, $|\psi|$ moves continuously from 0 as T drops below T_c, while in a first-order one, it jumps discontinuously.

Some simple examples are helpful. For the checkerboard $c(2\times 2)$ pattern discussed earlier $\psi = (N_A - N_B)/(N_A + N_B)$ where A and B denote the two "colors" of sites and N_I the total occupation of each (10.7). For the $(\sqrt{3}\times\sqrt{3})R30^\circ$ pattern on a triangular lattice (commonly observed in physisorption, as alluded to in Sect.11.2.4), in which there are three equivalent sites, we might write

$$\underline{\psi} = \frac{N_A\hat{x} + N_B\left(\frac{-\hat{x}+\sqrt{3}\hat{y}}{2}\right) + N_C\left(\frac{-\hat{x}-\sqrt{3}\hat{y}}{2}\right)}{N_A + N_B + N_C} \quad . \tag{11.8}$$

This vector is often more conveniently discussed in terms of two orthogonal components.

11.3.2 Universality Classes

Writing the free energy of the system as $F(T,\psi)$, we note that the minimum of F lies at $\psi = 0$ for $T > T_c$. In a second-order transition, this minimum moves infinitesimally away from $\psi = 0$ as T drops infinitesimally below T_c. If we expand $F(T \lesssim T_c, \psi)$ about $F(T \lesssim T_c, 0)$, we expect the lowest order term, in ψ^2, to be negative and the higher order terms positive. The explicit form of the expansion of F in powers of ψ thus determines the behavior of the system near T_c, i.e., it determines the critical properties. Each distinct expansion corresponds to a different universality class. The expansion obtained for a lattice gas system can also be obtained from some spin-spin interaction system which often bears no initially obvious relationship to the ordered overlayer pattern; but the analogy does give deeper understanding of how the overlayer disorders. The name of the magnetic system conventionally denotes the universality class.

The simplest magnetic interaction is of the form

$$\mathscr{H} = - J \sum_{<jk>} \underline{s}_j \cdot \underline{s}_k \equiv - J \sum_{<jk>} \sum_{i=1}^{n} s_j^{(i)} s_k^{(i)} \tag{11.9}$$

for n-d spins. For n = 1, 2, and 3, these systems are called Ising, XY (or planar rotor), and Heisenberg, respectively. From symmetry considerations alone one can write the expansion as

$$F(T,\psi) = F(T,0) + r \sum_{i=1}^{n} \psi_i^2 + u_2\left(\sum_{i=1}^{n} \psi_i^2\right)^2 + u_3\left(\sum_{i=1}^{n} \psi_i^2\right)^3 + \dots \tag{11.10}$$

where ψ_i is $\sum_j s_j^{(i)}$, and r is proportional to $T - T_c$ and so changes sign at T_c. For spins without full n-d rotational symmetry, there are further terms in the expansion. The general procedure for constructing these expansions using group theory is well developed but rather technical (the classic, detailed, semiscrutable exposition is [11.28]; also [11.29]). Landau gave three rules, couched in [group] representation language, for determining whether a particular expansion in order parameter can produce a second-order transition. Since this approach is in essence mean-field-like, one must check whether fluctuations destroy the claims. Indeed, for d = 2, the second Landau rule is violated (by the three- and four-state Potts models). We emphasize that these symmetry-based considerations indicate only whether a particular overlayer can have a continuous transition; whether it actually will depends on the expansion parameters in the expansion of F, and so, ultimately, on the lattice gas interactions E_i.

Most of the applications of these ideas to adsorption systems have been carried out by DOMANY, SCHICK, and co-workers [11.30]. SCHICK's excellent recent review [11.9] is uniquely readable from a surface science perspective. Most of the remainder of Sect.11.3 presents highlights and results of that treatment.

11.3.3 Landau Theory for Adlayers

The ordered state below T_c can be described by some 2-d wave vector $\underline{Q}$. However, if we include some Q, we must also admit all [say ℓ] independent members of its "star," i.e., all other wavevectors generated by applying the point-group operations of the lattice of adsorption sites (hereafter the "net"), that are not related by a reciprocal lattice vector g of the net,

$$\langle n_i\rangle = \theta + \sum_{s=1}^{\ell} \rho(\underline{Q}_s)\exp(i\underline{Q}_s \cdot \underline{R}_i) \quad . \tag{11.11}$$

If some $\underline{g}$ connects $\underline{Q}_s$ and $-\underline{Q}_s$, then $\rho(\underline{Q}_s)$ is real and

$$\langle n_i\rangle = \theta + \sum_{s=1}^{\ell} \rho(\underline{Q}_s)\cos(\underline{Q}_s \cdot \underline{R}_i) \quad , \tag{11.12}$$

otherwise $\rho(\underline{Q}_s) = C(\underline{Q}_s) + i\, S(\underline{Q}_s)$ with C and S real, and

$$\langle n_i\rangle = \theta + 2\sum_{s=1}^{\ell/2} C(\underline{Q}_s)\cos(\underline{Q}_s \cdot \underline{R}_i) + 2\sum_{s=1}^{\ell/2} S(\underline{Q}_s)\sin(\underline{Q}_s \cdot \underline{R}_i) \quad . \tag{11.13}$$

The first Landau rule states in essence that for a second- order transition there is just a *single* [star of] $\underline{Q}$ needed to describe the ordered state.

Now in the first case, the ρ's are the ensemble averages of the ℓ-component order parameter,

$$\psi_s = N^{-1} \sum_i n_i \cos(\underline{Q}_s \cdot \underline{R}_i) \qquad s = 1,\dots,\ell \quad , \tag{11.14}$$

while in the latter case it is

$$\psi_s = N^{-1} \sum_i n_i \begin{cases} \cos(\underline{Q}_s \cdot \underline{R}_i) & s = 1,\dots,\ \ell/2 \\ \sin(\underline{Q}_s \cdot \underline{R}_i) & s = 1,\dots,\ \ell/2 \quad (2^{nd}\ \ell/2 \text{ components}). \end{cases} \tag{11.15}$$

To illustrate the formalism, note that for the checkerboard $\underline{R}_i = (m,n)$ and $\underline{Q}_s = \pi(1,1)$, so we recover the single component $\psi = N^{-1} \sum n_i(-1)^{m+n}$. For the $\sqrt{3}$ pattern we have $\underline{R}_i = [(1/2)(m + n),\ (\sqrt{3}/2)(m - n)]$ and take $\underline{Q} = (4\pi/3)(1,0)$. Then $N^{-1} \sum n_i \cos[(2\pi/3)(m + n)]$ and $N^{-1} \sum_i n_i \sin[(2\pi/3)(m + n)]$ give the x and y components, respectively, of $\underline{\psi}$ as set down in (11.8).

The third Landau rule, also called the Lifshitz criterion, requires that $\underline{Q}$ be at a high symmetry point of the surface Brillouin zone (SBZ): either the origin or a symmetry point on the edge. This condition selects which commensurate overlayers can undergo continuous transitions; they must be relatively dense [11.31]. The actual criterion is somewhat more stringent. Continuous transitions are still possible to incommensurate states, which are not discussed here [11.28-30].

Landau theory also shows how to generate the free-energy expansion solely on the basis of symmetry. The mth-order term in the expansion contains products,

$$\rho(\underline{Q}_a)\rho(\underline{Q}_b)\ \ldots\ \rho(\underline{Q}_m)\quad . \tag{11.16}$$

In order for translation invariance to be satisfied, the sum of these m $\underline{Q}$'s must be a $\underline{g}$. Furthermore, the free energy (and hence each order in the expansion) must be invariant under the point group operations of the net. Since some of these operations interchange some of the Q's, the products must be symmetric under this interchange, or there must be other m^{th}-order products. Observe that this approach gives no information about the coefficients of the terms in the expansion.

11.3.4 Catalogue of Transitions

Table 11.1 summarizes the assignment of adatom overlayer transitions to universality classes. It is quite remarkable how few universality classes there are and also how few of the experimentally observed overlayer patterns [11.32] are able to undergo second-order transitions.

Table 11.1. Overlayer patterns, on nets with common symmetries, which can have continuous order-disorder transitions. Adapted from [11.9]

Universality class and exponents / Substrate symmetry	Ising	x-y with cubic anisotropy	3-state Potts	4-state Potts
	α 0(log) β 1/8 γ 7/4 ν 1	non universal	1/3 1/9 13/9 5/6	2/3 1/12 7/6 2/3
Skew (p1) or rectangular (p2 mm)	(2×1) (1×2) $c(2\times 2)$			
Centered rectangular (c2 mm) or square (p4 mm)	$c(2\times 2)$	(2×2) (1×2) (2×1)		
Triangular (p6 mm)			$(\sqrt{3}\times\sqrt{3})$	(2×2)
Honeycomb (p6 mm)	(1×1) $[p(2\times 2)]^a$			(2×2)
Honeycomb in a crystal field (p3 m1)			$(\sqrt{3}\times\sqrt{3})$	(2×2)

[a] Heisenberg with corner cubic anisotropy ($K_4 \lesssim 0$)

The most common class in the table is Ising. This class results whenever the order parameter has a single component; there is no possibility of extra terms in (11.10). In the rectangular case [e.g., {110} fcc], the three overlayer patterns correspond to $\underline{Q}$'s pointing to the center of either of the sides of the rectangular SBZ or to a corner. Since $\underline{g}$'s connect any one to any other member of its star, each star contains just one independent member. Similar remarks apply to the square lattice [e.g., {100} bcc, fcc] when $\underline{Q}$ points to a corner of the SBZ or to the centered rectangle [e.g., {110} bcc] when it points to the middle of the short edge. The $c(2\times2)$ overlayers are unusual in that the lattice gas Hamiltonians can actually be transformed into the Ising model by a familiar change of variables [that is, substitute $(s_i+1)/2$ for n_i (see Sect.10.3.1)]; thus, the correspondence holds not just near T_c but for all T. For the honeycomb, the situation is a bit more complicated since the lattice is not Bravais. The (1×1) pattern corresponding to $\underline{Q}=0$ has one of the two triangular sublattices occupied. SCHICK's careful treatment [11.9] contains numerous figures to illustrate these comments. Discussion of the $p(2\times2)$ overlayer, corresponding to a Heisenberg model with cubic anisotropy, is deferred to Sect.11.5.

In the square lattice case (that of the centered rectangle is quite similar), if $\underline{Q}$ points to the middle of an edge, we find

$$\begin{aligned} n_i &= \theta + (\underline{Q}_i)\cos(\underline{Q}_1\cdot\underline{R}_i) + \rho(\underline{Q}_2)\cos(\underline{Q}_2\cdot\underline{R}_i) \\ &= \theta + (-1)^m\rho(\underline{Q}_1) + (-1)^n\rho(\underline{Q}_2) \end{aligned} \tag{11.17}$$

if we write $\underline{R}_i=(m,n)$. This form describes a (2×2). If either $\rho(Q_1)$ or $\rho(Q_2)$ happens to vanish, we have a (2×1). Since $\ell=2$, we start with (11.10) with $n=2$, an XY model. There is an additional fourth-order term of the form $w[\psi_1^4+\psi_2^4]$, or equivalently $w'\psi_1^2\psi_2^2$, arising because the point group of a square is lower than full rotational. In magnetic language, a term $S_x^4+S_y^4$ is added to (11.9), favoring either the two coordinate axes or the diagonals depending on the sign of the coefficient. It is somewhat perversely referred to as "cubic anisotropy" since a square is a 2-d cube. This model is highly unusual in that its exponents are nonuniversal [11.33], depending on the particulars like interaction strengths and coverage. As this situation has no parallel in 3-d, it offers an unusual opportunity for surface science to advance insight into critical phenomena [11.9]. On the other hand, it seems bold if not foolhardy to investigate such a system before one has mastered the study of systems with firmly known exponents.

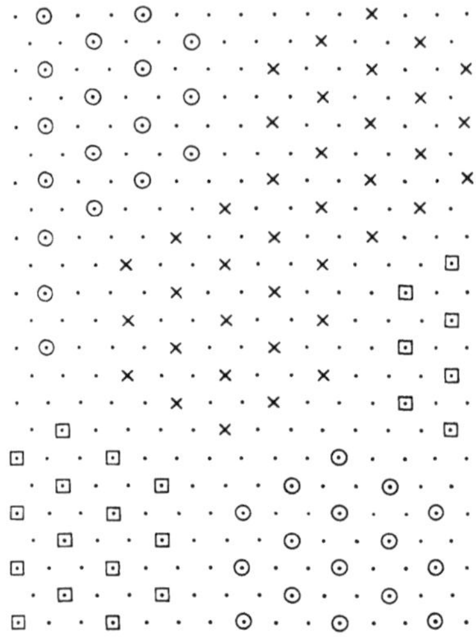

Fig.11.4. Several clusters of ordered $(\sqrt{3}\times\sqrt{3})R30^{\circ}$ regions, to motivate the correspondence to the three-state Potts model near T_c

Triangular lattices [e.g., atop binding on {111} fcc or {0001} hcp, centered sites on basal graphite] have as their SBZ a regular hexagon. As alluded to in Sect.11.3.3, a $(\sqrt{3}\times\sqrt{3})R30^{\circ}$ is produced by a $\underset{\sim}{Q}$ pointing to a vertex of the hexagon. Since its negative is not connected by a $\underset{\sim}{g}$, the slightly more cumbersome form of (11.15) was needed. The order parameter was seen to have two components. In addition to the terms of (11.10), there is a third-order term $\psi_1^3 - 3\psi_1\psi_2^2$ (violating the second Landau rule). This expansion is the same as that of the three-state Potts model [11.34], where the n-state Potts model [11.35] is defined by

$$\mathscr{H} = -J\sum_{\langle i,j\rangle}\delta_{s_i,s_j} \tag{11.18}$$

with s_i having n possible values (or directions or "colors"). Note that the two-state Potts model is just the Ising model. An intuitive grasp of the origin of this correspondence can be gained by examining Fig.11.4. Near T_c it is not the interaction of individual adatoms but rather those of large ordered clusters which determine the physics. In the $\sqrt{3}$ pattern there are three subsets of sites. There is one kind of domain wall (and resulting interaction energy) between clusters occupying the same subset. The walls between clusters on differing subsets do not depend on which two subsets are involved, but do depend on orientation. The figure shows the [two kinds of] antiphase boundaries that occur for strongly repulsive E_1. For weaker E_1, the domains approach each other closer. Computation of an effective cluster-cluster

Hamiltonian of the form (11.18) requires averaging over the many possible borders, often with some ab initio insight; such practice is implicit in the prefacing transformations [11.6b,8,36] mentioned in Chap.10.

A second star contains vectors pointing to the middle of the edges of the SBZ and generates (2 × 2) overlayers. Opposite $\underline{Q}$'s are related by $\underline{g}$'s. One finds that n = 3; the extra terms to be added to (11.10) are $w\psi_1\psi_2\psi_3$ and $v \sum_{i=1}^{3} \psi_i^4$, producing the expansion of the four-state Potts model. The logarithmic corrections associated with this model (see footnote 1) relate to the fact that, as a function of the (not necessarily integral) "number" of states, 4 is the largest at which a second-order transition occurs [11.37].

A honeycomb lattice is provided by the center (three-fold) sites of a close-packed layer of atoms. In physisorption this can be achieved by pre-plating based graphite, typically with Kr. The {111} faces of face-centered cubic crystals and {0001} faces of hcp crystals would also have this symmetry were it not for the second layer. The crystal field provides a binding-energy difference between those sites directly over atoms in the second layer (hcp sites) and the other half of the sites (fcc sites). This "staggered field" is expected to be small from solid-state experience.[2] From Sect.11.2.4 we recall that "well away" from T_c we can neglect this splitting.

While the honeycomb lattice has a triangular Bravais net, the two-site basis complicates the analysis. It turns out [11.30c] that the two-element star with Q's pointing to opposite vertices no longer satisfies the group theoretic expression of the Lifshitz condition.

There is, however, a new rich situation having the three-component star. The free energy has the extra term $w \sum_{i=1}^{n} \psi_i^4$. The ψ_i's are not just the $\rho(\underline{Q})$'s, but correspond to a 3-d (spin) vector. The extra term provides cubic (truly, this time!) anisotropy. For w positive, the six faces of the spin cube are favored. The adatoms form a (2 × 1) structure, occupying alternate rows in one of the three principal directions (for a total of six possible states). The transition has been recently shown to always be first order [11.39]. For w negative, eight concerns of the spin cube are favored. This case is discussed at the end of Sect.11.5.

2 We expect E_0 to be comparable to the stacking fault energy per atom for the substrate. This energy is well known to be small, on the order of tens of meV, and can vary by about an order of magnitude among metals [11.38].

11.3.5 Percolation

When all the E_i's of an overlayer are repulsive, the onset of long-range order occurs at roughly the same coverage for a wide range of temperatures (cf. R = 0 curve [11.14] in Fig.10.1). BINDER and LANDAU [11.14b] noticed that this coverage can be related to a (site-)percolation threshold: We imagine small planar objects with size determined by the area excluded on the lattice by repulsive E_i's. These objects are placed at random (but without overlap) on lattice sites. A cluster is defined as a set of these touching one another. As more of these objects are deposited, the size of the largest cluster starts to grow rapidly and suddenly girds the lattice (via periodic boundary conditions). In an infinite system, we say that the mean size of finite clusters diverges at a critical coverage θ_c, the percolation threshold.

Percolation has received much attention over recent years (a comprehensive review is in [11.40], also [11.41]). In spite of the absence of a Hamiltonian or a partition function [11.41c], it has been possible to draw analogies between functions describing the percolation problem and those describing a ferromagnet. Specifically, the mean size of finite clusters corresponds to the (zero-field) susceptibility, the mean number of clusters to the free energy, the fraction of sites in an infinitely large cluster ("percolation probability") to the magnetization, and the pair connectedness to the pair correlation function. The analogue of t is $\theta_c - \theta$. Following (11.1), we can define critical exponents. The percolation problem has been shown to correspond to the one-state Potts model [11.41], which has conjectured exponents [11.24b] $\nu = 4/3$, $\beta = 5/36$; thus, with the scaling equation (11.2), $\alpha = -2/3$, and $\gamma = 43/18$. It is not yet clear how fruitful such insights will prove to be for chemisorbed atoms .

11.4 LEED on Single Crystal Faces

11.4.1 Measurement of Exponents

In the remainder of this review, we shall specialize on the investigation of chemisorbed atoms on faces of metallic single crystals using LEED. Since it interacts strongly with matter, LEED is a natural probe of surface phenomena. There is no need for the enormous "surface" of exfoliated graphite required for neutron or (to date) X-ray scattering. Its resolution is at present as good as neutrons [11.42], and some new apparatuses are or will soon be capable of about a two-orders-of-magnitude improvement [11.43]. The technique is convenient. Most surface science laboratories have the needed equipment; there is no need to suffer the travails involved in experiments at a large multiuser facility. For over a decade, LEED has been used primarily for probing interatomc distances. This procedure involves an arduous effort due to the potent effect of multiple scattering [11.44]. In studying critical exponents, we use LEED for what it does best and most naturally: monitoring order on the surface. Many of the distinctions between chemisorption and physisorption were discussed in Sect.11.2.1. Because of the far greater binding energies, it is much easier to do LEED, making the delicate procedures developed by FAIN (Chap.9) unneccessary. Moreover, the greater strength of interactions leads to T_c's higher by half an order of magnitude; the size of the critical region, which is linked to reduced temperature, is scaled up accordingly in absolute temperature.

In LEED, in the kinematic limit, the intensity measured is the Fourier transform of the two-site correlation function,

$$\begin{aligned} I(k) &= \mathscr{F}\{\langle n_i n_j \rangle\} \\ &= \mathscr{F}\{\langle [n_i - \langle n_i \rangle][n_j - \langle n_j \rangle] \rangle + \langle n_i \rangle \langle n_j \rangle\} \end{aligned} \quad (11.19)$$

where the second term in braces is the product of the mean occupation of two sites and the first is the correlation of fluctuations (about these means) for the sites. Recalling (11.11), we note that for ordered overlayers we are interested in $n_j \exp(i\underline{Q} \cdot \underline{R}_j)$ rather than just n_j. Second, we must account for the fact that adatoms bind only to a particular lattice of sites with an associated reciprocal lattice. Third, the Fourier transform of the 2-d correlation function is Lorentzian. Thus, the intensity *due to the overlayer* is [11.20,45]

$$I(\underline{k}) \propto \sum_{\underline{Q}_s,\underline{g}} \left[\frac{\chi(\underline{Q}_s + \underline{g}, T)}{(\underline{Q}_s + \underline{g} - \underline{k})^2 + \xi^{-2}} + A\delta_{\underline{k},\underline{Q}_s+\underline{g}} |t|^{2\beta} \theta(-t) \right] \quad (11.20)$$

Thus, the LEED beams of interest are those at $\underline{Q}$ and its star, and all others related to them by reciprocal lattice vectors. In non- Bravais cases, the situation may be more complicated, e.g., glide planes can eliminate beams (Chap.9). We reemphasize that the $\underline{g}$'s are those due to the adsorption sites, not the atoms of the top layer of the substrate. For bridge sites on a square lattice, the lattices differ by $\sqrt{2}$ in lattice constant and 45° in orientation. For bridge sites on a triangular lattice or centered sites on a honeycomb net, even the symmetries are different (Kagomé and triangular, respectively).

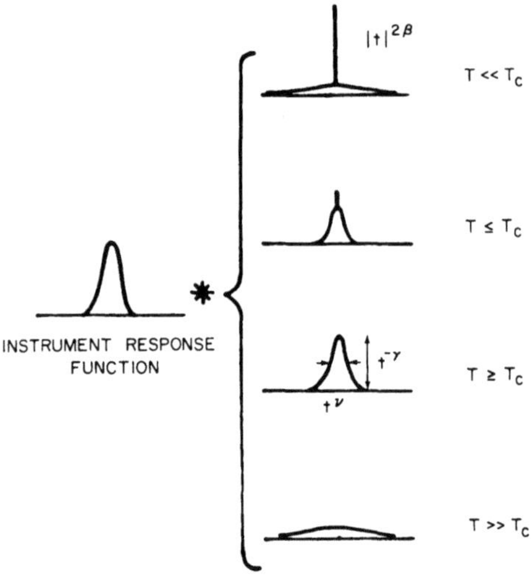

Fig.11.5. Schematic of adlayer-induced LEED spots vs k at various temperatures

The astute LEED practitioner may recognize that these beams of interest are just those that PARK long ago noted would be broadened by interference between antiphase domains [11.46]. The vectors connecting sites across the antiphase boundary are seen to be primitive vectors of the lattice of adsorption sites. Thus, in many cases, observation merely of which beams are broadened by critical scattering can identify the binding site and hence, in conjunction with the adlayer-induced spot pattern, predict the universality class of the transition.

Recalling (11.1) we see that the amplitude of the critical scattering goes as $t^{-\gamma}$ while its width goes as t^{ν}, as illustrated in Fig.11.5. As also shown, analysis of this behavior is confounded by the LEED instrument response function, which is convoluted with the ideal scattering of (11.20) in any measured profile. This function can be obtained from a low-temperature beam profile, where critical scattering is insignificant. Our procedure for removing the instrumental effects will be discussed in the next section, in conjunction with our case study [11.19,47]. (A preliminary account with some inaccuracies in the exponent analysis is in [11.6b].)

Heretofore many researchers, particularly those using weakly interacting probes (X-rays or neutrons), suspected that the strong interactions of electrons would preclude their use in studying critical exponents. The fear was that in addition to $\langle n_i n_j \rangle$ one would unavoidably collect multisite correlations like $\langle n_i n_j n_k \rangle$. Fortunately, the atomic scattering factor for light atoms below 50 eV is peaked in the backward and (more strongly) in the forward directions, with very little probability for right-angle scattering

[11.48]. If the incident beam is oriented normal to the surface, an electron is likely after (or before) scattering off an adatom to strike substrate atoms below the adatom. This sort of process leads to the multiple scattering that makes LEED I-V analysis so complicated. A process involving scattering off one adatom, then a second with subsequent escape requires a right-angle scattering followed by (for low-order spots) another nearly right-angle scattering. Such processes, which introduce the feared multiadatom correlations, are highly improbable and hence of very small amplitude. The effects of the preceding sort of process do not vary significantly near T_c since the substrate does not participate in the transition; accordingly, any modulation of the beam profile can be removed more or less in the pursuit of the known general form of the critical behavior.

11.4.2 Surface Defects

Finally, we discuss the nonideal nature of metallic crystal surfaces used in chemisorption. This problem involves not just the finite size of plateaus but also defects thereon. This sort of difficulty may well be intrinsically more stubborn here than in physisorption because the same chemical aspects of a surface promoting strong adsorption bonds renders it vulnerable to defects or distortions; the strong interplane bonding of planar materials like graphite make them poor chemisorbers [11.49].

The effect of the finite extent of perfectly flat areas on the surface is to raise T_c and to round (blur) the transition. Both effects are proportional to $L^{-1/\nu}$, where L is the linear dimension of a flat region; this dependence is useful for Monte Carlo but not of much use to the experimentalist unless the surface can be systematically damaged. The typical size of L is 100-200 Å in good samples; once ξ approaches this size, the effects become important. Note that probing of ξ is also limited by the instrument response function, which can be expressed in terms of a real-space window function through which the lattice is scanned [11.50]. With current instruments, the size of this window is roughly of the same size as L. Hence, recent improvements by orders of magnitude in the window width [11.43] may not be too helpful until better surfaces are available.

Local defects can be discussed in terms of their effect on the lattice gas Hamiltonian. If they create variations in the pair interactions, they couple to the energy density $|t|$. If α is nonnegative, they will beome important, with effects noticeable for $|t|$ less than of order $x^{1/\alpha}$, where x is the impurity concentration [11.51]. The transition may then be either smeared or sharp but with crossover to a new set of exponents associated with the impure system. The crossover exponent is α. The unusually large values of α for the three- and four-state Potts models will produce unusually large t ranges in which the effects of defects should be observable. If defects modify the single-site energies, the effect is more severe. By causing a particular site to be occupied or vacant, they can produce shifts and rounding comparable to finite size effects. Moreoever, they can, by locally changing the stability of one sublattice versus the others, change the nature of the transition [11.52]. In fact, in a recent study of a random-field Ising model, they were found to destroy order [11.53]. Further discussion of defects and finite size appeared in a recent review [11.8]; an earlier review presented renormalization group results [11.54].

Very recent work combining finite size scaling and Monte Carlo has illustrated how a small impurity concentration can dramatically change critical exponents. In studying the Baxter-Wu model (three-spin interaction on a triangular lattice [11.4b]. NOVOTNY and LANDAU [11.55] found that a small concentration of quenched impurities (viz., randomly omitted spins) changed the exponents ν, γ, and α from four-state Potts values to 1.00 ± 0.07, 1.95 ± 0.08, and $\lesssim 0$, respectively, while producing no detectable change in β. The effect of defects on lattice systems is under active theoretical investigation. There is naturally the difficulty of drawing correspondences between the sorts of defects a theoretician can readily handle to the kinds encountered in real systems, but many of the effects are expected to be independent of details [11.53]. In some cases, the defects may even change the effective dimensionality, leading to apparent violations of the [hyper]scaling laws of (11.2) [11.53].

11.5 Case Study: O/Ni{111}

Considering the vast number of ordered chemisorbed patterns [11.32], one would expect to have little difficulty obtaining realizations of nearly all the entries in Table 11.1. Life is not so simple. In many cases, the substrates are {100} bcc faces, which have been found to undergo spontaneous reconstructions and thus do not provide a passive net of sites (for recent reviews and references see [11.56]). In other cases, the transitions turned out to be first order.

For the system $c(2 \times 2)$ Na on Ni{100}, the adlayer-induced spot intensity (vs T) was fit to the ONSAGER solution [11.3], although no attempt was made to investigate exponents explicitly ([11.57], where a note added in proof raised the possibility that contaminants caused the order-disorder behavior). While such behavior is encouraging, a similar fit had been obtained earlier for $c(2\times2)$H or D on W{100}[11.58], a complicated substrate as just noted. Better evidence is needed to confirm "simple" 2-d Ising behavior. Recently, the system $c(2\times2)$ O on Ni{100} was investigated [11.59]. Unfortunately, at the temperature at which the O disordered, it was found to dissolve into the bulk. For the system $(\sqrt{3}\times\sqrt{3})R30^\circ$ CO on Ru{0001}, a comparable problem arises: the CO desorbs as it disorders [11.60]. In either case these behaviors preclude the observation of critical scattering above T_c. It is not clear how these complications affect critical scattering below T_c, which in any case is very difficult to isolate. Perhaps eventually it will give some insight into the fluctuations which lead to disordering, but it is not a priori clear whether equilibrium is achievable or what is being held constant (not θ, maybe μ); these problems should not be tackled until more nearly ideal systems are well understood.

The remainder of this section will be devoted to work at Maryland, investigating the order-disorder transition of $p(2\times2)$ O on Ni{111} [11.19, 47,61,62]. This study is the first to extract critical exponents, 1) with LEED and also 2) for a chemisorption system. The phase diagram and adsorption sites were described in Chap.10. Once again, the subtle vertical displacements [11.63] of the three Ni nearest neighbors of each O do not remove the 2-d nature of the transition but merely rather passively alters the lateral interactions.

The experiments were carried out in a UHV bell jar at total base pressures below 10^{-10} torr. A quadrupole mass analyzer was used to monitor the composition of gases in the chamber. A standard four-grid LEED optics system was used for all other measurements. Thus the equipment is available in any modern surface physics laboratory. The Ni samples were 99.995% pure. They were zone refined to eliminate mosaic spread and oriented to the <111> direction to better than 20 min. Measurements could be made down to 100 K. With a thermocouple-driven feedback circuit controlling the sample heater, temperature stability of ± 0.01 K was achieved. For further details of the apparatus and of sample preparation, see KORTAN and PARK [11.62].

A number of checks were performed to ascertain that the system was indeed on closed 2-d lattice gas. Monitoring the Auger peak height versus T showed that the oxygen did stay on the surface until 500 K [Ref.11.62, Fig.1]. At higher T it started to leave the surface; since no desorption was found, it must have dissolved into the bulk. To support the idea that the oxygen adsorbed in the same kind of bonding state regardless of whether the overlayer was ordered, the adsorption-induced work function change $\Delta\phi$ (along with Auger intensity) were monitored versus exposure [Ref.11.62, Fig. 2]. Translated into $\Delta\phi$ versus coverage, one finds that $\Delta\phi$ increases linearly up to 1/4 monolayer, not being affected by the onset of order near 1/4 [Ref.11.62, Fig.3].

The hallmark of a second-order transition as observed in LEED was described in Sect.10.3.3 (also [Ref.11.62,Fig.4]). The intensity of the adlayer-induced spot drops off smoothly from a plateau as T increases. The width of the spot seems constant out to about the temperature at which the intensity reaches half its lowest value or has its point of inflection; then it increases rapidly with increasing temperatures. This broadening, due to critical scattering [cf. (11.20) and Fig.11.5], also occurs below T_c, but is overshadowed in a casual observation by the Laue scattering (broadened by the instrument). The first-order transition of the $(\sqrt{3} \times \sqrt{3})R30^{\circ}$ overlayer is strikingly different: the beam essentially disappears above the transition and the intensity plummets [Ref.11.62,Fig.5]. Moreover, an intensity -versus- T plot shows strong hysteresis [Ref.11.62,Fig.6]. Operationally, life is not so clear-cut. While some of the intensity above T_c in [Ref.11.62,Fig.4] is due to critical scattering, much or most comes from finite size effects of the substrate and the LEED instrument response. Apparent hysteresis can be produced at a second-order transition by cycling T so fast that equilibrium is not achieved; at first-order transitions the hysteresis can in principal be eliminated by extremely slow cycling. Finally, for low-coverage first-order transitions due to island dissolution, there is also apparent beam broadening with increasing T as the island size shrinks [11.64]. (Here, however, the drop in intensity T is greatly stretched out and different in shape [11.65]). These practical problems are reminiscent of similar difficulties in Monte Carlo described in Chap.10.

In this experiment, the beam intensity was monitored versus k along a cut through an adlayer-induced spot. This novel measurement was achieved by projecting a real image of the diffraction beam, via an optical system outside the vacuum chamber, onto the aperture (1°) of a photon counter. The sweep was achieved by varying the incident electron energy rather than the position of the detector. To minimize extraneous I-V variations, the incident energy at the spot center was picked to give a maximum in the I-V plot. A check [11.47b] indicated that these effects were well within our level of uncertainty. Sixty profiles were recorded between 323 and 450 K. Figure 11.6 illustrates early samples of such profiles, far noisier than used in the

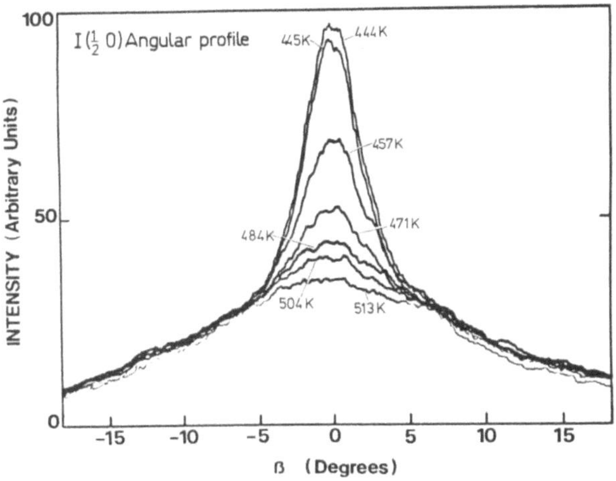

Fig.11.6. Angular profiles of an adlayer-induced LEED beam at several temperatures near T_c. Smoother versions of such profiles are the raw data for critical exponent analysis [11.61a]

analyses for exponents. Obviously the raw data are not very dramatic. In particular, it is hard to those locate T_c during the experiment. Preliminary processing included subtracting the uniform background due to thermal diffuse scattering and dividing out the effect of the Debye-Waller factor, as determined from the low-T dependence of the beam center.

The lowest T profile, more than 100 K below T_c, was used as the instrument response function. Two schemes were used to account for it. In the first, it was deconvoluted from the higher T profiles by dividing their 2-d Fourier transforms by its transform. From (11.20), we see that after this division the transformed profiles consist of a flat background proportional to $A|t|^{2\beta}$ and a peak at the origin (and at an array of real space points) with width proportional to ξ^{-2} and area proportional to χ. Operationally, there is much noise in the flat region due to uncertainty in the response function. The second approach, a fit scheme [11.66], takes advantage of (11.20) from the outset. That equation defines three temperature-dependent parameters which are determined (for each T) by convoluting with the instrument response function and optimizing the fit to the profile (at that T). Then each of these parameters are fit, again using a nonlinear least-squares routine, to the form of (11.3) where G(T) is the height of the delta-function contribution, the inverse amplitude of the Lorentzian (χ^{-1}), or the width of the Lorentzian; the resulting exponents λ are 2β, γ, and ν, respectively. While (11.3) involves a three-parameter fit, it is sometimes desirable to set T_c and fit

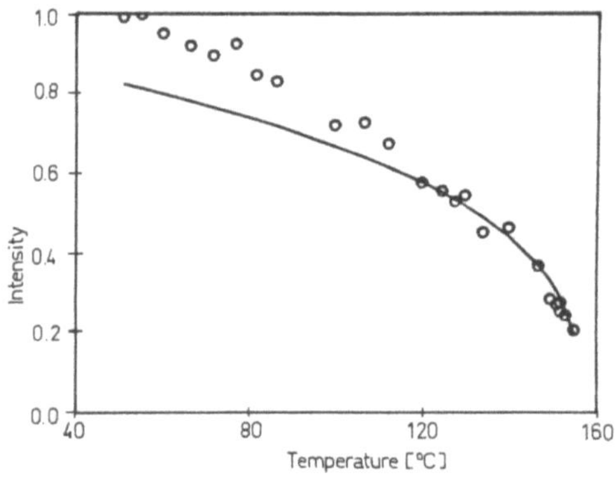

Fig.11.7. Analysis of the long-range-order intensity (at the center of the adlayer-induced beam) to determine the exponent β. The fitting range [120°C, 147°C] contains 7 data points [11.19]

just A and λ. The fitting scheme is the more expensive (time-consuming) method, but it is superior above T_c in that it avoids the persistent truncation oscillations of the first. Below T_c, the two methods give consistent results for β, while neither does well for the other two exponents because of noise after subtracting the long-range-order part. The choice of temperature range is important and nontrivial. The range should be as large as possible to enhance fitting statistics. It is limited at large $|t|$ by corrections to scaling and at small $|t|$ by finite size effects and incomplete deconvolution. In the present study, $0.015 \leq |t| \leq 0.06$ was used.

Plots of χ^{-1} [11.47a] and of width [11.47b] versus T for $T > T_c$ have already appeared in the literature. Both give T_c's slightly below 154°C. Due to overestimating T_c during the experiment (Fig.11.6), there were fewer data points below T_c. The analysis for β gave a T_c around 156°C. After reanalysis with T_c set at 154°C, β changed from 0.16 to 0.14, which seemed more trustworthy. Figure 11.7 illustrates this final analysis. We emphasize the sensitivity of the exponents to the value of T_c and hence the care that is required in the analysis (in an early crude analysis [11.47c], a T_c of 157°C was estimated and a β of 0.2 resulted).

A summary of the deduced exponents is presented in Table 11.2, along with the numbers for the Ising and the four-state Potts models. As noted in Chap.10, the expectation had been that the transition would fall into the four-state Potts universality class. This belief was based on the idea that the (2×2) pattern disordered into a triangular array of sites rather than a honeycomb array, i.e., into only one of the two kinds of three-fold sites. Monte Carlo results [11.19,47b] and fragmentary results on Heisenberg models with cubic anisotropy [11.67] suggested that the honeycomb cases exhibited only first-order transitions. The variety of complications discussed

Table 11.2. Exponent results, with model values for comparison (n.m.: not measured)

Exponent	Magnetic measurable	Ising	4-state potts	O/Ni{111} p(2 × 2)
α	Specific heat	0	2/3	n.m.
β	Magnetization	1/8	1/12	0.14 ± 0.02
γ	Susceptibility	7/4	7/6	1.9 ± 0.2
ν	Correlation length	1	2/3	0.94 ± 0.1

in Sect.11.2 were considered to try to understand the Ising-like exponents. The possibility of Fisher renormalization which would raise β, γ, ν by $(1-\alpha)^{-1}$, i.e., by a factor of 3 for four-state Potts was dismissed because the coverage was 1/4, a symmetry value for which the line of constant chemical potential would have no kink at T_c in a T -versus- θ plot (cf. Fig.11.3 with Fig.10.12). The effect of logarithmic corrections [cf. (11.5)] were evaluated in the asymptotic limit ($|t| \to 0$), in which the constant term is neglected. This assumption was required since the data did not warrant a four-parameter nonlinear least-squares fit. Explicitly, effective exponents were predicted by optimizing fits of $|t|^{\lambda_{eff}}$ over the experimental thermal range to the form $|t|^{\lambda}(-\ln|t|)^{\tau}$, where $\tau = -1/3$, 1/2, and 3/4 [11.21,68] for $\lambda = 1/12$, -2/3, and -7/6; the resulting λ_{eff}'s were 0.122, -0.80, and -1.37, respectively. All were thus substantially closer to the experimental numbers, but not by enough, particularly for γ. The effect of random surface defects or steps [11.69] might reduce the degrees of freedom for adatom fluctuations and thereby generate Ising-like behavior [11.70]. The work [11.55] on the impure Baxter-Wu model, cited at the end of Sect.11.4.2, found shifts of ν and γ to values consistent with the experimental results in Table 11.2. No corresponding shift of β could be detected. It is also not clear how such impurities would translate into the (2×2) lattice gas picture.

Another plausible explanation was suggested by SCHICK [11.71]. If E_0 is small [11.38], so that both kinds of three-fold sites are occupied, the (2×2) overlayer can be mapped into a Heisenberg model with cubic anisotropy, corners preferred. Each singly occupied cell of 8 sites in the overlayer goes into one site having a 3-d spin that can point to the 8 corners of a cube ($S_\alpha = \pm1$ for $\alpha = x,y,z$). The four fcc or hcp sites go to the four corners having the same value of $S_xS_yS_z$ (i.e., ±1); opposing vertices on the honeycomb go to (body-diagonally) opposing corners (Fig.11.8). This anisotropic Heisenberg system is a special case of the generalized Ashkin-Teller model studied very recently by GREST and WIDOM [11.72]; it contains a quadratic

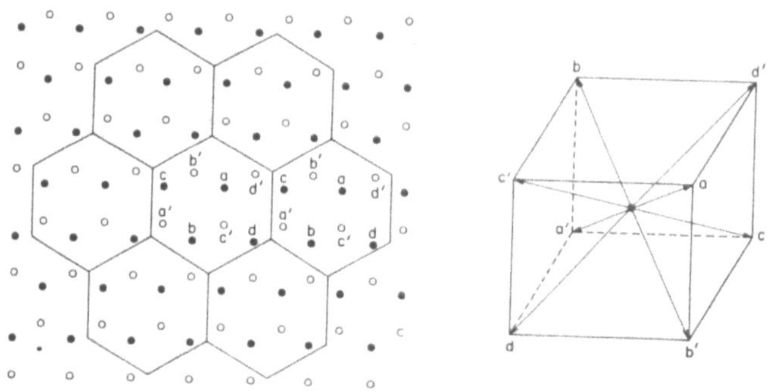

Fig.11.8. Correspondence between sites on a honeycomb lattice (with possible weak-energy differences between fcc and hcp sites) and the Heisenberg model with cubic anisotropy, corners preferred

term $K_2\, \underline{S}_i \cdot \underline{S}_j$ plus a quartic term with coefficient K_4, both terms coupling nearest neighbors. For $K_4 = 0$, the three spin directions decouple, each having Ising character. This behavior persists for small negative K_4. For $K_4 > 0$, the model predicts first-order transitions. With a prefacing transformation ([11.36,73], in the latter identical results were obtained by matching the low-energy excitations of the two models), the lattice gas interactions can be related to K_2 and K_4. This model predicts that, if the experiment could go to smaller $|t|$, one would observe crossover behavior to the four-state Potts model with g proportional to E_0 and ϕ being 13/8 [11.71]. Very preliminary results do not support this viewpoint as the explanation [11.74].

11.6 Conclusions and Exhortations

While there has been dramatic progress during the last decade in both the theory of 2-d phase transitions and in surface physics, close interaction between the two has been a relatively recent occurrence, particularly in the case of chemisorbed atoms. In physisorption the lateral interactions are far better characterized, and graphitic surfaces enabled study by traditional diffraction probes. Recent work on a complicated puzzle provided by the chemisorbed overlayer O on Ni{111} suggests that complete characterization of such systems will require a large arsenal of techniques. It is particularly remarkable that a result from theoretical study of an abstract model was needed to provide the clue to obtaining a set of lateral interactions that adequately explain all qualitative features. In short, chemisorption specialists cannot afford to be ignorant of new results in the theory of 2-d critical phenomena; correspondingly abstract theorists may find it gratifying to find that some models have realizations in the world, with complications that provide intriguing challenges.

For the 2-d statistical mechanic, many goals suggest themselves. Calculation of nonuniversal properties in terms of lattice gas parameters is crucial for a complete understanding of chemisorption systems. Detailed pictures of the correlation function, the size of critical region and the limiting corrections, and crossover behavior come readily to mind. The role of defects and finite plateaus also needs further attention; particularly important is some attempt to treat the actual defects, not just those that arise naturally in models.

Chemisorption experimentalists must identify more closed (no bulk absorption) systems and thoroughly characterize them. Critical exponents should be obtained at high- and low-symmetry regimes. Adlayer-induced spots should be examined above as well as below T_c. Efforts to probe smaller $|t|$ require both better LEED instruments (now becoming available) and better surfaces (a more difficult prospect). Close-packed surfaces should provide realizations of the three- and four-state Potts models [for {111} fcc or {0001} hcp faces] or the anisotropic XY model [for {110} bcc]. It would seem experience should be acquired on the former before tackling the less easily predictable latter. Also, a good $c(2\times 2)$ layer on a square lattice would permit comparisons with the most tractable theoretical model.

An important role of the surface theorist is to promote meaningful communication between the two preceding groups. Another is to advance our understanding of the interactions between adatoms, and between defects and adatoms, from a solid-state or chemical vantage. Such knowledge is vital to simplify the parameter set with which one approaches 2-d simulation and to gauge whether the output is physically sensible.

It is "a magnificant thing to make clearer and clearer to oneself something that seemed at first to be ... totally confused [11.75].

Acknowledgments. Close collaboration with N.C. Bartelt, A. Refik Kortan, Robert L. Park, and particularly Lyle D. Roelofs is gratefully and happily acknowledged. The last's contributions can be gauged by the number of figures taken from his thesis, of citations of his papers, and of references drawn from Chap.10 and [11.19]. I also benefited from many informative and enjoyable conversations with J. Bhattacharjee, S. Fishman, J.F. Nicoll, and M. Schick. I am especially indebted to the last for a draft of his review article [11.9].

References

11.1 Y. Imry: In *Chemistry and Physics of Solid Surfaces II*, ed. by R. Vanselow (CRC, Boca Raton 1979) p.461;
J.M. Kosterlitz: In *Phase Transitions in Surface Films*, ed. by G. Dash, J. Ruvalds (Plenum, New York 1980) p.193

11.2 B.I. Halperin, D.R. Nelson: Phys. Rev. Lett. *41*, 121, E515 (1978); D.R. Nelson, B.I. Halperin: Phys. Rev. B*19*, 2457 (1979)

11.3 L. Onsager: Phys. Rev. *65*, 117 (1944)

11.4a R.J. Baxter: J. Phys. A*13*, L61 (1980)

11.4b R.J. Baxter, F.Y. Wu: Phys. Rev. Lett. *31*, 1294 (1973); Aust. J. Phys. *27*, 357 (1974)
R.J. Baxter: Phys. Rev. Lett. *26*, 832 (1971); Ann. Phys. N.Y. *70*, 193 (1972)

11.5a T.H. Berlin, M. Kac: Phys. Rev. *86*, 821 (1952)

11.5b H.E. Stanley: Phys. Rev. *176*, 718 (1968)

11.5c E. Helfand: Phys. Rev.*183*, 562 (1969)

11.6a J.G. Dash, J. Ruvalds (eds.): *Phase Transitions in Surface Films* (Plenum, New York 1980) [Proceeding of Nato Advanced Study Institute, Erice, June 11-25, 1979]

11.6b S.K. Sinha (ed.): *Ordering in Two Dimensions* (North Holland, New York 1980) [Proceedings of the International Conference, Lake Geneva, Wisconsin, May 28-30, 1980]

11.7 Phase Transitions on Surfaces, University of Maine, Orono, August 3-7, 1981;
NATO Advanced Study Institute on Cooperative Phenomena in Two Dimensions, Banff, Summer, 1979

11.8 L.D. Roelofs: Appl. Surf. Sci. (to be published) [Proceedings of Second International Conference on Solid Films and Surfaces, College Park, Maryland, June 8-11, 1981]

11.9 M. Schick: Prog. Surf. Sci. (to be published)

11.10 C. Domb, M.S. Green (eds.): *Phase Transitions and Critical Phenomena* (Academic, London 1972-1976)

11.11 T.L. Einstein: In *Chemistry and Physics of Solid Surfaces II*, ed. by R. Vanselow (CRC, Boca Raton 1979) p.181

11.12 J.P. Muscat, D.M. Newns: Surf. Sci. *105*, 570 (1981);
J.P. Muscat: J. Chem. Phys. (to be published)

11.13 V. Hartung: Z. Phys. B*32*, 307 (1979);
J. Gallagher, R. Haydock: Surf. Sci. *83*, 117 (1979);
G. Theodorou: Surf. Sci. *81*, 379 (1979);
W. Kappus: Z. Phys. B*29*, 239 (1978); *38*, 263 (1980);
J. Lopez, G. Allan: Surf. Sci. *103*, 456 (1981)

11.14 K. Binder, D.P. Landau: Surf. Sci. *61*, 577 (1976); Phys. Rev. B*21*, 1941 (1980)

11.15 H.E. Stanley: *Introduction to Phase Transitions and Critical Phenomena* (Oxford University Press, New York 1971)

11.16 S.-K. Ma: *Modern Theory of Critical Phenomena* (W.A. Benjamin, Reading, MA 1976)

11.17 F.J. Wegner: In Ref.11.10, Vol.6 (1976) p.124

11.18 A. Aharony, M.E. Fisher: Phys. Rev. Lett. *45*, 679 (1980)

11.19 L.D. Roelofs: Ph. D. Thesis, University of Maryland (1980)

11.20 M.E. Fisher, R.J. Burford: Phys. Rev. *156*, 583 (1967)

11.21 M. Nauenberg, D.J. Scalapino: Phys. Rev. Lett. *44*, 837 (1980);
J.L. Cardy, M. Nauenberg, D.J. Scalapino: Phys. Rev. B*22*, 2560 (1980)

11.22 M.E. Fisher: Rev. Mod. Phys. *46*, 597 (1974) [and references therein];
E.K. Riedel, F.J. Wegner: Z. Phys. *225*, 195 (1969)

11.23 D.J. Amit, Y.Y. Goldschmidt, L. Peliti: Ann. Phys. N.Y. *116*, 1 (1978)
11.24a D.P. Landau, R.H. Swensen: Phys. Rev. Lett. *46*, 1437 (1981);
11.24b E.K. Riedel: Physica *106*A, 110 (1981) and references therein ;
11.24c A.N. Berker, S. Ostlund, F.A. Putnam: Phys. Rev. B*17*, 3650 (1978)
11.25 R.J. Birgeneau, G.S. Brown, P.M. Horn, D.E. Moncton, P.W. Stephens: J. Phys. C*14*, L49 (1981)
11.26 M.E. Fisher: Phys. Rev. *176*, 257 (1968)
11.27 A. Aharony: In Ref.11.10, Vol.6 (1976) p.357
11.28 G.Ya. Lyubarskii: *The Application of Group Theory in Physics* (Pergamon, Oxford 1960) Chap.7
11.29 D. Mukamel, S. Kransky: Phys. Rev. B*13*, 5065 (1976);
C. Rottman: Phys. Rev. B*24*, 1482 (1981)
11.30a E. Domany, M. Schick, J.S. Walker: Phys. Rev. Lett. *38*, 1148 (1977);
11.30b E. Domany, M. Schick, J.S. Walker, R.B. Griffiths: Phys. Rev. B*18*, 2209 (1978);
11.30c E. Domany, M. Schick: Phys. Rev. B*20*, 3828 (1979)
11.31 S. Ostlund: Phys. Rev. B*23*, 2235 (1981)
11.32 G.A. Somorjai, F.Z. Szalkowski: J. Chem. Phys. *54*, 389 (1971);
G.A. Somorjai, M.A. van Hove: *Adsorbed Monolayers on Solid Surfaces*, Structure and Bonding, Vol.38 (Springer, Berlin, Heidelberg, New York 1979)
11.33 J. José, L.P. Kadanoff, S. Kirkpatrick, D.R. Nelson: Phys. Rev. B*16*, 1217 (1977)
11.34 S. Alexander: Phys. Lett. A*54*, 353 (1975)
11.35 R.B. Potts: Proc. Cambridge Philos. Soc. *48*, 106 (1952)
11.36 A.N. Berker: Phys. Rev. B*12*, 2752 (1975); In Ref.11.6b, p.9
11.37 R.J. Baxter: J. Phys. C*6*, L445 (1973)
11.38 J.P. Hirth, J. Lothe: *Theory of Dislocations* (McGraw-Hill, New York 1968) pp.288ff, 764 [and references therein]
W.A. Harrison: *Pseudopotentials in the Theory of Metals* (Benjamin, New York 1966) p.207
11.39 B. Nienhuis, E.K. Riedel, M. Schick: To be published
11.40 J.W. Essam: In Ref.11.10, Vol.2 (1972) p.197
11.41a P.W. Kasteleyn, C.M. Fortuin: J. Phys. Soc. Jpn. Suppl. *26*, 11 (1969);
11.41b C.M. Fortuin, P.W. Kasteleyn: Physica Utrecht *57*, 536 (1972);
11.41c A.B. Harris, T.C. Lubensky, W.K. Holcomb, C. Dasgupta: Phys. Rev. Lett. *35*, 327 (1975)
11.42 J. Als-Nielson: In Ref.11.10, Vol.5a (1976) p.87
H. Dachs (ed.): In *Neutron Diffraction*, Topics in Current Physics, Vol.6 (Springer, Berlin, Heidelberg, New York 1978) p.1
11.43 M. Henzler: Appl. Surf. Sci. (to be published);
W.N. Unertl: Private communication
11.44 M.A. Van Hove, S.Y. Tong: *Surface Crystallography by LEED*. Theory, Computation and Structural Results. Springer Series in Chemical Physics, Vol.2 (Springer, Berlin, Heidelberg, New York 1979)
11.45 H.B. Tarko, M.E. Fisher: Phys. Rev. B*11*, 1217 (1975)
11.46 Robert L. Park: In *The Structure and Chemistry of Solid Surfaces*, ed. by G.A. Somorjai (Wiley, New York 1969) Chap.28
11.47a L.D. Roelofs, A.R. Kortan, T.L. Einstein, Robert L. Park: Phys. Rev. Lett. *46*, 1465 (1981)
11.47b L.D. Roelofs, A.R. Kortan, T.L. Einstein, Robert L. Park: J. Vac. Sci. Technol. *18*, 492 (1981)
11.47c L.D. Roelofs, A.R. Kortan, T.L. Einstein, Robert L. Park: In Ref. 11.6b, p.17
11.48 M. Fink, J. Ingram: At. Data *4*, 1 (1972);
M.B. Webb, M.G. Lagally: Solid State Phys. *28*, 301 (1973)
11.49 Robert L. Park: Private communication

11.50 L.D. Roelofs, Robert L. Park, T.L. Einstein: J. Vac. Sci. Technol. *16*, 478 (1979)
11.51 A.B. Harris: J. Phys. C*7*, 1671 (1974);
A.B. Harris, T.C. Lubensky: Phys. Rev. Lett. *33*, 1540 (1974);
T.C. Lubensky: Phys. Rev. B*11*, 3573 (1975)
11.52 A. Aharony, Y. Imry, S.-K. Ma: Phys. Rev. Lett. *37*, 1364 (1976) [and references therein]
11.53 S. Fishman: Private communications;;
E. Pytte, Y. Imry, D. Mukamel: Phys. Rev. Lett. *46*, 1173 (1981) [and references therein]
11.54 A. Aharony: J. Magn. Magn. Mat. *7*, 198 (1978)
11.55 M.A. Novotny, D.P. Landau: Phys. Rev. B*24*, 1468 (1981)
11.56 P.J. Estrup: In *Chemistry and Physics of Solid Surfaces II*, ed. by R. Vanselow (CRC, Boca Raton 1979);
P.J. Estrup, R.A. Barker: In Ref.11.6b, p.39
11.57 S. Andersson: Nobel Symp. *24*, 188 (1973)
11.58 P.J. Estrup: In *The Structure and Chemistry of Solid Surfaces*, ed. by G.A. Somorjai (Wiley, New York 1969) Chap.19
11.59 D.E. Taylor, Robert L. Park: Appl. Surf. Sci. (to be published) [Abstract]
11.60 D. Menzel: Private communication
11.61a A. Refik Kortan: Ph. D. Thesis, University of Maryland (1980)
11.61b A.R. Kortan, P.I. Cohen, R.L. Park: J. Vac. Sci. Technol. *16*, 541 (1979)
11.62 A.R. Kortan, Robert L. Park: Phys. Rev. B*23*, 6340 (1981)
11.63 T. Narusawa, W.M. Gibson, E. Törnqvist: Phys. Rev. Lett. *47*, 417 (1981)
11.64 J.C. Buchholz, M.G. Lagally: Phys. Rev. Lett. *35*, 442 (1975)
11.65 M.G. Lagally, G.-C. Wang, T.-M. Lu: In *Chemistry and Physics of Solid Surfaces II*, ed. by R. Vanselow (CRC, Boca Raton 1979) p.153
11.66 J. Als-Nielsen, O.W. Dietrich: Phys. Rev. *153*, 706 (1967)
11.67 M. Schick: Private communication
11.68 J. Bhattacharjee, D.J. Scalapino: Private communication
11.69 P. Kleban: Surf. Sci. *103*, 542 (1981);
P. Kleban, R. Flagg: Surf. Sci. *103*, 552 (1981)
11.70 J. Nicoll: Private communication
11.71 M. Schick: Phys. Rev. Lett. *47*, 1347 (1981)
11.72 G. Grest, M. Widom: Phys. Rev. B*24*, 6508 (1981)
11.73 N.C. Bartelt, T.L. Einstein: Unpublished
11.74 L.D. Roelofs, N.C. Bartelt, T.L. Einstein: Phys. Rev. Lett. *47*, 1348 (1981)
11.75 P. Ehrenfest to H.A. Lorentz, 1 January 1924, cited by M.J. Klein: Physica *106*A, 3 (1981)

Note added in proof:

I.F. Lyuksyutov, A.G. Fedorus: Sov. Phys. JETP *53*, 1317 (1981) [Zh. Eksp. Teor. Fiz. *80*, 2511 (1981)], recently reported the critical exponent β, obtained via LEED, for the disordering of the (2×1) and of the (2×2) overlayers of H chemisorbed on W{110}. Their analysis procedure is less sophisticated than that in Sect.11.5. These transitions lie in the class with nonuniversal exponents.

12. Structural Defects in Surfaces and Overlayers

M. G. Lagally

With 23 Figures

12.1 Introduction

The impact of structural defects on other surface properties has been evident, in some measure, since the inception of surface studies. From an understanding of the bulk properties of materials, it is easily recognized that a defect represents a loss of structural order and symmetry, has associated with it a free energy, and may have associated with it localized electronic states. Defects therefore may affect electronic and chemical properties, transport properties, thermodynamic properties, and mechanical properties.

The most extensive discussion, historically, of defects has centered on theoretical treatments of the thermodynamics and kinetics of crystal growth [12.1-6]. Calculations of the surface free energy have involved contributions of steps, terraces, and kinks (the terrace-ledge-kink model [12.5]). A surface roughening transition, a phase transition from a smooth to an atomically rough surface, has been theoretically predicted [12.2], although not, as yet, experimentally verified. Equilibrium shapes of crystals have been predicted on the basis of surface free energies [12.1-3]. In all these studies, surface defects (steps, kinks, ledges) are an inherent part of a physical phenomenon [12.6]. They play the role of structural entities with which is associated a free energy that affects growth kinetics and thermodynamics.

Defects have also been used deliberately to modify the properties of surfaces. The most familiar of this type of experiment is a surface intentionally cut at a particular orientation off singular to induce a known average step density. A large number of surface studies have been performed with such stepped surfaces. Examples include structural studies [12.7,8], reactivity studies [12.9], vibrational energy loss studies [12.10], electron-stimulated desorption studies [12.11], diffusion studies [12.12], and studies of electronic properties [12.13]. In these experiments, the major defect is assumed to be steps on the surface. From the results, it is obvious that steps markedly influence other surface properties. Little effort has been made in most of these experiments to identify other surface defects.

Surface defects are present to some extent in all surfaces and overlayers, even when they are not deliberately induced. They may be due to unintentional factors, such as sample preparation or the quality of the bulk material from which the sample is cut, or due to a natural phenomenon that is a result of the energetics and thermodynamics active at a surface. Examples of unintentional defects include steps or facets introduced by sample polishing and etching techniques, steps and inclusions or other point defects due to remanent sputter damage, steps due to cleavage, and mosaic at the surface due to subgrain boundaries in the bulk material. Natural phenomena that may lead to defects in surfaces or overlayers include surface reconstruction, finite-size effects in overlayers related to nucleation and growth or due to "surface melting" or other first-order phase transitions, multiple-phase formation on surfaces due to segregation or competitive adsorption, antiphase domain formation due to fluctuation phenomena in disordered layers just above their ordering temperature, and finite-size phenomena induced by metastability of various kinds.

Defects may be classified according to their dimension. Zero-dimensional or point defects include surface or overlayer vacancies, interstitial or substitutional atoms, or singly adsorbed atoms. Because of the difficulty in characterizing surface point defects, they have received little attention, but they can be quite important in providing sites at which nucleation of an overlayer phase takes place, in producing localized electronic states, in modifying the work function at a surface, and in affecting the reactivity of surfaces. One-dimensional or line defects include monatomic- or multiatomic-height steps, island and other two-dimensional phase boundaries, random strain fields associated with point or line defects, dislocations at a surface or in an overlayer, and projections of two-dimensional bulk defects, such as subgrain boundaries, onto the surface. Two-dimensional defects include the surface itself, of course, and additionally facets at a surface and misorientation of grains at a surface due to subgrain (mosaic) structure in the bulk.

Surface defects are most easily observed using a diffraction technique, because they modify the surface periodicity and thus affect the distribution of Fourier components that make up the diffracted intensity. Low-energy electron diffraction (LEED) [12.14,15], reflection high-energy electron difraction (RHEED) [12.16,17] grazing-angle X-ray diffraction (XRD) [12.18], and atomic beam diffraction [12.19] can all be used to study surface and overlayer defects. In all these cases, of course, the observation is indirect, i.e., one observes only a diffraction pattern and not an image. Direct observation of the influence of defects is possible in the field ion microscope [12.20]. Recently transmission electron microscopy has also been used to image directly steps and overlayer phase boundaries [12.21].

In this review we briefly discuss a variety of surface defects and their influence on the distribution of Fourier components in reciprocal space. That is, we consider the modification of the distribution of Fourier components for a perfect lattice by various defects, and the dependence of this distribution on reciprocal-space vectors, and thereby demonstrate that many defects have a unique effect on the angular distribution of diffracted intensity. This allows such defects to be identified and quantified with proper measurements. The discussion will initially be independent of the diffraction method used to measure the intensity distribution. Because LEED is presently the most commonly used technique for surface defect studies (and also because we are most familiar with it), we will discuss briefly its application to the measurement of surface defects and give a number of examples of experiments. The discussion could equally well be developed in terms of RHEED [12.22] or grazing-angle X-ray diffraction measurements by considering the appropriate Ewald constructions.

12.2 The Effect of Defects on the Intensity Distribution in Reciprocal Space

The ideal probe of the geometric structure of any object is radiation with wavelength of the order of the dimensions one is trying to resolve, which will be diffracted by the periodic arrangement of scatterers in the object. If one considers the elastic scattering of radiation with momentum $\underline{k}_0$ from a rigid three-dimensional crystal with lattice points

$$\underline{r}_i = m_1\underline{a} + m_2\underline{b} + m_3\underline{c} \quad , \tag{12.1}$$

where m_1, m_2, and m_3 are integers, with atoms located within each unit cell at positions

$$\underline{\rho}_n = u_n\underline{a} + v_n\underline{b} + w_n\underline{c} \quad , \tag{12.2}$$

where u_n, v_n, and w_n are fractions, the scattered amplitude at a given momentum transfer $\underline{S} = \underline{k} - \underline{k}_0$ is given by

$$A(\underline{S}) = \sum_{i,n} f_n(\theta,E)\exp[i\underline{S}\cdot(\underline{r}_i + \underline{\rho}_n)] \quad . \tag{12.3}$$

The sum is over lattice sites i and the atoms n within a unit cell. $f_n(\theta,E)$ is the atomic scattering factor of the nth atom, with θ half the scattering angle and E the energy of the radiation. $A(\underline{S})$ can be rewritten as

$$A(\underline{S}) = F(\theta,E) \sum_i \exp[i\underline{S}\cdot\underline{r}_i] \quad , \tag{12.4}$$

where $F(\theta,E)$ is called the structure factor and represents the scattering from one unit cell. $A(\underline{S})$ has its maximum value when $\underline{S} = \underline{G}_{hk\ell}$, where $\underline{G}_{hk\ell}$ is called a reciprocal-lattice vector. $\underline{G}_{hk\ell} = n2\pi/\underline{d}_{hk\ell}$, where $\underline{d}_{hk\ell}$ is the separation of $\{hk\ell\}$ planes. The intensity is

$$I(\underline{S}) = A(\underline{S})A^*(\underline{S}) = |F(\theta,E)|^2 \, \mathscr{I}(\underline{S}) \quad , \tag{12.5}$$

and for a particular reflection $\underline{G}_{hk\ell}$

$$I(\underline{G}_{hk\ell} + \underline{s}) = |F(\theta,E)|^2 \, \mathscr{I}(\underline{G}_{hk\ell} + \underline{s}) \quad , \tag{12.6}$$

where $\underline{s}$ is called the deviation parameter and $\mathscr{I}(\underline{G}_{hk\ell} + \underline{s})$ is called the interference function. It defines the reciprocal lattice for any crystal:

$$\mathscr{I}(\underline{G}_{hk\ell} + \underline{s}) = \frac{\sin^2[\tfrac{1}{2}N_1(\underline{G}_{hk\ell} + \underline{s})]\cdot\underline{a}}{\sin^2[\tfrac{1}{2}(\underline{G}_{hk\ell} + \underline{s})]\cdot\underline{a}} \cdot \frac{\sin^2[\tfrac{1}{2}N_2(\underline{G}_{hk\ell} + \underline{s})]\cdot\underline{b}}{\sin^2[\tfrac{1}{2}(\underline{G}_{hk\ell} + \underline{s})]\cdot\underline{b}} \cdot \frac{\sin^2[\tfrac{1}{2}N_3(\underline{G}_{hk\ell} + \underline{s})]\cdot\underline{c}}{\sin^2[\tfrac{1}{2}(\underline{G}_{hk\ell} + \underline{s})]\cdot\underline{c}} \quad , \tag{12.7}$$

$\mathscr{I}(\underline{G}_{hk\ell} + \underline{s})$ has its maximum value when $\underline{s} = 0$. The maxima have magnitudes proportional to $(N_1N_2N_3)^2$ and widths proportional to $1/N_1$, $1/N_2$, $1/N_3$ in the $\underline{a}$, $\underline{b}$, and $\underline{c}$ directions. It is obvious from the definition of $\underline{G}_{hk\ell}$ that $\mathscr{I}(\underline{G}_{hk\ell})$ for a defect-free crystal has the same value for any $\underline{G}_{hk\ell}$. For a three-dimensional infinite crystal ($N_1=N_2=N_3=\infty$), the reciprocal lattice is a three-dimensional array of points located at the ends of the reciprocal-lattice vectors $\underline{G}_{hk\ell}$. For a two-dimensional infinite crystal (i.e., a perfectly periodic single atomic layer, $N_1=N_2=\infty, N_3=1$) it can be seen from (12.7) that the intensity in the $\underline{c}$ direction is no longer modulated, but is constant, i.e., the reciprocal-lattice points have turned into reciprocal-lattice *rods* in the $\underline{c}$ direction (normal to the surface). Similarly diffraction from a single, infinite, perfect row of atoms ($N_1=\infty, N_2=N_3=1$) would produce reciprocal-lattice planes perpendicular to the row. Letting the crystal

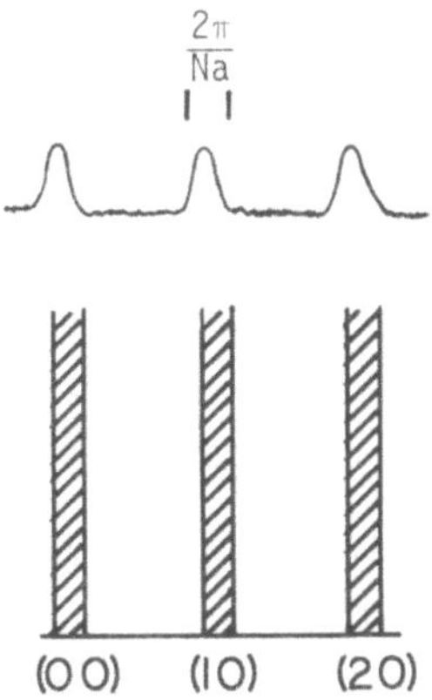

Fig.12.1. Reciprocal lattice in one dimension of a finite two-dimensional crystal. The dashed regions represent the full width at half-maximum (FWHM), $2\pi/Na$, of the intensity distribution. There are (N-2) additional maxima between the main rods that result in a background if N is large. Na is the size of the crystal in the $\underline{a}$ direction

shrink to one layer or one row demonstrates the ultimate finite-size effect. It illustrates the effect of reducing the size of the perfectly periodic structure by a boundary across which the phase is random (or at least different from the usual phase jump between adjacent atoms). The effect will be one of broadening of the reciprocal-lattice features by the introduction of different Fourier components into (12.7). Reducing the size of any dimension N_1, N_2, or N_3 causes the corresponding factor in (12.7) to be no longer a delta function but rather a function of the form $\sin^2 Nx/\sin^2 x$, which has a nearly Gaussian shape in its center, a full width at half maximum (FWHM) proportional to 1/N and wings higher than those of a Gaussian profile, with (N-2) sidebands that are distinct if N is small and unresolvable if N is large. For a finite, two-dimensional ($N_3=1$) crystal, N_1 and N_2 are finite; this causes a corresponding broadening of the reciprocal-lattice rods in the $\underline{a}$ and $\underline{b}$ directions. This is illustrated in Fig.12.1.

An example that illustrates more generally the broadening causes by defects can be given by considering a distorted crystal, where the position of the ith lattice site (or unit cell) is now

$$\underline{r}_i' = \underline{r}_i + \underline{R}_i \quad , \tag{12.8}$$

where $\underline{R}_i$ is the vector giving the displacement of the unit cell from its proper position $\underline{r}_i$. Around a particular reflection $\underline{G}_{hk\ell}$, (12.4) becomes

$$A(\underline{G}_{hk\ell} + \underline{s}) = F(\theta,E) \sum_i \exp[i(\underline{G}_{hk\ell} + \underline{s}) \cdot (\underline{r}_i + \underline{R}_i)] \quad . \tag{12.9}$$

The presence of the distortion thus introduces an added phase factor $\exp[i(\underline{G}_{hk\ell} + \underline{s}) \cdot \underline{R}_i]$. Consider $\underline{R}_i$ to be a sinusoidal variation in spacing in the $\underline{a}$ direction,

$$\underline{R}_i = \underline{\partial} \cos \frac{2\pi x_i}{\Lambda} \quad , \tag{12.10}$$

where $\underline{\partial}$ is very small compared to $\underline{a}$ and where $x_i = m_i a$ and Λ is the wavelength of the modulation. Then for the reflection $\underline{G}_{hk\ell}$,

$$A(\underline{G}_{hk\ell} + \underline{s}) = F(\theta,E) \sum_i \exp[i(\underline{G}_{hk\ell} + \underline{s}) \cdot \underline{r}_i] \exp\left[i\underline{G}_{hk\ell} \cdot \underline{\partial} \cos \frac{2\pi x_i}{\Lambda}\right] , \tag{12.11}$$

neglecting the term $(\underline{s} \cdot \underline{R}_i)$. Because $\underline{\partial}$ is small, the second exponential can be expanded to give

$$A(\underline{G}_{hk\ell} + \underline{s}) = F(\theta,E) \sum_i \left\{ \exp[i(\underline{G}_{hk\ell} + \underline{s}) \cdot \underline{r}_i] - \underline{G}_{hk\ell} \cdot \underline{\partial} \cos \frac{2\pi x_i}{\Lambda} \exp[i(\underline{G}_{hk\ell} + \underline{s}) \cdot \underline{r}_i] \right\} \quad . \tag{12.12}$$

The first term is just the Fourier transform of the perfect crystal, while the second leads to sidebands at positions $s_x = \pm 2\pi n/\Lambda$ away from the main peak. Figure 12.2 shows such sidebands for a two-dimensional crystal (N_3=1). If the crystal is finite also in $\underline{a}$ and $\underline{b}$, all the peaks will be broadened, and $\Delta s_x = 2\pi/N_1 a$, where Δs_x is the FWHM of the peaks in the $\underline{a}$ direction. If $\Lambda \gtrsim N_1 a$, the sidebands merge with the Bragg peak and cause broadening. The extent of the broadening increases with increasing h because of the factor $\underline{G}_{hk\ell} \cdot \underline{\partial}$. If $\Lambda \ll N_1 a$, the extra intensity goes into a background and leaves the main peak unchanged in width but reduced in intensity.

The effect on the intensity profile of several types of defects can be illustrated by considering their wavelength Λ. For example, most of the Fourier components of displacements due to thermal vibrations have Λ of the order of atomic dimensions, which is much smaller than the dimension of a typical ordered region on a crystal surface. Hence thermal vibrations cause a diffuse background intensity that reduces the Bragg intensity but causes no broadening. Similarly, point defects have Fourier components that have mostly small Λ, and they thus produce only a diffuse background. Random lattice strain, on the order hand, can have a much longer wavelength, of the size of the crystal, i.e., $\Lambda \lesssim N_1 a$. Hence random strain causes broadening of reflections that increases with increasing $\underline{G}_{hk\ell}$. This is illustrated for a two-dimensional lattice in Fig.12.3. Reflections with higher $\underline{G}_{||}$ but the same $G_{\perp}$ are broader if random strain is present. The distinct dependence of the broadening on $G_{||}$ makes it possible to identify random lattice strain at a surface uniquely [12.23].

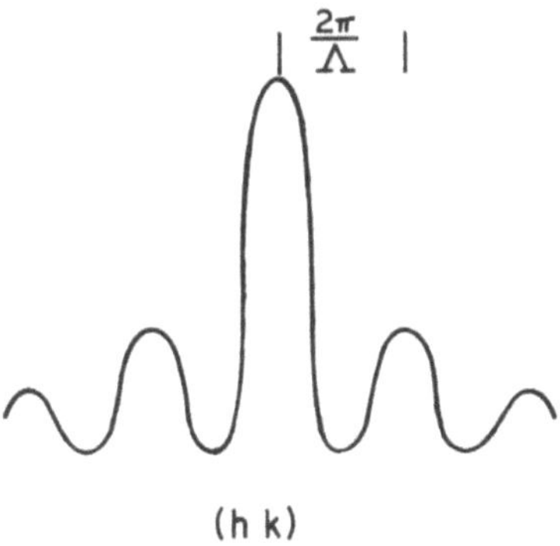

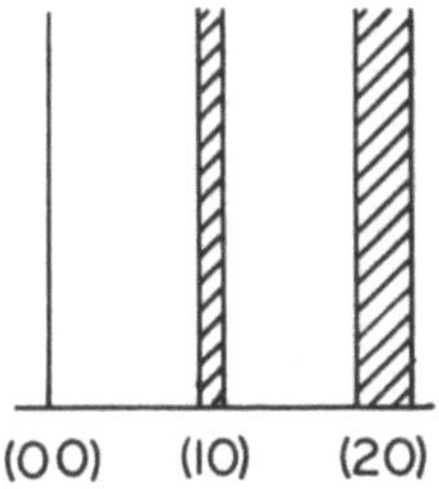

▲ Fig.12.3

◄ Fig.12.2

Fig.12.2. Intensity distribution in the vicinity of the (hk) rod for a finite two-dimensional crystal with a sinusoidal spacing variation in the a direction with wavelength Λ. Sidebands occurs at $s_x = \pm 2\pi n/\Lambda$

Fig.12.3. Reciprocal lattice in one dimension of an infinite two-dimensional crystal that contains random lattice strain in the a direction. The reciprocal lattice rods broaden with increasing order at the reflection h. The dashed regions represent the FWHM of the intensity distribution. The (00) rod is not sensitive to strain broadening parallel to the surface

Dislocations present a more complicated situation. The rapidly varying strains near the dislocation core give rise to background in the diffraction pattern, while the slowly varying long-range strain causes broadening as discussed above.

Another manifestation of finite-size effects at a surface can be explained in terms of a dislocation picture. Subgrain (mosaic) structure in a crystal, shown in Fig.12.4, manifests itself at the surface as finite-size domains that have small misorientations with respect to each other. Typical mosaic dimensions in well-grown crystals are of the order of 1 μm or larger, and misorientations are of the order of a tenth to several tenths of a degree. The magnitude of the misorientation can be expressed in terms of dislocations. A random array of edge dislocations in a crystal will give rise to a deviation from the average direction equal to b/ℓ, where b is the magnitude of the Burgers vector and ℓ is the average distance between dislocations. This misorientation manifests itself as a broadening of all reflections with $G_\perp$, again a unique dependence on diffraction parameters. This is illustrated in Fig.12.5 and is easily understood by considering each crystallite to have its own reciprocal lattice normal to its surface and recognizing that they must have a common origin. Misorientation of 0.1° is readily observable by making angular profile measurements at various energies (i.e., $G_\perp$) [12.23-25].

▲Fig.12.4. Schematic diagram of a crystal mosaic structure. The angles are vastly exaggerated. α is the mean misorientation angle

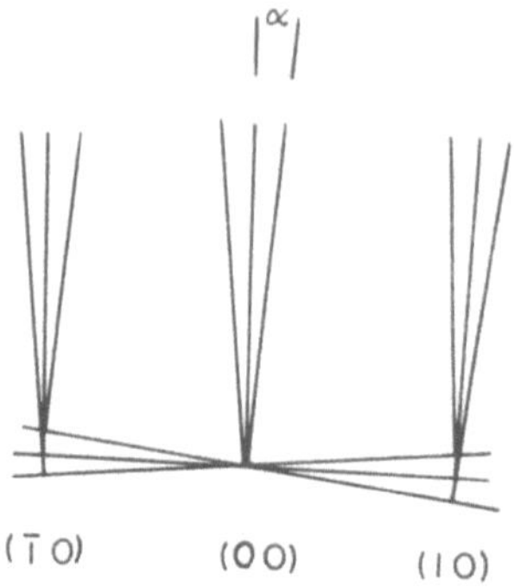

Fig.12.5. Reciprocal lattice in one dimension for a collection of large two-dimensional crystallites that have a small out-of-plane misorientation with respect to each other. The opening angle for the rods is twice the misorientation angle of the crystallites. The crystallite size is assumed to be large enough to produce very narrow rods from each ►

Surface steps represent an entirely different type of defect. Steps have been studied more extensively than any other surface defect [12.7,8,12,14,23, 26-31]. Many arrangements of steps are possible on a surface, with distinctive effects on the reciprocal lattice. The step arrangement for which the reciprocal lattice is most easily visualized is a monotonically increasing or decreasing step array with constant terrace size. As shown in Fig. 12.6, the reciprocal lattice can be considered as the product of the reciprocal lattice associated with the average surface and the terrace structure factor. This is equivalent to saying that the surface is the convolution of the single terrace unit with the step "lattice." The lattice points for this lattice represent the repeat units for the average surface, i.e., there is one lattice point associated with each terrace. Because the average surface consists of many lattice points, the reciprocal-lattice rods corresponding to the average surface will be sharp. Because the lattice points are far apart, the reciprocal-lattice rods are close together. Their separation is inversely related to the cosine of the angle of cut: the greater the deviation from singular, the farther apart these rods will be.

The terrace structure factor is just the reciprocal lattice associated with a single terrace. Because the terrace has finite dimensions N_1a and N_2b, the reciprocal-lattice rods will be broad, as discussed earlier. The greater the deviation of the average surface from singular, the smaller will be the terrace size and hence the greater will be the broadening. It should be evident that as the terrace structure factor gets broader (i.e., smaller terraces), the reciprocal-lattice rods of the average surface get farther apart (i.e., the repeat unit for the average lattice gets smaller). The proa)

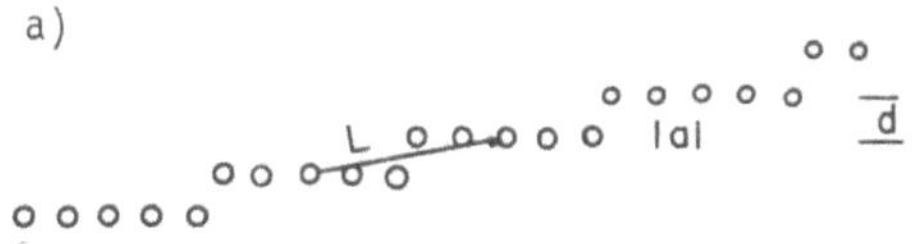

b)

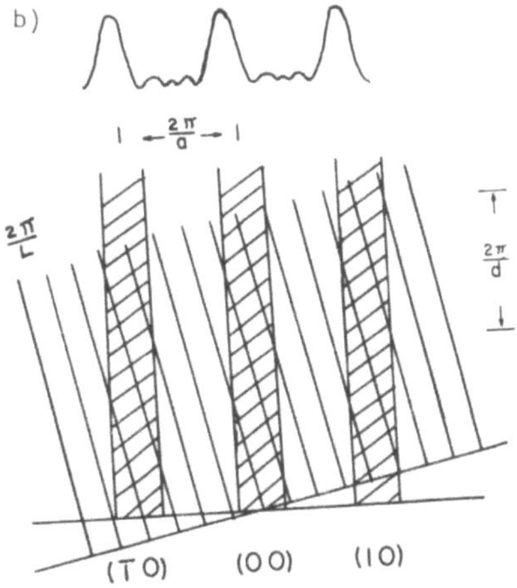

Fig.12.6a,b. Reciprocal lattice in one dimension for a vicinal surface with monotonically increasing steps and constant terrace size. (a) Surface, (b) reciprocal lattice. The sharp rods represent the reciprocal lattice of a lattice inclined to the average surface, with each terrace center (or edge) representing one lattice point. Their separation depends on the terrace size. The broad rods represent the FWHM $2\pi/Na$ of the terrace structure factor. The terrace structure factor, shown above the figure, has (N-2) subsidiary maxima. Reflections occur where the product of these two reciprocal lattices is nonzero. Thus there will be weak subsidiary reflections at some energies between the intense reflections. The latter will appear to oscillate between split and single reflections. The figure is drawn for a terrace size of 5 atoms

duct of these two factors, analogous to (12.4,6), is observed in reciprocal space, leading to rod-shaped regions of intensity inclined at the angle of the average surface.

Steps have been considered as a type of defect unique to surfaces, but have a three-dimensional analog, stacking faults, that is well known in transmission electron diffraction and microscopy [12.32]. Stacking faults are mistakes in the stacking of subsequent layers in a crystal. For example, in a <111> orientation of an fcc crystal, a stacking fault can consist of placing a layer into the hcp sites rather than the fcc sites, producing in part of the crystal a sequence ABCABABCABC rather than ABCABCABC. The diffraction from stacking faults can be viewed in the following way. The crystal can be considered as layers stacked on top of each other. Each layer will, by itself, produce an infinite rod in reciprocal space. Adding a number of layers in phase (i.e., proper stacking) reduces the length of this rod, until for an infinite crystal in the z direction the rods become spots again. If there are faults in the stacking, some of the reciprocal-lattice points will have streaks in them, the length of the streak increasing as the fault density increases (i.e., as the average thickness of the "perfect" crystal decreases). This can produce spots or streaks in the (transmission) diffraction pattern, depending on the orientation of the incident wave vector to the fault direction.

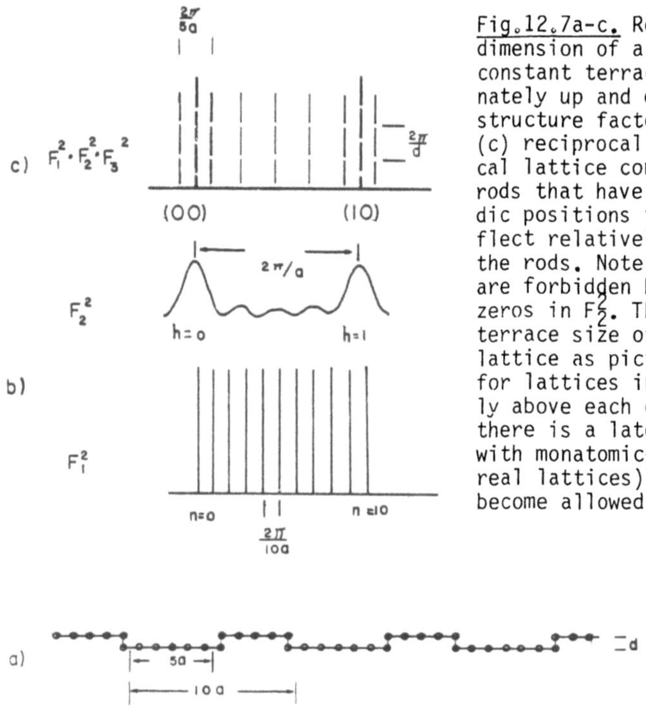

Fig.12.7a-c. Reciprocal lattice in one dimension of a stepped surface with constant terrace size and steps alternately up and down. (a) Surface, (b) two structure factor components, F_1^2 and F_2^2, (c) reciprocal lattice. The reciprocal lattice consists of delta-function rods that have zero intensity at periodic positions in $G_\perp$. Heavy lines reflect relatively greater intensity of the rods. Note that some subsidiary rods are forbidden because they occur at zeros in F_2^2. The figure is drawn for a terrace size of 5 atoms. The reciprocal lattice as pictured is obtained only for lattices in which atoms lie directly above each other at the step. If there is a lateral displacement (as with monatomic-step terraces in most real lattices), the forbidden rods become allowed

Stacking faults that occur parallel to the surface plane manifest themselves at the surface as antiphase boundaries that have an effect on specific reflections. These will be discussed below.

Step distributions different from the one that is monotonically varying with uniform terrace size produce different intensity distributions in reciprocal space [12.14]. For example, if the terrace size remains uniform, but the steps are alternately up and down, the reciprocal lattice consists of rods modulated in $G_\perp$ that are not inclined, because the average surface is flat. This is illustrated in Fig.12.7 and can be explained simply in the following manner. As shown in Fig.12.7a, the periodicity of the step structure is 2Na, where N is the effective number of scatterers in one terrace. The intensity can be described as the product of three structure factors, one that describes the periodicity with 2Na, one that reflects the scattering from one terrace of dimension Na, and one that gives the interference between terrace and trough. The first, F_1^2, is a reciprocal lattice consisting of rods that are delta functions (because this structure is infinite) spaced $2\pi/2Na$ apart, and oriented normal to the surface. The separation of adjacent rods, i.e., $n = 0$ and $n = 1$, reflects the size of one terrace plus one trough. F_2^2, the structure factor of one terrace of dimension Na, consists of an intensity function of form $\sin^2(N\underline{S} \cdot \underline{a}/2)/\sin^2(\underline{S} \cdot \underline{a}/2)$ with maxima spaced $2\pi/a$ apart (because a is the repeat unit in a terrace) and with N-2 subsidiary maxima. They are again oriented normal to the surface. The main maxima have a FWHM of $2\pi/Na$, and thus overlap three of the delta function rods. The minima of the terrace structure factor occur on alternate delta function rods, as do the subsidiary maxima. F_3^2 takes the difference in z spacing between terraces and troughs into account. This can easily be shown to be a function of the form $F_3^2 = 2[1 + \cos(h\pi)\cos(G_\perp d)]$, where $h = n/2N$ is the order of the reflection (i.e., the position of the nth delta function rod) and d is the height difference between terraces and troughs. If this function is evaluated at different values of $G_\perp$, it is found that its zeros occur alternately at the n = even and n = odd rods. The product of all three structure factors gives the reciprocal lattice shown in Fig.12.7c, with the rods with $n = 0$, $2N$ .... having maximum intensity, the adjacent ones having about half as much intensity, and the subsidiary rods having a few percent of the intensity of the main rods, depending on the size of a terrace. They are all, in principle, visible, although the subsidiary maxima may be quite weak.

The periodicity of the oscillation in $G_\perp$ between single and double spots is quite evidently related to the step height, as can be seen from the third structure factor. Physically this can be interpreted as constructive and destructive interference between terraces and troughs. When the interference is constructive, the diffraction does not recognize the existence of steps and only the reflections corresponding to an infinite lattice with lattice constant $\underline{a}$ appear, at $(2\pi h)/a$. At other $G_\perp$'s, all the other rods appear, and at characteristic $G_\perp$'s the rods at $(2\pi h)/a$ disappear while all the others are present. Each rod displays the same periodicity in $G_\perp$, but displaced in $G_\perp$ because the phase shift due to the parallel component $\underline{G}_{\parallel}$ is proportional to h. This periodicity in $G_\perp$ is reflected in all stepped structures that have a unique step height or multiples of such a unique height. Thus it is always possible in such cases to extract the step spacing, simply from the periodicity of the oscillation or modulation in width.

The reciprocal lattice shown in Fig.12.7c is directly applicable only to cases where atoms in the terrace lie directly above atoms in the trough. This can be realized for some double-height layers or other special cases. For most surfaces the existence of monatomic steps implies a lateral shift of the terrace atoms with respect to the substrate atoms by one half or

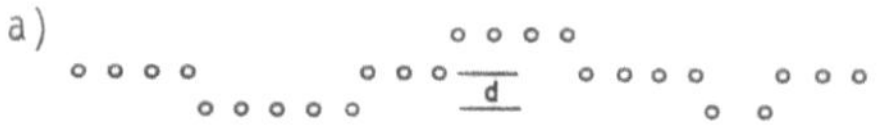

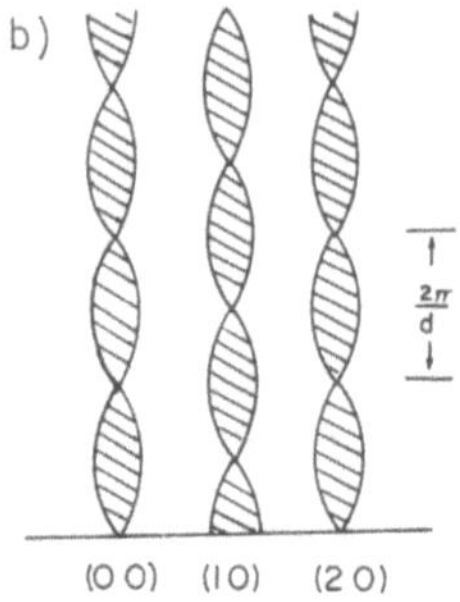

Fig.11.8a,b. Reciprocal lattice in one dimension for a surface with AB stacking and random up and down step edges occurring at random intervals. (a) Surface, (b) reciprocal lattice. The dashed regions represent the FWHM of the intensity distribution. The periodicity of the broadening is proportional to the step height and the FWHM (or the actual shape) reflects the average terrace size

one third of a lattice constant. This has the consequence that the maxima and minima of the terrace structure factor, F_2^2, no longer align with the delta functions of F_1^2. Thus there will be no structure-factor-forbidden reflections, and all the reflections become visible (at least in principle).

Uniform-size terraces that are alternately up and down may be physically realized in self-adsorption or surface reconstruction phenomena [12.33-35].

Introducing a distribution of terrace sizes causes broadening of reflections rather than a set of sharp delta functions. The most general case, shown in Fig.12.8, is the occurrence of random up and down step edges random intervals. This situation is approximated by most surfaces that are nominally flat but contain steps. The situation is similar to the alternate up and down terraces of uniform size, except that, because of the randomness, the original delta-function term, F_1^2, is absent and F_2^2 does not have subsidiary maxima but only a diffuse, monotonically changing intensity between the maxima. The reciprocal-lattice rods alternately broaden and narrow as $G_\perp$ is changed. The periodicity of the broadening is again related easily to the step height. At the conditions for which all the terraces scatter in phase, the diffraction spots will be sharp. At other $G_\perp$'s, there will be partial destructive interference and the spots will broaden. They will be broadest (for step distributions that contain only monatomic steps or a predominance of them) half way between the sharp spots.

If there is a predominance of steps of a height different from monatomic, the broadening will be different [12.25]. This is simply illustrated by considering all the steps to be double height, causing the periodicity in $G_\perp$ to halve. As the step-height multiplicity increases, the period of oscillation decreases. A superposition of steps with a range of multiple-step heights leads to a distribution in reciprocal space as shown in Fig.12.9.

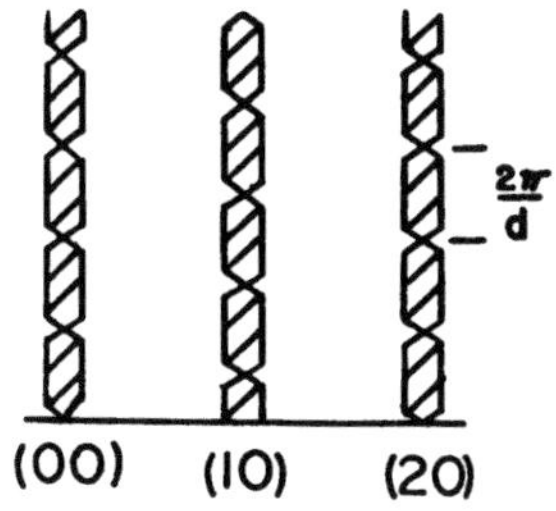

Fig.12.9 ▲

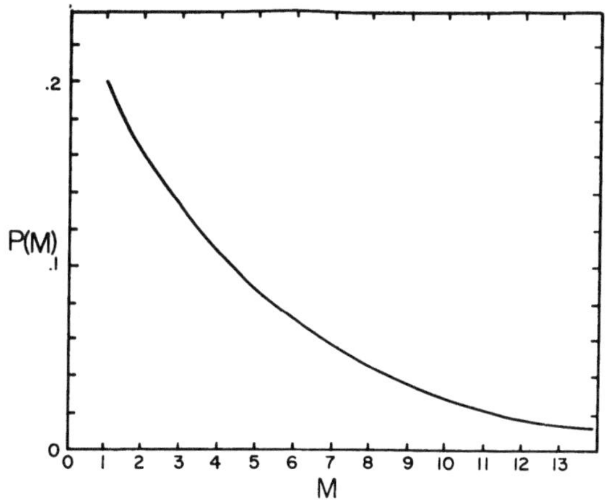

► Fig.12.10

Fig.12.9. Reciprocal lattice in one dimension for a surface with AB stacking that contains a wide distribution of multiatomic-height steps. The broadening is uniform over wide ranges of $G_\perp$, with sharp minima at the positions in $G_\perp$ of constructive interference

Fig.12.10. Form of the geometric distribution function. The fraction, P(M), of terraces with size Ma is plotted vs M, the number of lattice constants traversed in the $\underline{a}$ direction, for a given value of the probability γ of finding a step in going from any site to an adjacent site. $\gamma = 1/5$

The broadening becomes flat over most of the range of $G_\perp$, with very sharp minima in width occurring at the positions of constructive interference, which do not change. A physical analog of multiatomic steps is slip planes emanating at a surface. Slip in crystals can result, for example, during cleavage, and may result in step height of 20 to 500 lattice constants [12.36].

If the occurrence of step edges is random, an analytical form for the interference function can be developed [12.31]. For a random occurrence of steps the terrace size distribution is given by the geometric distribution function, shown in Fig.12.10 for a value of $\gamma = 0.2$ where γ is the probability of finding a displaced atom, i.e., a step, in going from one lattice site to the adjacent one. A certain density of multiatomic steps is included naturally. The probability of finding a double step is just the probability of finding two consecutive atoms displaced by a monatomic step height, i.e., $\gamma_1^2/2$, where γ_1 is the probability of finding a monatomic step. Similarly the probability of finding a triple step is $\gamma_1^3/4$. The form for the interference function for the geometric distribution is [12.31]

$$\mathscr{I}(\underline{G}_{hk\ell} + \underline{s}) = \frac{1 - \rho^2}{1 + \rho^2 - 2\rho\cos(\underline{G}_{hk\ell} + \underline{s})\cdot\underline{a}} , \qquad (12.13)$$

where ρ^2 is a "boundary structure factor" that is a function of γ. The reciprocal-space distribution was shown schematically in Fig.12.8. Equation (12.14) can be used to predict quantitatively the step density from the angular profiles resulting from a distribution of terraces with noninteracting step edges.

So far the discussion has been in terms of defects at clean surfaces. Most of the effects mentioned have analogs in adsorbed overlayers. The simplest results have already been presented indirectly. Equation (12.7) predicts the angular distribution of intensity for an adsorbed fractional monolayer. Several cases should be distinguished. The most commonly observed (if not the most common) form of adsorbed monolayer is one that is commensurate with the substrate (i.e., adsorbed in regular lattice sites of the substrate) and with a unit mesh larger than that of the substrate. Such layers produce additional, or superlattice, reciprocal-lattice rods. The positions of these superlattice rods can be established by substitution of the proper superlattice unit mesh vectors for $\underline{a}$ and $\underline{b}$ into (12.7), or else by an argument similar to that given by (12.9-13), which will demonstrate that the new Fourier components produce a "sideband" at the proper reciprocal-lattice positions. The reciprocal lattice for a complete, infinite, perfectly ordered monolayer with a double periodicity in the $\underline{a}$ direction [e.g., $p(2\times 1)$, $p(2\times 2)$, $c(2\times 2)$] is shown in Fig.12.11a. The intensity distribution consists of delta-function rods.

Another form of adsorbed layer is an "out-of-registry" layer, where the spacing of the overlayer atoms is slightly different from that of the substrate. If one considers an atom at an origin "lining up" with a substrate atom, then the $(N \pm 1)^{th}$ overlayer atom will again line up with the Nth substrate atom, leading to a superlattice with a large wavelength. The Fourier transform of this spacing causes the formation of sidebands, or satellite reciprocal-lattice rods, as illustrated in (12.13) and in Fig.12.2. The formation of satellite lines can be caused by an overlayer that may be rotationally or translationally out of registry, a periodic lattice distortion of the substrate caused by overlayer adsorption, or by similar effects (Chap. 9,[12.38,39]). If the distortion or displacement has a definite period, the satellite reflections will be sharp. Because the wavelength Λ of the distortion is generally large relative to a lattice constant, the satellite reflections will lie close to the position of the reflections for the undis-

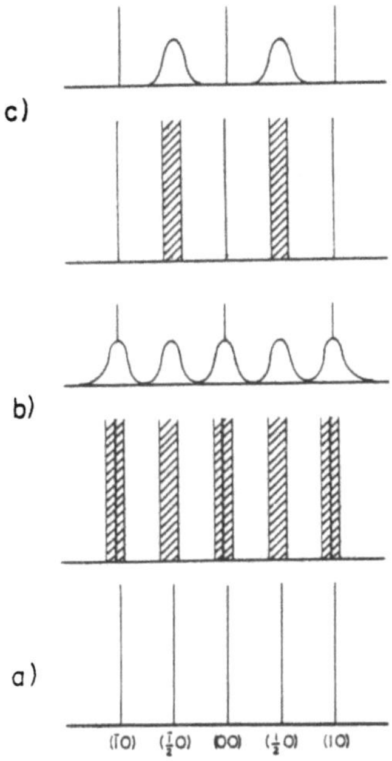

Fig.12.11a-c. Reciprocal lattices in one dimension for overlayer islands and domains with double spacing in the $\underline{a}$ direction adsorbed on an infinite, perfect substrate. (a) Infinite ordered overlayer. (b) Finite-size islands at low coverage. (The reciprocal lattice is the same for saturation coverage of an overlayer that forms both rotational and translational antiphase domains.) (c) Saturation coverage for an overlayer that forms translational antiphase domains only. In (b) and (c) the upper panel shows the dependence of the intensity on $\underline{G}_{\parallel}$ at a particular value of $G_{\perp}$ and the lower panel shows the WHM vs $G_{\perp}$, neglecting interference effects with the substrate. Teh broadening can be related quantitatively to the island or domain-size distribution [12.37]. For a p (1×1) overlayer that forms islands, the reciprocal lattice is as in (b) except that there are no superlattice reflections

torted or undisplaced lattice. There should be a multiplicity of sidebands, at values of $s_x = (2h\pi)/\Lambda$, with the one nearest the main peak being the most intense.

Any commensurately adsorbed monolayer that has a superlattice unit mesh larger than the substrate will form translational and possibly also rotational (depending on the symmetry of the unit mesh) antiphase domains. A fractional monolayer may exist in the form of *antiphase islands* [12.40] at some finite temperature if there is a net attractive interaction between the adsorbate atoms. There are a number of reasons why a fractional monolayer might exist as islands [12.37,41], the most important being kinetic limitations and substrate point or line defects that act as nucleation sites. For any distribution of islands at much less than saturation coverage [12.40], as well as at any coverage for overlayers that form in both translational and rotational antiphase domains [12.37], the reciprocal lattice is as shown in Fig.12.11b. The superlattice reciprocal-lattice rods will broaden while the fundamental reflections will consist of a broadened contribution, due to the overlayer islands or domains, and a delta function resulting from the periodicity imposed by the substrate. The substrate itself will make a contribution to the intensity of the delta function. Thus two layers have to be considered in this reciprocal-lattice construction if the relative intensities of the broadened contribution and the delta function need to be calculated.

If the overlayer forms only translational antiphase boundaries [e.g., p(2 × 2)] then at *saturation* coverage the fundamental reflections do not have a diffuse background as shown [12.37]. At low coverages, the reciprocal lattice is identical to that shown in Fig.12.11b. This is easy to understand. At low coverages, islands will be separated by a considerable amount of "sea," causing a randomization of phase. For overlayers that form rotationally inequivalent domains [e.g., p(2 × 1) and p(1 × 2)], these domains do not interfere; one always acts as sea for the other. If only translational antiphase boundaries exist, there is no sea at saturation coverage, and because the fundamental reflections are not sensitive to antiphase boundaries that occur at integral multiples of the fundamental spacing, these rods will be sharp, as shown in Fig.12.11c.

A third form of overlayer is one that adsorbs commensurately but does not form a superlattice. For such a p(1 × 1) lattice there obviously will be no superlattice reciprocal-lattice rods. At low coverages, there can still be island formation, and the reciprocal-lattice rods are identical to the fundamental reflections shown in Fig.12.11b. The delta-function contribution comes from both the overlayer and the substrate, and the diffuse scattering halo is from the overlayer alone. Thus if the scattering from the overlayer islands is sufficiently strong, it should be possible to determine the overlayer island size even for p(1 × 1) layers from the diffuse intensity [which is again given by (12.7)].

For a p(1 × 1) layer there can be no rotational antiphase domains or translational antiphase domains in the sense that we have so far described, namely occupation of the same type of sites but translationally displaced by a multiple of the substrate lattice constant. At saturation coverage the rods will therefore ordinarily be delta functions. However, a translational antiphase boundary of a different type can occur. This requires the occupation of two types of sites, e.g., hcp and fcc sites in an fcc {111} surface. The resulting antiphase boundaries (called twin or stacking-fault boundaries in bulk films or crystals) cause selective broadening of some of the reciprocal-lattic rods. Figure 12.12 shows a schematic diagram of an overlayer with a twin boundary and the resulting reciprocal lattice in one direction. The selective broadening of some rods can easily be understood physically be recognizing that the translational mistake at a twin boundary is typically

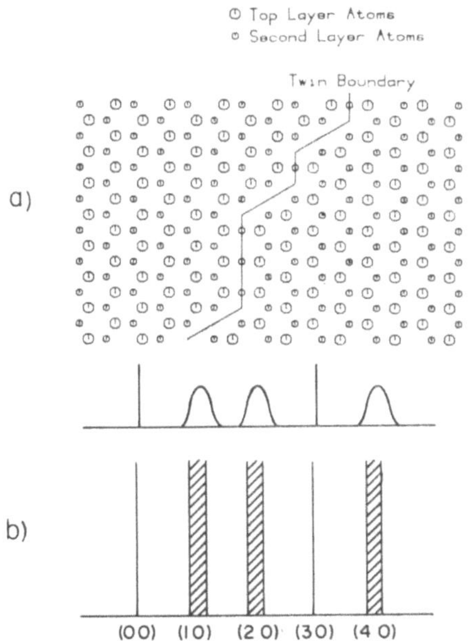

Fig.12.12a,b. Schematic diagram of a twin boundary (a) and the resulting reciprocal lattice (b). In this example, atoms occupy sites that have a translational displacement of 1/3 of a substrate lattice constant. Hence every third rod is not sensitive to the antiphase boundary, while the other rods are. The broadening reflects the domain size distribution. The upper panel in (b) shows the dependence of the intensity on $G_{\parallel}$ at a particular value of $G_{\perp}$. The lower panel shows the behavior with $G_{\perp}$, neglecting interference with the substrate

less than one lattice constant, thus causing broadening in all those peaks where the phase does not sum up to one, and no broadening where it does, while in an overlayer with a superlattice the translational mistake is one lattice constant or more [more, e.g., for a p(3 × 1) or p(4 × 2) lattice], causing broadening in the superlattice but not the fundamental reflections.

Twin boundaries can, of course, also occur for overlayers that form a superlattice. At saturation coverage in that case, the reciprocal lattice is a combination of those shown in Figs.12.11c,12b, i.e., some fundamental reflections and all superlattice reflections will broaden.

The selective broadening of superlattice reflections in different directions [12.39,42] can be used to distinguish between possible adsorption sites of the overlayer atoms. The position of the atoms determines the phase shift at a boundary in a given direction, causing selective broadening.

Finally, overlayers may form as completely incommensurate layers that, in a sense, form two-dimensional ordered rafts that have no definite phase relationship to the substrate on which they are adsorbed. The overlayer will have its own reciprocal lattice. At any coverage below a monolayer, this system behaves like a two-dimensional mosaic, i.e., there will be a random phase relationship between ordered islands, but if the substrate is flat, there will be no out-of-plane misorientation of the overlayer islands. If only trans-

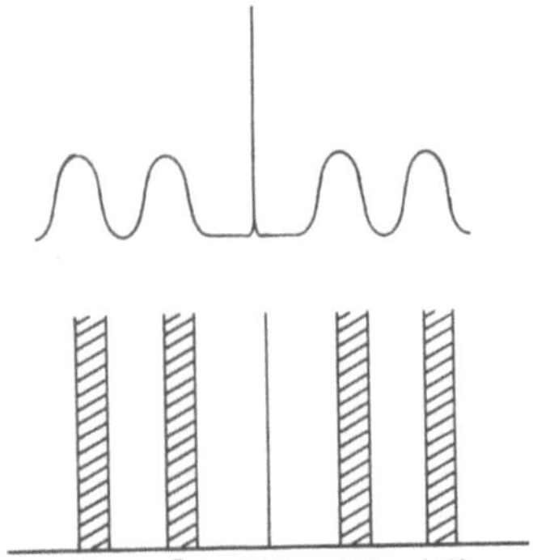

Fig.12.13. Reciprocal lattice in one dimension for a two-dimensional layer that consists of ordered regions that are translationally random in the $\underline{a}$ direction. The (00) beam does not broaden. The broadening of the other beams reflects the domain-size distribution. For a completely incommensurate two-dimensional layer, the rods will become rings because there will also be rotational randomness. The upper panel shows the dependence of the intensity on $\underline{G}_{\parallel}$ at a particular value of $G_{\perp}$. The lower panel shows the behavior with $G_{\perp}$, neglecting interference with the substrate

lational randomness exists, the reciprocal lattice is as shown in Fig.12.13. The (00) beam is a delta function because the specular reflection is not sensitive to lateral phase shifts. All the other beams will be uniformly broadened. This will be true for any flat mosaic, of course, but any out-of-plane misorientation among the crystallites results in broadening of the (00) beam also, except at $G_{\perp} = 0$, which cannot be observed. For the sake of clarity this point was ignored in the earlier discussion of crystal mosaic.

Because completely incommensurate layers will generally also involve rotational randomness, the rods turn into rings of finite width for all but the (00) beam. Rings have been obtained with LEED for a number of systems, e.g., from Kr on Ag [12.43] and from exfoliated graphite [12.44].

Islands adsorbed on a substrate should also cause modulation of the width of the fundamental reflections with $G_{\perp}$ for the same reasons that steps do, namely there occurs an interference between substrate and adsorbate layers. This is obvious for self-adsorbed layers, which form p(1 × 1) structures, because they are directly analogous to terraces. This phenomenon has been used to describe the island growth and ordering of W on W{110} [12.32]. The same reciprocal-lattice rod shapes described for step structures are possible for such systems. For other systems, such as O on W{110}, a weak modulation of the fundamental reflections should also exist, but it will no longer be simply oscillatory because in addition to the phase factor, due to differences in path length for the waves scattered from the two levels, there is a scattering phase difference because the atoms are different. However, with an accurate knowledge of the scattering factors of the atoms, it should be possible to determine to an accuracy of at least 0.1 Å the z spacing of any overlayer that scatters sufficiently strongly.

Finally, more complex situations are possible, involving defects in both substrate and overlayer. Islands on stepped surfaces [12.37,45] have been described. Another class of structures, not discussed so far, involves positional correlation between ordered regions that may be far apart [12.46]. So far we have considered only situations where ordered regions are Na apart, with N random. If some positional correlation exists, N will no longer be random. This can be the case if there are adatom interactions, regularly spaced defects that act as nucleation centers, or out-of-registry overlayers with slight lattice mismatch [12.38,39]. Either short-range or long-range positional correlations are possible. It has been found [12.46] that islands with short-range correlations give diffracted intensity profiles that differ negligibly from those obtained for a random arrangement of islands. Long-range correlations cause extra features to appear in the reciprocal-space distribution, which, however, cannot be resolved in a measurement unless a verry narrow detector is used [12.46].

In this section we have attempted to show that various extended defects at a surface produce distinct effects on the reciprocal lattice. They should thus be identifiable and quantifiable by scanning the reciprocal lattice with a diffraction technique. In the next section this is illustrated by means of examples of low-energy electron diffraction measurements of various surface defects.

12.3 Surface Defect Studies Using Low-Energy Electron Diffraction

In low-energy electron diffraction, a monochromatic beam of electrons in the energy range of 10 to 500 eV is used to scan the reciprocal lattice. Because the mean free path of electrons with this energy range is of the order of a few angstroms, it is appropriate to approximate the reciprocal lattice by rods, as has been done in the last section. In practice, several layers contribute to the diffraction, causing some *intensity* modulation along the rods, which does not, however, affect the angular distribution of intensity. The diffraction can be pictured simply by the superposition of an Ewald sphere on the reciprocal lattice, as shown in Fig.12.14. The superposition of such an Ewald sphere on any of the reciprocal lattices illustrated in the last section produces the appropriate diffraction pattern observed at the chosen diffraction parameters (energy, incident angle).

Numerous observations of diffraction patterns have been made that indicate spot splitting or spot broadening. In many cases, simply the number

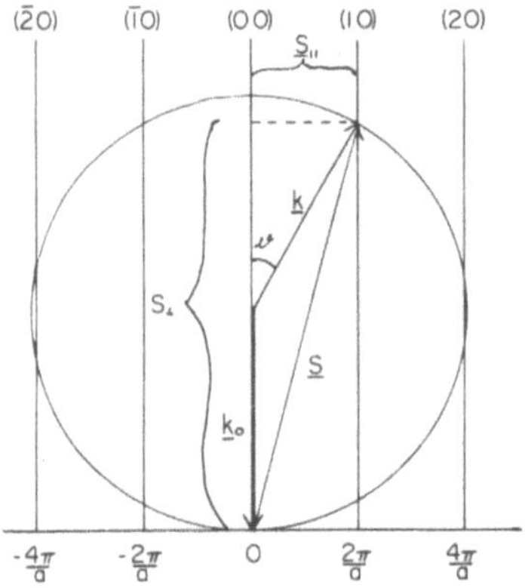

Fig.12.14. Reciprocal lattice and Ewald construction appropriate to LEED. k_0 and $\underline{k}$ are the incident and scattered wave vectors, and $\underline{S}$ is the momentum transfer

of spots, their position, or their variation in energy can be used to establish the type of finite-size effect on the surface, because of the characteristic effect many of them have on the beam profiles. Several such examples are given below. In order to determine quantitatively the size distributions of finite-size ordered regions, the existence or absence of reflections is insufficient. The angular distribution of intensity in the reflection (or at minimum their full width at half-maximum) must be known because it depends quite sensitively on the arrangement of ordered units and their sizes. To do this requires knowledge of the broadening effect of the instrument on a measurement. The effect of the instrument can be seen by recalling that any measurement is a convolution of the true intensity function $I(\vartheta,E)$ and the instrument response function $T(\vartheta,E)$,

$$J(\vartheta,E) = I(\vartheta,E)*T(\vartheta,E) \quad . \tag{12.14}$$

The important factors limiting the instrument response for LEED diffractometers have been outlined by PARK et al. [12.47]. Equation (12.14) shows that a $T(\vartheta,E)$ that is narrow in ϑ will perturb the angular distribution less than a broad $T(\vartheta,E)$. Additionally, the measurement accuracy is important in determining how precisely $I(\vartheta,E)$ can be extracted from a given measurement. The resolving power [12.48] of a LEED diffractometer can be defined in terms of the angular FWHM, b_T, of $T(\vartheta,E)$ and the measurement accuracy X. $T(\vartheta,E)$ is typically a Gaussian, and under those conditions a minimum angle of resolution ϑ_{min} can be defined by

$$\vartheta_{min} = [(b_T + Xb_T)^2 - (b_T - Xb_T)^2]^{\frac{1}{2}} = 2b_T(X)^{\frac{1}{2}} \quad . \tag{12.15}$$

With average instruments ($b_T \sim 1^o$) and average measurement accuracy (5%), the minimum angle of resolution is in the range of 0.4^o (depending on diffraction geometry), which corresponds for typical diffraction conditions to resolvable sizes of ordered regions of ~300-500 Å. Improved measurement accuracy increases this to ~1000 Å. Improved instrument response further increases this value. HENZLER [12.49] has recently demonstrated resolution of the order of 5,000 to 10,000 Å, using a magnetically focussed gun that reduces the contributions to $T(\vartheta,E)$ due to electron-beam size and divergence by about an order of magnitude. Low-energy beams with 10-μm diameter have recently been made with a field emission gun [12.50] that promises to produce an even narrower instrument response function.

Knowledge of the resolving power of the instrument has been used in several cases to provide a quantitative description of the size distribution of ordered regions on a surface. These are also discussed below.

The main components of a LEED diffractometer are an electron gun, a sample goniometer, and a detector (for reviews of progress in LEED instrumentation see [12.51]). A LEED apparatus [12.52] incorporating a vidicon camera to scan the diffraction pattern displayed on a fluorescent screen is shown schematically in Fig.12.15. Typical spectra obtained with this system are shown in Fig.12.16. Figure 12.16a is an angular scan taken of diffracted beams from a cleaved GaAs {110} surface, showing the (01) to (04) beams. Figure 12.16b is the same spectrum taken after the surface has been sputter etched. The angular broadening of the reflections results from the introduction of random steps, illustrated in Fig.12.8, with the Ewald sphere cutting the (01) to (04) rods at points of different broadening.

The measurement of the angular distribution of intensity is the most important one in determinations of finite-size effects. Frequently this measurement is made as a function of energy to scan through the reciprocal lattice. For thermodynamic or kinetic studies of overlayer ordering [12.39,41,54-58] it is made as a function of temperature and of coverage. In such studies, the peak intensity is frequently also measured, e.g., for order-disorder transitions or to determine roughly the coverage.

Figure 12.17 shows the results of Fig.12.16b plotted as a function of energy. The FWHM of the (01) beam profile is shown for several annealing conditions after sputtering [12.59]. It can be seen that the angular width of the profile oscillates as illustrated in Fig.12.8, and that the oscillation damps out when the surface is annealed. The minima occur at positions that correspond to a step height equal to the bulk interplanar distance in

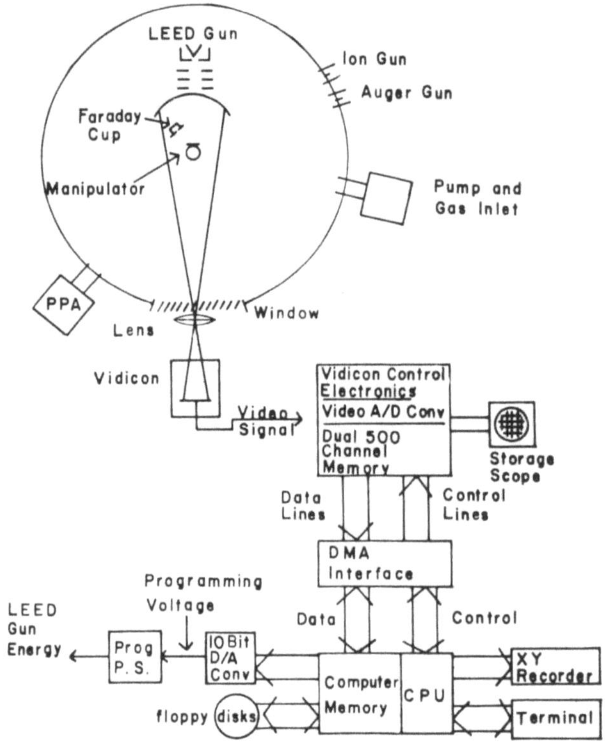

◄Fig.12.15

▼Fig.12.16

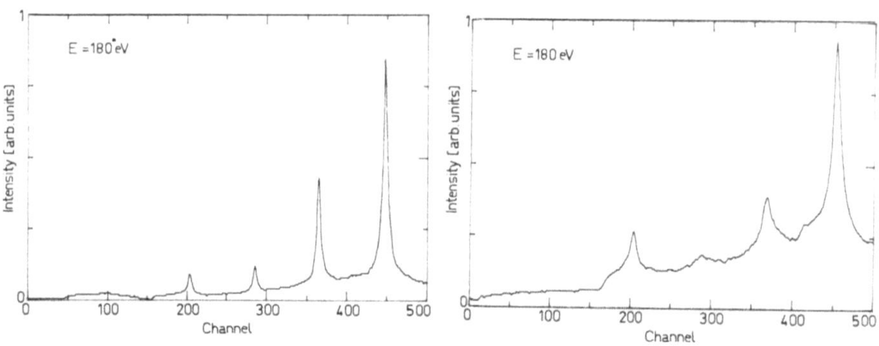

Fig.12.15. Schematic diagram of a LEED diffractometer with a vidicon camera detector. Angular profiles of diffracted beams are measured from a fluorescent screen by imaging the diffraction pattern on the vidicon. The angular resolution (size of collector "aperture") can be changed by adjusting the channel height and width of the vidicon raster scan [12.52]

Fig.12.16. Angular profiles of the (01) to (04) beams from the GaAs{110} surfaces measured by the vidicon system shown in Fig.12.15. (a) Cleaved in vacuum, (b) sputtered and partially annealed [12.53]

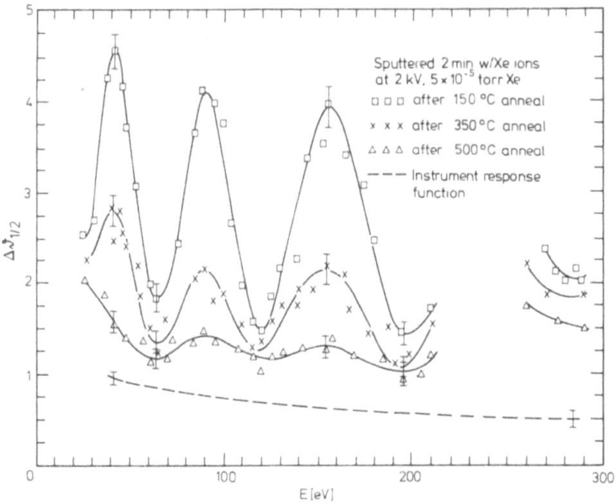

Fig.12.17. Full width at half-maximum of the (01) beam of GaAs{110} sputtered and annealed at three different conditions as a function of energy (or equivalently $S_\perp$). The corresponding terrace sizes are given in the text. Expressed in terms of widths normalized to the Brillouin zone boundary, the maximum oscillations correspond to $\Delta S_\parallel/[\Delta S_\parallel]_{B.Z.} \simeq 0.4$ and the instrument response function width corresponds to $\Delta S_\parallel/[\Delta S_\parallel]_{B.Z.} \simeq 0.05$, over the whole energy range [12.59]

GaAs {110} (this was already observed in [12.27]). The magnitude of the oscillation can be interpreted in terms of step densities or average terrace sizes. Using the random-probability model discussed earlier, average terrace sizes 10 ± 3, 7 ± 1, and 5 ± 1 rows are determined for the three annealing conditions.

Steps on a surface need not, of course, occur randomly. A nongeometric distribution implies interactions between steps. In particular, a short-range attraction or repulsion between steps eliminates small terrace sizes. The absence of some component of the size distribution in turn affects the angular profile. Attempts have recently been made to extract a range of interaction between steps by careful comparison of experimental angular profiles with calculated ones for random distributions and for distributions for which small terrace sizes are absent [12.53]. Preliminary results indicate that terraces smaller than 3 atomic rows are absent on the sputtered GaAs {110} surface.

The instrument response and measurement accuracy in the above experiment cause a minimum resolvable angular width of about 0.2°. Thus step densities that are low enough so that they cause oscillations in FWHM smaller than

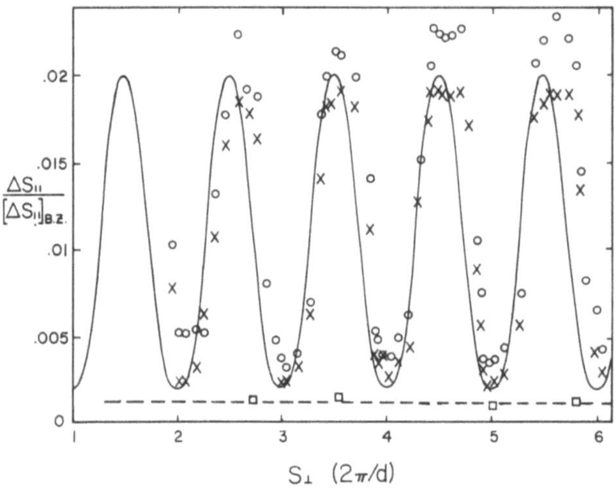

Fig.12.18. Full width at half-maximum of the (00) beam of Si{111} sputtered and annealed at 700°C as a function of $S_{\perp}$. The FWHM is normalized to the width of the Brillouin zone (i.e., the separation of reflections). (x) and (o) refer to measurements in two orthogonal directions on the surface. (□) is the instrumental limit. d is the Si{111} double-layer spacing [12.60]

this would not be resolvable, and the surface would appear flat. A better instrument can resolve smaller step densities (larger terraces). HENZLER has used his new high-resolution LEED system [12.49] to look at an ion-bombarded Si{111} surface annealed to 700°C. The results [12.60] are shown in Fig.12.18. The data are plotted in terms of the Brillouin zone width, and the oscillations correspond to a terrace width of 40 to 50 atomic rows, using the random probability model [12.31]. The resolving power in Figs. 12.16,17 is about the same as the maxima in Fig.12.18; hence these oscillations would be barely resolvable with a high-quality standard instrument.

A very interesting application of the model of alternating terraces and troughs shown in Fig.12.7 has recently been made to the reconstructed Si{111} - (7 × 7)-H surface [12.34]. With the adsorption of H, most of the intermediate reflections in the diffraction pattern of clean Si{111} - (7 × 7) disappear, leaving hexagonally symmetric "stars" of satellite spots around each fundamental reflection, with their intensity, however, showing only 3-fold symmetry, and this symmetry alternating with changes in energy (i.e., $G_{\perp}$) from a Y shape to an inverted Y shape. This is shown in Fig. 12.19a. A model of the surface analogous to that shown in Fig.12.7, but with triangularly shaped terraces and troughs, explains the experimental observations.

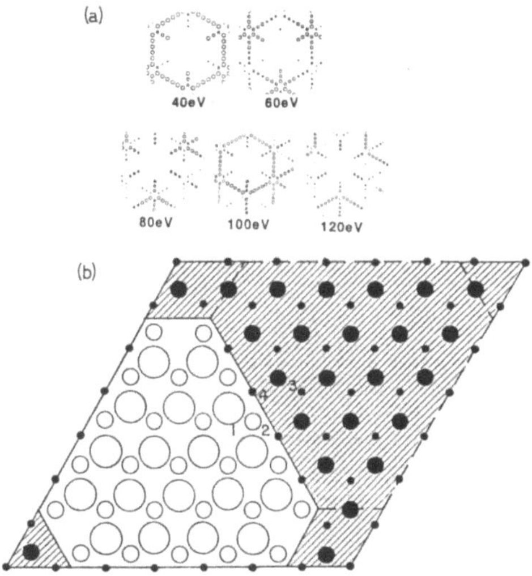

Fig.12.19a,b. Experimental LEED results and structural model for the Si{111} - (7 × 7)-H surface. (a) LEED pattern at different energies, showing the symmetry of the diffraction features. (b) Terrace-trough model for the surface, with the unit mesh divided roughly equally into terrace and trough. 1-4 label the Si layers; the terrace is one Si double layer above the trough [12.33]

The structure factor for triangular terraces is a 6-fold symmetric star. The interference terrace and trough as $S_\perp$ is changed results in the oscillation in intensity from the Y shape to the inverted Y shape. The period corresponds within 5% to the bulk double-layer spacing in Si{111}. The separation of satellite spots is related to the terrace size, as before. A roughly equal size of terrace and trough within the (7 × 7) unit mesh, as shown in Fig. 12.19b, produces agreement with experimental spot positions and intensity variations. The structure for the clean Si{111} - (7 × 7) surface is conjectured to be the same terrace-trough arrangement, but with periodic strain, or buckling, to produce the rest of the diffraction spots [12.34].

Surface defects that are projections of bulk defects have been observed in Ag{111} films [12.23,24] grown by electron-beam evaporation onto mica. These films give good diffraction patterns, but with broad beams, indicating that the films are well oriented but not single crystal. Some of the data are shown in Fig.12.20. It is clear that there is significant mosaic broadening. This can be quantified by replotting the data. A summary of the FWHM of the angular profiles of three beams is shown in Fig.12.21, plotted as a fraction of the Brillouin zone width (i.e., the separation of reflections) versus $S_\perp$, the perpendicular component of the momentum transfer. Review of the dependence of the broadening on $G_\perp$ and $G_{||}$ shown in Figs. 12.1,5 indicates that it is possible to determine misorientation of domains from the slope with $S_\perp$, and finite size by projecting this slope to $S_\perp = 0$. (In actuality, a straight line is not the proper extrapolation

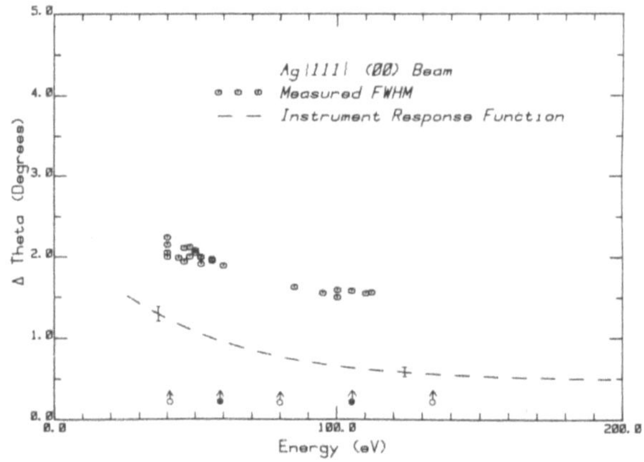

Fig.12.20

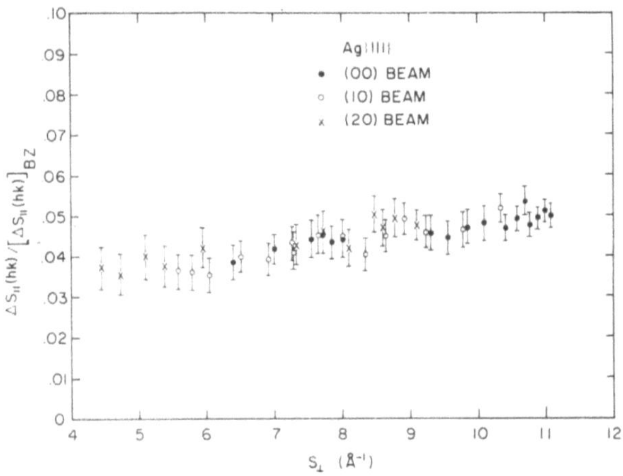

Fig.12.21

Fig.12.20. Full width at half-maximum of the (00) beam from a Ag{111} film grown on mica as a function of energy (or $S_\perp$). The dashed line gives the FWHM of the instrument function. The arrows indicate where maxima (o) and minima (•) in the angular width would occur if the surface contained steps with the bulk Ag{111} plane spacing [12.23]

Fig.12.21. Full width at half-maximum of the (00), (10), and (20) beams from a Ag 111 film as a function of $S_\perp$. The FWHM is normalized to the Brillouin zone width. The slope of a fit to the data gives the average misorientation, while the intercept at $S_\perp=0$ gives the average domain size. The fact that all three reflections have the same width within the uncertainty of the measurement indicates that random strain and twin boundaries are not significant at the surface of this film [12.24]

[12.23]). For this Ag film, the misorientation angle was determined to be $\alpha = 0.25 \pm 0.05^\circ$ and the average domain size 40 ± 15 atomic rows, depending on the distribution function of domain sizes that was chosen.

From Fig.12.3, it should be evident that another factor, strain broadening, can be extracted from these data. If strain is present, higher-order reflections are broader at the same $G_{\perp}$. Figure 12.21 shows that within the error bars, all three measured reflections have the same full width at half-maximum and, to an uncertainty of $\Delta\langle\varepsilon\rangle^2 = \pm\ 0.003$, no strain is present.

A similar statement can be made with regard to twin boundaries. If twin boundaries were present to a marked extent, the (01) and (02) beams should be broader than the (00) beam [12.23]. Within the error bars, all beams have the same width, and thus twin boundaries do not make a measurable contribution to the broadening.

Such analysis can be performed in conjunction with epitaxial growth. For Ge epitaxy on GaAs {110}, finite-size effects have been correlated with growth conditions. Low substrate temperatures yield a greater concentration of line defects, which can partially be annealed out after the film is deposited [12.25]. From the spot broadening, average domain sizes between $\sim$ 55 and $\sim$ 360 Å have been determined for epitaxy at different substrate temperatures and growth rates. Thus spot broadening can be used for quantitative determinations of the defect density at the surface of growing films.

A different example of the use of LEED profile analysis to study monolayer growth is shown in Fig.12.22, for the deposition of W onto W{110} [12.33]. In this system it has been possible to follow the growth and distribution of ordered overlayer islands by monitoring the changes in diffracted-beam profiles. Analysis of the LEED reflections indicates that the overlayer exists, depending on temperature, variously as round islands whose centers are spaced 10 to 15 Å apart, oblong islands with centers 19 Å apart, and terraced structures with terrace width between 9 and 40 Å. The surprising result of this work is that at a given temperature the separation of island centers apparently is independent of coverage, but that this separation is temperature-dependent [12.33]. This result may have implications about the strength of the potential at defects and its range.

An example of the quantitative determination of island size in an overlayer that forms a superlattice is shown in Fig.12.23. Oxygen chemisorbed on W{110} forms into a p(2×1) structure at low coverage, implying a net attractive interaction. By measuring the angular profiles of the superlattice reflections and modeling the width with various size distributions, it has been possible to determine limits on the island size as a function of coverage [12.37,54]. It is evident that even very small islands are observable. Interestingly, even considerable annealing did not cause an

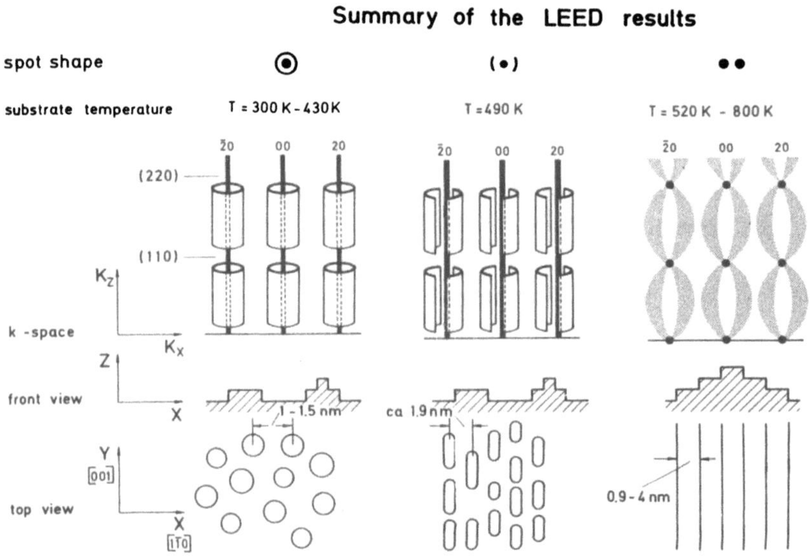

Fig.12.22. Summary of LEED results for monolayer growth of W on W{110}. The distribution of ordered domains depends on the temperature, giving different reciprocal lattices (top). Models for the overlayer ordering that give these reciprocal lattices are shown on the bottom [12.33]

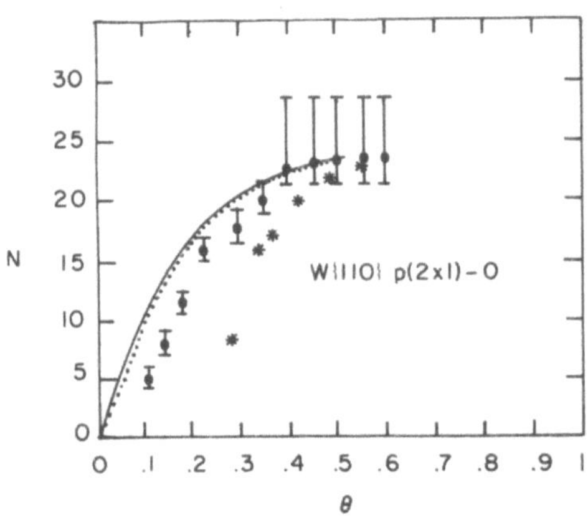

Fig.12.23. Equilibrium island size in a p(2×1) oxygen overlayer on W{110} as a function of coverage. N is the number of rows in an islands in the <11> direction. (•) island size at 300 K; (∗) island size at 425 K. The solid and dotted lines are island sizes predicted at 0 K and 300 K from a phase diagram calculated from LEED intensity vs temperature data for this system [12.54]

increase in size of the islands over those shown [12.41,54]. This can be explained with the hypothesis that substrate defects, e.g., steps, cause a high potential barrier for diffusion, such that the surface consists of a number of thermodynamic subsystems [12.41]. An alternative explanation, recently advanced, is that the symmetry of the layer causes ordering to be exceptionally slow [12.61]. Measurements at different temperatures show that it might be possible to determine a boundary energy contribution to the stability of islands by studying their size distribution with temperature and coverage [12.54].

A variety of other LEED studies of finite-size effects in surfaces and overlayers deal with spot splitting or with the existence or absence of reflections due to preferential ordering at or inhibition of ordering by substrate defects [12.62,63]. Recently it has been shown for W{100} that even steps spaced as closely as 5-10 atomic rows apart do not inhibit formation of the reconstructed surface phase [12.63].

Whereas LEED has been the most frequently used technique for studies of surface ordering phenomena and defect structures, it is not the only one that can be used in this way, as already noted. High-energy electron diffraction, grazing-angle X-ray diffraction, and atomic-beam scattering all have attributes that make them preferred under particular circumstances. These methods are briefly discussed in the next section.

12.4 Surface Defect Studies by Alternative Diffraction Techniques

The use of diffraction techniques other than LEED for surface defect analysis has been limited, but appears to be growing rapidly because each method has unique features to offer. For example, RHEED is applicable to high-temperature surface defect studies. Because of the small momentum transfers in RHEED, the Debye-Waller effect will be much smaller. This can be a distinct advantage for dynamic structural studies under epitaxial-growth conditions. In RHEED, the Ewald sphere has a much larger radius than in LEED (it can be approximated by a plane) and cuts the reciprocal lattice nearly parallel to the rods. Limited instrumental resolution and finite-size effects in the surface cause the characteristic streaks in RHEED patterns. All the effects discussed in the first section should be observable. There is the possibility that high resolving power can be obtained for RHEED more easily than it can for LEED, because precise electron beam definition is easier to achieve at higher energies. Analysis of the resolving power has, to our knowledge, not been performed for RHEED instruments. A treatment similar to that given for LEED should be applicable.

Several RHEED papers have dealt with defect studies. MATYSIK [12.16] has made an extensive study of the structure and nucleation kinetics of Cd on Ge{111}. He was able to observe diffraction features associated with Cd as it first began to order at a critical density and temperature. The Cd layer has hcp basal-plane symmetry, is commensurate with the Ge substrate, and has a lattice constant somewhat smaller than that of Ge. Critical nuclei estimated to have an average diameter of 10 Å were found to be either randomly arranged on the surface or spinning [12.16], i.e., broad rings formed in the diffraction pattern. These results could only be obtained dynamically, i.e., if the deposition of Cd was interrupted, the nuclei would immediately grow into three-dimensional structures.

The capabilities of LEED and RHEED have been directly compared in one study [12.64] of vicinal surfaces of Cu{111} and GaAs{100}. It was concluded that the RHEED instrument used in that study had a somewhat higher resolving power (i.e., was able to distinguish somewhat larger terrace sizes) than the LEED instrument. No estimates of the quality of either instrument were given, however. Results for terrace widths determined by LEED and RHEED agreed where they were compared.

A recent RHEED study has shown that terrace sizes on a vicinal GaAs{100} surface become more well defined and uniform in size if GaAs is grown on the substrate by molecular beam epitaxy, i.e., there appears to be an effective repulsive interaction between steps [12.65].

X-ray diffraction [12.66] using either a rotating anode or synchrotron radiation, can also be used for surface defect studies [12.18,67,68]. The Ewald construction looks similar to that for RHEED and the method should again be applicable at high temperatures. The advantages claimed for X-ray diffraction are that it involves no multiple scattering and that it is simpler. On the other hand, source intensity can be a problem. So far X-ray diffraction has been used mainly for phase transformation studies on surfaces [12.67,68] but there is clearly the possibility of observing finite-size effects.

Atomic-beam diffraction has also been used for surface structural studies. The Ewald sphere resembles that for LEED. The atomic beam probes the interaction potential between projectile and target. Insofar as this mirrors the positions of atoms (which is likely in most cases), atomic beams can be used to probe the reciprocal lattice in the same manner as electrons or X-rays. The terrace-trough model for Si{111} - (7×7)-H has been verified by atomic-beam scattering (Chap.7) [12.69]. Recently atomic-beam diffraction experiments from a stepped Pt{997} surface have been interpreted in terms of an upward bending of the terrace surfaces at the step edges [12.70].

12.5 Summary

In this review, an attempt has been made to describe a variety of surface and overlayer defects and to illustrate their effect on the distribution of intensity that would be observed in a diffraction experiment. From the discussion it should be apparent that many types of defects can be uniquely identified and quantified in this manner.

Most of the discussion has been in terms of a static description of the defects. However, the most interesting applications of surface defect studies are in terms of dynamic experiments related to nucleation, growth, chemical reactions, transport, and thermodynamics of overlayers. Few experiments of this nature have been done but it is likely that a major emphasis on such studies will occur as better instrumentation and quantitative models for the interpretation of the data become available. As a result, it should become possible to determine such parameters as the interaction of defects, the potential energy associated with defects, and rate laws for the growth of overlayers and films on crystalline substrates from diffraction experiments.

Acknowledgements. I would like to acknowledge useful discussions with M. Henzler, T.-M. Lu, D. Saloner, D.E. Savage, and M.B. Webb. I thank H. Clearfield, T.-M. Lu, and D. Saloner for criticlly reading the manuscript, and J. Bohlman and D. Jones for making the drawings. Those parts of the research discussed in this paper that we have performed have been supported in part by NSF Grant No. DMR-78-25754 and in part by ONR.

References

12.1 C. Herring: In *Structure and Properties of Solid Surfaces*, ed. by R. Gomer, C.S. Smith (U. Chicago Press, Chicago 1953)
12.2 W.K. Burton, N. Cabrera, F.C. Frank: Philos. Trans. R. Soc. London *243*A, 299 (1950)
12.3 J.P. Hirth: In *Energetics in Metallurgical Phenomena*, Vol.II, ed. by W.M. Mueller (Gordon and Breach, New York 1965)
12.4 H.J. Leamy, G.H. Gilmer, K.A. Jackson: In *Surface Physics of Materials*, Vol.I, ed. by J.M. Blakely (Academic, New York 1975)
12.5 I.N. Stranski: J. Phys. Chem. *136*, 259 (1928); *21*, 421 (1931)
12.6 B. Lewis, J.C. Anderson: *Nucleation and Growth of Thin Films* (Academic, London 1978)
12.7 W.P. Ellis, R.L. Schwoebel: Surf. Sci. *11*, 82 (1968)
12.8 M. Henzler: Surf. Sci. *19*, 159 (1970)
12.9 G.A. Somorjai: In *The Physical Basis for Heterogeneous Catalysis*, ed. by E. Drauglis, R.I. Jaffee (Plenum, New York 1975)
12.10 W. Erley, H. Ibach, S. Lehwald, H. Wagner: Surf. Sci. *83*, 565 (1979)
12.11 T.E. Madey: Surf. Sci. *94*, 483 (1980)

12.12 H. Wagner: "Physical and Chemical Properties of Stepped Surfaces", in *Springer Tracts in Modern Physics*, Vol.85, Solid Surface Physics (Springer, Berlin, Heidelberg, New York 1979)
12.13 S. Krueger, W. Mönch: Surf. Sci. *99*, 157 (1980)
12.14 M. Henzler: In *Electron Spectroscopy for Surface Analysis*, ed. by H. Ibach, Topics in Current Physics, Vol.4 (Springer, Berlin, Heidelberg, New York 1977)
12.15 M.G. Lagally, D.G. Welkie: In *Advanced Techniques for the Characterization of Microstructures*, ed. by F.W. Wiffen (TMS Warrendale PA, 1982)
12.16 K. Matysik: Surf. Sci. *46*, 457 (1974)
12.17 K. Matysik: J. Appl. Phys. *47*, 3826 (1976)
12.18 W.C. Marra, P. Eisenberger, A.Y. Cho: J. Appl. Phys. *50*, 6927 (1979)
12.19 G. Comsa, G. Mechtersheimer, B. Poelsema: Surf. Sci. *97*, L.297 (1980)
12.20 T.T. Tsong, P.L. Cowan: In *Chemistry and Physics of Solid Surfaces*, Vol.II, ed. by R. Vanselow (CRC, Boca Raton, FL 1979)
12.21 N. Osakabe, Y. Tanishiro, K. Yagi, G. Honjo: Surf. Sci. *102*, 424 (1981)
12.22 D. Saloner, D.E. Savage, M.G. Lagally: In preparation
12.23 D.G. Welkie: Ph. D. Dissertation, University of Wisconsin-Madison (1981)
12.24 D.G. Welkie, M.G. Lagally, R.L. Palmer: J. Vac. Sci. Technol. *17*, 453 (1980)
12.25 H.M. Clearfield, D.G. Welkie, T.-M. Lu, M.G. Lagally: J. Vac. Sci. Technol. *19*, 323 (1981)
12.26 J.E. Houston, R.L. Park: Surf. Sci. *21*, 209 (1970); *26*, 269 (1971)
12.27 M. Henzler: Surf. Sci. *22*, 12 (1970)
12.28 K. Besocke, H. Wagner: Surf. Sci. *53*, 351 (1975)
12.29 M. Henzler: Surf. Sci. *73*, 240 (1978)
12.30 W. Moritz, D. Wolf: In *Proceedings of the Fourth International Conference on Solid Surfaces and the Third European Conference on Surface Science*, ed. by P.A. Degras, M. Costa (Plenum, New York 1980)
12.31 T.-M. Lu, S.R. Anderson, M.G. Lagally, G.-C. Wang: J. Vac. Sci. Technol. *17*, 207 (1980);
T.-M. Lu, M.G. Lagally: Surf. Sci. (to be published)
12.32 P.B. Hirsch, A. Howie, R.B. Nicholson, D.W. Pashley, M.J. Whelan: *Electron Microscopy of Thin Crystals* (Buttersworth, Washington, D.C. 1965)
12.33 P.O. Hahn, J. Clabes, M. Henzler: J. Appl. Phys. *51*, 2079 (1981)
12.34 E.G. McRae, C.W. Caldwell: Phys. Rev. Lett. *44*, 1632 (1981)
12.35 H.M. Kramer, E. Bauer: Surf. Sci. *92*, 53 (1980); *93*, 407 (1980)
12.36 H.W. Hayden, W.G. Moffatt, J. Wulff: *The Structure and Properties of Materials*, Vol.III (Wiley, New York 1965)
12.37 T.-M. Lu, G.-C. Wang, M.G. Lagally: Surf. Sci. *107*, 494 (1981)
12.38 S.C. Fain, M.D. Chinn: J. Phys. *38*, Suppl. 10, C4-99 (1977)
12.39 S. Semancik, P.J. Estrup: J. Vac. Sci. Technol. *18*, 541 (1981);
R.A. Barker, P.J. Estrup: J. Chem. Phys. *74*, 1442 (1981);
P.J. Estrup, E.G. McRae: Surf. Sci. *25*, 1 (1971)
12.40 J.C. Tracy, J.M. Blakely: Surf. Sci. *15*, 257 (1969);
J.C. Tracy: Ph. D. Dissertation, Cornell University (1968)
12.41 M.G. Lagally, G.-C. Wang, T.-M. Lu: In *The Chemistry and Physics of Solid Surfaces*, Vol.II, ed. by R. Vanselow (CRC Boca Raton, FL 1979)
12.42 R.L. Park: In *The Structure and Chemistry of Solid Surfaces*, ed. by G.A. Somorjai (Wiley, New York 1969)
12.43 P.I. Cohen, J. Unguris, M.B. Webb: Surf. Sci. *58*, 429 (1976);
P.I. Cohen: Ph. D. Dissertation, University of Wisconsin-Madison (1975)
12.44 M.E. Schrader: J. Phys. Chem. *84*, 2774 (1980)
12.45 P. Kleban: Surf. Sci. *103*, 542 (1981)
12.46 T.-M. Lu, M.G. Lagally, G.-C. Wang, J.E. Houston: Surf. Sci. (to be published)

12.47 R.L. Park, J.E. Houston, D.G. Schreiner: Rev. Sci. Instrum. *42*, 60 (1971)
12.48 T.-M. Lu, M.G. Lagally: Surf. Sci. *99*, 695 (1980)
12.49 M. Henzler: To be published
12.50 J.A. Martin, M.G. Lagally: To be published
12.51 M.G. Lagally, J.A. Martin: Rev. Sci. Instrum.(to be published)
12.52 D.G. Welkie, M.G. Lagally: Appl. Surf. Sci. *3*, 272 (1979)
12.53 D.G. Welkie, T.-M. Lu, H.M. Clearfield, M.G. Lagally: 41st Annual Physical Electronics Conference, Bozeman, MT 1981 (to be published)
12.54 M.G. Lagally, T.-M. Lu, G.-C. Wang: In *Ordering in Two Dimensions*, ed. by S.K. Sinha (Elsevier, Amsterdam 1980)
12.55 G.-C. Wang, T.-M. Lu, M.G. Lagally: J. Chem. Phys. *69*, 479 (1978)
12.56 R.L. Park, T.L. Einstein, A.R. Kortan, L.D. Roelofs: In *Ordering in Two Dimensions*, ed. by S.K. Sinha (Elsevier, Amsterdam 1980);
A.R. Kortan, R.L. Park: Phys. Rev. B*23*, 6340 (1981)
L.D. Roelofs, A.R. Kortan, T.L. Einstein, R.L. Park: J. Vac. Sci. Technol. *18*, 492 (1981)
12.57 P.A. Bennett, M.B. Webb: Surf. Sci. *104*, 74 (1981)
12.58 P.J. Estrup, R.A. Barker: In *Ordering in Two Dimensions*, ed. by S.K. Sinha (Elsevier, Amsterdam 1980)
12.59 D.G. Welkie, M.G. Lagally: J. Vac. Sci. Technol. *16*, 784 (1979)
12.60 M. Henzler: Appl. Surf. Sci. (in press)
12.61 S.A. Safran: Phys. Rev. Lett. *46*, 1581 (1981)
12.62 M.K. Debe, D.A. King: Surf. Sci. *81*, 193 (1979)
12.63 G.-C. Wang: Private communication
12.64 F. Hottier, J.B. Theeten, A. Masson, J.L. Domange: Surf. Sci. *65*, 563 (1977)
12.65 J.M. Van Hove, P.I. Cohen: J. Vac. Sci. Technol. *20*, (in press)
12.66 W.W. Beeman, P. Kaesberg, J.W. Anderegg, M.B. Webb: In *Handbuch der Physik*, Vol.32, ed. by S. Flügge (Springer, Berlin, Heidelberg, New York 1957)
12.67 R.J. Birgenau, E.M. Hammons, P. Heiney, P.W. Stephens, P.M. Horn: In *Ordering in Two Dimensions*, ed. by S.K. Sinha (Elsevier, Amsterdam 1980)
12.68 P. Eisenberger, W. Marra, P. Fuoss: 41st Annual Physical Electronics Conference, Bozeman, MT 1981
12.69 M. Cardillo: Phys. Rev. B*23*, 4279 (1981)
12.70 G. Comsa, G. Mechtersheimer, B. Poelsema, S. Tomoda: Surf. Sci. *89*, 123 (1979)

13. Some Theoretical Aspects of Metal Clusters, Surfaces, and Chemisorption

R. P. Messmer

With 10 Figures

13.1 Intrinsic Properties of Metal Clusters

Before discussing what has been learned recently regarding the intrinsic electronic structure of metal clusters, it is useful to discuss a few general issues, so that a proper framework is established.

The first issue concerns the structure of metal particles. A primary problem which plagues all electronic structure calculations of clusters is the lack of experimental information on the geometrical structure (i.e., the positions of the atoms) of the clusters. For example, consider Fig.13.1, which shows two high-symmetry arrangements for 13 metal atoms. One polyhedron is a cubooctahedron, the other is an icosahedron. Which of these structures, if either, is the appropriate geometry for a cluster of 13 nickel atoms, for example? In most situations in molecular and solid-state physics, this structural information is known at the outset. However, for the case of clusters one is forced to assume "reasonable structures" in order to proceed. The usual wisdom is to assume highly symmetrical clusters (e.g., those of Fig. 13.1) most often reflecting the bulk structure (e.g., the cluster of Fig. 13.1a for the case of a face-centered-cubic metal) and to assume internuclear distances close to the bulk values. It is unlikely that these assumptions will be true for extremely small clusters, that is, on the order of ten atoms or less. Experimental information on structure is badly needed.

One might wonder why the electronic structure calculations are not used to determine the geometry of clusters. Unfortunately, at present, this appears to be an almost impossible task. The simple models which can treat reasonably large clusters ($\sim$25 atoms) are not accurate enough to reliably determine the small energy differences between various geometric configurations. On the other hand, the sophisticated models are so computationally complex that only a few atoms can be treated.

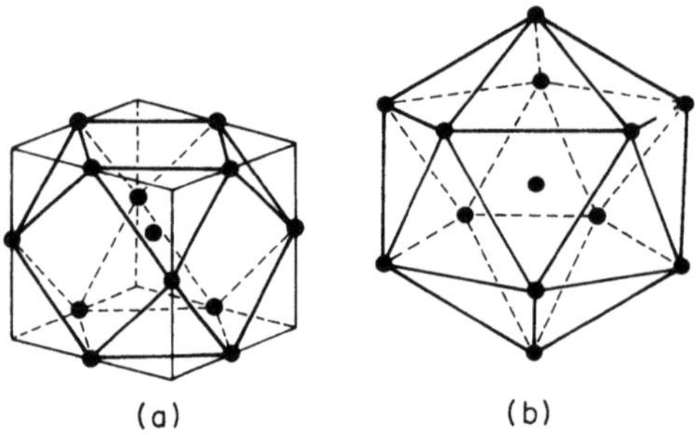

Fig.13.1a,b. Two 13-atom clusters. (a) Cubo-octahedral geometry. (b) Icosahedral geometry

A recent breakthrough in the experimental production of metal particles which have been mass analyzed from single atoms to clusters of over 100 atoms [13.1] is of fundamental importance. It promises to offer a source of metal clusters to experimentalists interested in the study of the intrinsic structure and properties characterization of such clusters. Such work will be essential to future progress in this field.

The second issue to be addressed is the form of molecular-orbital theory which one should use to calculate the electronic structure of metal clusters. It is the author's opinion that the best first-principles calculations possible should be carried out on clusters of 10-20 atoms to explore the properties of such clusters. With these results in hand, the simple semiempirical methods then can be properly parameterized to reproduce, as well as possible, the more accurate calculations for small clusters. These methods then could be logically extended to calculate the properties of much larger clusters.

The term "first-principles" methods means that such methods avoid, to the maximum extent possible, the introduction of approximations and parameters beyond the basic postulates of the methods. There are two such types of molecular-orbital methods in common use: 1) the Hartree-Fock method and 2) the density functional and $X\alpha$ methods. The former method has been the workhorse of quantum-chemical applications to molecules, but is not applied to the treatment of bulk metals via band theory because of its well-documented problems in the description of metals [13.2,3]. The $X\alpha$ method [13.4] (which is a particular form of density functional theory), on the other hand, has been widely used to treat the electronic structure of bulk metals by band theory and over the last ten years has been extended to the investigation of molecules.

Because of its proven ability to treat the electronic structure of bulk metals, the author believes that the $X\alpha$ method is the most reasonably suited first-principles technique for initial investigations of metal clusters. Thus far it has been the most widely applied method for the investigation of metal clusters and it is the results of this method which will dominate the discussion below.

13.1.1 Cluster Density of States

One of the most readily accessible theoretical characteristics of a metal cluster is its density of states (DOS). The DOS is a function which tells one about the number of energy levels per unit of energy as a function of energy. It is usually calculated for bulk metals from band theory computations. The ability to obtain such information from cluster calculations allows a comparison to be made between the DOS of clusters and that of the bulk metal.

The calculated energy-level diagram for a Ni_{13} cluster with the geometry of Fig.13.1a and a metal-metal distance appropriate to bulk Ni is shown in Fig.13.2. The method of calculation is the self-consistent-field $X\alpha$ scattered wave method [13.4,5]. Except for the change of boundary conditions to those appropriate to a cluster, the method is the same as the well-known Korringa-Kohn-Rostoker (KKR) method [13.6,7] of band theory. The calculations were carried out in a spin-polarized fashion, allowing different molecular orbitals for different spins. The spin-polarized energy levels of Fig.13.2 are displayed on two energy scales, one encompassing the entire d band and overlapping s, p levels and the other (to the right) resolving the upper part of the d band on a finer energy scale. The Fermi level E_F separates the unoccupied spin orbitals from the occupied ones. The Ni_{13} cluster has a net magnetic moment, and the moments of the central and peripheral atoms are aligned, i.e., the cluster is "ferromagnetic". This is particularly interesting as the bulk metal is known to be ferromagnetic. Even more striking is the fact that similar calculations for Pd_{13} and Pt_{13} clusters are *not* magnetic — again consistent with the bulk situation [13.8].

The complete set of levels in Fig.13. 2 can be characterized as a dense band of d levels through which the Fermi level passes at the upper end, bounded above and below with sp-like levels. Thus it is found that even for 13 atoms (actually, even for 4 atoms) the d band is totally overlapped by the sp band, as in the solid. Lest one think that such a cluster mirrors all the properties of the bulk, we hasten to point out that there are also considerable differences between the cluster and the bulk as discussed previously [13.8]. For example, the magnetic moment of the central and surface atoms are different (which is not surprising), and the surface atoms have a deficit of electron charge whereas the central atom has a surplus. Nonetheless, the number of qualitative similarities is rather striking.

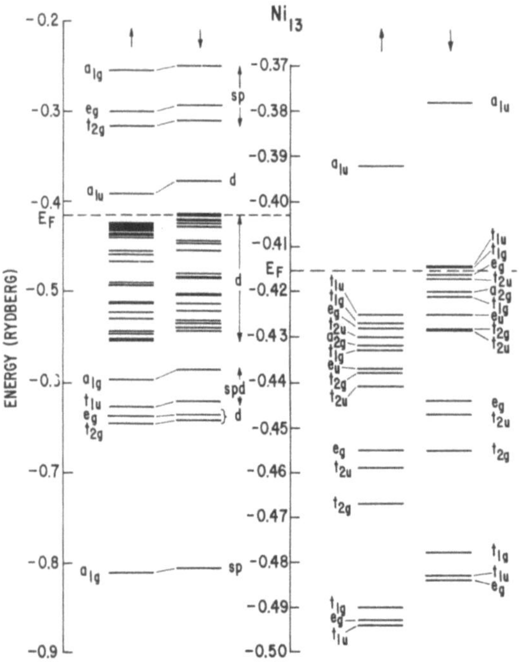

Fig.13.2. Spin-polarized SCF-Xα-SW electronic energy levels for a cubooctahedral Ni_{13} cluster with nearest neighbor internuclear distance equal to that of bulk crystalline nickel. The levels are labeled according to the irreducible representations of the O_h point group

A DOS may be generated from an energy-level diagram such as shown in Fig. 13.2 by centering a Gaussian of finite width at the position of each energy level. This has been done for a spin-restricted Ni_{13} calculation and compared with the bulk DOS of a band structure calculation as well as with two cluster calculations based on semiempirical molecular-orbital methods [13.8]. The band theory results were from on Xα-LCAO calculation [13.9], the semiempirical MO results were from extended Huckel (EH) cluster results [13.10], and "complete neglect of differential overlap" (CNDO) approximate MO cluster results [13.11]. From such a comparison [13.8] one can say that the CNDO and EH methods, with the parameterizations which have appeared in the literature, do not give DOS curves which resemble the main features of the bulk metal DOS. On the other hand the Xα-SW cluster results, while clearly different in detail, give an overall qualitative resemblance to the bulk DOS. This situation applies to other transition metals as well. It is the author's view that these differences in the cluster results are due to inappropriate and rather arbitrary parameterizations of the semiempirical methods (CNDO and EH), as pointed out previously [13.8]. This view is supported by two calculations which have been published using the EH method for Cu clusters [13.12,13]. The calculated d-band widths differ by an order of magnitude in the two calculations, which use different parameterizations. Thus, it would appear that until the semiempirical methods have found a consistent and reliable approach to parameterization, one must rely upon a nonempirical first-principles theory such as the Xα method to discuss the electronic structure of metal clusters.

The other type of nonempirical MO method, besides the density functional/Xα methods, is the Hartree-Fock approach mentioned above. This method has been applied to transition metal clusters only rather recently [13.14,15]. Both Ni [13.14] and Cu [13.15] clusters have been considered, with Ni_2 to Ni_6 investigated in the former case and in the latter case clusters up to Cu_{13}. One striking difference between the Hartree-Fock (HF) and Xα methods for such clusters is that, in contrast to the Xα method, the HF method produces s-like levels which are always above the d bands. That is, s-like character does *not* overlap the d bands as in the bulk or as in the Xα results for Ni and Cu clusters. Thus the studies based on the HF method conclude that such small clusters are fundamentally different from the bulk even in their qualitative characteristics. This raises a substantial conflict between the conclusions of the HF and Xα methods for metal clusters.

Although it is too early to say that there is a generally accepted resolution of this conflict, some insight may be obtained into how this situation might be resolved by looking at some previous conflicts between the conclusions of the two methods. Only three cases will be cited here. The first involves calculations on the model system $Cr_2(O_2CH)_4$. It is known experimentally that such molecules have substantial metal-metal bonding, yet a Hartree-Fock calculation [13.16] concluded that there was not net bonding between the metal atoms. Xα-SW calculations for the same system showed that there was metal-metal bonding in agreement with experiment [13.17]. Subsequent work [13.18,19] has shown that the usual Hartree-Fock method is inappropriate because it cannot take into account some important electronic correlation effects. When these effects are taken into account by a configuration interaction calculation, which introduces much additional computational complexity, a "completely different description of the bonding is achieved" [13.18,19]. That is, a Cr-Cr bond is found to exist, in agreement with the Xα results and experiment.

A second example is the ozone molecule. In this case the Hartree-Fock method gives the incorrect ground state and the wrong ordering of several excited states [13.20]. Again, if electronic correlation effects are taken into account, i.e., going beyond the HF molecular-orbital approximation, by generalized valence bond (GVB) and configuration interaction (CI) approaches, then a description consistent with experiment is achieved. However, it has been shown [13.21] that at the Xα molecular-orbital level one already obtains the correct ground state and a reasonable description of a number of excited states. Thus in these cases, as well as others, the Xα MO method appears to give a much better first-order description of the systems than achieved by the Hartree-Fock method. It should be clear that GVB and CI methods which incorporate correlation effects, if carried far enough, can provide very accurate descriptions of any system and descriptions far superior to the Xα method. However, that is not the point. The issue is whether the density functional/Xα or the Hartree-Fock description is the more appropriate one at the molecular-orbital level of approximation. This is particularly important for metal clusters since the GVB and CI methods are so computationally demanding in their present forms that they are unlikely to be applied to a cluster such as Ni_{13} for some time to come. Thus one will be confined to results at the MO level of approximation for the time being, and it is essential to know which approach bears the greater resemblance to the actual situation in metal clusters.

The last example to be discussed is the most relevant to the metal cluster problem, namely Ni_2. For the Xα results [13.22], one arrives at a situation similar to that found for Xα calculations on larger clusters (see, e.g., Fig.13.2), namely the d levels are bounded above and below by sp-like levels. That is, there is an overlap of s- and d-like excitations for Ni_2. For the Hartree-Fock case [13.14] on the other hand the sp-like levels are all above the d levels in energy. That is, there is no overlap of d- and sp-like levels in the HF calculations. However, for the Ni_2 molecule, extensive GVB and GVB-CI calculations have been performed in addition to HF calculations by UPTON and GODDARD [13.23]. They concluded that the states of Ni_2 and the ion states of Ni_2^+ cannot be appropriately described by Hartree-Fock calculations and that it was neccessary to go to a more sophisticated treatment such as the GVB method to obtain meaningful results. It is important to note that in their paper [Ref.23, Fig.6], they have shown that for the ionic states of Ni_2^+ that there *is* an overlap of the s- and d-type excitations in qualitative agreement with band theory and Xα results. It should be kept in mind that when angle-resolved photoelectron spectroscopy is used to determine energy-band dispersions [13.24] it is the ion states which are of relevance.

It is probably worthwhile reiterating a few points. First, when accurate calculations of the GVB or CI type can be carried out, they will always be more reliable than either Xα or HF results and contain more detailed information than either MO method is capable of giving. Second, as such accurate calculations will not be routine for larger transition metal clusters for sometime, one will probably have to be content temporarily with results at the MO level of approximation. Third, on the basis of known examples, three of which are discussed above, the density functional/Xα results are in better qualitative agreement with the more accurate calculations then are the Hartree-Fock results. Thus, one is led to conclude from the information presently available that the density functional/Xα methods are likely to give the most reliable information on the electronic structure of metal clusters of any of the presently available molecular orbital theories.

13.1.2 Cluster Magnetism

One of the fascinating aspects of investigating transition metal clusters is the prospect of finding unusual magnetic effects—again due to the large surface-to-volume ratio of such clusters. Of the cluster calculations carried out to date using various methods, only the Xα-SW method has been used to investigate the magnetic properties of various metal clusters.

The first Xα-SW work on transition metal clusters [13.8] was concerned with Ni_8, N_{13}, Cu_{13}, Pd_{13}, and Pt_{13}. It was found that Ni_8 and Ni_{13} clusters were magnetic while Pd_{13} and Pt_{13} were not. For the Ni_8 cluster the calculated average spin magneton number per atom was 0.25, whereas for the Ni_{13} cluster the value was 0.46. These may be compared to the bulk value of 0.54. As pointed out above, the spin density on the central atom is quite different from that found on the 12 surface atoms.

Although it is rather remarkable that the 13-atom clusters of Ni, Pd, and Pt which all have the same number of valence electrons turn out to be magnetic for Ni and nonmagnetic for Pd and Pt, and thus minimic the behavior found in the bulk metals, one cannot expect that this will be a general phenomena. In fact, recent calculations [13.25] on a 15-atom Cr cluster have shown that such clusters carry a net magnetic moment, while it is known that the bulk metal is antiferromagnetic. Nonetheless, the Cr_{15} cluster results mimic the antiferromagnetic behavior to a large degree, but due to surface effects there is not an exact cancellation of moments of opposite spin orientation. It is interesting to consider the Cr_{15} results in a broader context, however—namely, to study the magnetic properties of clusters of three different elements, which have the same crystal structure in the bulk (body-centered cubic), and exhibit three different types of magnetic ordering. The metals are vanadium (nonmagnetic), chromium (antiferromagnetic), and iron (ferromagnetic).

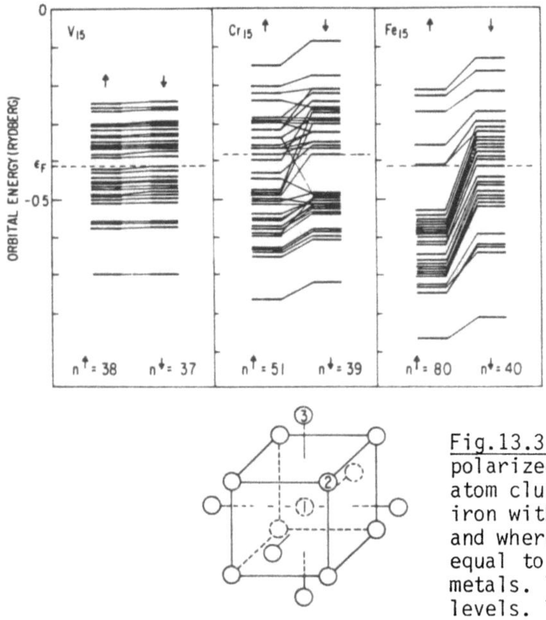

Fig.13.3. Orbital eigenvalues from spin-polarized SCF-Xα-SW calculations on 15-atom clusters of vanadium, chromium, and iron with geometry shown at the bottom and where the internuclear distances are equal to that of the respective bulk metals. The dashed lines mark the Fermi levels. The numbers of spin-up and spin-down valence electrons are given at the bottom of each panel

Figure 13.3 shows the orbital energies from spin-polarized Xα-SW calculations for V_{15}, Cr_{15}, and Fe_{15}. The cluster geometry is shown at the bottom of the figure and consists of 15 atoms representing the bcc structure with a central atom, the eight nearest neighbors at the corners of a cube of side a (appropriate to the observed bulk internuclear distance), and six second neighbors in the faces of a cube of side 2a. The lines connecting up-spin and down-spin energy levels of Fig.13.3, merely relate the same symmetry types in sequential order.

The calculations were carried out fully self-consistently, allowing a flow of electrons between the two spin manifolds such that at convergence Fermi statistics are satisfied. In the case of V_{15}, the added degree of freedom provided by spin polarization has a negligible effect. The converged result has a single extra majority spin electron and the exchange splitting is very small. Thus V_{15} mimics, as far as possible for a system with an odd number of electrons, the nonmagnetic situation of bulk vanadium.

For Cr_{15} an approximate calculation of the magnetic moment [13.25] for the various atoms yields $-0.7\mu_B$ for the central atom, $+4.1\mu_B$ for the first neighbors, and $-3.4\mu_B$ for the second neighbors, which may be compared to $\pm 0.7\mu_B$ for bulk antiferromagnetic chromium as determined from neutron diffraction data [13.26]. The near-perfect agreement of the central atom value with experiment must be considered largely fortuitious. Thus while the alternation in sign characteristic of bulk chromium is well represented by the results, they also indicate that the 15-atom cluster in the bulk geometry should have a permanent magnetic moment, and also that the moments for the surface atoms may be quite different from those of atoms in the interior. Adding further shells of atoms will undoubtedly lead to equivalent spin-up and spin-down potentials in the interior, but the overall magnetic ordering will remain, as it is largely determined by local short-range interactions.

Turning finally to the case of iron, it can be seen from Fig.13.3 that Fe_{15} shows a very large spin polarization and has an excess of 40 majority spin electrons. Neglecting the contribution to the magnetization from orbital angular momentum (which is small for bulk iron [13.27]), this yields an average magnetic moment per iron atom of $2.7\mu_B$ which compares to the experimental value of $2.2\mu_B$ for bulk iron. However, the three groups of symmetry equivalent iron atoms in the Fe_{15} cluster are in different environments, and one would not expect that the charge and spin densities for them would be the same. An approximate calculation [13.25] of magnetic moments yields $1.1\mu_B$ for the central atom, $2.7\mu_B$ for the first neighbors, and $2.8\mu_B$ for the second neighbors; thus the cluster is ferromagnetic. A detailed discussion of the Fe-cluster results can be found elsewhere [13.28]. Thus it is seen again that while the qualitative features of magnetism of bulk iron are reasonably well interpreted with the results of an Fe_{15} cluster, there are definite effects which arise from the finite size of the cluster (surface effects).

Summarizing cluster magnetism for 15-atom clusters of three bcc transition metals, one finds that the clusters appear to be sufficiently large to exhibit magnetic ordering which is qualitatively similar to that shown by the bulk metals, and hence that insight into certain aspects of bulk properties may be obtained. However, the clusters also exhibit characteristic differences as compared to the bulk, and these intrinsic differences are important and of interest in their own right.

13.2 The Interaction of CO with Cu Clusters

The nature of the multiple lines observed in the X-ray photoelectron spectroscopy (XPS) of core levels of adsorbates on metals has received considerable attention recently. For the case of CO chemisorbed on Cu, two experimental studies have been reported [13.29,30]. The first studied C 1s and O 1s core hole spectra of CO on polycrystalline Cu [13.29], and the second studied the C 1s core hole spectrum of CO on Cu{100}[13.30]. Although initially the possibility of multiple adsorption sites was considered as an explanation of the observed multiple lines in the core hole spectra, it is now generally agreed that the multiple peaks arise from a single adsorption site [13.29,30].

In this review we will discuss both the O 1s hole spectrum and the C 1s hole spectrum for CO chemisorbed on Cu{100}, although it is only the latter case for which single crystal data has been reported. However, it should be noted that the reported C 1s spectra for CO on polycrystalline Cu [13.29] and on Cu{100}[13.30] appear very similar; both exhibit a characteristic three-peak spectrum.

The first theoretical treatment to consider the C 1s hole spectrum as arising from a single adsorption site for CO on Cu was the work of GUNNARSSON and SCHONHAMMER [13.31]. They employed a simple model Hamiltonian approach and concluded that the shape of the valence density of states (DOS) of the metal (Cu) can dramatically influence the form of the XPS core spectrum of the adsorbate. A second approach to the problem and a rather different interpretation of the origin of the three peak spectrum was given in a preliminary communication [13.32]. In that communication a molecular-orbital cluster method was employed. Specifically, self-consistent-field Xα scattered wave calculations [13.4,5] were presented for a Cu_5CO cluster in which the CO was in a fourfold site.

The third and most recent theoretical study [13.33] has employed a molecular cluster approach using the self-consistent-field Hartree-Fock method to study a Cu_5CO cluster, in which the CO is in a onefold site on the metal cluster. The physical model obtained from the latter study is rather different from either of the two previous studies.

Because of the rather different physical models arrived at by the three studies and the fact that a description of the physical processes is fraught with a variety of semantic difficulties, it is important to review the similarities and differences found in the three works. A synopsis will be presented of the salient features of the three models from a common viewpoint as well as the viewpoint of the individual studies. This, hopefully, will remove some of the problems which are a matter of semantics and point out the differences which are a matter of physics. The actual experimental spectrum which one is trying to explain should be kept in mind.

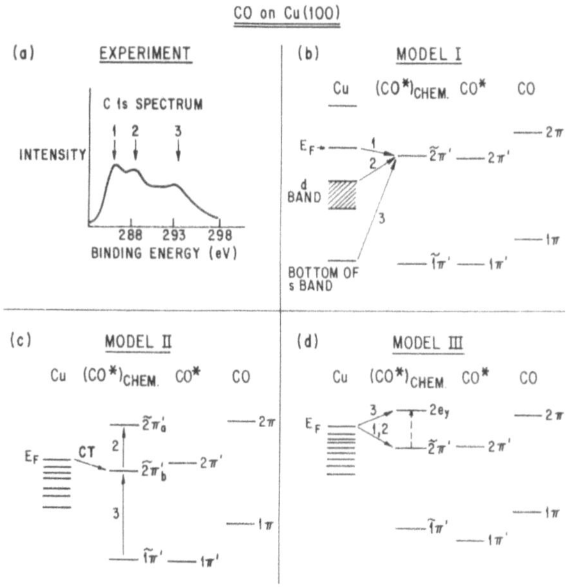

Fig.4.4a-d. Schematic representation of three interpretations of the C 1s photoelectron spectrum for CO chemisorbed on Cu{100} surface. (a) The experimental spectrum; (b) model proposed by [13.31]; (c) model proposed by [13.32]; (d) model proposed by [13.33]. In (b-d) the right-most column represents the 1π and 2π orbital energies of isolated CO; the column labeled CO^* represents the energies of these orbitals after ionization of a C 1s electron. The column labeled Cu is a representation of levels near the Fermi level, and the column labeled $(CO^*)_{chem.}$ shows the levels of the chemisorbed CO with a C 1s electron ionized. The transitions associated with the peaks in the experimental spectrum are labeled according to the interpretations of the three models

In Fig.13.4a the C 1s XPS experimental spectrum [13.30] for CO chemisorbed on Cu{100} is shown. The three peaks in the spectrum are labeled to facilitate the discussion below. Naively, in a one-electron picture two extreme situations can be imagined as a qualitative guide to the understanding of the three peaks. In the first, one can imagine that the photoionization of a C 1s electron from chemisorbed CO yields three final ion states with different probabilities in which orbitals with roughly the same energies but different hole-screening capabilities are occupied in each state. These differences in hole-screening capabilities would lead to the observed differences in the C 1s binding energies. At the other extreme, one might imagine the situation where three different orbitals with roughly the same hole-screening capabilities are occupied in the three final states but that these orbitals have considerably different orbital energies. Thus it would be these differences in orbital energies which would be reflected in the experimental spectrum. It should be clear however that the actual situation, expressed in one-electron terms, will most likely be a combination of these two limiting situations.

In Fig.13.4b, a schematic energy-level diagram is given which is useful in the discussion of the results of the GUNNARSSON-SCHONHAMMER (GS) model [13.31]. At the right of Fig.13.4b, the 1π and 2π levels of the isolated CO molecule are shown. The 1π is completely occupied with four electrons and the 2π level is empty. The occupied σ levels of CO are not shown as they are not relevant to the present discussion. At the left of Fig.13.4b is a schematic representation of the occupied portion of the energy-level structure of Cu metal, showing the wide sp band overlapping the narrow 3d band. On introduction of a core hole in the CO molecule (labeled CO^*) the 1π and 2π levels are shifted to lower energies in response to the increased positive charge on the core hole site. Orbitals which have changed in response to a core hole are denoted with a prime. The chemisorbed CO levels are denoted by a tilde ($\tilde{\ }$) over the orbital designation, and thus the $2\tilde{\pi}'$ level (which is the level of the chemisorbed molecule in the presence of the core hole) is pulled down below the Fermi level (E_F) of Cu so that a charge transfer from the metal can take place which will help to screen the core hole on CO.

The three-peak structure in Fig.13.4a is explained by the GS model in the following manner. All three peaks are the result of the transfer of a substrate valence electron to the $2\tilde{\pi}'$ level of the CO molecule. Peak 1 results from the transfer of an sp-like electron which is initially close to E_F into the $2\tilde{\pi}'$ level of chemisorbed CO. Peak 2 corresponds to an electron at the top of the d band, $\sim$2 eV below E_F, being transferred into the $2\tilde{\pi}'$ level. Finally, peak 3 corresponds to an sp-like electron close to the bottom of the sp-band "tunneling" into the $2\tilde{\pi}'$ level. The three labeled arrows in Fig.13.4b thus schematically show the origin of the three peaks of Fig.13.4a, as determined by the GS model.

It is interesting to note that this model provides an explanation for the three-peak structure which is rather closely related to the second of the simple explanations mentioned above. That is, the screening orbital is the same in each case, and the positions of the peaks are related to the energy-level positions in the metal from which the electron is transferred, hence the conclusion of GS, that the valence DOS can dramatically influence the form of the XPS core spectrum of an adsorbate.

This particular explanation of GS for CO on Cu is rather different from previous work of these authors [13.34] in which multipeak structures in core level spectra was discussed. In previous work, a two-peak structure was discussed in terms of a screened and a nonscreened peak arising from an unoccupied adsorbate level being pulled below the Fermi level on creation of a core hole [13.34].

A schematic representation of the model deduced in [13.32] to explain the experimental spectrum of Fig.13.4a, is shown in Fig.13.4c. Again at the right are the 1π and 2π levels of the isolated CO molecule, the next column to the left shows the levels of CO with a core hole. The interaction of the 2π level of CO with Cu results in a mixing between metal and 2π producing two levels $2\tilde{\pi}_a$ and $2\tilde{\pi}_b$ (antibonding and bonding, respectively). The $2\tilde{\pi}_a$ level, which is higher in energy than the $2\tilde{\pi}_b$, has far more CO 2π character than the $2\tilde{\pi}_b$ level. However, when a core hole is introduced into the chemisorbed CO producing the levels $2\tilde{\pi}_a'$ and $2\tilde{\pi}_b'$ shown in Fig.13.4c, it is found that the character of these orbitals is considerably different from those of $2\tilde{\pi}_a$ and $2\tilde{\pi}_b$. In fact, $2\tilde{\pi}_b'$ becomes more strongly CO 2π-like and $2\tilde{\pi}_a'$ becomes more strongly Cu sp-like. This situation will be discussed latter. The $2\tilde{\pi}_b'$ level which is strongly CO 2π in character is partially occupied with one electron. Thus the first peak in the experiment spectrum can be attributed to a transition between the ground state of the neutral chemisorbed system and a final state in which a core hole on CO is produced together with a transfer of an electron from Cu to the $2\tilde{\pi}_b'$ orbital. This $2\tilde{\pi}_b'$ orbital containing very significant 2π CO character contributes to the screening of the core hole. It is the main contributing factor to the extramolecular screening of the core hole. This final state, i.e., the final state associated with peak 1 is the calculated ground state of the chemisorbed core hole ion system.[1]

If one chooses this ion state as the zero of energy for discussing the spectrum of Fig.13.4a, then the other two peaks represent shake-up states as they are given by excitations from this core hole ion ground state. Thus peak 2 can be viewed as a transition from this ground-state ion to an excited-state ion by virtue of an excitation of an electron from the $2\tilde{\pi}_b'$ orbital to the $2\tilde{\pi}_a'$ orbital. Considering the character of these orbitals as discussed above, the overall transition resulting in peak 2 may be described rather well as the absence of a net charge transfer from the substrate to the chemisorbed species. Peak 3 is described as a one-electron excitation from the $1\tilde{\pi}'$ level to the $2\tilde{\pi}'$ level. This is the analog of the $1\tilde{\pi}' \rightarrow 2\tilde{\pi}'$

---

1 Referring to Fig.13.4, it should be noted that peak 1 of the spectrum (a) corresponds to the charge transfer (CT) shown schematically in (c). Peak 2 corresponds to CT followed by shake-up transition labeled 2 in (c) and peak 3 corresponds to charge transfer (CT) followed by shake-up transition labeled 3 in (c).

shake-up in isolated CO, which is found at ~8 eV above the main peak in the molecular core hole spectrum [13.35].

As a consequence of the choice of the zero of energy above, we refer to peaks 2 and 3 as shake-up peaks. If, however, one were to choose a zero of energy based on the isolated Cu and CO^* states (first and third columns of Fig.13.4c), one might describe the spectrum differently, although the physics remains unchanged.

Consider the consequences of basing the zero of energy on isolated Cu and CO^*. The Cu cluster remains neutral; the isolated molecule is ionized, thus its $2\tilde{\pi}'$ level is lower than the Cu Fermi level. The final state of the interacting cluster-molecule system which most closely resembles this defined zero-of-energy state is peak 2. In this configuration the electron donated from the Cu to the $2\tilde{\pi}'_b$ has been transferred to the $2\tilde{\pi}'_a$, which also has predominantly metallic character. Then peak 1, as it is lower in energy than peak 2, must be considered a shake-down state. This peak is made possible by the increased screening derived from the transfer of an electron from the substrate to the molecule. Clearly the nature of the final states resulting in peaks 1 and 2 are the same regardless of how the zero-of-energy reference point is chosen. Thus whether we refer to peak 2 as the main peak and peak 1 as a shake-down peak, or alternatively refer to peak 1 as the main peak and peak 2 as a shake-up peak is purely a matter of semantics, not of physics.

BAGUS and SEEL [13.33] (BS) have recently discussed a third theoretical model to explain the core level spectrum of CO on Cu. They employed a Cu_5CO cluster and Hartree-Fock theory to discuss CO chemisorbed on a onefold Cu site. A schematic representation of the BS model is shown in Fig.13.4d. The two columns at the right are the same as for the two previous models discussed. When Cu and CO^* are combined as in the second column from the left an electron is transferred to the $2\tilde{\pi}'$ level of CO^*, resulting in a single electron occupying the $2\tilde{\pi}'$ level of the combined system. Because BS chose their reference point as the isolated Cu and CO^*, they referred to this transfer of charge from the Cu to CO as a shake-down process. They assigned both peaks 1 and 2 of Fig.13.4a to this shake-down process. They assigned peak 3 as the "main peak", as the final state in this case ($2\tilde{\pi}'$ empty and one electron in $2e_y$) is almost entirely Cu in character and hence is very similar to their reference point of isolated Cu and CO . Note however that with a change in reference point one might call the first peak the main peak and the third peak a shake-up (arising from the transition shown by the dotted arrow in Fig.13.4d). The BS model does not really explain the full three-peak spectrum of Fig.13.4a, because it does not differentiate between peaks 1 and 2. Indeed these authors have only discussed the experiments in terms of a broad spectrum whose structure extends over an energy range which correlates with the energy difference between the two states considered in their calculations.

One point of agreement between the three models, in terms of the physics, is the nature of the final state which is responsible for peak 1. In all three cases this peak is said to arise from a core hole on the chemisorbed CO molecule, with an electron occupying a CO 2π-like orbital which has been transferred from the Cu sp states near the Fermi level. This is in spite of the fact that the language used to describe the situation in each of the models appears to be rather different.

However, beyond this point the agreement in terms of the basic physical situation vanishes. BS [13.33] have assumed that peak 2 has the same origin as peak 1, and the MLS [13.32] and GS [13.31] models propose two further explanations. Likewise, for peak 3, the three models provide three separate explanations. Only the MLS studies have considered the possible importance of excitations involving the $1\tilde{\pi}'$ orbital.

13.2.1 Cu_9CO Calculations [13.36]

The calculations for Cu_9CO with CO at the onefold site (cf. Fig.13.5c) at a Cu-C distance of 1.9 Å (the distance determined by LEED) will be discussed first. Shown in Fig.13.6a are the ground- state orbital energies for this cluster. The orbital energies are measured with respect to the highest occupied orbital (E_F) as the zero. This highest occupied orbital and the one below it are Cu sp-like in character. The unoccupied levels are denoted by dashed lines. The first unoccupied level is the $2\tilde{\pi}_b$ which is a mixture of Cu sp and CO 2π components; a contour plot of it is shown in Fig.13.7. About 2 eV below E_F there is a group of closely spaced levels which are strongly Cu 3d-like in character with some admixture of Cu sp character. Below this group of closely spaced levels which constitute the cluster analog of the Cu d band, there are several levels between -5 eV to -6.5 eV which are Cu sp-like. Finally, starting at $\sim$-8 eV are the levels associated with the CO molecule. A contour plot of the $1\tilde{\pi}$ orbital is also shown in Fig.13.7. The splitting between the $1\tilde{\pi}$ and $5\tilde{\sigma}$ orbitals of the chemisorbed system tends to be quite exaggerated by the muffin-tin approximation to the potential. Another orbital of interest is the $2\tilde{\pi}_a$ orbital (a contour plot is given in Fig.13.7) which is unoccupied and is not

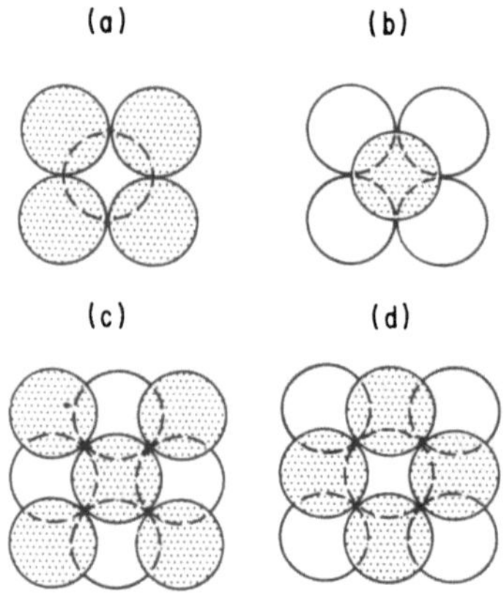

◄ Fig.13.5a-d. Cluster geometries for Cu_5 and Cu_9 calculations. CO is positioned C-end down, perpendicular to the page at the center of each cluster. The shaded atoms denote "surface" atoms. Cluster (a) represents a fourfold adsorption site; (b) a onefold site, (c) a onefold site and (d) a fourfold site

▼ Fig.13.6

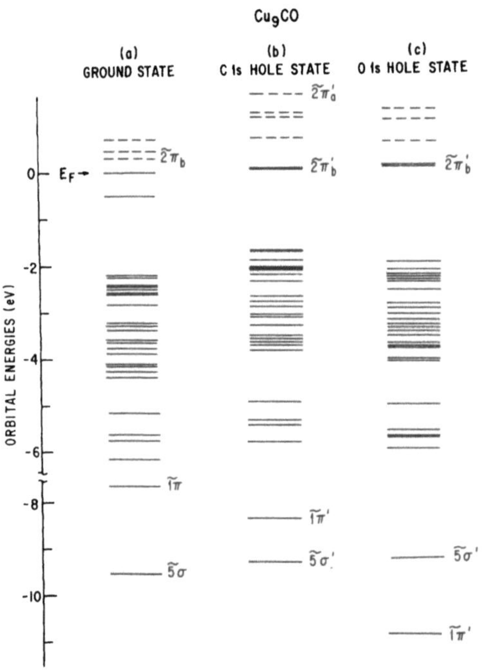

▼ Fig.13.7

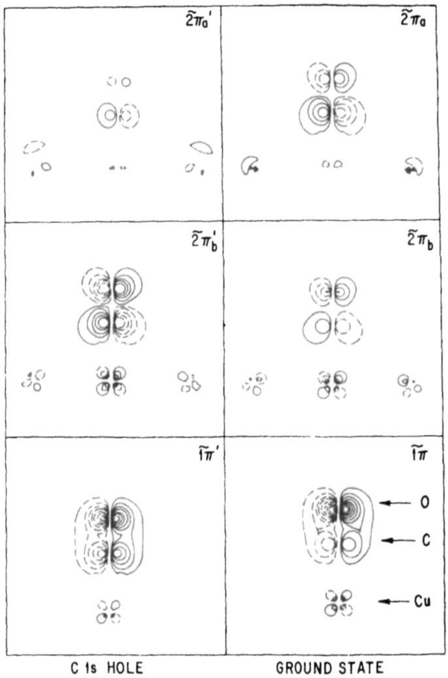

Fig.13.6a-c. Orbital energy-level diagrams determined from Xα scattered wave calculations for Cu_9CO cluster. CO is positioned above the central atom of Cu cluster shown in Fig.13.5c. The energy levels of each of the three cases have been rigidly shifted so as to align the highest occupied levels with $E_F = 0$. Dashed lines represent unoccupied levels. (a) Ground-state orbital energies of Cu_9CO; (b) orbital energies of the C 1s hole state; (c) orbital energies of the O 1s hole state. The $2\tilde{\pi}_b'$ levels in (b) and (c) contain one electron

Fig.13.7. Caption see opposite page

shown on the orbital energy level of Fig.13.6 because it is too high in energy.

On the introduction of a C 1s core hole in the chemisorbed CO molecule, the originally unoccupied $2\tilde{\pi}_b$, is pulled down in energy such that it becomes partially occupied. Note that the $1\tilde{\pi}$ orbital is also stabilized. These CO levels are pulled down in energy relative to the Cu levels as a result of the localized core hole produced on the CO molecule. It is the π orbitals of chemisorbed CO which exhibit the most dramatic response to the creation of a core hole. This is shown clearly in Fig.13.7, where contour plots of the $1\tilde{\pi}$, $2\tilde{\pi}_b$, and $2\tilde{\pi}_a$ orbitals are given for before and after the introduction of the core hole. The response of the $1\tilde{\pi}$ orbital when a C 1s electron is ionized is to shift toward the carbon atom in order to screen the core hole. The response of the $2\tilde{\pi}_b$ orbital is quite dramatic. In the ground state it is unoccupied and has some CO content, but is mainly Cu in character. In the C 1s hole state it is occupied and is now very largely CO 2π in character. As the $2\tilde{\pi}_a'$ orbital must remain orthogonal to the $2\tilde{\pi}_b'$ orbital, there is also a significant change in the $2\tilde{\pi}_a$ orbital on going from the ground state to the C 1s hole state. In fact in the hole state, this orbital ($2\tilde{\pi}_a'$) is now almost exclusively Cu sp-like in character.

Qualitatively, a very similar situation occurs for the case of the O 1s hole state, with the exception of course that the $1\tilde{\pi}$ orbital shifts to the oxygen end in the O 1s hole state in order to screen the core hole. However, the qualitative changes in the $2\tilde{\pi}_b$ and $2\tilde{\pi}_a$ orbitals are similar to those discussed for the C 1s case.

In Fig.13.6, Fermi statistics apply in each case. For the ground state, the Fermi level is determined by a metal-like $7b_2$ level which contains one electron. For the O 1s hole state, the $2\tilde{\pi}_b'$ level contains one electron and determines E_F. However, for the C 1s hole state, Fermi statistics are only satisfied if the $2\tilde{\pi}_b'$ contains a small fraction of an electron beyond an integer occupancy. The differences in wave functions, energies, etc. are rather slight between the cases of integer occupancy of $2\tilde{\pi}_b'$ and the nonin-

Fig.13.7. Contour plots of the $1\tilde{\pi}$, $2\tilde{\pi}_b$, and $2\tilde{\pi}_a$ orbitals of chemisorbed CO on the onefold site of Cu_9 cluster shown in Fig.13.2c. The orbitals at the right are for the ground state of the cluster before C 1s ionization; the orbitals at the left are for the C 1s hole state which results after a C 1s electron is ionized. The positions of the C, O, and Cu nuclei are shown in the lower right panel. They are in the same relative orientation in the other panels. The Cu atoms show very few contours because the Cu character is mostly diffuse s and p teger occupancy of this orbital. As a consequence of this and the ease of dealing with integral occupancy configurations, only those calculations which assume integer occupation numbers will be discussed here.

Thus the ground state of the core hole ion has the configuration I: $(1\tilde{\pi}')^4(2\tilde{\pi}_b')^1(2\tilde{\pi}_a')^0$ and the various shake-up states (since we are choosing this state as our zero of energy to discuss other final states) will be obtained by electronic excitations from this configuration. As discussed below, the only transitions which have any appreciable intensity are those involving these three orbitals, the $1\tilde{\pi}'$, $2\tilde{\pi}_b'$, and $2\tilde{\pi}_a'$. The lowest excited ion state has the configuration II, $(1\tilde{\pi}')^4(2\tilde{\pi}_b')^0(2\tilde{\pi}_a')^1$, and the other excited ion state has the configuration III, $(1\tilde{\pi}')^3(2\tilde{\pi}_b')^2(2\tilde{\pi}_a')^0$. These are the three final states which give rise to the spectrum of Fig.13.4a.

In order to determine the energy separations between these three states, spin-polarized transition-state calculations were carried out. Taking configuration I as the zero of energy, we obtained for the C 1s core hole case energy separations of 2.3 eV and 8.1 eV for configurations II and III respectively. Likewise for the O 1s core hole, we obtained energy separations of 1.9 eV and 12.5 eV. For the case of configuration III, there are three open shells (including the core shell) which lead to many possible states — as a result of the various allowed spin couplings. Within the context of $X\alpha$ theory the various states cannot be resolved since one does not generate proper eigenstates of the total spin. As a consequence, the numbers quoted above are based on the high-spin configuration. Although within the context of ab initio calculations such spin couplings can be rigorously treated, the work of BS [13.33] on Cu_5CO has not considered this problem.

In Table 13.1 the results for the shake-up intensities are presented; only those shake-up transitions are given which have calculated relative intensities greater than one percent. Again we note that the energies and intensities are given relative to the calculated first peak, i.e, the core hole ion ground state. Only two transitions have significant intensities; they are the ones previously referred to as giving rise to the two satellite peaks in the spectrum of Fig.13.4a. Although the calculated relative intensities for the two satellites are not quantitative, they do have the correct behavior of decreasing in intensity with increasing binding energy, which contrasts with the behavior found in the recent Hartree-Fock cluster calculations [13.33]. We note that the $1\tilde{\pi}' \rightarrow 2\tilde{\pi}_b'$ peak position for the O 1s hole is considerably higher in energy than the corresponding peak for the C 1s hole. This is completely analogous to the situation found for the isolated CO molecule [13.36].

Table 13.1. Results for Cu_9CO onefold site core level spectra. Energies and intensities are given relative to the core hole ion ground state

	O 1s core hole		C 1s core hole	
Transition	E[eV]	Intensity[%]	E[eV]	Intensity[%]
$12a_1' \rightarrow 13a_1'$	-	-	(1.36)	1.5
$8e' \rightarrow 2\tilde{\pi}_b'$	(2.53)	2.2	(2.03)	1.6
$2\tilde{\pi}_b' \rightarrow 2\tilde{\pi}_a'$	1.86	22.9	2.26	80.4
$1\tilde{\pi}' \rightarrow 2\tilde{\pi}_b'$	12.5	11.9	8.09	9.1
$1\tilde{\pi}' \rightarrow 2\tilde{\pi}_a'$	(12.8)	1.7	(9.96)	1.1

In Table 13.1 the energies have been determined by the transition-state procedure for the intense transitions. For the weak transitions the energies have been estimated from the ion ground-state orbital energies; these values are given in parentheses.

For the case of the C 1s hole spectrum a further calculation was performed. In this calculation the CO-to-Cu distance was increased from 1.9 Å to 2.4 Å. Then E and the relative intensity of the $2\tilde{\pi}_b' \rightarrow 2\tilde{\pi}_a'$ transition were calculated. The results for this larger distance were found to be 2.55 eV and 127% as compared to the values (at 1.9 Å) of 2.26 and 80.4% in Table 13.1. Thus as the Cu-CO distance is increased the intensity of peak 2 becomes larger than that of peak 1. The significance of this result will be discussed in the next section.

Results have also been obtained for CO chemisorbed in a fourfold site using a Cu_9 cluster (cf. Fig.13.5d), with the Cu- C distance of 1.43 Å as employed in the previous Cu_5CO study [13.32]. It is probably useful only to discuss these results in a qualitative way, pointing out the differences among the Cu_9CO (onefold site), the Cu_9CO (fourfold site) and the previous study of Cu_5CO.

The most important point, however, is that qualitatively the results of all three calculations are very similar. The same basic picture of the physics involved in producing the satellite structure applies. The differences of interest are the following. Comparing the fourfold-site cases of Cu_5CO and Cu_9CO (4), one finds in the former case [13.32] that both the $1\tilde{\pi}' \rightarrow 2\tilde{\pi}_b'$ and $1\tilde{\pi}' \rightarrow 2\tilde{\pi}_a'$ transitions have roughly comparable intensities; however, in the latter case the $1\tilde{\pi}' \rightarrow 2\tilde{\pi}_b'$ transition has a calculated intensity about an order of magnitude greater than for the $1\tilde{\pi}' \rightarrow 2\tilde{\pi}_a'$ transition. Thus the fourfold-site Cu_9CO result is similar to the onefold-site results for Cu_9CO (1)

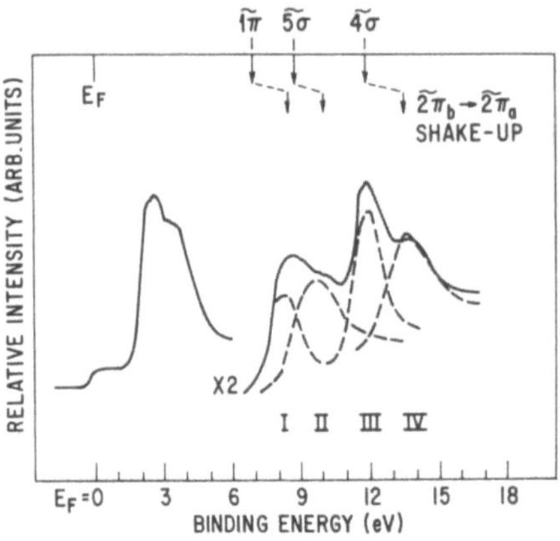

Fig.13.8. Comparison of experimental ultraviolet photoelectron transition energies [13.29] with calculated spectrum based on Xα scattered wave calculations for Cu_9CO. The calculated values were rigidly shifted so as to align the calculated $4\tilde{\sigma}$ ionization energy with peak III of the spectrum

shown in Table 13.1. A second difference between the Cu_9CO (1) and Cu_9CO (4) results and the Cu_5CO results is that in both of the former cases the $2\tilde{\pi}_b$ orbital is unoccupied before the core hole is introduced. In all cases the $2\tilde{\pi}_b$ orbital on becoming the $2\tilde{\pi}_b'$ orbital (on introduction of the core hole) looses considerable Cu character and gains CO 2π character. This effect is more dramatic for the Cu_9CO cases.[2]

All of the results presented thus far have been concerned with the satellite structure found in the core level spectroscopy of CO chemisorbed on Cu. However, there is also experimental information for the valence region of CO chemisorbed on Cu using ultraviolet photoelectron spectroscopy [13.29,37]. The experimental data are shown in Fig.13.8. NORTON et al. [13.29] suggested that the structure in the valence region of the chemisorbed CO should be considered to consist of four peaks which have been labeled with Roman numerals in the figure. However, ALLYN et al. [13.37] considered peaks I and II as one peak with two components. The latter study used angle-resolved photoelectron spectroscopy to investigate CO on Cu(100). They showed that peak III of Fig. 13.8 was due to the ionization of the $4\tilde{\sigma}$ orbital of CO, and that the structure in the region of I and II was due at least in part to the $1\tilde{\pi}$ and $5\tilde{\sigma}$ ionizations of the chemisorbed CO molecule. They assigned peak IV to a shake-up (the nature of which was not specified) associated with the $4\tilde{\sigma}$ ionization.

2 The total charge on CO from the $2\tilde{\pi}_b'$ orbital is about the same in both the Cu_5CO and Cu_9CO calculations. However, in the former case this orbital is occupied by three electrons, whereas in the latter case it contains only one electron. Hence, the change in the orbital on going from $2\tilde{\pi}_b$ to $2\tilde{\pi}_b'$ is less dramatic for Cu_5CO than for Cu_9CO [13.32].

From transition-state calculations for the Cu_9CO cluster we have calculated the ionization energies of the chemisorbed CO $1\tilde{\pi}$, $5\tilde{\sigma}$, and $4\tilde{\sigma}$ orbitals. By shifting the calculated values so as to match the $4\tilde{\sigma}$ ionization energy with the position of peak III, one arrives at the positions and assignments shown at the top of Fig.13.8. It should be noted that although the calculated $5\tilde{\sigma}$ binding energy is found to fall in the proper region the $1\tilde{\pi}$ is found at a binding energy which is too low. We have discovered from our many calculations on these systems that the $1\tilde{\pi}$ orbital energy and ionization energy are particularly sensitive to errors which arise from the muffin-tin nature of the potential used in these scattered wave calculations. The $1\tilde{\pi}$ binding energy would undoubtedly be much closer to that of the $5\tilde{\sigma}$, if the muffin-tin errors did not occur.

The only shake-up transition calculated to have any appreciable intensity is the $2\tilde{\pi}_b \rightarrow 2\tilde{\pi}_a$, which was also important in understanding the core region.[3] The calculated shake-up energy positions (relative to their parent orbital binding energies) are shown by the lower set of arrows in Fig.13.8. Taking into account the likelihood that the $1\tilde{\pi}$ position should be closer to the $5\tilde{\sigma}$ position, we arrive at the following assignment of the structure in the spectrum. Peaks I and II arise from the primary ionizations from the $1\tilde{\pi}$ and $5\tilde{\sigma}$ orbitals together with a $2\tilde{\pi}_b \rightarrow 2\tilde{\pi}_a$ shake-up which accompanies the primary ionization. Peak III is due to the ionization from the $4\tilde{\sigma}$ orbital, and peak IV is due to the $2\tilde{\pi}_b \rightarrow 2\tilde{\pi}_a$ shake-up which accompanies the ionization from the $4\tilde{\sigma}$ orbital. Thus one may understand the satellite structure in the valence as well as core regions from a simple unified point of view.

13.2.2 Discussion of Core Level Spectra

In order to assess the validity and generality of the interpretation which emerges from the calculations presented in the previous section, it is worthwhile considering the core level satellites of CO in a more general context; that is, to make a comparison of core level spectra for isolated CO, molecular carbonyls, and chemisorbed CO. Such a comparison has been given by FREUND and PLUMMER [13.38] and Fig.13.9 is adapted from their work. We consider the O 1s core region of CO in different environments, free, molecular, and chemisorbed. For the free molecule spectrum at the bottom of Fig.

3 The calculated intensities of the $2\tilde{\pi}_b \rightarrow 2\tilde{\pi}_a$ shake-ups relative to the primary ionization are 13.9%, 10.6% and 7.4% for the $4\tilde{\sigma}$, $1\tilde{\pi}$ and $5\tilde{\sigma}$ primary ionizations respectively. The next largest intensities are for the $1e \rightarrow 14e$ and $13e \rightarrow 14e$ shake-ups from the $4\tilde{\sigma}$ primary ionization which are 1.4% and 1.3% respectively.

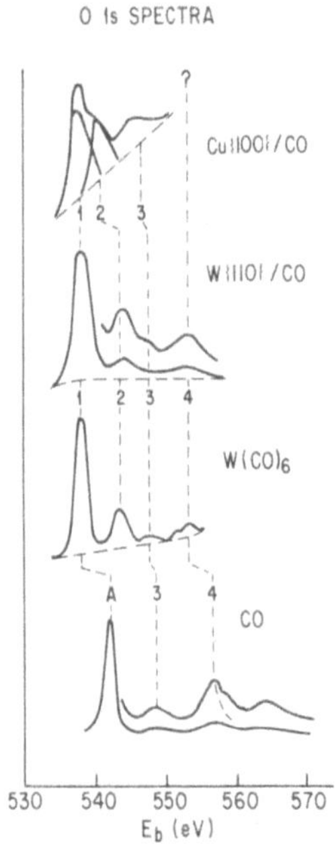

Fig.13.9. Comparison of experimental O 1s spectra for isolated CO molecule, $W(CO)_6$ molecule, CO chemisorbed on W{110} and CO chemisorbed on Cu{100} [13.38]

13.6, the three peaks of interest are labeled A, 3, and 4. Peak A arises from the ionization of the O 1s electron; peaks 3 and 4 are shake-ups which accompany the O 1s ionization and arise from the transition $1\pi \rightarrow 2\pi$. There are two peaks from this transition because two independent doublet states are created.

When the CO molecule interacts with a metal atom as in the case of the $W(CO)_6$ molecule, the 2π orbital of CO interacts with the metal atom so as to produce two new orbitals—a bonding $2\tilde{\pi}_b$ and an antibonding $2\tilde{\pi}_a$ orbital. The stronger the interaction between the metal and CO, the larger the energy separation between the $2\tilde{\pi}_b$ and $2\tilde{\pi}_a$ levels. On ionizing the O 1s electron, the $2\tilde{\pi}_b$ orbital should become partially occupied and pick up considerable CO 2π character (see Fig.13.7 for the Cu_9CO case). This contributes screening to the core hole and results in peak 1 being at lower binding energy as compared to peak A of isolated CO. It is also seen from Fig.13.9, that a new peak arises, peak 2, which does not exist in the free CO molecule. This peak can exist in the interacting case because the core hole ion ground state has the $2\tilde{\pi}'_b$ orbital partially occupied (the $2\tilde{\pi}'$ orbital is empty in the CO case). Further, due to the energy splitting produced between the $2\tilde{\pi}'_b$ and $2\tilde{\pi}'_a$ orbitals, a transition (shake-up) can occur between the $2\tilde{\pi}'_b$ and $2\tilde{\pi}'_a$ orbitals. The shake-up transition is of the nature of a ligand to metal change transfer (see Fig.13.7 for Cu_9CO case). Peaks 3 and 4 are shifted to lower binding energies due to the screening effects of the partial occupancy of the $2\tilde{\pi}_b'$ orbital. They arise from $1\tilde{\pi}' \rightarrow 2\tilde{\pi}_b'$ shake-ups, which are direct analogs of the molecular shake-up transitions.

For the case of CO on W{110}, the explanation should be virtually identical to that of $W(CO)_6$, and one finds that the observed spectra are very similar.

Considering the situation for Cu{100}/CO, the peaks 1 and 2 are less separated than for the two previous cases. This undoubtedly arises from the fact that the Cu-CO interaction is considerably weaker than the W-CO interaction leading to a smaller splitting between the $2\tilde{\pi}_b$ and $2\tilde{\pi}_a$ orbitals. Thus the shake-up transition $2\tilde{\pi}_b' \rightarrow 2\tilde{\pi}_a'$ has a lower energy. Peak 3 arises from a $1\tilde{\pi}' \rightarrow 2\tilde{\pi}_b'$ shake-up. An interesting question arises in this case: if the spectrum had been recorded to higher binding energies would there be a peak 4? Clearly the systematics observed here indicate that this should be the case.

A comment with regard to the assignment of BAGUS and SEEL [13.33] is in order at this point. They maintained that peaks 1 and 2 of the Cu{100}/CO spectrum are due to shake-down and that peak 3 corresponds to peak A of free CO. We know of no reason why peak A of free CO should shift to *higher* binding energies when interacting with a metal. All experience has been that the shift should be to lower binding energies due to screening effects. BAGUS and SEEL provided no explanation for this rather curious situation, which is a consequence of their assignments.

Another interesting way to gain some insight into the relationship between the isolated CO spectrum and that for the chemisorbed case is through Fig.13.10, which has been adapted from [13.38]. Figure 13.10 shows a schematic representation of the changes in the spectrum as a function of the Cu-CO distance. Curve e represents the essential features of the isolated CO molecule (cf. Fig.13.9). Basically, peaks 3 and 4 which arise from $1\tilde{\pi}' \rightarrow 2\tilde{\pi}_b'$ shake-ups should be in roughly the same positions for all distances. For the case of CO on Cu, peak 4 has not yet been observed, and hence it is designated in curves a-d as a dashed-line portion of the curves.

When the isolated CO molecule interacts with the metal, peak A is split into two peaks, peaks 1 and 2 of curves a-d. The splitting arises from the Cu-CO 2π interaction producing the bonding $2\tilde{\pi}_b'$ and antibonding $2\tilde{\pi}_a'$ orbitals with the $2\tilde{\pi}_b'$ orbital being occupied in the case of peak 1 and the $2\tilde{\pi}_a'$ orbital being occupied in the case of peak 2. As the $2\tilde{\pi}_b'$ orbital is largely CO

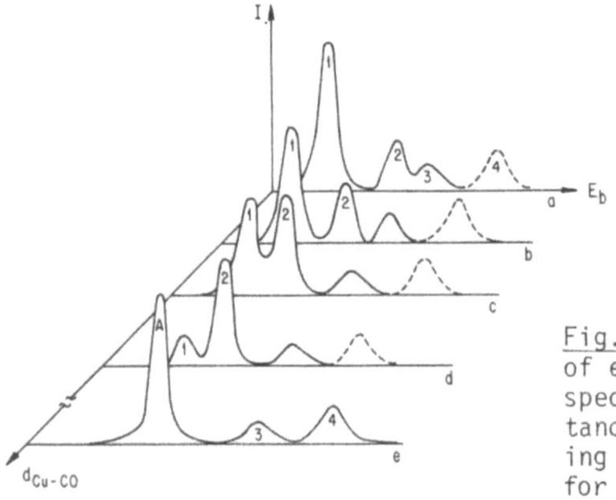

Fig.13.10. Schematic representation of expected behavior of core hole spectra as a function of Cu-CO distance. Curves a-d are for increasing metal- CO distances, curve e for the isolated CO molecule

2π in character and the $2\tilde{\pi}'_a$ orbital largely metal (see Fig.13.7) one may view peaks 1 and 2 in curves a-d roughly as the screened and unscreened counterparts, respectively, of peak A in curve e.

The binding-energy separation between peaks 1 and 2 reflects the separation between the $2\tilde{\pi}_b$ and $2\tilde{\pi}_a$ levels. The stronger the interaction between the metal and CO (i.e., the shorter the bond length curve a), the larger the separation. Conversely, the weaker the interaction and longer the bond length, the smaller the separation (curve d). The trend in intensities of peaks 1 and 2 may also be understood in simple terms. At large distances (curve d), the overlap of metal orbitals with the CO $2\tilde{\pi}'$ orbital is not very significant, thus the probability that an electron will be transferred from the metal to CO is small and the intensity of peak 1 is consequently small. However, at short distances (curve a), the overlap is quite significant and thus the probability that an electron is transferred is significantly increased, resulting in a larger intensity of peak 1.

It is interesting to note that for the Cu_9CO calculations with the Cu-CO distance at 1.9 A the results presented in the preceeding section suggest a qualitative situation somewhat intermediate between curves b and c. The results for the C 1s spectrum at a Cu-CO distance of 2.4 A suggest a qualitative situation intermediate between curves c and d. Thus the calculated results clearly conform to the schematic picture represented in Fig. 13.10.

We believe that the model and interpretation of experimental results which emerges from our present and previous work has the advantage over other models of providing a coherent framework with which to view the core hole spectrum of CO in various environments. Its viewpoint is very similar in spirit to that espoused by FREUND and PLUMMER [13.38] and consistent with the experimental information which they have discussed. Furthermore, we believe it will provide a conceptual framework with which to discuss the features of core level spectra in other systems.

References

13.1 K. Sattler, J. Muhlbach, E. Recknagel: Phys. Rev. Lett. *45*, 821 (1980)
13.2 C. Kittel: *Quantum Theory of Solids* (Wiley, New York 1963)
13.3 H.J. Monkhorst: Phys. Rev. B*20*, 1504 (1979)
13.4 J.C. Slater: Adv. Quantum Chem. *6*, 1 (1972)
13.5 K.H. Johnson: Adv. Quantum Chem. *7*, 143 (1973)
13.6 J. Korringa: Physica *13*, 392 (1947)
13.7 W. Kohn, N. Rostoker: Phys. Rev. *94*, 1111 (1954)
13.8 R.P. Messmer, S.K. Knudson, K.H. Johnson, J.B. Diamond, C.Y. Yang: Phys. Rev. B*13*, 1396 (1976)
13.9 J. Callaway, C.S. Wang: Phys. Rev. B*7*, 1096 (1973)
13.10 A.B. Anderson, R. Hoffmann: J. Chem. Phys. *61*, 4545 (1974)
13.11 G. Blyholder: J. Chem. Phys. *62*, 3193 (1975)
13.12 R.C. Baetzold, R.E. Mack: J. Chem. Phys. *62*, 1513 (1975)
13.13 R.C. Baetzold: J. Phys. Chem. *82*, 738 (1978)
13.14 H. Basch, M.D. Newton, J.W. Moskowitz: J. Chem. Phys. *73*, 4492 (1980)
13.15 C. Bachmann, J. Demuynck, A. Veillard: Faraday Symp. Chem. Soc. *14*, 170 (1980)
13.16 C.D. Garner, I.H. Hillier, M.F. Guest, J.C. Green, A.W. Coleman: Chem. Phys. Lett. *41*, 91 (1976)
13.17 F.A. Cotton, G.G. Stanley: Inorg. Chem. *16*, 2668 (1977)
13.18 M. Benard, A. Veillard: Nouv. J. Chim. *1*, 97 (1977)
13.19 M. Benard: J. Chem. Phys. *71*, 2546 (1979)
13.20 P.J. Hay, T.H. Dunning, Jr., W.A. Goddard, III: J. Chem. Phys. *62*, 3912 (1975)
13.21 R.P. Messmer, D.R. Salahub: J. Chem. Phys. *65*, 779 (1976)
13.22 N. Rösch, T.N. Rhodin: Phys. Rev. Lett. *32*, 1189 (1974)
13.23 T.H. Upton, W.A. Goddard, III: J. Am. Chem. Soc. *100*, 5659 (1978)
13.24 E. Dietz, D.E. Eastman: Phys. Rev. Lett. *41*, 1674 (1978)
13.25 D.R. Salahub, R.P. Messmer: Surf. Sci. *106*, 415 (1981)
13.26 J.C. Slater: *Quantum Theory of Molecules and Solids, Vol.4*, The Self-Consistent Field for Molecules and Solids (McGraw-Hill, New York 1974)
13.27 C.G. Shull, H.A. Mook: Phys. Rev. Lett. *16*, 184 (1966)
13.28 Chiang Y. Yang, K.H. Johnson, D.R. Salahub, J. Kaspar, R.P. Messmer: Phys. Rev. B*24*, 5673 (1981)
13.29 P.R. Norton, R.L. Tapping, J.W. Goodale: Surf. Sci. *72*, 33 (1977)
13.30 J.C. Fuggle, E. Umbach, D. Menzel, K. Wandelt, C.R. Brundle: Solid State Commun. *27*, 65 (1978)
13.31 O. Gunnarsson, K. Schönhammer: Phys. Rev. Lett. *41*, 1603 (1978)
13.32 R.P. Messmer, S.H. Lamson, D.R. Salahub: Solid State Commun. *36*, 265 (1980)
13.33 P.S. Bagus, M. Seel: Phys. Rev. B*23*, 2065 (1981)
13.34 K. Schönhammer, O. Gunnarsson: Solid State Commun. *23*, 691 (1977)
13.35 T.A. Carlson, M.O. Krause, W.E. Moddeman: J. Phys. Paris *32*, C4-76 (1971)
13.36 R.P. Messmer, S.H. Lamson, D.R. Salahub: Phys. Rev. B*25* (1982)
13.37 C.L. Allyn, T. Gustafsson, E.W. Plummer: Solid State Commun. *24*, 531 (1977)
13.38 H.-J. Freund, E.W. Plummer: Phys. Rev. B*23*, 4859 (1981)

14. The Inelastic Scattering of Low-Energy Electrons by Surface Excitations; Basic Mechanisms

D.L. Mills and S.Y. Tong

With 5 Figures

14.1 Introduction

The task of developing experimental probes that provide information on the outermost atomic layer or two of crystals, and on atoms or molecules adsorbed there, is a formidable one. Techniques used commonly to deduce the geometry or properties of elementary excitations in bulk crystalline materials employ probes such as neutrons or X-rays that interact with each atomic constituent weakly. As a consequence, it is relatively straightforward to develop a quantiative theoretical description of their interaction with various forms of condensed matter, but the probes penetrate deeply into the crystal interior. Information on the near vicinity of the surface is thus a very small fraction of the total signal received at the detector.

Surface sensitive probes such as low-energy electrons, ions or atoms, sample only the outermost few atomic layers by virtue of the fact that they interact very strongly with the atomic constituents. While the data contain information on only the surface region, the strong interaction between the probe and crystal requires one to develop complex theories that do not treat the probe-crystal interaction as a weak perturbation. Without such theories, the data can be interpreted in only a schematic and qualitative fashion.

The article will be devoted to the discussion of one probe that has come into rather wide use during the past decade: high-resolution electron energy loss spectroscopy. The method is for surface physics an analog of the thermal neutron scattering experiments that have played a central role in the study of elementary excitations in condensed matter. A highly monoenergetic and well-collimated electron beam, with energy $E^{(I)}$ and wave vector $\underline{k}^{(I)}$, is directed onto the crystal surface. The electrons at these energies (a few electron volts typically) are backscattered off the surface. As they do so, a certain fraction are inelastically scattered by the elementary excitations of the surface region. If we consider a surface geometry with periodicity in the two directions parallel to the surface plane (clean surface, or one with an ordered commensurate layer of adsorbate present), then the elementary excitation may be characterized by a two-dimensional wave vector $\underline{Q}_{\parallel}$ parallel to the surface, and an excitation energy $\hbar\omega_s(\underline{Q}_{\parallel})$. The electron emerges with energy $E^{(s)} = E^{(I)} \pm \hbar\omega_s(\underline{Q}_{\parallel})$, with the lower sign chosen for the case where the elementary excitation is emitted in the scattering process, and the upper sign for the case where it is absorbed. There is clearly an absence of translational symmetry in the direction normal to the surface. One consequence is that there is no condition on conservation of wave vector in the direction normal to the surface. The two-dimensional periodicity parallel to the surface demands, in a manner familiar from solid-state physics, that components of wave vector parallel to the surface are conserved to within a reciprocal-lattice vector. Thus, if $\underline{Q}_{\parallel}$ lies within the first Brillouin zone of the two-dimensional unit cell appropriate to the surface structure, and $\underline{G}_{\parallel}$ is a reciprocal-lattice vector we have $\underline{k}_{\parallel}^{(s)} = \underline{k}_{\parallel}^{(I)} \pm \underline{Q}_{\parallel} + \underline{G}_{\parallel}$, where $\underline{k}_{\parallel}^{(s)}$ and $\underline{k}_{\parallel}^{(I)}$ are the projections onto the surface plane wave vectors of the scattered and incident electron wave vector onto the plane parallel to the surface.

If the direction and energy of the scattered electron are measured, then one may deduce both $\underline{Q}_{\parallel}$ and $\omega(\underline{Q}_{\parallel})$, i.e., one may explore the dispersion relation of the elementary excitations in the surface, in a manner familiar from thermal neutron scattering. Thus, in contrast to both low-energy electron diffraction and the various forms of photoelectron spectroscopy, high-resolution electron energy loss spectroscopy is a probe of the dynamical properties of the surface region. In much of the literature the focus has been placed on use of the method to explore vibrational motions of atoms and molecules on the surface. The method also proves to be useful as a quantitative probe of electronic excitations which lie in the infrared frequency regime. In Sect.14.2 we shall place considerable emphasis on use of the method to explore the electronic degrees of freedom of the crystal surface.

While the remarks above draw an analogy between electron energy loss spectroscopy and thermal neutron scattering, of course the two experiments differ in many important qualitative senses. For one thing, the electron impact energy is several electron volts in a typical experiment, and if we consider vibrational spectroscopy for the moment, $\hbar\omega_s(\underline{Q}_{\parallel})$ is typically a few tens of millivolts (1 meV $\simeq$ 8 cm^{-1}). The fractional change in energy, $\hbar\omega_s(\underline{Q}_{\parallel})/E^{(I)}$ is thus very small, and the task of constructing suitable monochromators that pass sufficient current is formidable. In the recent literature, one finds experiments carried out with systems that provide an energy resolution in the range of 4 meV, so the resolution offered by electron energy loss spectroscopy is now in the range where frequencies well below the maximum phonon frequency of metallic substrates may be explored. Recent experiments have resolved well defined structures below the maximum substrate phonon frequency for the $c(2\times 2)$ and the $p(2\times 2)$ oxygen overlayers on the Ni{111} [14.1] and the Ni{100} [14.2] surface. The data has interesting implications for the structure of these overlayers, for the {100} surface particularly [14.3].

To outline the motivation of the present paper, it is necessary to discuss the nature of the scattering mechanisms which play a role in these experiments.

The angular distribution of electrons scattered inelastically by, say, a vibrational quantum on the surface, frequently has the following form. One finds a narrow peak of electrons which emerge very close to the specular direction. This (relatively) intense peak will contain roughly one tenth of a percent of the electrons backscattered from the surface, for scattering off a typical monolayer of adsorbed atoms. The near specular peak is superimposed on a background of inelastically scattered electrons that is distributed over the entire solid angle, but two orders of magnitude weaker. The near specular peak is observed not only for scattering off vibrational excitations, but for scattering from low-lying electronic excitations as well.

The origin of the near specular peak is now well established to arise from the time-dependent electric dipole moment associated with the excitation responsible for the scattering; it is true also that the principal features of the phenomenon follow from a simple argument, which proceeds as follows. Consider a sheet of dipoles, parallel to the surface, with dipole moment that varies with position as $\exp(i\underline{Q}_{\|}\cdot\underline{x}_{\|})$, with $\underline{x}_{\|}$ a vector parallel to the surface. In the vacuum above the crystal, there necessarily exists an electrostatic potential of the form $\Phi(z)\exp(i\underline{Q}_{\|}\cdot\underline{x}_{\|})$, i.e., the electrostatic potential must also be proportional to $\exp(i\underline{Q}_{\|}\cdot\underline{x}_{\|})$. The form of $\Phi(z)$ is fixed completely to within an overall multiplicative constant, unimportant here, by the requirement that Laplace's equation to be obeyed in the vacuum above the crystal. This gives $\Phi(z) = \Phi_0\exp(-Q_{\|}z)$.

From the result just quoted, one sees that as $\underline{Q}_{\|}\rightarrow 0$, an array of dipoles arranged along the surface with dipole moment proportional to $\exp(i\underline{Q}_{\|}\cdot\underline{x}_{\|})$ sets up an electric field that extends very far out into the vacuum, a distance the order of $\ell_0(Q_{\|}) = Q_{\|}^{-1}$ above the crystal. As the electron approaches the surface, it sees this electric field for a very long time, with the consequence that it couples strongly to the surface excitation. Hence, the small $\underline{Q}_{\|}$ normal modes which, from the kinematical considerations given earlier, scatter electrons very near the specular direction (if $\hbar\omega_s(Q_{||}) \ll E^{(I)}$) have a large cross section because of the long-ranged nature of the electric dipole field.

One may carry the argument one step further, and isolate the characteristic deflection angle which eneters the problem. Let an electron with momentum $p^{(I)}$ strike the crystal by approaching along the normal, and suppose it suffers an angular deflection $\Delta\theta$ as it backscatters from the surface. One has $\hbar Q_{||} \simeq p^{(I)}\Delta\theta$, for $\Delta\theta \ll 1$. The electron also sees the surface excitation for the time $\Delta t = 2/v^{(I)}Q_{||}$, with $v^{(I)}$ as velocity. Here we use the fact that $\ell_0(Q_{||}) \simeq Q_{||}^{-1}$, and the factor of two enters because the electron can scatter from the surface excitation either as it approaches or exits from the crystal Now if ω_s is the frequency of the excitation, when $\omega_s\Delta t \ll 1$, the time the electron spends bathed in the electric field is so short that the excitation receives only a very brief impulse from the electron. Here the excitation efficiency is small. Similarly when $\omega_s\Delta t \gg 1$, the electric field seen by the electron varies slowly with time, and the adiabatic theorem of quantum mechanics insures that no transitions are induced, so there can be no scattering. The condition for maximum excitation probability is thus $\omega_s\Delta t \sim 1$ which, with the above formula for Δt gives $\Delta\theta = \Delta\theta_c \simeq \hbar\omega_s/E^{(I)}$ as the characteristic deflection angle. Evaluation of $\Delta\theta_c$ for frequencies typical of vibrational losses, and $E^{(I)} \simeq 5$ eV shows $\Delta\theta_c$ to be a few tenths of a degree.

It is important to realize that when $\Delta\theta \simeq \Delta\theta_c$, the electron is quite far above the crystal surface, when the inelastic scattering event takes place. The argument just presented gives $\ell_0(Q_{||}) \simeq v^{(I)}/2\omega_s$ as the average distance above the surface where the scattering takes place. One has $\ell_0(Q_{||}) \simeq 100$ Å, for deflection angles $\Delta\theta \sim \Delta\theta_c$. The fact that the electron is quite far above the crystal when it scatters allows one to develop a theory of small-angle dipole scattering which, though fully quantum mechanical in nature [14.4,5], ultimately expresses the cross section in terms of certain general response functions closely related to the dielectric response of the material [14.6-8]. In essence, the electron scatters from time-dependent electric field fluctuations in the vacuum above the surface. By use of the appropriate form of the fluctuation-dissipation theorem, the amplitude and frequency spectrum of these fluctuations may be related to the dielectric properties of the substrate, and far from the surface it is the macroscopic properties of the substrate that control the strength of the electric field fluctuations [14.6-8].

In the small angle, dipole-dominated regime, the theory yields analytic formulae for the electron scattering efficiency, and in a number of instances, these may be placed in quantitative contact with the data. The argument above also shows that when $\Delta\theta \gg \Delta\theta_c$, the length $\ell_0(Q_\parallel)$ becomes microscopic in magnitude. With $\ell_0(Q_\parallel)\Delta\theta$ $\hbar/p^{(I)}\Delta\theta$, a 5-eV electron incident on the surface and $\Delta\theta = 20^\circ$, we have $\ell_0(Q_\parallel) \simeq 2$ Å, a distance considerably shorter than the penetration depth of the electron into the crystal. In this large-angle deflection regime (and, as we shall see, for excitations which have symmetry such that an appropriate dipole moment does not exist), it is necessary to resort to a fully microscopic theory of the scattering process. While at first glance, to carry through a full microscopic theory of inelastic scattering of electrons from a surface seems a formidable task, there is a wealth of information on microscopic aspects of the surface that may be obtained from such an analysis if proper contact between theory and experiment can be achieved. We have initiated a theoretical program [14.9-11] which explores large-angle deflections of electrons by scattering from surface vibrations, and an initial series of calculations explore the angle and energy variation of the cross section, for two systems of current experimental interest [14.10,11].

These remarks outline the motivation of the present paper. Sect.14.2 reviews the description that has been developed for small-angle, dipole-dominated scatterings, and a number of experiments that explore dipole scattering by both vibrational and electronic excitations of the surface. Then in Sect.14.3 we turn our attention to the microscopic theory of inelastic scattering from dipole inactive modes.

14.2 Small-Angle Dipole Scattering

As remarked in Sect.14.1, oscillating electric dipole moments in and near the crystal surface lead to electric field fluctuations in the vacuum off which the incoming electron scatters. This is the basic mechanism that gives rise to the relatively intense, near-specular lobe. A few years ago, it was found that the scattering efficiency could be related to the frequency dependent dielectric constant of the substrate [14.6]. The expression obtained in this analysis, when allowance is made for the differences in dielectric response of the surface region of the crystal, describes a diverse variety of experimental data. In this section, we summarize a discussion presented elsewhere [14.8], which employs this expression to examine a number of particular cases.

Suppose the average charge density of the crystal is denoted by $\rho_0(\underline{x})$. There will be thermodynamic fluctuations in charge density, so at time t, the instantaneous charge density will be $\rho_0(\underline{x}) + \rho_1(\underline{x},t)$, where the time average of $\rho_1(\underline{x},t)$ vanishes by definition; and at each time considerations of charge neutrality give

$$\int d^3x\rho_1(\underline{x},t) = 0 \quad . \tag{14.1}$$

The charge fluctuation $\rho_1(\underline{x},t)$ sets up electric field fluctuations in the vacuum above the crystal, and in the limit that the deflection of the electron by the fluctuations is small, the scattering efficiency per unit solid angle $d\Omega(\hat{k}_s)$ per unit energy $d\hbar\omega$ may be written [14.8] when certain assumptions are met,

$$\frac{d^2S}{d\Omega(\hat{k}_s)d\hbar\omega} = \frac{2m^2e^2v_\perp^4}{\pi\hbar^5\cos\theta_I}\left(\frac{k_s}{k_I}\right)\frac{P(\underline{Q}_\parallel,\omega)}{[v_\perp^2Q_\parallel^2 + (\omega - \underline{v}_\parallel\cdot\underline{Q}_\parallel)^2]^2} \quad . \tag{14.2}$$

Here $\underline{v}_\parallel$ and $v_\perp$ are the velocity of the incoming electron parallel and perpendicular to the surface and θ_I is the angle of incidence, the energy transfer is $\hbar\omega$ in the scattering process and $\underline{Q}_\parallel$ is the wave vector change of the electron parallel to the surface. Then $|R_I|^2$ is the reflectivity of the surface, in the specular direction, and finally $P(Q_\parallel,\omega)$ contains information on the dynamics of the charge fluctuations near the surface. In terms of $\rho_1(\underline{x},t)$, one has

$$P(\underline{Q}_\parallel,\omega) = \int d^2x_\parallel \int_{-\infty}^{+\infty} dt\, \exp[i(\underline{Q}_\parallel\cdot\underline{x}_\parallel - \omega t)]$$

$$\times \int_{-\infty}^{0+} dz' \int_{-\infty}^{0+} dz''\, \exp[Q_\parallel(z' + z'')]\langle\rho_1(\underline{x}_\parallel z'',t)\rho_1(0z';0)\rangle_T \tag{14.3}$$

where the angular brackets denote an average of the quantities enclosed over the appropriate finite temperature thermodynamic ensemble. In earlier publications, one can find explicit relations between $\underline{Q}_\parallel$, $\hbar\omega$, and an appropriately defined pair of angles that describe the direction of the outgoing electron beam [14.4,8].

The general formulae are displayed above to illustrate one key point. We saw in Sect.14.1 that when the electron suffers a scattering in which its wave vector change, projected onto the plane of the surface, is $\underline{Q}_\parallel$, then the electron is above the crystal a distance roughly equal to $\ell_0(Q_\parallel) = 1/Q_\parallel$ when the scattering event takes place. Small-angle scatterings are thus produced by electric field fluctuations that extend quite far into the vacuum, as emphasized earlier. Now (14.3) shows that in this same scattering event, the electron "looks down" into the substrate the same distance $1/Q_{\parallel}$. This can be many atomic layers, so in the dipole scattering regime, one can probe bulk as well as surface parameters of the material. It is intriguing to note that $Q_{\parallel}$ varies strongly with scattering angle and energy transfer $\hbar\omega$, so that with sufficient angular resolution, one has the possibility of a continuously tunable depth probe. As discussed elsewhere [14.8] electron energy loss spectroscopy may prove a powerful probe of semiconductor surfaces, with either depletion or accumulation layers whose thickness is quite close to $1/Q_{\parallel}$. An initial experimental study of a material with a depletion layer [14.12], and one with an accumulation layer [14.13] have appeared quite recently. The results indeed show that this will be a most fruitful area of experimental study.

To proceed further, one requires an explicit expression for $P(Q_{\parallel},\omega)$, which contains information on the nature of the charge fluctuations near the surface. The following model proves useful, in a variety of circumstances [14.6,8]. Consider the substrate to be modeled simply as a semi infinite dielectric, with dielectric constant (complex) $\varepsilon(\omega)$. Then suppose on the surface, we have a thin layer of thickness d which is assumed to be described by the dielectric constant $\varepsilon_s(\omega)$. The thin surface layer can be used to model a layer of adsorbates, with infrared active adsorbate vibrations, electronic surface state bands with oscillator strength confined to the outermost few atomic layers, or electrons in an accumulation layer of a semiconductor.

For the model just described, one may show [14.6]

$$P(Q_{\parallel},\omega) = \frac{2\hbar Q_{\parallel}}{\pi}[1 + \bar{n}(\omega)] \,\mathrm{Im}\left\{\frac{-1}{1 + \tilde{\varepsilon}(Q_{\parallel},\omega)}\right\} \quad , \tag{14.4}$$

where $\tilde{\varepsilon}(Q_{\parallel},\omega)$ is an effective frequency- and wave-vector-dependent dielectric function given by

$$\tilde{\varepsilon}(Q_{\parallel},\omega) = \varepsilon_s(\omega)\left[\frac{1 + \Delta(\omega)\exp(-2Q_{\parallel}d)}{1 - \Delta(\omega)\exp(-2Q_{\parallel}d)}\right] \quad . \tag{14.5}$$

where $n(\omega) = [\exp(\hbar\omega/k_BT)-1]^{-1}$ is the Bose-Einstein function, and

$$\Delta(\omega) = \frac{\varepsilon(\omega) - \varepsilon_s(\omega)}{\varepsilon(\omega) + \varepsilon_s(\omega)} \quad . \tag{14.6}$$

Note the two limits

$$\lim_{Q_{\parallel}\to 0} \tilde{\varepsilon}(Q_{\parallel},\omega) = \varepsilon(\omega) \quad , \tag{14.7a}$$

$$\lim_{Q_{\parallel}\to 0} \tilde{\varepsilon}(Q_{\parallel},\omega) = \varepsilon_s(\omega) \quad . \tag{14.7b}$$

We now consider application of the above result to various cases. First suppose $d \to 0$, so we consider a clean material, with electronic structure uniform right up to the surface. Then there are two interesting applications of the resulting expression for $P(Q_{\parallel},\omega)$:

$$\lim_{d\to 0} P(Q_{\parallel},\omega) = P_{\infty}(Q_{\parallel},\omega) = \frac{2Q_{\parallel}}{\pi} [1 + \bar{n}(\omega)]\mathrm{Im}\left\{\frac{1}{1+\varepsilon(\omega)}\right\} \quad . \tag{14.8}$$

a) *Scattering from Interband Transitions in Semiconductors*

FROITZHEIM and co-workers [14.6] have applied (14.8) to describe the exitation of interband transitions by the incoming electron in the frequency regime where transitions from the valence band to the conduction band make the dominant contribution to the dielectric function. The results give an excellent quantitative account of the data, for scattering off the Si{111} surface, the Ge{111} surface, and also from GaAs. At small energy transfers, surface state contributions are present in the spectra; the surface layer may be added, and the oscillator strength deduced from the electronic energy loss data compares quite favorably with that deduced from earlier optical studies of the surface state transitions [14.14].

b) *Scattering from Surface Optical Phonons on Ionic Crystal Surfaces (Fuchs-Kliewer Modes)*

For scattering from the surface of an ionic crystal, where optically active surface phonons are the dominant contributor to the infrared optical response, we may let

$$\varepsilon(\omega) = \varepsilon_{\infty} + \frac{\Omega_p^2}{\omega_{to}^2 - \omega^2 - i\omega\gamma} \tag{14.9}$$

with ε_{∞} the frequency-independent contribution to the low-frequency dielectric constant from the electronic excitations (ε_{∞} assumed real here), ω_{to} is the transverse optical phonon frequency (we confine our attention to a cubic material, with two atoms per unit cell), $\Omega_0^2 = 4\pi n e^{*2}/\mu$, with n the number of unit cells per unit volume, e^* is the transverse effective charge, and μ is the reduced mass of the ions in the unit cell. Finally, γ is the damping constant.

After some algebra, one finds

$$P(Q_{\|},\omega) = \frac{2Q_{\|}}{\pi}[1 + \bar{n}(\omega)] \frac{\Omega_p^2}{(1+\varepsilon_\infty)^2} \frac{\omega\gamma}{(\omega_s^2 - \omega^2)^2 + \omega^2\gamma^2} \tag{14.10}$$

where $\omega_s^2 = \omega_{to}^2[(\varepsilon_s + 1)/(\varepsilon + 1)]$ is the frequency of a surface optical phonon that exists on the surface of an ionic crystal. This mode is commonly called the Fuchs-Kliewer optical phonon.

In the limit of small damping constant γ, (14.10) describes a well-defined loss peak produced by scattering of the electron from the Fuchs-Kliewer surface optical phonon. A very detailed experimental study of these modes, with attention to quantitative comparison of a number of features in the data with theory, was presented a number of years ago by IBACH [14.15]. The integrated strength of the energy loss line, obtained by combining (14.10) with (14.2), then integrating over wave vector transfer and energy transfer, agrees with a result given some years ago by Lucas and Sunjic, who employed a fully classical description of the electron's motion. A quantum-mechanical derivation was given shortly afterward by EVANS and MILLS [14.4], with attention to multiphonon processes [14.5].

c) *Scattering from low-energy electron-hole excitations in metals and semiconductors*

By simply taking the appropriate limit in (14.10), one obtains a description of scattering by the electronic excitations (intraband excitations) in metals, or from free carriers in a semiconductor. One lets $\omega_{to} \to 0$, and views Ω_p^2 as the electron plasma frequency. One then finds that (14.10) remains as written, but $\omega_s^2 = \Omega_p^2/(1 + \varepsilon_\infty)$, the surface plasmon frequency of the material.

The result in (14.10), modified as just described, gives an account of the scattering of electrons by surface plasmons on doped semiconductors and metals. In semiconductors, the surface plasmon frequency lies in the infrared typically, and the assumptions used to obtain the expressions above (the energy and wave vector change of the electron upon scattering is a small fraction of the incident wave vector) are comfortably satisfied for near-specular scattering. For metals, the surface plasmon frequency is generally several electron volts, so care must be exercised in quantitative application of the formula.

A most intriguing application of (14.10) has been made recently by IBACH [14.8]. He observed a structureless "background" signal, when electrons scatter from the surface of a clean Ni surface. Equation (14.10) describes this background quantitatively. The result is most illuminating if it is rewritten somewhat. For energy transfers in the infrared, we have $\omega \ll \Omega_p$, and also $\omega \ll \gamma$ for small energy transfers. Then also with $\bar{n} \ll 1$, and if we recall that the electrical conductivity of the metal is $\sigma = ne^2/m\gamma$ in the present notation, (14.10) becomes

$$P(Q_{\|},\omega) \simeq \frac{Q_{\|}}{2\pi^2}\frac{\omega}{\sigma} \quad . \tag{14.11}$$

From (14.11), we see that dipole scattering from low-energy electronic excitations gives a featureless background proportional to the *electrical resistivity* of the substrate. The same result may be obtained directly from (14.8) by writing $\varepsilon(\omega) = \varepsilon_\infty + 4\pi i\sigma(\omega)/\omega$, with $\sigma(\omega)$ the frequency-dependent conductivity, then noting that at low frequencies one has $\varepsilon(\omega) \simeq 4\pi i\sigma(\omega)/\omega \gg 1$. It is then $\sigma_R(\omega)$, the real part of the frequency-dependent conductivity, that appears in (14.11).

A useful and simple expression for the background intensity is obtained [14.8] if one integrates (14.2) over solid angle $d\Omega(\hat{k}_s)$ to obtain the scattering efficiency per unit energy transfer, $dS/d\hbar\omega$. One finds

$$\frac{dS}{d\hbar\omega} = \int d\Omega(\hat{k}_s)\,\frac{d^2S}{d\Omega(\hat{k}_s)d\hbar\omega} = \frac{1}{2\pi}\,\frac{|R_I|^2}{\hbar a_0 k_I\,\cos\theta_I}\,\frac{[1+\bar{n}(\omega)]}{\sigma_R(\omega)} \quad , \tag{14.12}$$

where a_0 is the Bohr radius and k_I the wave vector of the incident electron.

IBACH has compared (14.12) with the background signal that he has measured for scattering from a Ni surface, to find remarkable quantitative agreement between theory and the data [14.8]. Most particularly, as the temperature is increased, as expected from (14.12) with $\bar{n}(\omega) \ll 1$. One may use the background as a means of calibrating the intensity of vibrational losses to obtain a more reliable account of their absolute intensity.

Next we turn our attention to the case where a thin layer of material is present on the surface. If the thickness of the surface layer is small, then $P(Q_{\|},\omega)$ may be expanded in powers of d, to yield the first surface correction to $P_\infty(Q_{\|},\omega)$. We have

$$P(Q_{\|},\omega) = P_\infty(Q_{\|},\omega) + P_s(Q_{\|},\omega) \quad , \tag{14.13}$$

where a short calculation gives

$$P_s(Q_{\|},\omega) = \frac{2\hbar Q_{\|}^2 d}{\pi}\left(\mathrm{Im}\left\{\frac{\varepsilon^2(\omega)}{[1+\varepsilon(\omega)]^2}\,\frac{-1}{\varepsilon_s(\omega)}\right\} + \mathrm{Im}\left\{\frac{1}{[1+\varepsilon(\omega)]^2}\right\}\right) \quad . \tag{14.14}$$

Under many circumstances at small energy transfers, the substrate dielectric constant is very large, $|\varepsilon(\omega)| \gg 1$. This is true in two situations encountered commonly. In a metal, for frequencies well below the electron plasma frequency, $|\varepsilon(\omega)| \simeq \Omega_p^2/\omega^2 \gg 1$, for frequencies ω that lie in the infrared. Also, in most semiconductors, the low-frequency dielectric constant is the order of 12, or possibly even larger. The first term in (14.14) then dominates, to give

$$P_s(Q_{\|},\omega) \simeq \frac{2\hbar Q_{\|}^2 d}{\pi} \operatorname{Im}\left\{\frac{-1}{\varepsilon_s(\omega)}\right\} \quad . \tag{14.15}$$

The result in (14.15) has a simple physical interpretation. The electron energy loss spectrum will contain loss structures at these frequencies ω where the thin surface layer possesses excitations with fluctuating electric dipole moment *normal* to the substrate surface. To see this, consider a thin slab of material with dielectric constant $\varepsilon_s(\omega)$. A fluctuating electric dipole moment normal to the slab surface generates an electric field $\underline{E}$ inside the slab, and normal to the surface. There is also a $\underline{D}$ field normal to the surface, and also $\underline{D} = \varepsilon_s(\omega)\underline{E}$. In the absence of free charges, we must have $\nabla \cdot D = 0$ everywhere, including at the slab surfaces. This condition is satisfied only if $\underline{D} \equiv 0$ everywhere, and with $E \neq 0$, we are led to require $\varepsilon_s(\omega) \equiv 0$ as the condition on the frequency of such an excitation. A zero in $\varepsilon_s(\omega)$ produces a peak in $\operatorname{Im}\{-1/\varepsilon_s(\omega)\}$; the loss spectrum described by (14.15) thus contains structure at those frequencies for which the surface layer possesses excitations with electric dipole moment normal to the surface.

One may argue that the second term in (14.14) describes the contribution to the loss spectrum from elementary excitations with electric dipole moment *parallel* to the slab surfaces. These contributions are supressed in the limit which leads to (14.15). We thus have a derivation of the dipole selection rule introduced by EVANS and MILLS [14.4] that is of substantially greater generality than the particular context in which it was introduced.

If we consider a surface covered by a layer of vibrating adsorbate atoms, then we may describe their contribution to the loss cross section by choosing for $\varepsilon_s(\omega)$ the form

$$\varepsilon_s(\omega) = \varepsilon_s^{(\infty)} + \frac{\Omega_p^{(s)2}}{\omega_0^2 - \omega^2 - i\omega\gamma} \quad . \tag{14.16}$$

It may be shown that (14.15,16), in combination with (14.2), lead to results that generalize those obtained by EVANS and MILLS, whose model ignores the influence of $\varepsilon_s^{(\infty)}$ (this has its physical origin in the electronic polarizability of the adsorbate atoms or molecules), and the difference between the vibration frequency ω_0 of an isolated adatom, and the peak in $\mathrm{Im}\{-1/\varepsilon_s(\omega)\}$, shifted away from ω_0 by local field effects. Studies of the coverage dependence of the intensity and frequency of the CO stretching vibration on Ru [14.16] and on Cu [14.17] provides compelling evidence for the importance of both effects. The simple model just outlined accounts for the data very well in both cases, in an internally consistent fashion.

In electron-electron energy loss studies of electronic transitions on semiconductor surfaces, the surface state contributions may also be described by combining (14.15,16) with (14.2). Here ω_0 is the interband excitation frequency, and $\Omega_p^{(s)2}$ is related to the oscillator strength of the surface state interband transition. The oscillator strengths deduced from the electron energy loss spectra are in good accord with those deduced from optical studies of the same features [14.6].

This section has presented a brief outline of the description of small-angle electron energy losses produced by coupling of the incoming electron to a variety of elementary excitations at crystal surfaces. An extended and complete discussion along these lines has been given elsewhere [14.8].

We conclude with one final application of the expression given in (14.2, 4). Quite frequently free carriers are trapped on semiconducting surfaces to form a thin, conducting surface layer. An inversion layer is an example of such a situation, where an insulating region of bulk material lies behind the thin conducting surface layer. Quite recently, experimental studies of such inversion layers by electron energy loss spectroscopy have appeared. Also, IBACH has made the intriguing observation that a surface conducting layer is present on the (7×7) reconstructed form of a semiconductor surface.

The scattering of electrons from a thin conducting surface layer on an otherwise insulating material may be described by replacing $\varepsilon(\omega)$ by ε_B, assumed real. Then $\varepsilon_s(\omega) = \varepsilon_B - \Omega_p^2/\omega(\omega + i\gamma)$, with Ω_p the plasma frequency of the electrons in the thin layer, and γ the inverse of the scattering time. An interesting limit to consider is that where $Q_{||}d \ll 1$ (satisfied in the experiments described above) and $\omega \ll \Omega_p$. The loss function $P(Q_{||},\omega)$ assumes the form

$$P(Q_{||},\omega) \quad \frac{2\hbar Q_{||}}{\pi}\frac{1}{(1+\varepsilon_B)}\frac{\omega^2(Q_{||})\omega\gamma}{[\omega^2(Q_{||})-\omega^2]^2+\omega^2\gamma^2} \quad , \tag{14.17}$$

where $\omega^2(Q_{||}) = \Omega_p^2 Q_{||} d/(1+\varepsilon_B)$ is the dispersion relation of the low-lying plasmons of the two-dimensional electron gas. These low-lying modes have electric fields *parallel* to the conducting layer, and for this geometry no surface depolarization field can be developed. As a consequence, as $Q_{\parallel} \to 0$, the frequency of the plasmons vanishes.

The scattering of an electron from these low-lying two-dimensional plasmons has been studied recently, as remarked above. If one collects all electrons scattered near the specular, with angular resolution insufficient to actually measure the dispersion relation, then the scattering from these modes produces a strongly frequency-dependent inelastic "wing" centered about zero frequency, in the loss spectrum. It is possible to work out the shape of this wing, to obtain a simple result, if we allow ourselves the further assumption $\omega > \Omega_p^2(d/v_0)/(1+\varepsilon_B)$, in addition to $\omega \ll \Omega_p$, with v_0 the velocity of the incident electron. If $\Omega_p = 500\ \mathrm{cm}^{-1}$, $d = 3$ Å and the incident electron energy is 5 eV, this condition is satisfied for all energy losses greater than 1 cm^{-1}; then we find for scattering from the low-energy excitations

$$\frac{dS}{d\hbar\omega} = \int d\Omega(\hat{k}_s)\, \frac{d^2S}{d\Omega(\hat{k}_s)\, d\hbar\omega}$$

$$= \frac{2|R_I|^2}{a_0 k_I \cos\theta_I} \frac{d}{v_0} \frac{[1+\bar{n}(\omega)](1+\cos^2\theta_I)}{(1+\varepsilon_B)^2} \left(\frac{\Omega_p}{\omega}\right)^2 \quad . \tag{14.18}$$

In his study of scattering from the conducting layer on the 7×7 Si{111} surface, IBACH saw a low-energy tail qualitatively similar to that described by (14.18). The form displayed holds only when $\omega \ll \Omega_p$, and the full expression for $P(Q_{\parallel},\omega)$ must be used to describe the whole loss spectrum. When $\omega \sim \Omega_p$, the surface contribution to $P(Q_{\parallel},\omega)$ is described well by (14.15), so there is a loss peak (present in Ibach's spectrum) produced by the plasmon of the two-dimensional layer with electric field *normal* to the surface. The collective excitations of the two-dimensional electron gas have a two-branch spectrum, with one branch approaching Ω_p (electric field normal to the slab surfaces), and the second approaching $\Omega_p(Q_{\parallel}d)^{\frac{1}{2}}/(1+\varepsilon_B)^{\frac{1}{2}}$, as $Q_{\parallel} \to 0$. The high-frequency branch gives rise to the plasmon line described by (14.15), while the low-lying excitations produce the "wing" described by (14.18).

This completes our discussion of small-angle inelastic electron scattering by the dipole mechanism. The principal message is that a wide variety of low-lying elementary excitations of the crystal surface can and have been studied by the method, and the results can be interpreted with the description provided by (14.2,4). We now turn to the discussion of inelastic scatterings which involve large-angle deflections . Here both the theoretical and the experimental literature have explored losses produced by scattering from the vibrational normal modes of adsorbate layers.

14.3 Inelastic Electron Scattering from Surfaces with Large Deflection Angles; The Scattering by Dipole-Inactive Surface Vibrations

The remarks in Sect.14.1 show that, when the electron is deflected from the specular direction by an angle $\Delta\theta$ very large compared to $\Delta\theta_c = \hbar\omega_s/E^{(I)}$, the electric field produced by the oscillating electric dipole moments in the surface extends out into the vacuum only one or two angstroms, a distance quite small compared to the depth of penetration of the electron beam into the substrate. Thus, under these conditions, the electron will interact with any particular excitation primarily during the time in which it is inside the crystal. A similar statement applies also to small-angle deflections from excitations which fail to generate an electric dipole moment when $Q_{\parallel} \simeq 0$, so the long-ranged field described earlier is absent in the vacuum above the crystal (more precisely, it will be very weak, since for $Q_{\parallel}$ small and in a general direction, there will be a small electric dipole moment proportional to the magnitude of $Q_{\parallel}$). Thus, when we consider the inelastic scattering of electrons under conditions where the deflection angle is not small, or from surface excitations that are dipole inactive when $Q_{\parallel} = 0$, we must resort to a fully microscopic theory, since the detailed form of the electron wave function in the crystal is required.

The present authors and collaborators have developed a formalism which describes the large-angle inelastic scattering of low-energy electrons from the vibrational motions of surfaces, for substrates upon which ordered layers of adsorbate molecules or atoms are present [14.9]. We have applied the formalism to an analysis of the energy and angle variation of the cross section for scattering from the various normal modes of the $c(2\times2)$ overlayer of CO adsorbed on the Ni{100} surface [14.10], and more recently to hydrogen adsorbed on the W{100} surface. This section describes the formalism used in this work, and presents a summary of some key results.

We have seen that in the dipole-dominated regime, a selection rule operates that selects out those normal modes of the system which possess a dynamic dipole moment normal to the surface. In the regime of large-angle deflections, this selection rule no longer operates, so that in principle one may probe all normal modes of the system through proper choice of the scattering geometry. When the energy loss spectrum measured at large-angle deflections is compared with the near-specular dipole loss spectrum, one may then sort out the symmetry of the various modes. Those with symmetry such that a dipole moment normal to the surface is generated at $\underline{Q}_{||} \simeq 0$ will show a loss cross section that increases dramatically as the specular direction is approached, while the dipole-inactive modes will have a cross section that varies smoothly as the specular is approached, or which, for reasons we shall appreciate shortly, drops to zero here. HO et al. used this behavior of the cross section to discriminate between the vibrational modes of H on W{100} which are parallel to the surface, and the mode polarized normal to it [14.17]. Similarly, BARO et al. [14.18] have identified the symmetry of the normal modes of hydrogen on Pt{111} by using this principle.

There are selection rules which operate in the large-angle-deflection regime, and for scattering from dipole-inactive modes, however. We have discussed thse completely elsewhere [14.10], so here we will be content to cite one or two particular examples. Suppose we consider the scattering geometry illustrated in Fig.14.1, where the incident electron wave vector $\underline{k}^{(I)}$, the scattered electron wave vector $\underline{k}^{(s)}$, and the surface normal all lie in a single plane we call the scattering plane. Then consider a simple adsorbate atom on a high-symmetry site, so its vibrational normal mode spectrum consists of modes polarized either normal to the surface, or parallel to it. Finally, suppose the scattering plane is oriented so that it is a plane of reflection symmetry for the problem. Now the wave functions $\psi^{(I)}$, $\psi^{(s)}$ of the incident and scattered electrons are both *even* under reflection through the scattering plane, while $\partial V/\partial u$, the derivative of the crystal potential with respect to the displacement u of the adsorbate, is odd under the same reflection operation for the vibrational mode polarized *normal* to the scattering plane (and parallel to the surface). It follows that the matrix element for scattering from this mode vanishes identically for these conditions. One must swing the direction of the scattered electron out of the scattering plane to observe such a mode.

There is a second selection rule that operates for the parallel modes, but this one has an origin that is more subtle, along with a constraint that comes from an assumption that must be injected into the argument to prove the rule. Consider the geometry outlined in the previous paragraph, and scattering from a mode polarized parallel to *both* the surface and the scattering plane, i.e., the displacement is parallel to the x axis of Fig.14.1. Now let the electron emerge right along the specular direction, and for the moment we ignore the difference in energy between the incident and the scattered electron (this is the assumption referred to earlier in this paragraph). A combination of reflection in the *yz* plane *and* time reversal leaves the scatter-

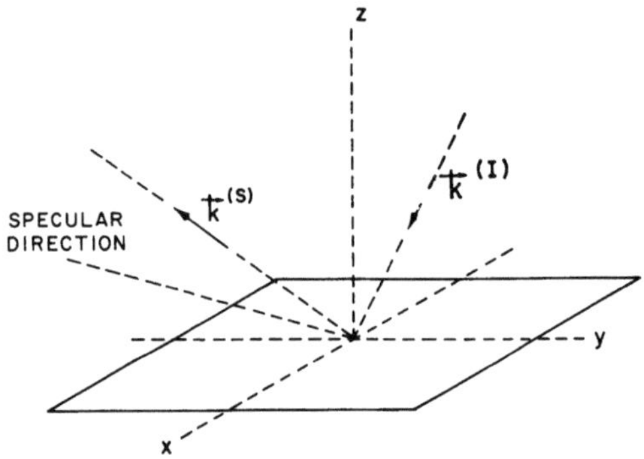

Fig.14.1. A scattering geometry in which the incident electron wave vector $\underline{k}^{(I)}$, the scattered electron wave vector $\underline{k}^{(s)}$, and the normal to the surface lie in one plane, referred to in the text as the scattering plane

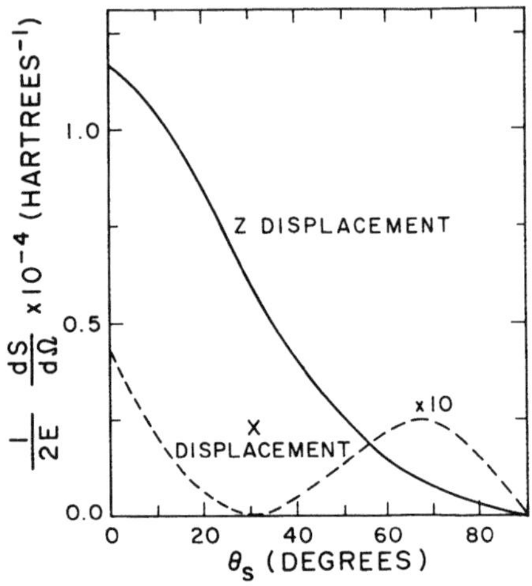

Fig.14.2. The angular variation of the scattering cross section for scattering from two normal modes of a c(2 × 2) overlayer of CO on the Ni{100} surface. The figure shows scattering from the CO "stretch" mode, polarized parallel to the $\hat{z}$ direction, and from a parallel mode polarized parallel to $\hat{x}$, which is parallel to both the surface and the scattering plane. The zero illustrates a selection rule discussed in the text

ing geometry invariant, but once again changes the sign of $\partial V/\partial u$, so the matrix element for scattering vanishes when the electron emerges along the specular. This selection rule is illustrated in Fig.14.2, where we present calculations of the angular variation of the scattering efficiency for scattering off two of the normal modes of CO molecules for a c(2 × 2) layer of CO on the Ni{100} surface [14.10]. The z polarized mode is the CO stretching vibration, polarized normal to the surface. For this mode, as outlined earlier, we should have the near-specular dipole lobe for deflections very near the specular direction, but this has been omitted from these calculations for reasons discussed earlier [14.9]. Then the intensity of scattering from the CO stretching vibration varies smoothly with scattering angle, when the scattered beam lies in the scattering plane and is swept through the specular direction. The curve labeled "x polarized" describes the angular variation of scattering from one of the CO vibrational modes polarized parallel to the x axis of Fig.14.1. Here we see a dip in the scattering intensity as the scattered beam sweeps through the specular direction; this illustrates the operation of the selection rules.

The previous two paragraphs outline two particular selection rules that operate in the large-angle deflection regime, and for dipole-inactive modes. There are several other rules that apply here for various geometries. These have been examined in a more complete discussion given elsewhere [14.10]. It must be kept in mind that some, but not all, of the proofs of the selection rules require the assumption that the wave function $\psi^{(s)}$ of the scattered electron may be replaced by a wave function of an outgoing electron at the incident energy without introducing serious error. At first glance this assumption does not seen serious since the energy change of the electron is a very small fraction of the impact energy in nearly all electron energy loss studies. However, at low impact energies, it is not uncommon to encounter "fine structure resonances" in the LEED intensity curves. In such regimes of energy and impact angle, the amplitude for *elastic* scattering from the surface, and hence necessarily the electron wave function, are affected strongly by rather small variations in energy. The selection rules may well break down in such regimes of energy and impact angle; we presently have calculations underway that will incorporate the image potential experienced by the electron as it approaches and exits from the surface. This will introduce the fine structure resonances into the calculations, and enable us to check whether or not the selection rules continue to operate in such a regime.

Our calculations for the c(2 × 2) CO overlayer on the Ni{100} surface illustrate one more point that is of general interest.

In the small-angle, dipole-dominated scattering regime the inelastic scattering efficiency decreases as the electron energy is increased for impact energies sufficiently large that the deflection angle $\Delta\theta_c$ is sufficiently small that all electrons in the "dipole lobe" are collected by the apparatus. There are two reasons for this. First of all, we see from (14.2) that the scattering efficiency $d^2S/d\Omega(\hat{k}_s)d\hbar\omega$ is proportional to $|R_I|^2$, the intensity of the specular beam. As the electron energy is lowered from, say, a few tens to a few electron volts, the specular intensity rises dramatically. It is also true that, when (14.2) is integrated over solid angle and energy transfer, then $|R_I|^2$ is multiplied by a "dipole form factor" which decreases with increasing energy. As illustrated by the elementary example of Rutherford scattering of electrons by nuclei, coulomb cross sections are decreasing functions of impact energy. These considerations show that low impact energies, in the range of 2-5 eV are favored in the dipole regime.

In the large-angle deflection regime we find, in contrast to the behavior just described, that the excitation probability *increases* with impact energy,

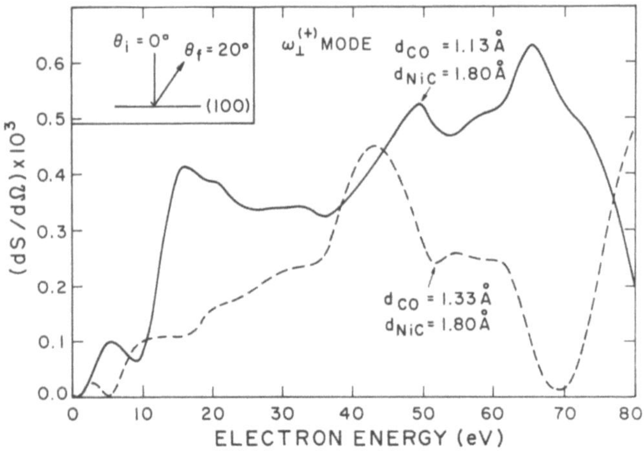

Fig.14.3. The energy variation of the scattering efficiency for excitation of the CO stretching vibration, for the c(2 × 2) overlayer of CO on the Ni{100} surface. The figure presents calculations for two values of the CO bond length to illustrate structural sensitivity of the energy variation. See [14.10] for further details

on the average. This is illustrated in Fig.14.3, where the scattering efficiency per unit solid angle is given for scattering from the CO stretching vibration for energies considerably higher than those used in present-generation, small-deflection-angle studies. The inelastic scattering intensity, considered as a function of impact energy, contains structure which is sensitive to details of the bonding geometry. One sees this from the two curves in Fig.14.3 calculated for different values of the CO bond length. One should also notice that, on the average, the cross section for inelastic scattering increases with impact energy for the regime explored in Fig.14.3, while at the same time the dipole contribution to the cross section falls off. We estimate that for an impact energy of 50 eV the dipole cross section has fallen off, and the "impact" contribution has risen to the point where the two are comparable in magnitude.

The remarks above suggest that large-deflection-angle studies should be carried out at energies very considerably higher than those employed in present spectrometers. We understand it is difficult to construct a practical spectrometer which can *scan* a wide range of energies to measure the inelastic cross section of wide range of impact energies. However, it may prove possible to construct a spectrometer which operates at a selected narrow range of impact energies in the 50-100 eV range. Our calculations suggest the construction of such a spectrometer would prove of great value in vibrational spectroscopy. We are unaware of any apparatus under construction at this time aimed at this regime of energy, and we would very much like to see experimentalists focus their attention on this possibility.

While the calculations reported above have been carried out for the c(2 × 2) CO overlayer on Ni{100} surface, as yet no experimental data is available on this system, in the regime of large-angle deflections. Also, scattering from modes polarized parallel to the surface have not been detected. We have recently explored scattering from the c(2 × 2) overlayer of hydrogen on the W{100} surface [14.11]. This system has been studied by HO et al. [14.17]. Here we present a brief summary of the results of this study, with emphasis on the principal qualitative conclusions that may be drawn from the analysis.

On the W{100} surface, the H atom sits on a bridge site between two tungsten atoms, so the local site symmetry is C_{2v}. There are thus three normal modes of vibration, each of which has been identified in the electron energy loss spectra reported by HO and co-workers [14.17]. One is the vibration of the H atom normal to the surface, which can be identified in the data by the presence of the "dipole lobe", a two-orders-of-magnitude increase in the scattering intensity as the specular direction is approached. There is then the antisymmetric stretching vibration, polarized parallel to the surface and to the plane formed by the hydrogen and the two tungsten atoms to which it is bonded, and finally a wagging mode, where the hydrogen vibrates parallel to the surface, but perpendicular to the plane of the triangle.

In Fig.14.4 we show the variation of scattering efficiency for excitation of the wagging mode with exit angle for two angles of incidence and the following scattering geometry. The incident and scattered beam both lie in a plane normal to the surface, and this plane (referred to earlier as the scattering plane) is aligned to contain the <100> axis which lies in the surface. In these calculations, we ignore the influence of reconstruction of the W atoms. As the reader will appreciate shortly, we have explored the effect of shifts in the W positions on the calculated excitation efficiencies, but these shifts have little qualitative effect on the dependence of the scattering efficiency on exit angle.

The most striking feature present in Fig.14.4 is the zero in the excitation efficiency along the specular direction. This is a manifestation of the selection rules discussed earlier in the present section. From the point of view of planning future experiments, the most important feature of the results displayed in Fig.14.4 is the large scattering efficiencies calculated for exit angles *above* the specular direction. In each case, the cross section calculated for a geometry where the exit angle *below* the specular is very

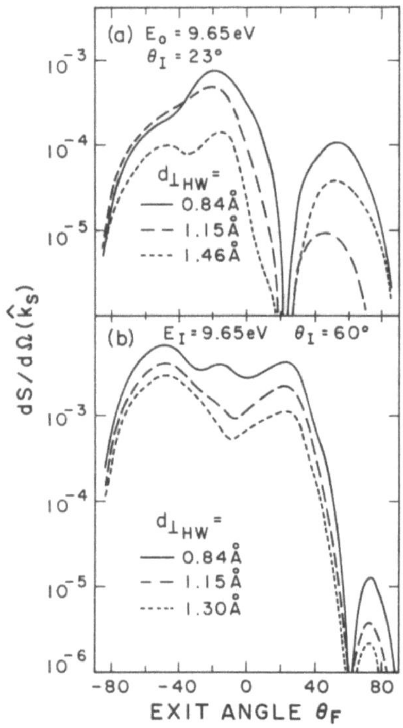

Fig.14.4a,b. The magnitude and angular variation of the scattering efficiency per unit solid angle, dS/d (k_S), for scattering from the "wagging mode" of a c(2 × 2) array of hydrogen atoms on the W{100} surface. The effect of reconstruction of the W atoms has been ignored, and we present calculations for an angle of incidence of (a) 60° and (b) 23°. The scattering plane contains the {100} axis in the surface

much smaller than a configuration in which the electron exists near the surface normal. In the work of HO et al. [14.17]on H on W{100}, and also that of BARO et al. [14.18] which examines H on Pt{111}, rather large angles of incidence were employed, and the electrons which emerge *below* the specular direction were sampled. Thus, both experiments were carried out in a most unfavorable geometry, and substantially larger signals should result if the scattered electrons which emerge above the specular are sampled. Electron energy loss studies which sample large-angle deflections are very difficult to carry out successfully, largely because the signals are very weak, for scattering from dipole inactive modes, and well outside the dipole lobe of dipole active vibtations. We will be most anxious to see if experiments can be performed in a more favorable scattering geometry.

It is intriguing to inquire if the angular variation of the excitation energy is sensitive to adsorbate-induced shifts of the substrate atoms. The case of H on W{100} is a particularly interesting system from this point of view, since it is now well established that the adsorbate induces reconstruction of the {100} surface at room temperature. Since it is clear from the work of HO and colleagues that all the hydrogen vibrations may be seen

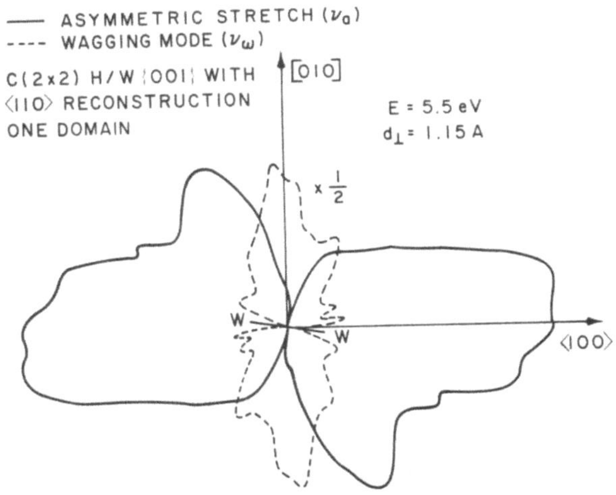

Fig.14.5. The φ plot, described in the text, for electrons normally incident on a W{100} surface with a c(2 × 2) layer of H present. The exit angle is 70° from the crystal normal. We show a plot for scattering from the asymmetric stretching vibration (ν_a), and the wagging mode (ν_w). The calculation assumes the electron reflects from a single domain of the adsorbate layer

in the electron energy loss spectrum, we have the possibility of inferring the nature of the shift in position of the substrate atoms from further studies of this system.

As remarked earlier, our calculations show that the dependence of the excitation efficiency on exit angle is not sensitive to displacements of substrate atoms in its qualitative features, though its quantitative details are. We find that the dependence of the excitation efficiency on the azimuthal orientation of the scattering plane is strikingly sensitive to the nature of the surface reconstruction, particularly for scattering from a single domain of the adsorbate layer. We consider a scattering geometry when the incident beam, the scattered beam, and the normal to the surface lie in one plane. Then, with incident and scattered beam fixed, the crystal is rotated about the normal to the surface, and the scattered intensity is monitored as a function of the azimuthal angle between the scattering plane and the <100> direction. We call data generated by this procedure "φ plotts".

In Fig.14.5 we show a "φ plot" for scattering from two of the three normal modes of H on W{100}, the asymmetric stretch (ν_a) polarized parallel to the surface, and the wagging mode (ν_w), also parallel to the surface. Figure 14.5 is calculated for a model of the surface reconstruction which supposes the W atoms to be displaced along the <110> direction. The calculations are carried out for the case where the electron beam scatters from a single domain of the absorbate layer. What is striking is that the φ plot patterns do not have symmetry under reflection through either the <100> or the <010> directions, while such reflection symmetry is present if the displacement is along <100>.

In this section, we have outlined some of the principal results we have obtained in our theoretical studies of inelastic scattering of electrons from adsorbate vibrations, in the regime where the angular deflection is sufficiently large that the dipole description of the scattering process breaks down, and a full microscopic theory is required. We see that a series of selection rules apply in this regime, and to the description of near-specular scattering by dipole-inactive vibrations. In addition, the calculations suggest new ways of generating information from electron energy loss spectroscopy; important, and possibly difficult modifications of present day apparatus will be required to implement some of these suggestions, while it should be possible to implement others with possibly minor alterations of present procedures. This chapter provides only a brief outline of our results, and we refer the reader to detailed discussions presented elsewhere for full details [14.8-11].

Acknowledgments. The results described in this article obtained recently have emerged through a research program supported by the U.S. Department of Energy, through Grant Nos. DE-AC02-79ER10431 and DOE AT03-79-ER10432.

References

14.1 H. Ibach, D. Bruchmann: Phys. Rev. Lett. *44*, 36 (1980)
14.2 S. Lehwald, H. Ibach: Conference on Vibrations at Surfaces, Namur, Belgium, 1980 (to be published)
14.3 Talat S. Rahman, J.E. Black, D.L. Mills: Phys. Rev. Lett. *46*, 1469 (1981)
14.4 E. Evans, D.L. Mills: Phys. Rev. B*5*, 4126 (1972)
14.5 E. Evans, D.L. Mills: Phys. Rev. B*7*, 853 (1972)
14.6 H. Froitzheim, H. Ibach, D.L. Mills: Phys. Rev. B*11*, 4980 (1970)
14.7 D.L. Mills: Surf. Sci. *48*, 59 (1975)
14.8 H. Ibach, D.L. Mills: *Electron Energy Loss Spectroscopy and Surface Vibrations* (Academic, New York forthcoming)
14.9 S.Y. Tong, C.H. Li, D.L. Mills: Phys. Rev. Lett. *44*, 407 (1980); C.H. Li, S.Y. Tong, D.L. Mills: Phys. Rev. B*21*, 3057 (1980)
14.10 S.Y. Tong, C.H. Li, D.L. Mills: Phys. Rev. B*24*, 806 (1981)
14.11 B.M. Hall, C.H. Li, S.Y. Tong, D.L. Mills: To be published
14.12 R. Matz, H. Luth: Phys. Rev. Lett. *46*, 1652 (1981)
14.13 A. Mang, I. Wagner, A. Rosenthal, J.I. Gersten, Y. Goldstein: Phys. Rev. Lett. *46*, 1658 (1981)
14.14 G. Chiarotti, S. Nannarone, R. Pastore, P. Chiaradia: Phys. Rev. B*4*, 3398 (1971)
14.15 H. Ibach: Phys. Rev. Lett. *24*, 1416 (1970)
14.16 S. Andersson, B.N.J. Persson: Phys. Rev. Lett. *45*, 1421 (1980); to be published
14.17 W. Ho, R.F. Willis, E.W. Plummer: Phys. Rev. Lett. *40*, 1463 (1978); W. Ho, N. DiNardo, E.W. Plummer: J. Vac. Sci. Technol. *17*, 134 (1980)
14.18 A.M. Baro, H. Ibach, H.D. Bruchmann: Surf. Sci. *88*, 384 (1979)

15. Electronic Aspects of Adsorption Rates

O. Gunnarsson and K. Schönhammer

With 8 Figures

15.1 Introduction

Recently there has been much interest in the energy transfer between a surface and an atom or a molecule approaching a surface. It has been shown that inelastic scattering of rare-gas atoms can give information about the elementary excitations of the surface [15.1]. The energy transfer processes also determine the probability that the incoming particle loses its kinetic energy and is adsorbed at the surface, the so-called sticking probability. Since adsorption is the first step in heterogeneous catalysis and since it represents a simple example of a surface reaction, the dynamics of adsorption are of considerable interest.

There are several possible mechanisms for the energy transfer. In the past it has often been assumed that the dominating process is the excitation of substrate phonons by the impinging adsorbate [15.2]. In the case of metal substrates, however, there has recently been an increasing interest in the excitation of electron-hole pairs [15.3-11]. Molecular adsorbates can also lose kinetic energy to internal degrees of freedom of the molecule, such as rotational and vibrational excitations [15.12]. Although there is no doubt that the phonon mechanism can be very important, we shall focus here on the electronic excitations since the phonon mechanism has already been thoroughly reviewed in the literature [15.2].

Because of the large mass of the adsorbate nucleus it is usually assumed that nuclear motion can be treated classically, i.e., that the adsorbate nucleus follows a classical trajectory. The motion of the adsorbate nucleus can then be viewed as an external perturbation, and we are interested in the response of the electrons of the system. The response of a fermion system to a localized time-dependent perturbation is also of great interest in other situations, for instance for spectroscopies such as X-ray absorption and emission [15.13] and X-ray photoemission [15.14]. Although much of our discussion will be fairly general and could be applied to other situations, we shall keep the adsorption problem in mind. In particular, we shall develop methods suitable for slow perturbations, while in spectroscopies the perturbation is usually very fast.

Because of the slow nuclear motion of the adsorbate one might expect the deviations from the adiabatic limit to be small, in which case the electronic contribution to the energy transfer would be insignificant. However, if the adiabatic state changes rapidly with the position of the adsorbate, the nonadiabatic effects could nevertheless be very important. NØRSKOV and LUNDQVIST [15.4] have proposed that this should be the case for adsorbates which have an affinity level close to the Fermi energy of the substrate. As the adsorbate approaches the surface the level is generally shifted downwards [15.15], and it may move below the Fermi level. This leads to a partial filling of the affinity level in the adiabatic state. If the affinity level is narrow when it crosses the Fermi level, the system may not have time to follow the changes in the adiabatic state, and nonadiabatic effects become important [15.4] (Fig.15.1). For instance, in the case of adsorption of halogen molecules on a sodium surface and oxygen molecules on an aluminium or magnesium surface, nonadiabatic effects have been observed directly in the form of the emission of photons and electrons [15.16]. This shows that there is a small probability that the adsorbate level is filled through a radiative process or through an Auger process involving an electron with sufficient energy to leave the system. However, the dominating nonadiabatic effect should be the creation of low-energy electron-hole pairs, which gives a much larger contribution to the sticking probability, but which is not so easily observed directly.

A central quantity for the calculation of the sticking probability is the probability $P_{E'}(E)$ that an incoming particle with the energy E' looses the energy E during one round trip in the potential well (Fig.15.2). If the substrate temperature T_s is zero, the sticking probability is

$$s = \int_{E'}^{\infty} P_{E'}(E)\, dE \qquad (15.1)$$

since the particle can not leave the potential well if it has lost an energy $E > E'$. If $T_s > 0$ a particle which is trapped in the potential well during the first round trip may later obtain enough energy from the substrate to leave the surface. ICHE and NOZIERES [15.17] have proposed a master equation approach to calculate the sticking coefficient in this case.

The loss function $P_{E'}(E)$ is usually calculated under the assumption that the electrons are independent particles. This is not unreasonable, since

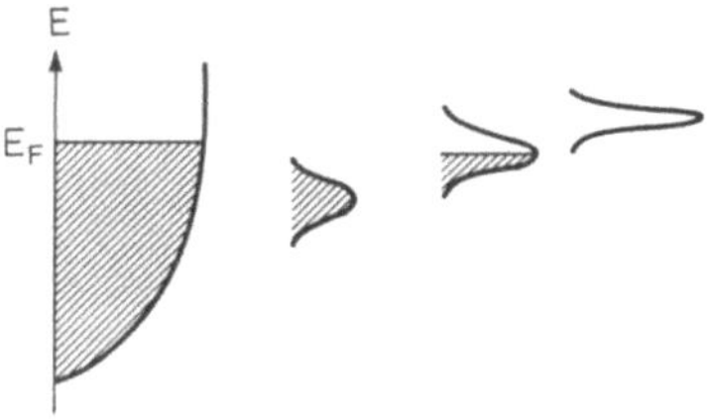

Fig.15.1

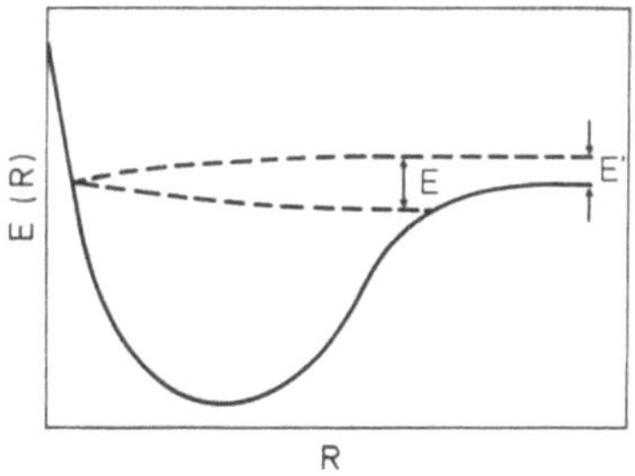

Fig.15.2

Fig.15.1. To the left the unperturbed substrate density of states is shown. The adsorbate-induced density of states for different adsorbate positions is shown to the right. The dashed area shows the occupancy for the adiabatic states

Fig.15.2. Schematic representation of the adiabatic energy of the system as a function of the separation between the adsorbate and the substrate. The dashed curve indicates how in the nonadiabatic treatment an adsorbate with the initial kinetic energy E' loses the energy E to the substrate

almost all the created excitations have low energies and may be viewed as Landau quasiparticles. The function $P_{E'}(E)$ can then be obtained by solving the time-dependent Schrödinger equation for noninteracting electrons. This is in principle a straightforward numerical problem. In practice, however, the slow nuclear motion of the adsorbate makes such a direct approach difficult and nontransparent. Instead, the low-energy excitations can be treated as independent bosons. This leads to a time-dependent version of the displaced harmonic oscillator problem, which is fairly straightforward to solve. This approach is in certain situations exact. The boson approach was first used in this context by MÜLLER-HARTMANN et al. [15.18], who obtained an expression for the weak coupling limit. We have later developed a formalism for strong but slow perturbations [15.7,8].

There has been a number of calculations [15.3] of the so-called friction coefficient, which is directly related to the center of gravity of $P_{E'}(E)$. The friction coefficient gives a good idea of the size of the nonadiabatic effects, but it does not, in general, give enough information about $P_{E'}(E)$ [15.8].

There are a few examples of a quantum-mechanical treatment of the motion of the adsorbate nucleus. For systems where the nonadiabatic effects are not too strong it is convenient to use a method similar to BORN and HUANG's treatment of coupled nuclear and electronic systems [15.19]. The (adiabatic) electronic eigenstates are calculated as a function of the adsorbate position (Fig.15.3), and the wave function for the total system, including the nuclei,

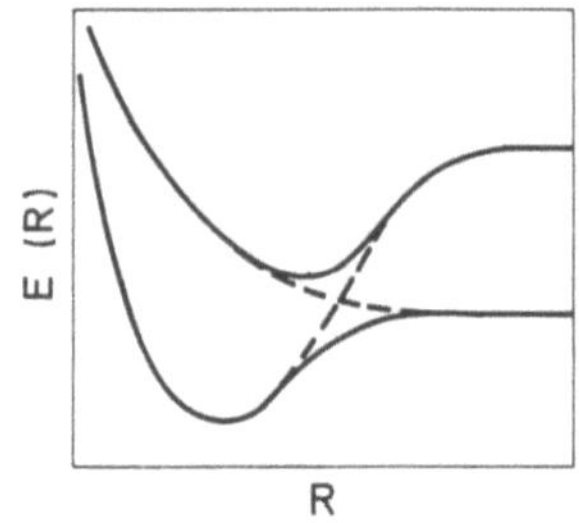

Fig.15.3. Schematic drawing of the adiabatic (——) energy curves used in the Born-Huang type approach and the diabatic (---) energy curves used by GADZUK and METIU [15.11]

is expanded in the electronic eigenstates. Such an approach was first used in the present context by BRIVIO and GRIMLEY [15.5].

They calculated the sticking coefficient of hydrogen atoms on cubium and found results in the range 10^{-3} to 10^{-1}. If the coupling between the adsorbate and the substrate electrons is weak or if the nonadiabatic effects are small the fairly complicated formalism can be simplified substantially [15.9].

GADZUK and METIU [15.11] have taken a different approach and focused on the diabatic curves of the coupled adsorbate-substrate system corresponding to different adsorbate states (Fig.15.3). As the adsorbate approaches the substrate two diabatic curves may cross; there is then a certain probability that the system changes from one diabatic state to another. This can be associated with a large and sudden change of the adsorbate electronic structure, which may excite electron-hole pairs in the substrate. METIU and GADZUK presented a formalism for treating this mechanism but explicit calculations have not been published.

In Sect.15.2 we derive formulas for the function $P_{E'}(E)$, and obtain results in terms of boson excitations. The crucial approximation in Sect.15.2 is discussed in Sect.15.3 and shown to be exact in certain cases. In Sect. 15.4 we discuss the general features of $P_{E'}(E)$ and explicit results are given for certain models. A stochastic description of sticking for nonzero substrate temperatures is given in Sect.15.5. In Sect.15.6 we calculate $P_{E'}(E)$ for the scattering of He from a metal surface using a quantum-mechanical treatment of the nuclear motion.

15.2 The Energy Distribution Function

In this section the adsorbate nuclear motion is treated classically. The probability $P_{E'}(E)$ that an incoming particle with the energy E' loses the energy E during one round trip is calculated. We start from the time-dependent Schrödinger equation and treat the electrons as fermions. Two approximations are introduced which lead to an expression for $P_{E'}(E)$ in terms of linear combinations of excitations that obey boson statistics. This derivation has the advantage that it allows us to see which approximations are needed to obtain a boson description. For systems where the boson treatment would not be accurate enough, the derivation provides a more accurate and complicated formula. In Sect.15.3 we provide justification for the crucial approximation and show that the formalism is exact in certain situations. The treatment closely follows [15.10].

Before the adsorbate approaches the surface the substrate is in the electronic state $|E_m\rangle$ with the probability $p(E_m) = \exp(-E_m/kT_s)/Z$, where T_s is the substrate temperature and Z is the partition function. The time evolution of the states of the substrate is described by a time-dependent Hamiltonian $H(t) = H_0 + V(t)$, where $V(t)$ describes the perturbation due to the adsorbate. As discussed in the introduction, $H(t)$ is assumed to be a one-particle operator. We introduce the states $|\phi_m(t)\rangle$ which describe the time evolution of $|E_m\rangle$. Thus

$$i \frac{\partial}{\partial t} |\phi_m(t)\rangle = H(t)|\phi_m(t)\rangle \tag{15.2}$$

with the initial condition $|\phi_m(-\infty)\rangle = |E_m\rangle$.

The energy distribution function is given by

$$P_{E'}(E) = \sum_{m,n} |\langle E_n|\phi_m(\infty)\rangle|^2 p(E_m)\delta(E-[E_n-E_m]) \equiv p_0\delta(E) + \tilde{P}(E) \tag{15.3}$$

where p_0 is the probability for elastic scattering. The E' dependence enters via $V(t)$. In the following we suppress the E' label. For finite substrate temperatures $P(E)$ contains processes where the adsorbate loses $(E > 0)$ as well as gains $(E < 0)$ energy.

Since we consider slow perturbations, it is convenient to use the adiabatic states $|E_m(t)\rangle$, defined by

$$H(t)|E_m(t)\rangle = E_m(t)|E_m(t)\rangle \quad . \tag{15.4}$$

The time evolution of $|\phi_m(t)\rangle$ can then be described in terms of the amplitudes

$$c_{nm}(t) = \exp[i\theta_n(t)]\langle E_n(t)|\phi_m(t)\rangle \quad , \tag{15.5}$$

where $\theta_n(t) = \int_{-\infty}^{t} E_n(t')\,dt'$. Using the Schrödinger equation (15.2), we find that

$$\frac{\partial c_{nm}(t)}{\partial t} = \sum_{n'} \left\langle \frac{\partial}{\partial t} E_n(t)|E_{n'}(t)\right\rangle \exp\{i[\theta_n(t) - \theta_{n'}(t)]\}c_{n'm}(t) \quad . \tag{15.6}$$

Equation (15.6) can be integrated in time, which leads to an integral equation which can be iterated once, using the zeroth-order result $c_{nm}^{(0)} = \delta_{nm}$ to obtain

$$c_{nm}(t) = \int_{-\infty}^{t} \left\langle \frac{\partial}{\partial t'} E_n(t') | E_m(t') \right\rangle \exp\{i[\theta_n(t') - \theta_m(t')]\} \, dt' + \dots \quad n \neq m \quad . \tag{15.7}$$

We now differentiate $\langle E_n(t)|[H(t) - E_n(t)] = 0$ with respect to time and multiply with $|E_m(t)\rangle$ from the right. This gives

$$\left\langle \frac{\partial}{\partial t} E_n(t) | E_m(t) \right\rangle = [E_n(t) - E_m(t)]^{-1} \langle E_n(t)|\dot{H}(t)|E_m(t)\rangle \tag{15.8}$$

where $\dot{H}(t) = \partial H(t)/\partial t$. Since $\dot{H}(t)$ is a one-particle operator, it can be expressed in terms of the creation and annihilation operators $\psi_\alpha^+(t)$ and $\psi_\alpha(t)$ of the adiabatic one-particle states $|E_\alpha(t)\rangle$,

$$\dot{H}(t) = \sum_{\alpha\alpha'} \langle E_\alpha(t)|\dot{H}(t)|E_{\alpha'}(t)\rangle \psi_\alpha^+(t) \psi_{\alpha'}(t) \quad . \tag{15.9}$$

For a system of noninteracting electrons, the states $|E_m(t)\rangle$ are constructed from the operators $\psi_\alpha^+(t)$ in exactly the same way as the states $|E_m\rangle$ are constructed from the operators $\psi_\alpha^+ \equiv \psi_\alpha^+(t = -\infty)$. Thus

$$\langle E_n(t)|\psi_\alpha^+(t)\psi_{\alpha'}(t)|E_m(t)\rangle = \langle E_n|\psi_\alpha^+\psi_{\alpha'}|E_m\rangle \quad . \tag{15.10}$$

Furthermore, for a semiinfinite substrate the energy of an extended one-particle state is not changed by a localized perturbation, i.e., $E_\alpha(t)$ is independent of t. The energy difference $E_m(t) - E_n(t)$ between two states which are connected by the operator $\psi_\alpha^+(t)\psi_{\alpha'}(t)$ is therefore also independent of time and equal to $E_\alpha - E_{\alpha'}$. This is a drastic simplification, since the difference $\theta_n(t') - \theta_m(t')$ in (15.7) is now trivial to calculate. Combining this result with (15.7,8) we obtain

$$c_{nm}(t = \infty) = \sum_{\alpha\alpha'} v_{\alpha\alpha'} \langle E_m|\psi_\alpha^+\psi_{\alpha'}|E_n\rangle \tag{15.11}$$

where

$$v_{\alpha\alpha'} = \int_{-\infty}^{\infty} \frac{\langle E_\alpha(t)|\dot{H}(t)|E_{\alpha'}(t)\rangle}{E_\alpha - E_{\alpha'}} \exp[i(E_\alpha - E_{\alpha'})t] \, dt \tag{15.12}$$

and an unimportant phase factor has been suppressed. We can then write the inelastic part of P(E) as

$$\tilde{P}(E) = \sum_{m \neq n} |c_{nm}(t = \infty)|^2 p(E_m)\delta[E - (E_n - E_m)]$$

$$= \sum_{m \neq n} \sum_{\alpha_1\alpha_2} \sum_{\alpha_3\alpha_4} v_{\alpha_1\alpha_2} v_{\alpha_3\alpha_4} <E_m|\psi^+_{\alpha_1}\psi_{\alpha_2}|E_n><E_n|\psi^+_{\alpha_3}\psi_{\alpha_4}|E_m>$$

$$\times \delta[E - (E_{\alpha_3} - E_{\alpha_4})]p(E_m)$$

$$= \sum_{\alpha_1\alpha_2} \sum_{\alpha_3 \neq \alpha_4} v_{\alpha_1\alpha_2} v_{\alpha_3\alpha_4} <\psi^+_{\alpha_1}\psi_{\alpha_2}\psi^+_{\alpha_3}\psi_{\alpha_4}> \tag{15.13}$$

where < > denotes a thermal average. For $\alpha_3 \neq \alpha_4$ we have

$$<\psi^+_{\alpha_1}\psi_{\alpha_2}\psi^+_{\alpha_3}\psi_{\alpha_4}> = f(E_{\alpha_1})[1 - f(E_{\alpha_2})]\delta_{\alpha_1\alpha_4}\delta_{\alpha_2\alpha_3} \tag{15.14}$$

where f(E) is the Fermi function. The inelastic part of P(E) is then given by

$$\tilde{P}(E) = \sum_{\alpha_1\alpha_2} |v_{\alpha_1\alpha_2}|^2 f(E_{\alpha_1})[1 - f(E_{\alpha_2})]\delta[E - (E_{\alpha_1} - E_{\alpha_2})] \quad . \tag{15.15}$$

The elastic scattering probability follows from the sum rule $\int P(E)\, dE = 1$, and we obtain

$$P(E) = \frac{1}{2\pi}\int_{-\infty}^{\infty} dt \Big\{1 - \sum_{\alpha_1\alpha_2} |v_{\alpha_1\alpha_2}|^2 f(E_{\alpha_1})[1 - f(E_{\alpha_2})]$$

$$[1 - \exp[-i(E_{\alpha_1} - E_{\alpha_2})t]]\Big\} e^{iEt} \tag{15.16}$$

where we have expressed the delta function as a time integral. Expression (15.16) is valid for weak inelastic scattering, while for large values of $v_{\alpha 1\alpha 2}$ it leads to the physically meaningless result that p_0 is negative. A natural way to avoid this difficulty and to obtain an expression of broader validity is to view the braces in (15.16) as the first two terms of the expansion of an exponential. Then we find

$$P(E) = \frac{1}{2\pi}\int_{-\infty}^{\infty} dt\, \exp\Big\{- \sum_{\alpha_1\alpha_2} |v_{\alpha_1\alpha_2}|^2 f(E_{\alpha_1})[1 - f(E_{\alpha_2})]$$

$$[1 - \exp[-i(E_{\alpha_1} - E_{\alpha_2})t]]\Big\} e^{iEt} \quad . \tag{15.17}$$

This is one of the two main approximations in this approach. Although this approximation may seem fairly questionable, we show in the next section that it is exact in certain cases. For $T_s = 0$, (15.17) is identical to a result we have obtained earlier Ref.15.7, Eq.(2.15) .

In general, the states $|E_\alpha\rangle$ can be labeled by energy and additional quantum numbers σ. Thus we make the replacement $|E_\alpha(t)\rangle \to |E,\sigma,t\rangle$ in the following. We use the fact that the perturbation from the adatom is slow, which means that mainly low-energy electron-hole pairs are created. If $\bar{E}$ is the range over which the matrix elements in (15.12) vary appreciably, we assume that the perturbation is slow on the time scale $\hbar/\bar{E}$ and that $T_s \ll \bar{E}$. Then in (15.12) we can use the states at the Fermi level

$$V_{\alpha_1\alpha_2} \approx \frac{1}{E_1 - E_2}\int_{-\infty}^{\infty} dt \langle E_F\sigma_1 t|\dot{H}(t)|E_F\sigma_2 t\rangle \exp[i(E_1 - E_2)t]$$

$$\equiv \frac{\lambda_{\sigma_1\sigma_2}(E_1 - E_2)}{E_1 - E_2} \quad . \tag{15.18}$$

This is our second major approximation, which is valid only for slow perturbations. In Sect.15.4 we test this approximation numerically. The argument in the exponent in (15.18) now takes the form

$$-\sum_{\sigma_1\sigma_2}\int_{-\infty}^{\infty} dE_1 \int_{-\infty}^{\infty} dE_2 \left|\frac{\lambda_{\sigma_1\sigma_2}(E_1 - E_2)}{E_1 - E_2}\right|^2 f(E_1)[1 - f(E_2)]$$

$$\times \{1 - \exp[-i(E_1 - E_2)t]\}$$

$$= -\int_{-\infty}^{\infty} dE' \sum_{\sigma_1\sigma_2}\left|\frac{\lambda_{\sigma_1\sigma_2}(E')}{E'}\right|^2 A(E')(1 - e^{-iE't}) \quad , \tag{15.19}$$

where

$$A(E) = \int_{-\infty}^{\infty} dE_1 \int_{-\infty}^{\infty} dE_2\, f(E_1)[1-f(E_2)]\delta[E-(E_1-E_2)] = E[1+n(E)] \tag{15.20}$$

with

$$n(E) = 1/[\exp(E/kT_s) - 1] \tag{15.21}$$

being the Bose function. The function $|\lambda_{\sigma_1\sigma_2}(E')|^2$ is an even function of E', and the integral in (15.19) over all E' can be replaced by an integral over positive E' only. This leads to our final result

$$P(E) = \frac{1}{2\pi}\int_{-\infty}^{\infty} dt \exp\Big(-\int_0^{\infty}\frac{dE'}{E'}\sum_{\sigma_1\sigma_2}|\lambda_{\sigma_1\sigma_2}(E')|^2\{1 + 2n(E')$$

$$- [1 + n(E')]e^{-iE't} - n(E')e^{iE't}\}\Big) \quad . \tag{15.22}$$

This is the desired expression for P(E) in terms of boson excitations. From (15.20) we can see that each excitation effectively describes *all* the possible electron-hole pairs of the same energy.

The meaning of the term in (15.22) can be understood if we expand the exponent in powers of $|\lambda_{\sigma_1\sigma_2}(E')|^2$. The term proportional to exp(-iE't) leads to a delta function $\delta(E - E')$, which describes a loss process, where the incoming particle excites a boson. This term has the right factor 1 + n(E) for stimulated emission of bosons. The term proportional to exp(iE't) describes a gain process, where the adsorbate gains energy due to the deexcitatin of a boson. Finally, the factor 1 + 2n(E') in (15.22) is needed to preserve the normalization of P(E).

To obtain a better understanding of (15.22) we consider the case when the state is labelled by energy only, so that no extra quantum numbers σ are needed. We define a phase shift

$$\delta_E(t) = \text{Im}\{\ln (\det[1 - G_0(E)V(t)])\} \tag{15.23}$$

where the Green's function is defined as

$$G_0(E) = (E - i0 - H_0)^{-1} \quad . \tag{15.24}$$

By using the diagonal representation of $G_0(E)$ one can easily see that the unperturbed density of states is given by

$$\rho_0(E) = -\frac{1}{\pi}\frac{d}{dE}\,\text{Im}\{\ln \det G_0(E)\} \tag{15.25}$$

with a similar result for the perturbed density. The perturbed Green's function satisfies a Dyson's equation

$$G(E) = G_0(E) + G_0(E)\ V(t)\ G(E) \quad . \tag{15.26}$$

Thus

$$\Delta\rho(E) = \rho(E)-\rho_0(E) = -\frac{1}{\pi}\frac{d}{dE}\,\text{Im}\{\ln \det G(E)G_0(E)^{-1}\} = \frac{1}{\pi}\frac{\partial}{\partial E}\,\delta_E(t) \quad . \tag{15.27}$$

This result is the justification for calling the quantity in (15.23) a phase shift.

We now use the identity det A = exp(Tr ln A) and differentiate $\delta_E(t)$ with respect to time,

$$\begin{aligned}\dot{\delta}_E(t) &= -\mathrm{Im}\{\mathrm{Tr}([1 - G_0(E)\ V(t)]^{-1}\ G_0(E)\ \dot{V}(t)\} \\ &= -\mathrm{Im}\{\mathrm{Tr}([E - H(t)]^{-1}\ \dot{V}(t))\} \\ &= -\pi\ \mathrm{Tr}\ \delta(E - H(t))\ \dot{V}(t)\} \\ &= -\pi\langle E(t)|\dot{V}(t)|E(t)\rangle \quad . \end{aligned} \tag{15.28}$$

Thus we have

$$\lambda(E) = -\frac{1}{\pi}\int_{-\infty}^{\infty} \dot{\delta}(t)\ e^{iEt}\ dt \tag{15.29}$$

where $\dot{\delta}(t) \equiv \dot{\delta}_{E_F}(t)$, describes the rate of creation of new states below the Fermi level. Since a finite time is needed to fill these states, there are large nonadiabatic effects if $\dot{\delta}(t)$ is large. This is closely related to the intuitive picture [15.4] presented in the introduction (Fig.15.1).

15.3 Derivation of a Boson Formalism

In the previous section we introduced excitations obeying boson statistics. These excitations effectively describe all electron-hole pairs of the same energy, which brings the method of TOMONAGA [15.20] to mind. To justify the step between (15.16) and (15.17) in the previous section we therefore give a *derivation* in the spirit of TOMONAGA for the case when the eigenstates can be labelled by energy only.

Following [15.8] we write the Hamiltonian as

$$H(t) = \int_0^B E\psi_E^+(t)\psi_E(t)\ dE \quad , \tag{15.30}$$

where the energy zero is taken at the bottom of the band, B is the bandwidth, and the Fermi operators $\psi_E^+(t)$ and $\psi_E(t)$ satisfy the anticommutation relations

$$\{\psi_E(t),\ \psi_{E'}^+(t)\} = \delta(E - E') \quad . \tag{15.31}$$

For simplicity, we first discretize the energy variable

$$E_n = nB/N = n\Delta E \qquad (n = 0,1,\dots,N) \quad , \tag{15.32}$$

and later we consider $N\to\infty$. The Hamiltonian is now given by

$$H(t) = \Delta E \sum_{n=0}^{N} n\psi_n^+(t)\psi_n(t) \quad , \tag{15.33}$$

and (15.31) is replaced by

$$\{\psi_n(t), \psi_{n'}^+(t)\} = \delta_{nn'} \quad . \tag{15.34}$$

Following TOMONAGA [15.20] and SCHOTTE and SCHOTTE [15.13] we introduce a coherent particle-hole excitation operator ($n > 0$),

$$b_n(t) = \frac{1}{\sqrt{n}} \sum_{m=0}^{N-n} \psi_m^+(t)\, \psi_{m+n}(t) \quad , \tag{15.35}$$

which annihilates electron-hole pairs with the energy $n\Delta E$. Using (15.34) the commutation relations for $b_n(t)$ can be derived in a straightforward manner,

$$[b_n(t), b_{n'}(t)] = [b_n^+(t), b_{n'}^+(t)] = 0 \tag{15.36}$$

$$[b_n(t), b_n^+(t)] = \frac{1}{n} \sum_{m=0}^{n-1} \psi_m^+(t)\psi_m(t) - \frac{1}{n} \sum_{m=N-n+1}^{N} \psi_m^+(t)\psi_m(t) \quad , \tag{15.37}$$

with a similar relation for $[b_n(t), b_{n'}^+(t)]$. If we only consider states which have no holes in the n lowest states and no particles in the n highest states (Fig.15.4), we can replace (15.37) by

$$[b_n(t), b_{n'}^+(t)] = \delta_{nn'} \quad . \tag{15.38}$$

This is justified in the limit of very slow perturbations, where we are only interested in states with low-energy electron-hole pair excitations and operators $b_n(t)$ and $b_{n'}^+(t)$ with small n and n'.

The commutator of $b_n(t)$ with $H(t)$ can easily be calculated,

$$[b_n(t), H(t)] = n\Delta E\, b_n(t) \quad . \tag{15.39}$$

This indicates that the Hamiltonian can be written as

$$H(t) = \Delta E \sum_{n=1}^{N} n\, b_n^+(t)\, b_n(t) + E_0(t) \tag{15.40}$$

where $E_0(t)$ is the ground-state energy of $H(t)$. This was proved explicitly by TOMONAGA [15.20]. We now want to find a time-dependent equation for $b_n(t)$. We write

$$H(t + dt) = \dot{H}(t) + H(t)\, dt + O(dt^2) \tag{15.41}$$

where

$$\dot{H}(t) = \sum_{m,m'=0}^{N} [\dot{H}(t)]_{mm'} \psi_m^+(t)\psi_{m'}(t) \tag{15.42}$$

Fig.15.4. The commutation relation (15.30) is valid for states of this type if $n,n' \leq 3$

and $[\dot{H}(t)]_{mm'} = \langle E_m(t)|\dot{H}(t)|E_{m'}(t)\rangle$. This matrix element is now replaced by its value for the states at the Fermi level, which is given by the phase shift $-\dot{\delta}(t)/\pi$ (15.28). First we consider the effect of this replacement on the diagonal part $m = m'$ of $\dot{H}(t)$. The approximation is inaccurate only for levels which are sufficiently deep (high) to be occupied (unoccupied) for all states of interest. Thus for the diagonal part this replacement just adds a constant to the Hamiltonian. The approximate diagonal part obtained in this way contains the total number operator, which is a constant of motion. Thus for the states of interest the diagonal part can be treated as a constant. Similarly the replacement of the matrix elements in the nondiagonal part of the Hamiltonian does not change the results for these states. Thus we find

$$\dot{H}(t) = -\frac{\dot{\delta}(t)}{\pi}\sum_{m\neq m'}\psi^+_{m'}(t)\psi_m(t) + \langle E_0(t)|\dot{H}(t)|E_0(t)\rangle \tag{15.43}$$

where the constant is determined by the requirement that the expectation value of $\dot{H}(t)$ in the adiabatic state should be correct. However, this operator can be expressed in terms of the boson operators

$$\dot{H}(t) = -\frac{\dot{\delta}(t)}{\pi}\sum_{n>0}\sqrt{n}[b_n(t) + b^+_n(t)] + \dot{E}_0(t) \tag{15.44}$$

where we have also used the Hellmann-Feynman theorem. We now let $N\to\infty$ and obtain

$$b_E(t) = \int_0^{B-E}\psi^+_{E'}(t)\psi_{E'+E}(t)\,dE' \tag{15.45}$$

$$[b_E(t), b^+_{E'}(t)] = \delta(E - E') \tag{15.46}$$

$$H(t) = \int_0^B E\, b^+_E(t)\, b_E(t)\, dE + E_0(t) \tag{15.47}$$

and

$$H(t + dt) = H(t) - \frac{\dot{\delta}(t)}{\pi}\int_0^B \sqrt{E}[b_E(t) + b^+_E(t)]\, dE + \dot{E}_0(t) \quad . \tag{15.48}$$

We can then extract an equation for $b_E(t)$ because $H(t + dt)$ can also be expressed in the form (15.47) with t replaced by t + dt,

$$b_E(t + dt) = b_E(t) - \dot{\delta}(t)\, dt/(\pi\sqrt{E}) \qquad (15.49)$$

or

$$b_E(t) = b_E - \frac{1}{\pi\sqrt{E}} \int_{-\infty}^{t} \dot{\delta}(t')\, dt' \qquad (15.50)$$

where $b_E = b_E(t = -\infty)$. Using (15.47) we obtain

$$H(t) - E_0(t) = \int_0^B E\, b_E^+ b_E\, dE - \frac{\delta(t)}{\pi} \int_0^B \sqrt{E}(b_E^+ + b_E)\, dE + B\left[\frac{\delta(t)}{\pi}\right]^2 . \qquad (15.51)$$

Thus we have been able to express our Hamiltonian exactly in terms of bosons in the part of the Hilbert space which is relevant for slow perturbations. To obtain the energy distribution we introduce the time evolution operator U(t), which obeys the Schrödinger equation

$$i \frac{d}{dt} U(t) = H(t)U(t) \qquad (15.52)$$

and the Hamiltonian in the Heisenberg representation

$$\tilde{H}(t) = U^+(t)\, H(t)\, U(t) \quad . \qquad (15.53)$$

Then the energy distribution is given by

$$P(E) = \langle\phi_0|\delta(E - [\tilde{H}(\infty) - E_0(\infty)])|\phi_0\rangle \qquad (15.54)$$

where $|\phi_0\rangle$ is the ground state. Thus we need the operators $\tilde{b}_E(t)$ in the Heisenberg picture. They obey the equation of motion

$$i \frac{d\tilde{b}_E(t)}{dt} = E\, \tilde{b}_E(t) - \sqrt{E}\, \frac{\delta(t)}{\pi} \qquad (15.55)$$

with the solution

$$\tilde{b}_E(t) = e^{-iEt}\left[b_E + i \int_{-\infty}^{t} \sqrt{E}\, \frac{\delta(t')}{\pi}\, e^{iEt'}\, dt'\right] \quad . \qquad (15.56)$$

A partial integration gives

$$\tilde{b}_E(t) = e^{-iEt}\left[b_E + \frac{1}{\sqrt{E}} \frac{\delta(t)}{\pi}\, e^{iEt} + \frac{1}{\sqrt{E}}\, \lambda(E,t)\right] \qquad (15.57)$$

where

$$\lambda(E,t) = -\frac{1}{\pi} \int_{-\infty}^{t} \dot{\delta}(t)\, e^{iEt'}\, dt' \quad . \qquad (15.58)$$

Straightforward algebra now yields

$$\tilde{H}(t) - E_0(t) = \int_0^B [E b_E^+ b_E + \sqrt{E}\, b_E^+ \lambda(E,t) + \sqrt{E}\, b_E \lambda^*(E,t) + |\lambda(E,t)|^2]\, dE \quad . \tag{15.59}$$

If the temperature T_s is small, only initial states of the restricted Hilbert space considered above contribute with an appreciable statistical weight and the boson description remains valid. The calculation of P(E) is then straightforward [15.18] (see also [Ref.15.21]) and yields

$$P(E) = \frac{1}{2\pi}\int_{-\infty}^{\infty} dt\, e^{iEt} \exp\left(-\int_0^B \frac{dE'}{E'} |\lambda(E',t=\infty)|^2 \{1 + 2n(E') - [1+n(E')] \times e^{-iE't} - n(E')\, e^{iE't}\}\right) \quad . \tag{15.60}$$

This is identical to our result (15.22).

Thus we have shown that in the limit of a slow perturbation, (15.22) in Sect.15.2 is exact if the states can be labelled by energy only. If additional quantum numbers σ are needed, (15.22) is in general not exact. However, it can be shown that it gives the two first moments exactly for slow perturbations. Equation (15.17) is in addition exact for weak perturbations with an arbitrary time dependence. Thus we expect (15.17) to be of broader validity than (15.22). Actually, comparisons to exact numerical calculations show that for strong and rapid perturbations (15.17) often gives a qualitatively correct behavior, while (15.22) in this limit fails [15.9].

15.4 General Features of the Energy Distribution Function

In this section we discuss some general aspects of P(E) in (15.22). First we notice that the probability for elastic scattering is given by

$$p_0 = \exp\left\{-\sum_{\sigma_1\sigma_2}\int_0^{\infty} dE \frac{|\lambda_{\sigma_1\sigma_2}(E)|^2 [1 + 2n(E)]}{E}\right\} \quad . \tag{15.61}$$

For the case of an adsorbate hitting the surface and leaving again the interaction is switched off for large times. A partial integration of (15.18) shows that in this case $\lambda_{\sigma_1\sigma_2}(E) \sim E$ for small E and the integral in (15.61) converges. This is in contrast to, for instance, the case of X-ray photoemission where the interaction is not switched off. Then $\lambda_{\sigma_1\sigma_2}(E) \to$ constant for small E and the integral in (15.61) diverges. Thus, in this case $p_0 = 0$ and P(E) instead exhibits the well-known infrared singularity $P(E) \sim E^{\alpha-1}$ for small

E [15.14]. MÜLLER-HARTMANN et al. [15.18] emphasized the relation between p_0 and the time scale T, over which the perturbation is switched on to a given value and switched off again. In particular they considered the limit of large T. Let us write

$$<E_1\sigma_1 t|\dot{H}(t)|E_2\sigma_2 t> = \frac{1}{T}\, f(t/T) \quad . \tag{15.62}$$

A variable substitution in (15.18) gives

$$\lambda_{\sigma_1\sigma_2}(E) = g_{\sigma_1\sigma_2}(ET) \tag{15.63}$$

for some $g_{\sigma_1\sigma_2}(E)$. From (15.61) we can see that for $T_s = 0$ the elastic scattering probability is independent of how slowly the perturbation is varied. For the substrate temperature $T_s > 0$ we even find that $p_0 \to 0$ as $T \to \infty$. Thus, as stressed by MÜLLER-HARTMANN et al. [15.18], the process is never strictly adiabatic since the probability p_0 that the substrate remains in its ground state is always less than one. On the other hand, for $T_s = 0$ the width of the distribution goes to zero as $T \to \infty$. Thus in this limit the substrate always ends up in a state which in energetically very close to the ground state.

From (15.22) we can also calculate the first moment

$$\mu_1 \equiv \int_{-\infty}^{\infty} EP(E)\; dE = \sum_{\sigma_1\sigma_2} \int_0^{\infty} dE|\lambda_{\sigma_1\sigma_2}(E)|^2 \tag{15.64}$$

and the second moment

$$\mu_2 = \int_{\infty}^{\infty} E^2P(E)\; dE = \sum_{\sigma_1\sigma_2} \int_0^{\infty} dE\; E\; |\lambda_{\sigma_1\sigma_2}(E)|^2[1 + 2n(E)] + \mu_1^2 \tag{15.65}$$

of P(E). In Fig.15.5 a typical example of P(E) is shown. As the substrate temperature T_s is increased the elastic scattering probability p_0 is decreased. The weight goes into gain as well as loss processes in such a way that μ_1 stays constant, while μ_2, of course, increases.

We shall now discuss a model which includes the mechanism emphasized by NØRSKOV and LUNDQVIST [15.4], namely that a level on the adsorbate crosses the Fermi level as the adsorbate approaches the surface,

$$H(t) = \sum_k E_k n_k + E_a[Z(t)]n_a + \sum_k V_{ak}[Z(t)]\psi_a^+\psi_k + \text{H.c.} \tag{15.66}$$

This Hamiltonian includes the substrate levels E_k, the adsorbate level E_a, and the coupling between them V_{ak}. The adsorbate follows a prescribed classical trajectory Z(t), which determines the time dependence of H(t). It is useful to introduce

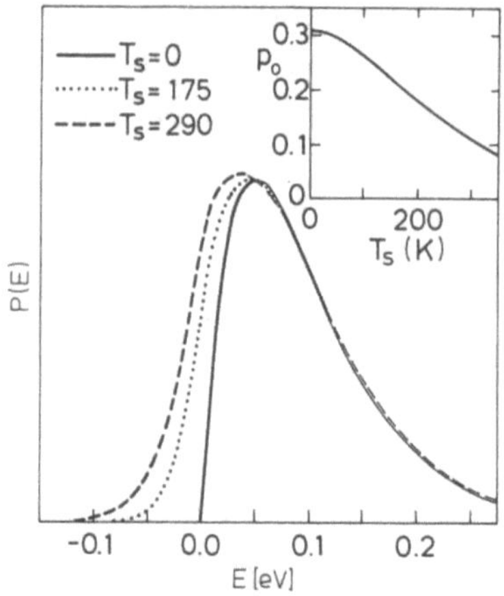

Fig.15.5. The energy distribution function P(E) according to (15.22,69,70). We have used $\Delta t = 20\ eV^{-1}$ and $T = 30\ eV^{-1}$. The curves show P(E) for $T_s = 0$ K (———) $T_s = 175$ (·····) and $T_s = 290$ (-----). The insert shows the elastic scattering probability p_0 as a function of the substrate temperature T_s

$$\Gamma(t) = \sum_k \frac{|V_{ak}(Z(t))|^2}{E_F - i0 - E_k} \tag{15.67}$$

which leads to

$$<E_F t|\dot{H}(t)|E_F t> = \frac{1}{\pi}\frac{d}{dt} \arctan \frac{\mathrm{Im}\{\Gamma(t)\}}{E_F - E_a[Z(t)] - \mathrm{Re}\{\Gamma(t)\}} \quad . \tag{15.68}$$

The matrix element tends to be large if $|E_F - E_a[Z(t)] - \mathrm{Re}\{\Gamma(t)\}| \lesssim \mathrm{Im}\{\Gamma(t)\}$. If the adsorbate level crosses the Fermi level at time t_0 when the adsorbate moves towards the surface and at $T + t_0$ when it moves out, we can use the approximation

$$<E_F t|\dot{H}(t)|E_F t> = f(t) - f(t + T) \tag{15.69}$$

where

$$f(t) = -\frac{1}{\pi}\frac{\Delta t/2}{(t - t_0)^2 + (\Delta t/2)^2} \tag{15.70}$$

and Δt is the time the adsorbate resonance needs to cross the Fermi level. This leads to

$$|\lambda(E)|^2 = 2(1 - \cos ET)e^{-E\Delta t} \quad . \tag{15.71}$$

For $T_s = 0$ we can calculated P(E) analytically and obtain [15.8]

$$P(E) = \frac{(\Delta t)^2}{(\Delta t)^2 + T^2}[\delta(E) + T^2 E\ e^{-E\Delta t}] \quad . \tag{15.72}$$

The inelastic part of this expression was later also obtained by NØRSKOV [15.22] in a semiclassical calculation. The first moment of (15.72) is given by

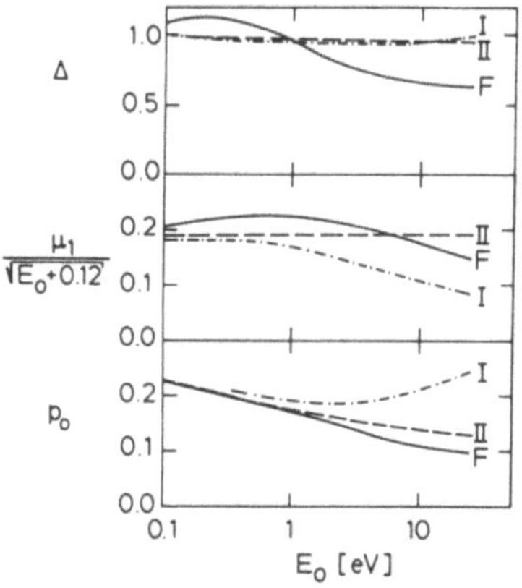

Fig.15.6. Results for an H_2 molecule for the elastic scattering probability p_0, the center of gravity μ_1 of P(E), and the relative width Δ. The exact numerical calculation is denoted F and the approximate calculations I (15.17) and II (15.22). The adsorbate initial energy is given by E_0; we have used $dE_a(Z)/dZ = 0.25$ eV/a_0. For the details of the model see [15.9]

$$\mu_1 = \frac{2}{\Delta t}\frac{1}{1+(\Delta t/T)^2} \quad . \tag{15.73}$$

For $T_s \neq 0$ we calculate P(E) numerically and find the results [15.10] shown in Fig.15.5. For $T_s = 0$ and large values of T the relative width Δ of P(E) is given by

$$\Delta \equiv (\mu_2 - \mu_1^2)^{\frac{1}{2}}/\mu_1 = 1/\sqrt{2} \quad , \tag{15.74}$$

and for $T_s \neq 0$ Δ is even larger. Actually, we generally find for the electronic mechanism that $\Delta \gtrsim 1$ if the coupling is monotonously switched on and then monotonously switched off again [15.8]. If the adsorbate moves parallel to the surface so that the coupling goes through many maxima this argument is, however, not valid.

Since the model (15.66) can also be solved exactly using straightforward numerical integration, we can check the accuracy of our boson approximations. Figure 15.6 shows results for a case where Z(t) describes how the adsorbate is accelerated by the attractive well and then reflected as it hits the cores of the substrate. We have assumed that $E_a(Z)$ varies linearly with Z and crosses the Fermi level when the adsorbate approaches the surface. At the same time $V_{ak}(Z)$ grows exponentially. If we use a coupling of the form $V_{ak}(Z) = V(Z)A_k$, the states in this model can be labelled by energy only, so that the only approximation in the boson model is the assumption that the perturbation is slow. This is illustrated in Fig.15.6 which shows that both (15.17) and (15.22) approach the exact results when the initial adsorbate energy is reduced [15.9]. In addition, these approximations remain good for adsorbate energies which are larger than what we are usually interested in.

Finally, we discuss semiquantitatively the case when the adsorbate level stays above the Fermi level. First the average energy transfer (15.64) is rewritten as

$$\mu_1 = \frac{1}{\pi}\int_{-\infty}^{\infty} [\dot{\delta}(t)]^2\, dt \tag{15.75}$$

where Parseval's theorem and (15.29) have been used. The phase shift $\delta(t)$ is assumed to grow linearly to a maximum value $\delta_{max} = \pi\,\Delta n$ over a time T, where n is the occupancy of the adsorbate level, and (15.27) has been used. The adsorbate is assumed to move with a velocity $v = \sqrt{2E/M}$, where E is an estimate of the kinetic energy of the adsorbate in the potential well. We now relate T to a typical distance d via T = dv, which gives

$$\Delta E = \pi\,\Delta n^2\, 2\,\frac{1}{d}\sqrt{\frac{2E}{M}} \quad . \tag{15.76}$$

Similar estimates have been obtained by, for instance, d'AGLIANO et al. [15.3] and NOURTIER [15.3]. If we express the energies in eV, the distance in a_0, and use the proton mass as unit for M, we find

$$\Delta E \approx \frac{1}{d}\sqrt{\frac{E}{M}}\,(\Delta n)^2 \quad . \tag{15.77}$$

It is interesting to compare this with a very simple model of a phonon mechanism, where the adsorbate is assumed to hit a substrate atom with the mass M_s in a head-on collision. Neglecting the interaction between the substrate atoms we obtain in the limit $M \ll M_s$

$$\Delta E_{phonon} \approx 4\,\frac{M}{M_s}\,E \quad . \tag{15.78}$$

Comparing (15.77) and (15.78) we can see that the electronic mechanism should be of importance for light (M small), chemically active (Δn large) atoms for not too large energies E.

15.5 Stochastic Description of the Sticking Process

For zero substrate temperature T_s, P(E) alone determines the sticking probability. For $T_s > 0$, however, there is a possibility that the adsorbate loses its initial kinetic energy during the first round trip in the adsorbate well but that at a later time, before it has fallen all the way to the bottom of the adsorption well, it regains enough energy from the substrate to leave the surface. To describe this situation we follow ICHE and NOZIERES ([15.17], also [15.10]). Thus we assume that there is no statistical correlation between two oscillations in the potential well, i.e., that the sticking process is Markovian. This assumption is discussed in Appendix A. We consider the case when kT_s is much smaller than the depth of the potential well. Then we can treat the potential well as being infinitely deep and even neglect that an adsorbate which has arrived at the bottom of the well will eventually be given enough energy from the substrate to leave the surface. In addition, we assume that the initial kinetic energy is small compared with the depth of the well. Then the adsorbate is close to the top of the well for all processes of interest, and the classical trajectory describing the adsorbate is essentially independent of the energy E' of the adsorbate. When this approximation breaks down, the adsorbate is so deep in the well that it is unlikely that it can leave the surface. Thus we take P(E) to be independent of E'.

Let a particle approach the surface with the energy E'. After one round trip it has the energy E with the probability P(E'-E). If $E > 0$, the adsorbate leaves the system, and if $E < 0$, it starts a new round trip with the sticking probability S(E). The function S(E) obeys the integral equation [15.17]

$$S(E') = \int_{-\infty}^{0} P(E' - E)S(E)\, dE \tag{15.79}$$

which is solved with the boundary condition $S(E) \to 1$ when $E \to -\infty$. In (15.79) the assumption that the potential well is infinitely deep is used once more. We introduce

$$r(E') = 1 - S(E') \quad , \quad E' < 0 \quad , \tag{15.80}$$

which obeys an inhomogeneous integral equation

$$r(E') = \int_{0}^{\infty} P(E' - E)\, dE + \int_{-\infty}^{0} P(E' - E)r(E)\, dE \qquad E' < 0 \tag{15.81}$$

with the boundary condition $r(E') \to 0$ when $E' \to -\infty$. The numerical solution of this equation is straightforward.

Figure 15.7 shows a typical example of the sticking probability as a function of T_s for two different values of the initial energy E_{in} [15.10]. For E_{in}=0.005 the dependence on T_s is quite weak. As T_s is increased the probability that the adsorbate loses at least the energy E_{in} increases due to the factor 1+n(E') in (15.22), determining the probability for stimulated emission. On the other hand, there is also an increase in the probability that an initially trapped adsorbate later obtains enough energy to leave the surface

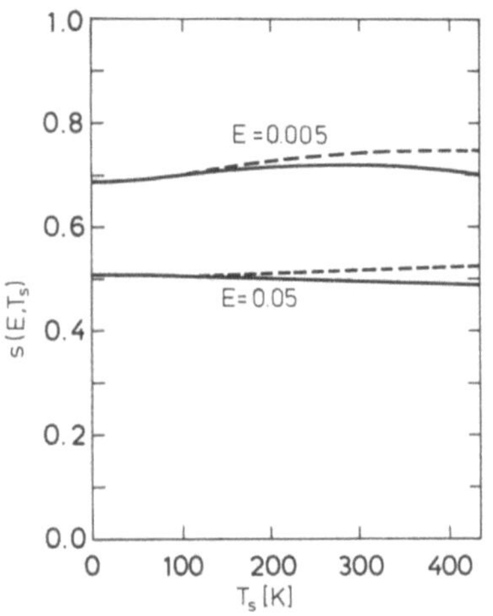

Fig.15.7. The sticking probability S(E) as a function of the substrate temperature T_s, obtained from the energy distribution P(E) in Fig.15.5. We have considered the initial kinetic energies 0.005 eV and 0.05 eV. The solid curve shows the calculated value for S(E), and the dashed curve the probability that the adsorbate is trapped after the first round trip. The difference between the two curves gives the probability that a particle which is initially trapped later obtains enough energy to leave the substrate

again. In Fig.15.7 these two effects largely cancel each other [15.10]. However, if the elastic scattering probability is small the latter effect tends to dominate and the sticking probability tends to decrease with T_s.

15.6 Quantum-Mechanical Treatment of the Adsorbate Motion

In the previous section we have assumed that the adsorbate nucleus follows a classical trajectory. In this section we use a quantum-mechanical treatment of the nuclear motion. If the system is close to the adiabatic limit, it is convenient to start from the adiabatic electronic eigenstates and calculate the nonadiabatic corrections. Such a formalism was given by BORN and HUANG [15.19] and has been applied to the sticking problem by BRIVIO and GRIMLEY [15.5]. Here we essentially follow [15.9].

We assume that the ions of the substrate are rigid and write the Hamiltonian as

$$H = -\frac{\hbar^2}{2m}\sum_i \nabla_i^2 + V(r,R) - \frac{\hbar^2}{2M}\nabla_{\underset{\sim}{R}}^2 \equiv H_E + T_N \tag{15.82}$$

where $\underset{\sim}{r}_i$ is the coordinate of electron i, $\underset{\sim}{R}$ the coordinate of the adsorbate nucleus, and $\underset{\sim}{r} = (\underset{\sim}{r}_1,\underset{\sim}{r}_2,\dots)$. The electron and nuclear masses are given by m and M, respectively. We first treat $\underset{\sim}{R}$ as a parameter and calculate the electronic eigenfunctions $\phi_m(\underset{\sim}{r},\underset{\sim}{R})$ and eigenvalues $E_m(\underset{\sim}{R})$ of H_E for fixed $\underset{\sim}{R}$. These quantities are then used to expand the wave function of the combined electronic and nuclear system

$$\chi(\underset{\sim}{r},\underset{\sim}{R}) = \sum_m \psi_m(\underset{\sim}{R})\phi_m(\underset{\sim}{r},\underset{\sim}{R}) \quad . \tag{15.83}$$

This leads to an equation for $\psi_m(\underset{\sim}{R})$,

$$[T_N + E_n(\underset{\sim}{R}) - E]\psi_n(\underset{\sim}{R}) = \sum_m A_{nm}(\underset{\sim}{R},\nabla_{\underset{\sim}{R}})\psi_m(\underset{\sim}{R}) \tag{15.84}$$

where

$$A_{nm}(\underset{\sim}{R},\nabla_{\underset{\sim}{R}}) = \frac{\hbar^2}{2M} \int d^3r\phi_n^*(\underset{\sim}{r},\underset{\sim}{R})\left[\nabla_{\underset{\sim}{R}}^2\phi_m(\underset{\sim}{r},\underset{\sim}{R}) - \phi_m(\underset{\sim}{r},\underset{\sim}{R})\nabla_{\underset{\sim}{R}}^2\right] \quad . \tag{15.85}$$

We introduce a Green's function

$$\left[-\frac{\hbar^2}{2M}\nabla_{\underset{\sim}{R}}^2 + E_n(\underset{\sim}{R}) - E\right]G_n(\underset{\sim}{R},\underset{\sim}{R}',E) = \delta(\underset{\sim}{R} - \underset{\sim}{R}') \tag{15.86}$$

and impose outgoing scattered wave boundary conditions. For the inelastic part this yields

$$\psi_n(\underset{\sim}{R}) = \sum_m \int d^3\underset{\sim}{R}'G_n(\underset{\sim}{R},\underset{\sim}{R}',E)A_{n,m}(\underset{\sim}{R}')\psi_m(\underset{\sim}{R}') \quad . \tag{15.87}$$

From the asymptotic behavior of $\psi_n(\underset{\sim}{R})$ we can obtain the $m \rightarrow n$ inelastic scattering probability. This equation is in general very hard to solve. It is, however, drastically simplified if the inelastic scattering is weak so that it is sufficient to solve (15.87) iteratively in the spirit of the distorted wave Born approximation. First we find the $n = 0$ solution $\psi_0^0(\underset{\sim}{R})$ of (15.84), neglecting the term A_{nm} and imposing the appropriate boundary condition, and then (15.87) is iterated once,

$$\psi_n^{(1)}(\underset{\sim}{R}) = \int d^3\underset{\sim}{R}'G_n(\underset{\sim}{R},\underset{\sim}{R}',E)A_{n,0}(\underset{\sim}{R}')\psi_0^0(\underset{\sim}{R}') \quad . \tag{15.88}$$

The second major complication is the calculation of $A_{nm}(\underset{\sim}{R})$, which requires the calculation of the adiabatic wave functions as a function of $\underset{\sim}{R}$. If the adsorbate can be viewed as a fairly weak pseudopotential, we may formally treat it in lowest order perturbation theory. If the electrons are treated as noninteracting, the one-particle orbitals are given by

$$\phi_\mu(\underset{\sim}{x},\underset{\sim}{R}) = \phi_\mu(\underset{\sim}{x}) + \sum_{\nu\neq\mu} \frac{\phi_\mu(\underset{\sim}{x}) \int d^3x'\phi_\mu^*(\underset{\sim}{x}')v(\underset{\sim}{x}' - \underset{\sim}{R})\phi_\nu(\underset{\sim}{x}')}{E_\mu - E_\nu} \tag{15.89}$$

where $v(\underset{\sim}{x})$ is the adsorbate pseudopotential. Since $v(\underset{\sim}{x})$ often is of fairly short range, higher order effects can be included through the replacement

$$\int d^3x\phi_\mu^*(\underset{\sim}{x})v(\underset{\sim}{x} - \underset{\sim}{R})\phi_\nu(\underset{\sim}{x}) \rightarrow \phi_\mu^*(\underset{\sim}{R})\phi_\nu(\underset{\sim}{R})t_0 \tag{15.90}$$

where t_0 is the t matrix of the pseudopotential for a plane wave with $\underset{\sim}{k} = 0$. The calculation of $A_{nm}(\underset{\sim}{R})$ is then straightforward.

To compare a classical and quantum-mechanical treatment we use a simple model where the adiabatic potential-energy curves are approximated by

$$E_n(\underset{\sim}{R}) = \begin{cases} E_n & Z > Z_0 \\ \infty & Z < Z_0 \end{cases} \quad . \tag{15.91}$$

We consider the probability of scattering an electron from the state $|E_\mu>$ to the state $|E_\nu>$, where the wave functions in the region $Z > Z_0$ are given by

$$\phi_i(\underset{\sim}{x}) = \exp(-K_i Z)\exp(i\underset{\sim}{k}{}^i_{\|}\underset{\sim}{x}_{\|}) \quad , \quad i = \mu,\nu \quad . \tag{15.92}$$

In the classical treatment we obtain

$$P^{Cl}_{\mu\to\nu} \sim \left[\frac{\hbar\alpha v^i_\perp}{(\hbar\alpha v^i_\perp)^2 + (E - \hbar\underset{\sim}{k}_{\|}\underset{\sim}{v}{}^i_{\|})^2}\right]^2 \tag{15.93}$$

where E and $\underset{\sim}{k}_{\|}$ is the energy and momentum transferred to the electron, $\alpha = K_\mu + K_\nu$, and $v^i_\perp(v^i_{\|})$ is the perpendicular (parallel) velocity of the adsorbate in the initial state. In this classical treatment we could also use the velocities $v^f_\perp$ and $v^f_{\|}$ in the final state. The quantum-mechanical treatment gives

$$P^{QM}_{\mu\to\nu} \sim \frac{v^f_\perp}{v^i_\perp}\left[\frac{\hbar\alpha v^i_\perp}{\left(\frac{\hbar\alpha^2}{2M}\right)^2 + \frac{1}{2}\hbar^2\alpha^2(v^{i2}_\perp + v^{f2}_\perp) + \left[\varepsilon - \frac{1}{2}\hbar\underset{\sim}{k}_{\|}\cdot(\underset{\sim}{v}{}^i_{\|} + \underset{\sim}{v}{}^f_{\|})\right]^2}\right]^2 \quad . \tag{15.94}$$

If the energy and momentum transfers are small the classical description is valid if the first term in the denominator can be neglected, i.e., if

$$v_\perp \gg \frac{\hbar\alpha}{2M} \tag{15.95}$$

or if

$$\lambda \equiv \frac{2\pi\hbar}{Mv_\perp} << \frac{4\pi}{\alpha} \quad . \tag{15.96}$$

Thus the classical description is valid if the adsorbate wave function has many oscillations over the range where the particle interacts with the substrate. For instance, for a helium atom with normal incidence and the energy 0.025 eV,

$$\frac{2Mv_\perp}{h\alpha} \sim 6 \tag{15.97}$$

if $\underset{\sim}{k}_{\|} = 0$ and the condition (15.95) is fairly well satisfied. The classical treatment becomes even better if the nuclear mass is larger or the kinetic energy is larger. In Fig.15.8 the classical and quantum- mechanical treatment of the system mentioned above are compared [15.9].

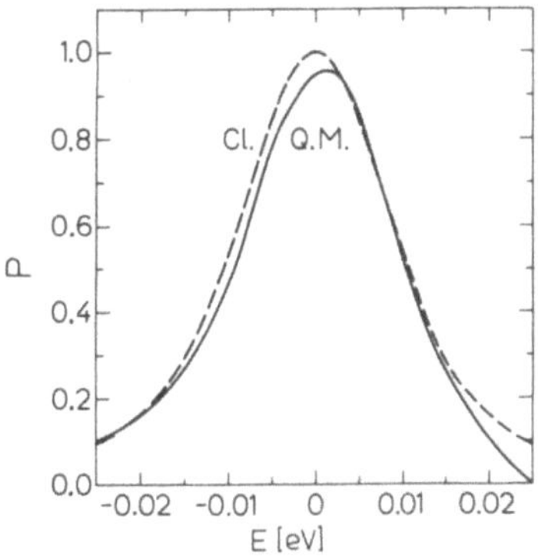

Fig.15.8. The probability of exciting an electron-hole pair with energy E as a function of E. The results of a quantum mechanical (Q.M) (15.94) and a classical (Cl.) (15.93) treatment of the nuclear motion are compared. We have considered a helium atom with the initial energy 25 meV and assumed normal incidence without parallel momentum transfer. The curves have been normalized so that the classical result for E = 0 is unity

We have also considered a more realistic treatment of the scattering of rare-gas atoms from metal surfaces. We use the calculation of ZAREMBA and KOHN [15.23] to obtain the adiabatic potential-energy curve for He interacting with Cu. The electronic structure of Cu is described in the step barrier model, where the electrons are assumed to feel the potential

$$v(\underset{\sim}{x}) = \begin{cases} 0 & Z < 0 \\ E_F + \phi & Z > 0 \end{cases} \qquad (15.98)$$

where E_F is the Fermi energy and ϕ is the work function. The He-electron t matrix was obtained from the work of JORTNER et al. [15.24]. We considered He atoms with the energy 22.6 meV, an incident angle of 65°, and a substrate temperature of 15 K as in the work of MASON and WILLIAMS [15.1]. The inelastic scattering probability is found [15.9] to be 10^{-5}, which is much smaller than the experimental value. We have also obtained results for the energy accomodation coefficient

$$\alpha = \frac{E_r - E_i}{E_s - E_i} \qquad (15.99)$$

where E_i is the energy of the incident atoms, E_r the average energy of the scattered atoms, and $E_s = k_B T_s$ the energy of atoms in thermal equilibrium with the surface. This leads to $\alpha \approx 10^{-6}$, which is about four orders of magnitude smaller than experiment [15.2]. Thus the electronic mechanism should be of little importance for scattering of rare-gas atoms.

15.7 Summary

In this chapter we have reviewed methods for calculating the energy transfer between an adsorbate and the electron-hole pair excitation of the substrate. GADZUK and METIU [15.11] have presented an approach based on diabatic states. This should be good starting point when the nonadiabatic effects are very strong. In most other treatments it is assumed that the nonadiabatic effects are, in a certain sense, not too large. For instance, a quantum-mechanical treatment based on the BORN-HUANG [15.19] formalism has been used by BRIVIO and GRIMLEY [15.5] and by us [15.9]. Similary, when the adsorbate nuclear motion is treated classically, the nuclear trajectory is usually calculated from the adiabatic energy surface. In theories assuming a classical treatment of the adsorbate nucleus the fact that in important cases the substrate excitations can be treated as bosons [15.6-10,18] is a significant simplification. We have shown that this approach is exact for slow perturbations if the levels of the substrate can be labelled by energy only [15.8]. In most cases the assumption about a slow perturbation should be fairly well justified [15.9].

In the boson formalism the nonadiabatic effects are directly related to $\dot{\delta}(t)$, the time derivative of the phase shift at the Fermi level due to the perturbation, which gives the rate of creation of new states below the Fermi level in the adiabatic state. If this rate is high the system may not have time to adjust to the changes in the adiabatic state and nonadiabatic effects are important. This should in particular be the case if an adsorbate level crosses the Fermi level when the adsorbate approaches the surface.

NØRSKOV and LUNDQVIST [15.4] have argued that this effect is very important for the adsorption of halogen molecules on, e.g., sodium. In this case the adsorbate resonance should be very narrow when the Fermi level is crossed which would lead to rapid changes in the adiabatic states. The opposite extreme is a rare-gas atom on a surface, for which the adsorbate levels are far from the Fermi level. We have shown that for, e.g., He on a metal surface the electronic mechanism gives a contribution which is several orders of magnitude smaller than the experimental results. In this case the phonon mechanism should be the dominanting one [15.9]. Other adsorbate systems are probably in between these extremes. For instance, for molecules such as CO, N_2, and H_2 the interaction between the substrate and the adsorbate leads to a partial filling of the adsorbate affinity level. Model calculations [15.8,22] indicate that the electronic mechanism is of importance for such light, chemically reactive adsorbates. In these cases it may be necessary to consider both the electronic and phonon mechanisms.

Appendix A

In Sect.15.5 we assumed that there is no statistical correlation between two round trips in the potential well. To test this assumption we let the particle follow a classical trajectory which leads to two round trips. If these round trips are independent the energy distribution function should be the self-convolution of the function P(E) for one round trip. We use the same model as in Sect.15.4 (15.69) and assume

$$\langle E_F t|\dot{H}(t)|E_F t\rangle = f(t) - f(t+T) + f(t+T+\tau) - f(t+2T+\tau) \tag{15.100}$$

where f(t) is given by (15.70). This equation describes how the adsorbate level goes through the Fermi level at the times t_0 and t_0+T during the first round trip and at the times $t_0+T+\tau$ and $t_0+2T+\tau$ during the second round trip. The perturbation is essentially switched off during a time between the two round trips when the adsorbate is far from the surface.

It is straightforward to show that the energy distribution function is the self-convolution of (15.72) if $\tau \gg T$. The elastic scattering probability, for instance, is given by

$$P_0 = \left[\frac{(\Delta t)^2}{T^2 + (\Delta t)^2}\right]^2 \frac{[(\tau + T)^2 + (\Delta t)^2]^2}{[(\tau + 2T)^2 + (\Delta t)^2]\,[\tau^2 + (\Delta t)^2]} \quad . \tag{15.101}$$

For $\tau \gg T$ this is the square of the elastic scattering probability for one round trip (15.72). Thus, if τ is sufficiently large the system loses its memory of the previous round trip, and the round trips are statistically independent. The assumption $\tau \gg T$ may be reasonable if the adsorbate is close to the top of the potential well. Then τ is determined by the motion far from the surface, where the kinetic energy is very small, while T is determined by the faster motion when the adsorbate interacts strongly with the adsorbate.

References

15.1 J.M. Horne, D.R. Miller: Phys. Rev. Lett. *41*, 511 (1978);
G. Brusdeylins, R.B. Doak, J.P. Toennies: Phys. Rev. Lett. *44*, 1417 (1980); *46*, 437 (1981);
W. Allison, B. Feuerbacher: Phys. Rev. Lett. *45*, 2040 (1980);
B.F. Mason, B.R. Williams: Phys. Rev. Lett. *46*, 1138 (1981) [and references therein]
15.2 R.M. Logan: In *Solid State Surface Science*, Vol.3 (Decker, New York 1973) p.1;
F.O. Goodman, H.Y. Wachman: *Dynamics of Gas-Surface Scattering* (Academic, New York 1976)
15.3 E.G. d'Agliano, P. Kumar, W. Schaich, H. Suhl: Phys. Rev. B*11*, 2122 (1975);
W.L. Schaich: Surf. Sci. *49*, 221 (1975);
K.P. Bohnen, M. Kiwi, H. Suhl: Phys. Rev. Lett. *34*, 1512 (1975);
A. Blandin, A. Nourtier, D.W. Hone: J. Phys. Paris *37*, 369 (1976);
A. Noutier: J. Phys. Paris *38*, 479 (1977)
15.4 J.K. Nørskov, B.I. Lundqvist: Surf. Sci. *89*, 251 (1979)
15.5 G.P. Brivio, T.B. Grimley: Surf. Sci. *89*, 226 (1980); J. Phys. C*10*, 2351 (1977)
15.6 R. Brako, D.M. Newns: Solid State Commun. *33*, 713 (1980)
15.7 K. Schönhammer, O. Gunnarsson: Z. Phys. B*38*, 127 (1980)
15.8 K. Schönhammer, O. Gunnarsson: Phys. Rev. B*22*, 1629 (1980)
15.9 O. Gunnarsson, K. Schönhammer: Phys. Rev. (in press)
15.10 K. Schönhammer, O. Gunnarsson: Phys. Rev. B*24*, 7084 (1981)
15.11 J.W. Gadzuk, H. Metiu: Phys. Rev. B*22*, 2603 (1980);
H. Metiu, J.W. Gadzuk: J. Chem. Phys. *74*, 2641 (1981)
15.12 R.L. Palmer, H. Saltsburg, J.N. Smith: J. Chem. Phys. *50*, 4661 (1969);
F. Frenkel, J. Häger, W. Krieger, H. Walther, C.T. Campbell, G. Ertl, H. Kuipers, J. Segner: Phys. Rev. Lett. *46*, 152 (1981)
15.13 P. Nozières, C.T. de Dominicis: Phys. Rev. *178*, 1097 (1969);
K.D. Schotte, U. Schotte: Phys. Rev. *182*, 479 (1969); *185*, 509 (1969)
15.14 S. Doniach, M. Sunjić: J. Phys. C*3*, 285 (1970)
15.15 N.D. Lang, A.R. Williams: Phys. Rev. Lett. *37*, 212 (1976);
O. Gunnarsson, H. Hjelmberg, J.K. Nørskov: Phys. Scr. *22*, 165 (1980)
15.16 B. Kasemo, E. Törnqvist, J.K. Nørskov, B.I. Lundqvist: Surf. Sci. *89*, 554 (1979);
J.K. Nørskov, D.M. Newns, B.I. Lundqvist: Surf. Sci. *80*, 179 (1979) [and references therein]
15.17 G. Iche, P. Nozières: J. Phys. Paris *37*, 1313 (1976)
15.18 E. Müller-Hartmann, T.V. Ramakrishnan, G. Toulouse: Phys. Rev. B*3*, 1102 (1971)
15.19 M. Born, K. Huang: *Dynamical Theory of Crystal Lattices* (Clarendon, Oxford 1954)
15.20 S. Tomonaga: Prog. Theor. Phys. *5*, 544 (1950)
15.21 D.C. Langreth: Phys. Rev. B*1*, 471 (1970)
15.22 J.K. Nørskov: J. Vac. Sci. Technol. *18*, 420 (1981)
15.23 E. Zaremba, W. Kohn: Phys. Rev. B*15*, 1769 (1977)
15.24 J. Jortner, N.R. Kestner, S.A. Rice, M.H. Cohen: J. Chem. Phys. *43*, 2614 (1965)

16. Thermal Desorption

D. Menzel

With 9 Figures

6.1 Introduction

Thermal desorption, the removal of adparticles from a surface by thermal excitations induced by heating the substrate, is one of the basic surface kinetic processes. It has therefore been investigated in great detail since the days of Langmuir. In most of this work, the integrated effect of desorption has been examined. Recently, the use of molecular beams both as source and probe has led to additional differential information. While this is a very promising avenue for more detailed microscopic understanding, the present survey will be limited to the classical type of integral experiments. It will be shown that even here new and interesting possibilities arise from better experimentation and careful interpretation.

The same subject has been reviewed before by this author about 7 years ago [16.1]. The present survey will be based on this review and try to update it. No detailed description of the basic facts and approaches, therefore, will be given here; the reader is referred to [16.1] for this purpose. Some of the aspects discussed in the following have also been treated by KING [16.2] and in a recent review by YATES [16.3].

16.2 Critical Examination of the Usual Procedures

16.2.1 Short Description

Thermal desorption can be viewed as a simple chemical reaction where one or more particles A bound to surface sites S are removed into the gas phase to leave empty surface sites behind:

$$\text{S-A} \rightarrow \text{S} + \text{A}_{\text{gas}} \quad \text{or} \quad 2\ \text{S-A} \rightarrow 2\text{S} + \text{A}_{2\text{gas}} \tag{16.1}$$

or more complicated. The rate of such processes can be described by the change of particle number n on the surface or in the gas phase which must be equal by mass balance

$$- \frac{dn_{ad}}{dt} = \frac{dn_{gas}}{dt} \quad . \tag{16.2}$$

Usually the desorption rate R_d is normalized to the surface area A; it can also be expressed in terms of fractional coverage Θ,

$$- \frac{dn_{ad}}{A\,dt} = - \frac{dN_{ad}}{dt} = R_d = - N_s \frac{d\Theta}{dt} \quad . \tag{16.3}$$

where N_s is the number of sites per area and Θ is that of adparticles per available surface sites (e.g., adatoms/surface metal atoms; but other normalizations can of course also be used).

As the usual procedure of chemical kinetics, isothermal measurement, is difficult to do for desorption (see however below), the most widely used method is temperature-programmed desorption (TPD; earlier called "flash desorption"), mostly with a linear temperature sweep. This makes measurements easy as the decrease in rate by depletion of the surface layer is counteracted by the temperature increase. Together, these influences lead to one (or more) peak(s) of the rate as a function of T. The measurement is usually performed by monitoring the pressure rise in the system while increasing T (as discussed in [16.1], Δp is proportional to R_d in a pumped system provided the temperature is not raised too fast). Several facts account for the popularity of this kind of measurement. For example, the integral of the peak area is proportional to the starting coverage, so that relative coverages and sticking coefficients can be obtained. The number of peaks lends itself to a qualitative discussion of the layer in terms of "states". The shapes of the peaks and their behavior with changing initial coverage can be used to derive an "order" of desorption. This derives from the use of a Polanyi-Wigner equation,

$$R_d = N_{ad}^x k_o \exp(-E_d/RT) \tag{16.4}$$

as the basic ansatz to interpret measurements. If $T = T_o + \beta t$, it can be readily shown ([16.1] and references therein) that for constant E_d and k_o the peak temperature T_p is independent of initial coverage N_o for $x = 1$ (first order), and moves to lower T with increasing N_o for $x = 2$, and to higher T for $x = 0$, in a characteristic way. So-called first- or second-order plots can then be used to extract E_d and k_o from measurements of T_p with changing β [16.1]. Another method of more detailed analysis is curve fitting, i.e., the selection of $E - k_o$ pairs, possibly changing with Θ, such that a calculated curve matches the measured one.

The aim of obtaining a value for the order x is to derive information about the mechanism [e.g., for associative desorption $2A \rightarrow A_2$ (this will be termed a (2-1) desorption) x = 2 would be expected in the simplest models if the rate-determining step is the association of adsorbed atoms, and x = 1 if desorption of the resulting molecule is rate limiting]. The resulting values of the kinetic parameters k_o and E are then usually discussed in terms of transition state theory [16.4]. For a (1-1) desorption (for instance desorption of molecularly adsorbed CO) we obtain [16.1]

$$R_d = \kappa k^{\ddagger} N^{\ddagger} = \kappa (kT/h) N_{ad} \exp(-\Delta F^{\ddagger}/RT)$$

$$= \kappa (kT/h) N_{ad} \exp(-\Delta E^{\ddagger}/RT) \exp(\Delta S^{\ddagger}/RT) \quad \text{or} \tag{16.5}$$

$$k_o^{(1)} = \kappa (kT/h) (\mathcal{F}^{\ddagger}/\mathcal{F}_{ad}) = \kappa (kT/h) (f_{tr}^2 f_r f_v / \mathcal{F}_{ad}) \tag{16.6}$$

where the fs are partition functions of individual degrees of freedom (for tr: translation, r: rotation, v: vibrations) and the $\mathcal{F}$'s products thereof, κ is the transmission coefficient, the superscript $\ddagger$ denotes the activated complex, and the other symbols have their usual meaning [16.1]. If the transition state is assumed to be a two-dimensional gas and the adsorbate layer is fully delocalized (mobile in the statistical sense), $F^{\ddagger} \approx F_{ad}$ and $k_o \approx 10^{13}\ s^{-1}$.

Correspondingly, for a simple (2-1) desorption, with second order and for mobile adlayer, $k_o^{(2)} \approx 10^{-2}\ s^{-1}\ \text{molecule}^{-1}/\text{cm}^2$. These values are often termed the "normal" values and are frequently assumed to apply without further questioning.

16.2.2 Critique

Close examination of the basic approaches outlined above reveals a number of approximations, some of which are not acceptable in most of the encountered cases. This in turn can be used to improve the overall interpretation of measurements. In the following some of these points will be discussed.

1) A basic assumption always implicit made is that the rate R_d is *only* a function of coverage and temperature, which means that the layer is always in *internal* equilibrium. On the other hand, the state of the layer at a certain coverage can well change with temperature, e.g., in a phase transition.

2) The use of transition state theory (TST) implies that the layer is always in thermal equilibrium with the *substrate*. In terms of the phenomenological approach of KRAMERS [16.5], applied to surface processes by SUHL et al.

[16.6] and others [16.7], this means that the adparticle-substrate (heat-bath) coupling expressed by Kramers' friction coefficient should be neither very small nor very large. In the first case the Boltzmann tail of excited particles would fly off at first to be only slowly refilled at the quasi-stationary rate, leading to a decrease of the rate; in the second, energetic particles about to leave the surface could be recaptured during their depart by the high friction, which would also lead to a rate decrease compared to TST. A hydrodynamic treatment of R_d [16.7] suggests that under most conditions expected for surfaces no problems would arise in this respect. On the other hand, recent microscopic theoretical treatments [16.8] suggest that strong deviations from the equilibrium treatment may well occur. Even in cases where an overall Arrhenius form results, the energy E_{eff} obtained from its slope can differ from the binding energy E_{eq}. Experimentally, no clear proof of such effects has so far been given to my knowledge.

3) The use of a Polanyi-Wigner equation as the basic formula is sensible only if the *total* coverage dependence is contained in the N^x term, and at least the overwhelming temperature dependence is contained in the exponential term, leading to E_d and k_o (or in TST language: the activation energies and entropies) being independent of both T and Θ. The T dependence of E_d can most probably be expected to be negligible over a not too large T range (it is not zero since it must be connected to the difference of the specific heats of layer and transition state); that of k_o will not be negligible (as T is contained in the partition functions) but small compared to the exponential. Therefore an Arrhenius form with a weak T dependence in the preexponential can be defended in most cases as a first approximation. However, the assumption of a *constant order*—which, as outlined, is equivalent to coverage-independent E_d and k_o—will practically never be realized given the knowledge now accessible about coverage-dependent interactions in adsorption layers. This would make it necessary to make k_o and E (and/or alternatively x) coverage dependent which would lead to a breakdown of the concept of order contained in the Polanyi-Wigner equation. It is proposed, therefore, to evaluate data in terms of an Arrhenius equation

$$R_d = R_d^o(T)\ \exp(-E/RT) \tag{16.7}$$

where R_d^o is still somewhat temperature dependent; with more detailed knowledge this T dependence can be included thus making the E values even better. One can then further *define* normalized preexponentials by dividing by N^m,

$$k_d^{(1)} \equiv \frac{R_d^o}{N_{ad}} \quad \text{for} \quad (1\text{-}1) \text{ desorption}$$

$$k_d^{(2)} \equiv \frac{R_d^o}{N_{ad}^2} \quad \text{for} \quad (2\text{-}1) \text{ desorption} \tag{16.8}$$

where m is the expected molecularity of desorption. This puts all coverage dependence into k_o and E and makes the interpretation clearer. Such an approach should *not* be confused with imposing the concept of a *fixed order*.

4) The entire previous discussion only applies rigorously for *direct* desorption, i.e., if the desorbing particle follows a smoothly varying potential curve from the adsorbed to the gas phase. It is well known, however, that in many cases a *precursor* state [16.9,10], i.e., a weakly adsorbed state, can play an important role in *ad*sorption kinetics. It should be expected, then, from microscopic reversibility that such states can also be important in *de*sorption kinetics. Apart from earlier considerations [16.10], the consequences have been examined in detail only in the last 5 years, using either a picture of consecutive and parallel reactions [16.11], or the probabilistic approach of KISLIUK [16.12,13]. The results must, of course, be equivalent [16.14], especially if the realistic assumption is made that the (intrinsic [16.13]) precursor above empty sites is not identical in properties to the (extrinsic) precursor above occupied adsorbate site. Not surprisingly, quite drastic effects of precursor properties on TPD traces can result, as has been shown in simulation calculations [16.2,13]. Even in the most simplified version [16.11], the equation for R_d

$$R_d = \frac{k_d k_D \Theta N_s}{k_d + k_a(1-\Theta)} \tag{16.9}$$

still contains 6 parameters (k_a, k_d, k_D — see Fig.16.1 for their definition — each consist of a preexponential and an exponential) which cannot possibly be all determined from desorption measurements, as detailed as they may be. One approach is to try and determine at least one parameter independently; e.g., (E_d-E_a) can be determined from the temperature dependence of the sticking coefficient from the gas phase into the precursor if the condensation into the latter is T independent, which is quite unlikely, and if further assumptions can be reasonably made (e.g., k_d^o, k_a^o, k_D^o given by TST). More generally, the precursor properties can be determined from detailed sticking coefficient measurements.

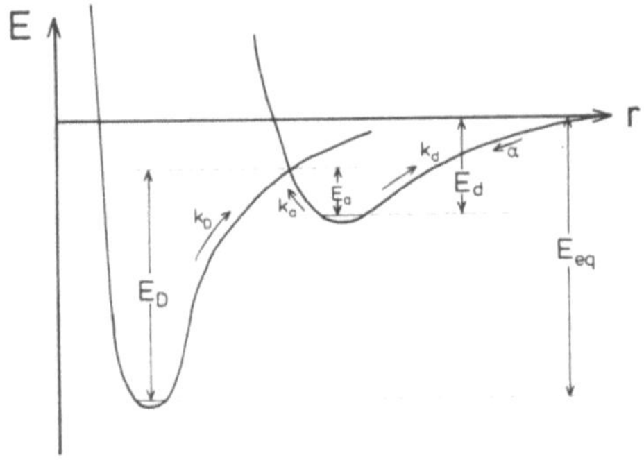

Fig.16.1. Schematic one-dimensional potential-energy diagram for a simplified (1 - 1) precursor-mediated desorption process, with definition of rate constants and energies of [16.9]

The application of such results to desorption measurements is still not simple, though. Both *ad*sorption and *de*sorption measurements are usually done irreversibly, so that sticking coefficients are taken at low, and desorption data at high temperatures. This may well lead to different ad/desorption paths in both cases (remember that the adparticle really follows a three-dimensional trajectory, not a one-dimensional one), and microscopic reversibility breaks down.

Another approach is to try and eliminate influences of the precursor by examining limiting cases and looking for conditions where these may be realized. For the simple case of (16.9) two limiting cases are obvious:

a) $k_d \gg k_a(1 - \Theta)$

$$\rightarrow \quad R_d = k_D \Theta N_s \qquad (16.10)$$

$$E_{eff} = E_D < E_{eq}$$

and the transfer into the precursor is rate limiting;

b) $k_d \ll k_a(1 - \Theta)$

$$\rightarrow \quad R_d = \frac{k_d k_D}{k_a} \frac{\Theta}{1 - \Theta} N_s \qquad (16.11)$$

$$E_{eff} = E_D + E_d - E_a = E_{eq} \quad \text{for not too large } \Theta \quad ,$$

and desorption from the precursor becomes rate limiting, with the precursor being in a (temperature-dependent) equilibrium with the adsorbed layer.

It can be seen even for this very simple case that E_{eff}, the effective energy derived from the temperature dependence of desorption rates, can be different from E_{eq}, the equilibrium adsorption energy (even after correction for the temperature dependences of partition functions contained in the k_0's).

As mentioned, some recent microscopic theories of desorption [16.8] have also shown that the effective desorption energy can differ from E_{eq}; in these cases the causes are less obvious in macroscopic terms but are contained in the microscopic mechanism (e.g., if the adsorption well contains only one vibrational level or if the "climbing of the ladder" of vibrational states becomes asymmetric on going up or down). Such effects would manifest themselves in deviations from the TST behavior; it would be difficult, however, to detect it in classical desorption experiments.

5) All these problems can be avoided if one works under equilibrium conditions from the start [16.7,15]. Then the chemical potential of adlayer and gas are equal and set by p,T(gas). Also, the rates of adsorption and desorption are equal, so that for (1-1) ad/desorption

$$R_a = sN_g v_T = N_{ad} k_o^{(1)} \exp(-E_{eff}/RT) = R_d \tag{16.12}$$

where s is the overall sticking coefficient and $v_T = \sqrt{\pi kT/2m}$ is the thermal velocity of the 3D gas. On the other hand, equilibrium implies that

$$N_g/N_{ad} = [\mathscr{F}_g(3D)/\mathscr{F}_{ad}] \exp(-E_{eq}/RT) \tag{16.13}$$

where $\mathscr{F}_g(3D)$ is the partition function for the gas phase. Then

$$[k_o^{(1)} \exp(-E_d/RT)]/sv_T = (\mathscr{F}_g/\mathscr{F}_{ad}) \exp(-E_{eq}/RT) \tag{16.14}$$

and

$$k_o^{(1)} = (s\mathscr{F}_g v_T/\mathscr{F}_{ad}) \exp[-(E_{eq} - E_d)/RT] \quad . \tag{16.15}$$

When $E_{eq} = E_d$, then

$$k_o^{(1)} = s\mathscr{F}_g v_T/\mathscr{F}_{ad} \quad . \tag{16.16}$$

As

$$\mathscr{F}_g(3D) = f_{tr}^3 f_r f_v \quad ,$$

$$v_T \mathscr{F}_g(3D) = (kT/h) f_{tr}^2 f_r f_v \quad . \tag{16.17}$$

Obviously, this treatment becomes fully equivalent to the TST treatment of (16.5,6) if

$$s(\Theta) = \kappa \quad . \tag{16.18}$$

In this way, the full information about the mechanism as well as all non-equilibrium effects are put into the sticking coefficient. Apart from it (and from a possible discrepancy between E_{eq} and E_{eff}), the preexponential of desorption

$$k_o^{(1)} = s(\Theta)\,\frac{kT}{h}\,\frac{f_{tr}^2 f_r f_v}{\mathcal{F}_{ad}} \tag{16.19}$$

contains information on the equilibrium statistical mechanics of the adlayer. Reversing the argument, if sticking coefficient measurements are available or at least if it is known that $s(\Theta)$ is well behaved (i.e., does not vary strongly) in the region of interest, desorption kinetics measurements can be used to derive information on the (equilibrium) entropy of the adlayer. Again, however, the cautioning remarks apply which were made above about the combination of s- and R_d-information with the implication of detailed balance.

16.3 Experimental Difficulties and Advances

As discussed in Sect.16.2.1, the usual method of thermal desorption kinetics is temperature-programmed desorption monitored through the pressure rise in the pumped system (Δp-TPD).

Although it is now standard in such experiments to use a mass spectrometer so that only the wanted molecular species is monitored, such measurements suffer from a number of limitations. Since all particles appearing in the gas phase upon heating of the sample are monitored, contributions from the less well characterized parts of a single crystal such as edges, rounded parts, etc., from the support, thermocouple wires, etc., fully contribute. All such effects will broaden the spectra which are especially detrimental for curve fitting (see below). Also, the gas particles interact with the walls, pumps, pressure measuring devices, etc., to change the gas composition. In extreme cases of reactive particles (e.g., NO) this can lead to virtual disappearance of the signal despite the fact that originally such molecules do desorb. Also, the possible variation of heating rates—so important to determine good kinetic parameters—is very limited. β cannot be made very small as this makes the signal difficult to distinguish from the background; it cannot be made too large because of the limitation imposed by the pumping speed. Also, data evaluation is problematic. As discussed in Sect.16.2.2, the assumption of constant order ("order plots") is usually not warranted. Alternatively, E and k_o values are often obtained by curve-fitting procedures. Because of the compensatory effects of these two parameters, however, such procedures are quite ambiguous as has been shown by several authors [16.16], especially in view of

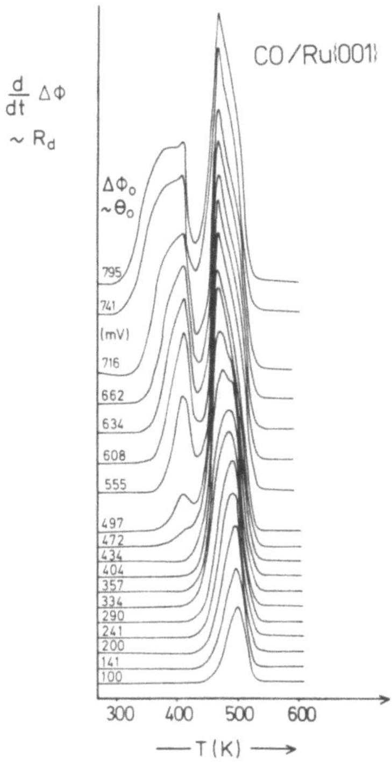

Fig.16.2. A set of desorption spectra obtained by the $\Delta\phi$ method for the system CO/Ru{001} [16.19] at a heating rate of 4 K/s^{-1}. The parameter is initial coverage given as $\Delta\phi_0$. Together with the values of d $\Delta\phi$/dt, those of $\Delta\phi$ are also recorded, so that $R_d(T,\Theta)$ can be obtained directly

the broadening effect of any disturbance. In β variation, the range is rather small in Δp-TPD because of the limitations on both ends as discussed.

With rather simple means, considerable improvements are possible. The detrimental effect of nonoptimally ordered surface contributions and of interactions with walls and pumps can be minimized by putting a glass cap with a small aperture around the mass spectrometer ion source [16.17]. This leads to a rather dramatic improvement of resolution, and even to detection of active gases. Even better results are obtained if, instead of detecting the desorbing molecules in the gas phase, the changing residual coverage on the surface is monitored. The best parameter for doing this is the work function change $\Delta\phi$ caused by adsorption, which is appreciable for many cases. Of course it has to be calibrated carefully against coverage, but once the $\Delta\phi(\Theta)$ relationship is known, it can be used for measuring R_d (via d$\Delta\phi$/dt) as well as Θ (via $\Delta\phi$) at the same time. We have used an especially suited vibrating capacitor with high reproducibility (< 1 mV), low noise (0.1 mV at 1 Hz, 1 mV unfiltered) and drift, and high time resolution (< 20 ms) for this purpose [16.18]. There are many advantages to this method. For instance, resolution is much better and noise much lower (see Fig.16.2 for an example); the sampled surface area is determined by the size of the counterelectrode and can be made small and undisturbed by edge effects, etc.; the range of β variation can be made much larger (we have used a range from 0.01 to 30 K s^{-1}, i.e., 3½ orders of the magnitude); isothermal desorption as well as TPD is possible; and the determination of desorption isosters (see below) is very simple as the $\Delta\phi$ value recorded along with its time derivative gives the coverage for any $R_d(T)$ value. Also, once the $\Delta\phi(\Theta)$ dependence—and if necessary, its dependence on T—is known, it can also be used to measure sticking coefficients and to set the coverage in equilibrium measurements. For the latter, isobar measurements which can then be converted to isotherms or isosters by the appropriate cuts, are much to be preferred over direct isosters (obtained by adjusting T and p such that a constant Θ is obtained). For an example, see Fig.16.3. The $\Delta\phi$ method can, of course, not be used if the ϕ change is small and/or changes its direction with coverage, or if several desorbing species exist. Then the improved Δp-TPD method would appear as the method of choice.

Even if the TPD traces obtained by i-Δp-TPD are much better than the conventional results and those by $\Delta\phi$-TPD are still better [16.17], shape analysis is still to be rejected because of its basic ambiguity. Comparison of calculated and measured traces should only be used as a negative check: they *have* to agree for the kinetic parameters (determined differently) to be correct. One trustworthy method of evaluation of TPD results is β variation. As has been shown by FALCONER and MADIX [16.21], even for arbitrary "order",

$$\ln(T_p^2/\beta) = E_d/RT_p + \ln(E_d/k_0R) + (\text{"order" dependence}) \tag{16.20}$$

so that the effective activation energy for the coverage at T_p can be taken from a plot of $\ln(T_p^2/\beta)$ vs 1/Tp. Figure 16.4 shows an example of such a plot from $\Delta\phi$-TPD for CO/Ru over a range of 3 orders of magnitude. To take such measurements at many coverages is very tedious, however. A simpler evaluation is the construction of *desorption* isosters [$R_d(T)$ for fixed Θ] which can be taken from a family of TPD curves by construction of R_d (Θ) curves for fixed T and taking cuts at fixed Θ values. The resulting plots of $R_d(T)$ for constant Θ can then be directly evaluated according to (16.7) to yield *independent* values of E_d and $k_0(\Theta)$. Δp-TPD traces have to be backintegrated to each wanted point to obtain the coverage, which is again time consuming (Fig.16.5). One can also use the procedure of BAUER [16.22] (Fig.16.6). For $\Delta\phi$-TPD traces, isoster determination is very easy as the $\Delta\phi$ values can be

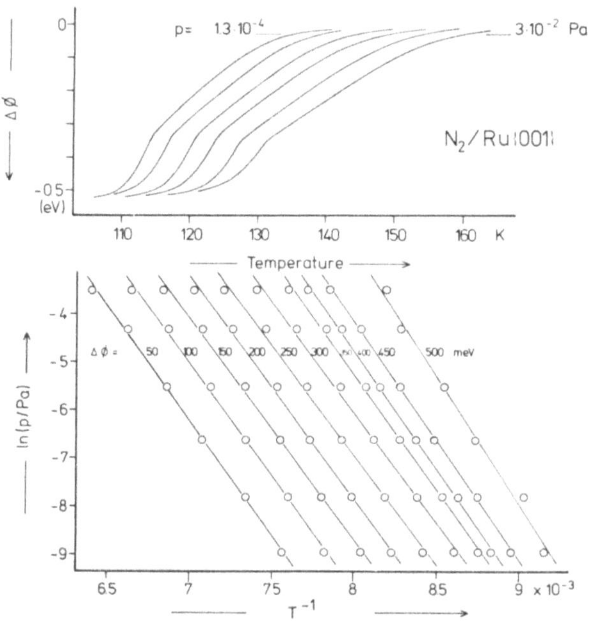

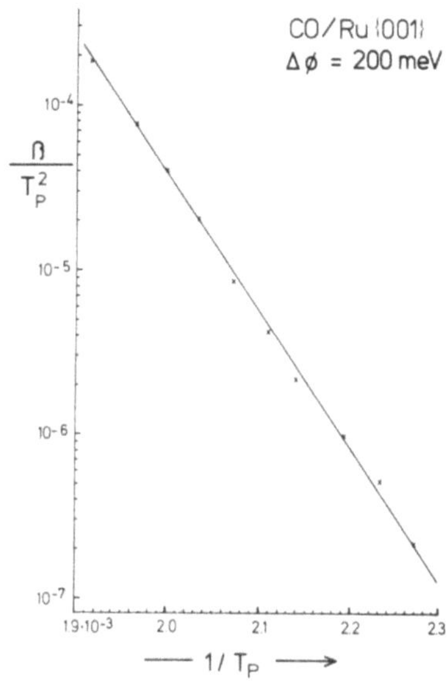

◄Fig.16.3 ▲Fig.16.4

Fig.16.3. (a) Equilibrium isobars (parameter:p) obtained by using $\Delta\phi$ as coverage indicator for molecular N_2 on Ru{001}, and (b) derived equilibrium isosters (parameter: $\Delta\phi \sim N_{ad}$), from which $E_{eq}(\Theta)$ and $S_{eq}(\Theta)$ can be derived independently. The change of slope in the isobars around $\Delta\Phi$ = 0.3 eV is due to a phase transition [16.20]

Fig.16.4. Example for evaluation of the dependence of T_p on heating rate β to derive E_d, for the system CO/Ru{001} [16.19]. Data have been taken by $\Delta\phi$TPD (see Fig.16.2) over 3 orders of magnitude

recorded along with the $\Delta\phi$ values in each trace and the rearrangement of data is easily done by computer.

Isothermal desorption measurements are easy with $\Delta\phi$ as coverage parameter also; they are usually run by introducing a jump in T, with an initial *irreversible* coverage at a base temperature, and starting the measurement of $R_d(\Theta)$ [i.e., d $\Delta\phi/dt(\Delta\phi)$] as soon as the temperature has settled to a constant value. A family of such curves (see Fig.16.7) can then be evaluated simply by cuts along certain $\Delta\phi$ values to yield $R_d(T)$ for this fixed coverage; from these, independent $E_d(\Theta)$ and $K_o(\Theta)$ values can again be taken.

For complicated systems, for instance in which more than one species is evolved upon heating, the combination of i-Δp-TPD and $\Delta\phi$ can yield additional information. While the $\Delta\phi(T)$ traces cannot now be evaluated in the same quantitative detail as simple systems, the additional information provided

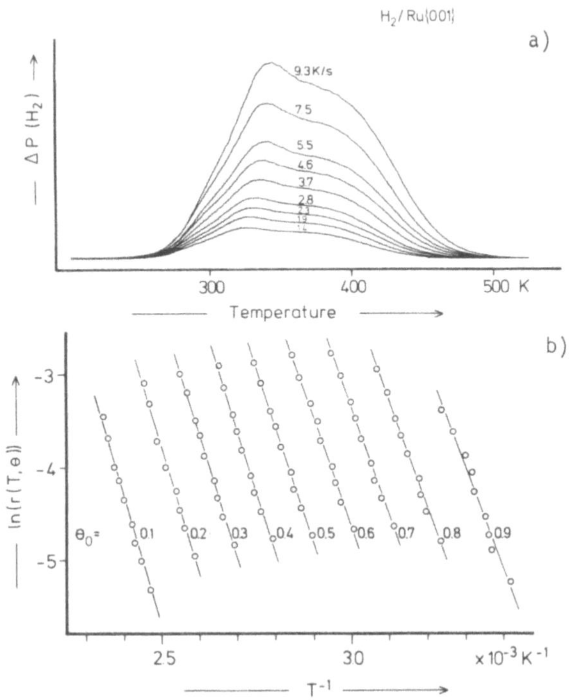

Fig.16.5a,b. Example for the construction of desorption rate isosters from Δp-TPD (in this case the family of curves shown in (a) have been obtained for constant N_0 with β variation). Coverage is obtained by backintegration of TPD traces; $R_d(T)$ values at constant Θ are then combined to yield rate isosters (b). In this system (H_2/Ru{001})$\Delta\phi$-TPD cannot be taken [16.20]

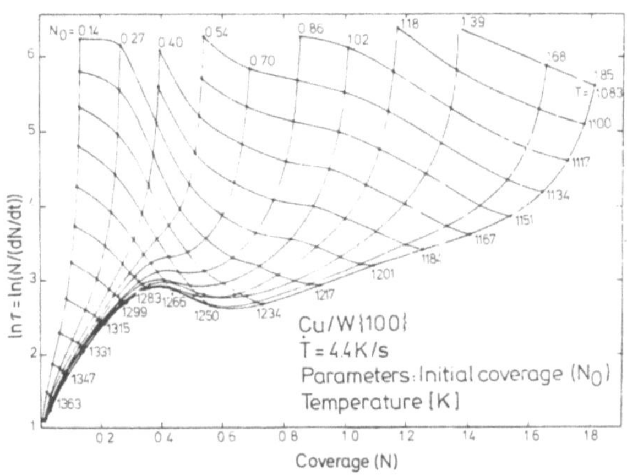

Fig.16.6. Example for the construction of rate isosters and isotherms from TPD spectra with the method in [16.22]

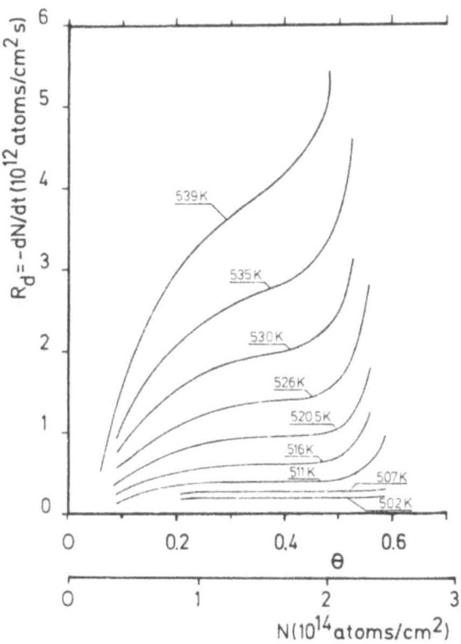

Fig.16.7. Isothermal desorption runs in the system oxygen/Ag{110}, taken with $\Delta\phi$ as coverage indicator [16.23]

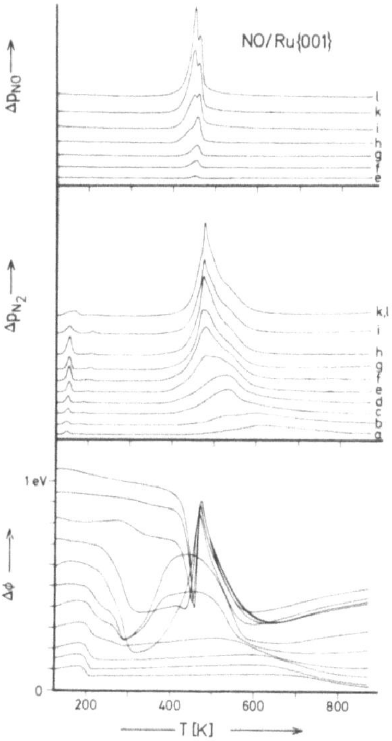

Fig.16.8. Correlation of Δp-TPD for the two desorbing species with $\Delta\phi(T)$, for the system NO/Ru{001}. For a discussion of the conclusions see [16.24]

by it can make conclusions about the evolution of adsorption states more unambiguous. As an example, Fig.16.8 shows a set of data for NO/Ru{001} [16.24].

16.4 Some Results and Discussion

This shows that a number of independent measurements is possible which can then be evaluated in more than one way each to yield $E_{eq}(\Theta)$ or $E_{eff}(\Theta)$ and $k_0(\Theta)$ values which can be used to check each other. This has been done most extensively for the system CO/Ru{001} [16.19]; a number of the results are shown in Fig.16.9. The good agreement of the very different independent methods and evaluations, especially regarding the striking behavior of $k_0(\Theta)$, was proof that no artefacts are the cause, and that no disturbance is brought about by precursor effects. For the latter conclusion the acceptable agree-

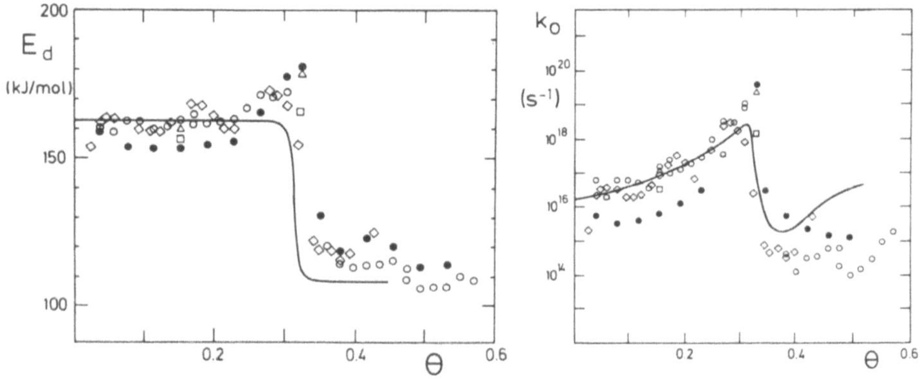

Fig.16.9. Collection of (a) E(Θ) and (b) k_0 (Θ) values for CO/Ru{001}, obtained by different independent measurements and evaluations [16.19]. Open symbols: values derived from kinetic measurements of various kinds (see [16.19] for details). Solid symbols: equilibrium energy E_{eq}; k_0 derived from E_{eq} and sticking coefficient. The solid lines give the result of the generalized quasi-chemical approximation [16.25], which uses next-neighbor repulsion only

ment of E_{eff} and E_{eq} is most important (the systematic deviation between E_{eq} values and all the various E_{eff} values is probably due to a disturbance at high pressure of the E_{eq} measurement). CO is known to adsorb on Ru{001} via a precursor, but the sticking coefficient is well behaved except at very high coverages (where the highest scatter occurs in the data). Similar, though not as extensive, results have been obtained for O/Ag{110} [16.23], N_2/Ru{001} [16.20], and H_2/Ru{001} [16.20]. For the much more complicated case of NO/Ru{001}, where NO and N_2 desorb and there is competition of desorption and dissociation of two molecularly adsorbed states, the parallel measurement of Δp-TPD of both products and of Δϕ (Fig.16.8), made a quite detailed assignment possible [16.24].

We shall limit the detailed discussion to the best-documented system, CO/Ru{001}. At the time of first publication, the high k_0 values and especially their very strong Θ dependence appeared remarkable (first order k_0 values of 10^{16} s^{-1} had been observed before— see references in [16.19, 21]—but had been debated by others even though such a possibility had been realized very early [16.10]). Since then another set of similar data, obtained by E_{eq}+ sticking measurements, has been reported for CO/Ni{111} [16.26]. The explanation for the observed behavior is quite straightforward, so that by now it appears to be accepted. We shall sketch it in the following.

CO is adsorbed on the hexagonal Ru{001} face in on-top sites [16.27] initially; island growth of the $\sqrt{3}$ structure and a second-order phase transition [16.28] indicate weak attractive interactions between the $\sqrt{3}$ sites and explain the slight increase of E with Θ up to $\Theta = 1/3$. The $\sqrt{3}$ structure is optimal at this coverage and disappears above it; it persists up to desorption around $\Theta = 1/3$. At this coverage, the layer is very uniform and homogeneous, as shown by the narrow linewidth of IR absorption of the C-O stretch [16.27]. The step in E at this coverage is indicative of strong repulsive interactions between CO molecules approaching each other more closely. As no influence of precursor kinetics is to be expected (see above), the k_o values can be taken to give direct information on the entropy of the adsorbed layer. The high values of k_o show that the layer must be localized ("immobile" in the statistical sense). As derived in (5.19),

$$k_o^{(1)} = s(\Theta)(kT/h)(f_{rg}^2 f_{rg} f_{vg})/(f_{ra} f_{va} f_{conf}) \qquad (16.21)$$

where the second subscripts g and a denote gas and adsorbate, respectively, and f_{conf} is the configurational partition function of the layer. This neglects a factor $f_{ph}(\text{clean})/f_{ph}(\text{cov})$, the ratio of the partition functions of the surface phonons for the clean and covered surfaces which is equal to 1 only for $\Theta \to 0$. Because its dependence on Θ is not known, we set it equal to 1 throughout.

For CO, f_{rg} (two degenerate modes) $\approx 2 \cdot 10^2$ at 500 K, and $f_{vg} = 1$. On the other hand, the admolecule does not rotate ($f_{ra} = 1$), and of the adsorbate vibrations, the frustrated rotations ($h\nu \approx 400\ \text{cm}^{-1}$ [16.29]) and translations ($h\nu \approx 40\ \text{cm}^{-1}$ [16.29] are nonnegligible, which results in $f_{ra} \approx 10^2$. For an immobile layer at $\Theta = 0$, $f_{conf} = N_S$, the number of surface atoms [16.30], so that we indeed get $k_o^{(1)} \approx 10^{16}\ \text{s}^{-1}$ for low coverages.

This treatment neglects the temperature dependences of the partition functions, because they are weak and partly cancel. Taking them into account fully as suggested by IBACH et al. [16.26] would increase the accuracy somewhat.

The only remaining coverage-dependent factor is f_{conf}. For an ideal lattice gas, we would get [16.30]

$$k_o(\Theta) = k_o(0)\,\frac{1}{1 - \Theta}$$

where now Θ would pertain to the coverage of the lattice gas ($\Theta = 1$ for the $\sqrt{3}$ structure). This would yield a divergence at $\Theta = 1/3$ in Fig.16.8

[16.19]. However, such a treatment does not only exaggerate repulsion (as it sets them infinite beyond $\sqrt{3}$) and neglect attraction, but also gives a wrong count of available sites: for the first CO molecule, *all* surface sites are accessible; however each adsorbing molecule excludes *7* such sites from further adsorption; finally in the $\sqrt{3}$ structure every *third* site is occupied. The normal quasi-chemical approximation (QCA) which works with pairwise interactions can therefore not be used (the Bragg-Williams approximation has the same drawback and does not affect entropy anyway). A generalized QCA has been developed by LEUTHÄUSSER [16.25] which works with triangular clusters and avoids this difficulty. The fat lines in Fig.16.8 are the result of this treatment in which the step in E has been used for determination of parameters and no further adjustments were necessary to obtain the entropy, i.e., $k_0(\Theta)$.

The attractive interactions have been neglected so far; they would introduce only minor modifications for desorption. This is an example of a case where kinetic measurements can be interpreted in terms of equilibrium statistics of the adlayer because of the well-behaved (i.e., high and weakly coverage-dependent) sticking coefficient. In terms of the limiting cases discussed above, this must be the case of rate-limiting k_D, i.e., E_{eff} should be smaller than E_{eq}. The opposite deviation shown by the experiment is probably due to a systematic error, as discussed above. We conclude that it will be difficult to unequivocally determine such deviations [16.31]. Also, the reader is reminded of the remarks about the deviation from true equilibrium and detailed balance conditions in the *measurements* which have been given above. These would well lead to errors of the same magnitude, making conclusions even more difficult.

On the other hand, any *true* nonequilibrium effects in the *mechanism* (very large or very small friction, etc., see above) will be immediately obvious in the sticking coefficient. Arguing in reverse, such effects can be excluded for all cases of adsorption with high sticking coefficients; if they seem to show up in desorption only, there must be an error.

We have not touched at all on the question if the energy transfer between adsorbate and substrate is mediated by *vibrational* or *electronic* coupling. This is a question which is being actively investigated at present from the side of theory (see Chap.15 and the references given in it); as it is so far not accessible to experiment, it has been omitted.

16.5 Conclusions

Within the scope of classical desorption measurements, simple refinements of measurement techniques as well as evaluation procedures have led to a considerable improvement in resolution and accuracy of desorption results.

A number of effects can be envisaged which would lead to deviation from local equilibrium in adsorption or desorption. However, such effects would show up most readily in the sticking coefficients; as long as these are well behaved, desorption can be discussed in terms of equilibrium statistics of the adlayer. This becomes most obvious if the measurements themselves are obtained via the combination of equilibrium isosters and sticking coefficients. In cases where this cannot be done easily, desorption measurements can be used instead. The comparison of desorption and equilibrium measurements can be employed to check the reliability of the data. To find out details about nonequilibrium behavior, angle- and energy-resolving measurements will be important which have not been within the range of this discussion. But even without them, desorption measurements – if done carefully – yield very interesting results about interactions in adlayers, apart from the more straightforward information on coverages and overall energetics for which they have long been used. To show this was the purpose of this survey.

Acknowledgements. I acknowledge with great pleasure the very excellent and pleasant cooperation of my former and present collaborators H.A. Engelhardt, P. Feulner, and H. Pfnür, as well as many illuminating discussions with W. Brenig and U. Leuthäusser. The reported experiments have been supported financially by the Deutsche Forschungsgemeinschaft through Sonderforschungsbereich 128.

References

16.1 D. Menzel: In *Interactions on Metal Surfaces*, ed. by R. Gomer, Topics in Applied Physics, Vol.4 (Springer, Berlin, Heidelberg, New York 1975) p.101
16.2 D.A. King: In *Chemistry and Physics of Solid Surfaces II*, ed. by R. Vanselow (CRC, Boca Raton 1979) p.87
16.3 J.T. Yates, Jr.: To be published
16.4 S. Glasstone, K.J. Laidler, H. Eyring: *The Theory of Rate Processes* (McGraw-Hill, New York 1941)
16.5 H.A. Kramers: Physica *7*, 284 (1940)
16.6 E.G. d'Agliano, W.L. Schaich, P. Kumar, H. Suhl: Nobel Symp. *24*, 200 (1973)
16.7 W. Brenig, H. Müller, R. Sedlmaier: Phys. Lett. *54*A, 109 (1975); W. Brenig, K. Schönhammer: Z. Phys. B*24*, 91 (1976)

16.8 H. Müller, W. Brenig: Z. Phys. B*34*, 165 (1979);
Gopa Sarkar De, U. Landman, M. Rasolt: Phys. Rev. B*21*, 3256 (1980);
S. Efrima, K.F. Freed, C. Jedrzejek, H. Martin: Chem. Phys. Lett. *74*, 43 (1980);
Z.W. Gortel, H.J. Kreuzer, D. Spaner: J. Chem. Phys. *72* , 234 (1980)
J.C. Tully: Surf. Sci. *111*, 461 (1981)
16.9 I. Langmuir: Chem. Rev. *6*, 451 (1929);
J.B. Taylor, I. Langmuir: Phys. Rev. *44*, 423 (1933)
16.10 G. Ehrlich: J. Phys. Chem. *59*, 473 (1955); J. Phys. Chem. Solids *1*, 3 (1956)
16.11 R. Gorte, L.D. Schmidt: Surf. Sci. *76*, 559 (1978)
16.12 P.J. Kisliuk: J. Phys. Chem. Solids *3*, 95 (1957); *5*, 5 (1958)
16.13 D.A. King: Surf. Sci. *64*, 43 (1977)
A. Cassuto, D.A. King: Surf. Sci. *102*, 388 (1981)
16.14 K. Schönhammer: Surf. Sci. *83*, L633 (1979)
16.15 G. Iche, Ph. Nozières: J. Phys. Paris *37*, 1313 (1976)
16.16 C. Pisani, G. Rabino, F. Ricca: Surf. Sci. *41*, 277 (1974)
16.17 P. Feulner, D. Menzel: J. Vac. Sci. Technol. *17*, 662 (1980)
16.18 H.A. Engelhardt, P. Feulner, H. Pfnür, D. Menzel: J. Phys. E *10*, 1133 (1977)
16.19 H. Pfnür, P. Feulner, H.A. Engelhardt, D. Menzel: Chem. Phys. Lett. *59*, 481 (1978); Ned. Tijdschr. Vacuumtech. *16*, 232 (1978); to be published
16.20 P. Feulner, D. Menzel: To be published
16.21 J.L. Falconer, R.J. Madix: Surf. Sci. *48*, 393 (1975)
16.22 E. Bauer, F. Bonczek, H. Poppa, G. Todd: Surf. Sci. *53*, 87 (1975)
16.23 H.A. Engelhardt, D. Menzel: Surf. Sci. *57*, 591 (1976)
16.24 P. Feulner, S. Kulkarni, E. Umbach, D. Menzel: Surf. Sci. *99*, 489 (1980)
16.25 U. Leuthäusser: Z. Phys. B*37*, 65 (1980)
16.26 H. Ibach, W. Erley, H. Wagner: Surf. Sci. *92*, 29 (1980)
16.27 H. Pfnür, D. Menzel, F.M. Hoffmann, A. Ortega, A.M. Bradshaw: Surf. Sci. *93*, 431 (1980);
G. Michalk, H. Pfnür, D. Menzel: Verh. Dtsch. Phys. Ges. *16*, 976 (1981); To be published
16.28 H. Pfnür, D. Menzel: Verh. Dtsch. Phys. Ges. *16*, 977 (1981); To be published
16.29 N.V. Richardson, A.M. Bradshaw: Surf. Sci. *88*, 255 (1979)
16.30 T.L. Hill: *Introduction to Statistical Thermodynamics* (Addison-Wesley, Reading 1960) Chaps.7,14
16.31 Very recently, a case [Xe/W(110)] has been reported for which $E_{eff} < E_{eq}$:
R. Opila, R. Gomer: Surf. Sci. *112*, 1 (1981)

17. Field Desorption and Photon-Induced Field Desorption

J.H. Block

With 15 Figures

17.1 Introduction

More than 30 years have passed since E.W. Müller first introduced the field ion microscope which now allows scientists to observe single atoms on surfaces. Relatively few laboratories have started or continued to develop and use field ion microscopy, despite the enormous possibilities for investigating the many urgent open questions in surface science offered by this technique.

There are, perhaps, practical as well as fundamental reasons for only having at present a limited utilization of field ion microscopy and related techniques. For example, field ionization suffers from the requirement that the sample must have a special shape. The preparation of field emitters with a radius of curvature of only some 100 nm is thought to be an art and contributes to the difficulties of field ion microscopy. For many materials, however, preparation of an emitter tip is no longer such an enormous task.

Furthermore, rather costly instruments are required if the application of field ionization is not to be restricted to just looking at field ion images. A serious problem with field ionization in surface science is that the fundamentals of the process are still very obscure. Field ionization is taken to be the interaction of a particle primarily with a high electric field; the surface is only considered as an afterthought. The interaction of matter with high electric fields still represents an area of extended ignorance. Therefore, one may be reluctant to use this technique for other purposes in studying solid surfaces. Nonetheless, field ion microscopy remains a technique whereby many problems of scientific importance may be directly studied.

There is one area where field ionization has been developed extensively and has found a diverse number of applications. This is field ionization mass spectrometry. The growth of this technique to its present maturity was mainly initiated by the pionerring work of H.D. Beckey. Field ionization mass spectrometry has many applications in analytical chemistry including the analysis of drugs or medical tissue. The properties of the field ionization process makes this technique very valuable in various analytical problems. The advantageous characteristics of field ionization include the frequently minimal ion fragmentation and the possibility of investigating nonvolatile high molecular weight compounds in the field desorption mode.

The development of field ionization mass spectrometry opens new possibilities in surface science which may have been neglected by many scientists. Among the well-known techniques in field ion mass spectrometry is E.W. Müller's atom probe. Using this technique the mass of individual surface atoms can be analyzed which are first imaged to locate their position on the crystal surface. There are, however, other developements in field ion mass spectrometry which are also of great importance to surface science:

1) Field ion appearance spectroscopy allows one, in addition to mass separation to measure the onset in energy distribution of ions and provides very valuable information on interaction energies at surfaces.

2) Field pulse methods have been improved and can now be used to study not only the chemical composition of certain reaction intermediates, but also their kinetic data.

3) Photon pulse methods have been recently introduced to field ion mass spectrometry, leading to photon-induced field desorption and permitting studies of surface diffusion and surface reactivity.

The purpose of this contribution is to show that field ionization – or, more precisely, field desorption mass spectrometry – is a valuable method to investigate the chemistry and physics at solid surfaces.

17.2 Field Desorption and Thermal Desorption

Field desorption is a process in which an atom or molecule leaves a solid surface due to the influence of a high electric field. The atom or molecule usually departs the surface in a charged state as an ion. Field desorption was first observed by MÜLLER [17.1] and has been subsequently discussed in several review articles [17.2-5].

Let us first discuss field desorption in comparison with thermal desorption. Both processes are used for analyzing surface layers and for studying chemical reactions at solid surfaces. The phenomenon of thermal desorption has been studied [17.6-10] and theoretically investigated extensively [17.11-13].

In a simple description of thermal desorption, a particle is bound to the surface by an energy E_A, the adsorption energy. The desorption process of the particle from the surface occurs when the interaction between the solid surface and the particle is diminished by highly energetic vibrations such that the bonding to the surface is severed with the particle overcoming the energy E_A at a critical distance x_c from the surface. The energy required to overcome E_A must be obtained from statistical energy fluctuations. The probability of accumulating this energy at the critical bond is given by a Boltzmann distribution $p \propto \exp(-E_A/kT)$ and a preexponential factor. If we include this in the form of a vibrational frequency ν_0, we obtain a rate equation

$$k = \nu_0 \exp(-E_A/kT) \tag{17.1}$$

similar to the Frenkel equation [17.14]. In practice, the models used to interprete thermal desorption data are complicated. Many factors need to be considered in thermal desorption, such as desorption via precursor states, interactions between adsorbed molecules, coverage-dependent variations in ν_0, etc. There are, as well, questions of internal energy fluctuations, energy accumulation, and repartition in the thermal desorption process, as are pertinent in other kinetic processes (see Chap.16).

For field desorption, a minimum energy is also required to cleave the bond to the surface. The desorbing particle transfers one or more electrons to the solid while leaving the surface.

At a field strength F, the binding energy of a neutral adparticle is slightly changed as a result of polarization forces. For nonpolar particles "the field binding energy" [17.15] is given by

$$E = \frac{\alpha}{2} F^2 \tag{17.2}$$

where α is the polarizability. However, the value of $(\alpha/2)F^2$ is usually only of the order of 0.1 eV at field strengths of $1 < F < 10$ V/nm and is neglected in Fig.17.1. With neutral polar molecules in electric fields, the behavior is more complicated. The energetics of the field desorption process can be described by the potential energy diagram shown in Fig.17.1 [17.2]. The critical distance x_c is defined as the minimum distance from the surface for ionization. For $F = 0$, far away from the surface at $x \gg x_c$, the potential energy of the singly charged ion is determined by the ionization potential I_p of the particle less the work function ϕ. The reduction by the amount of the work function results from the gain in energy when the

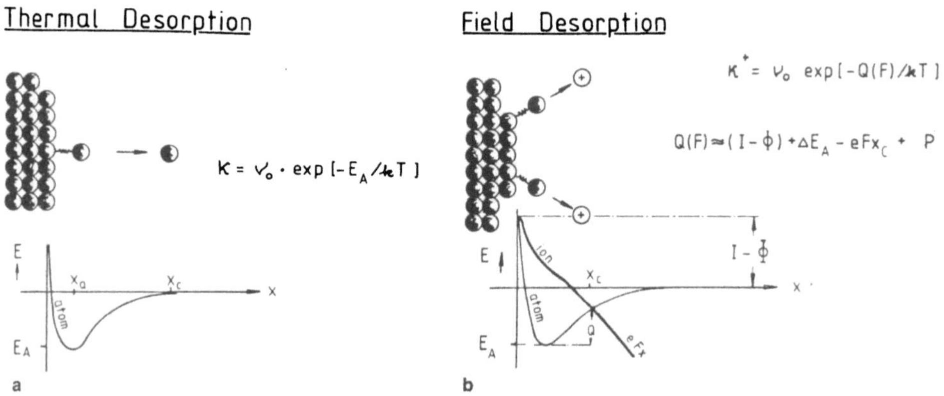

Fig.17.1. (a) Thermal desorption as thermally activated process. The rate K is determined by a frequency factor ν_0 and a Boltzmann distribution function. In the potential energy diagram E_A is the heat of adsorption, x_a the equilibrium distance, x_c the critical distance. (b) Field desorption of ionic species. The rate of desorption K^+ is determined by the activation energy Q. Q as well as x_c decrease with increasing field strength. P is a polarization energy; details in (17.3)

electron from the ionized desorbing particle enters the metal. With an applied field F, the potential energy curve of the ion is given by the superposition of the field-free ion potential curve and the potential given by the electric field eFx as represented in Fig.17.1. When the distance x exceeds x_c in this model, ionic species are more stable than neutral ones. At the crossing point of the ionic and neutral potential energy curves, ions can be formed as described in the charge exchange model [17.16]. Alternatively, the field desorption process may be described by the image hump model [17.17], which includes an activation barrier of the charged particle prior to desorption.

Both thermal desorption and field desorption are thermally activated processes. The activation energy of thermal desorption E_A (Fig.17.1) may be compared to the value Q for field desorption of ions. From potential-energy considerations, the value of Q for a singly charged ion can be calculated [17.4]. It is important to note that Q depends on the electric field strength F (Fig.17.1),

$$Q(F) = (I - \Phi) + E_A - eFx_c + P \qquad (17.3)$$

where P is an energy term due to polarization and image potential terms.

Field desorption is very sensitive to surface site and geometry, due to the inhomogeneity of the electric field strength near steps and kinks as well as to differences in binding energy, geometry, and work function at different planes or surface sites.

Thermally desorbed neutral particles display an angular and energy distribution in accord with a cosine law and the Maxwell-Boltzmann distribution, respectively. Exceptions have, however, been reported [17.18]. The energy distribution of field-desorbed charged particles reflects the position of ion generation. The potential evaluated at $x = x_c$ usually determines the maximum kinetic energy (appearance energy) desorbing ions will gain, rather than the potential precisely at the surface $x = x_0$. From the experimental appearance energy the minimum ionization distance, x_c, can be evaluated with great precision [17.19]. With increasing field strength the crossing point of neutral and ionic potential-energy curves shifts to lower energy values and the critical distance x_c decreases. The field dependence of x_c can be determined from the appearance energy of desorbing ions as a function of the field. The shape of the potential energy curve for adsorbed neutral particles can be evaluated (for $x_a < x < \infty$) from the field dependence in the energy distribution of desorbing ions. Only the displacement due to polarization forces has to be taken into account.

A special case of field desorption occurs when lattice atoms from the substrate surface are desorbed themselves. This process is called field evaporation. In principle, field evaporation does not differ from the ionization mechanism described in the potential-energy diagram for field desorption.

In thermal desorption rapid heating of a substrate (flash desorption) is used to identify different binding states with different binding energies E_A. We can investigate a chemical surface reaction which preceeds desorption and which is activated by a temperature increase by using temperature-programmed reaction spectroscopy as described by MADIX [17.20].

A temperature rise as well as an increase in field strength may enhance field desorption. Chemical reactions preceeding field desorption, either temperature or field induced, can be investigated as well. The principles of these measurements have been discussed before in connection with the chemical reactions on surfaces in the presence of high electric fields [17.21,22].

Surface species may be chemically identified using various types of mass spectrometers (magnetic sector field, quadrupole, time-of-flight (TOF), or a combination of the latter with an FIM in the "atom probe") in field desorption [17.23,24]. Furthermore, energy analyzers or retarding fields may be used for measurements, either differential or integral, of the energy distribution.

The sensitivity of the analyses of surface species has reached the ultimate limit for field desorption in the atom probe. Individual surface atoms or molecules can first be imaged and then analyzed for chemical composition in a TOF measurement which has a transmission coefficient of better than 0.5. In thermal desorption, a portion of the neutral desorption products is counted following ionization in a mass spectrometer or a pressure gauge some distance away from the surface. Present gas analyzers detect low partial pressures with relatively high scanning speeds, so that with a gas pressure of 10^{-11} torr the entire gas composition can be monitored within a few milliseconds. Thus in thermal desorption a surface coverage of $\theta < 0.1\%$ can readily be detected. Thermal desorption is useful for investigating an ensemble of surface sites, including steps and kink sites, while in field desorption a single individual site may be investigated. However, it is well known that the behavior of chemisorption structures on an ideal single crystal plane differ from the behavior on the surfaces presented by the half-spherical shape of a field emitter. Therefore, we cannot always expect field desorption and thermal desorption to provide the same result.

Surface diffusion occurs during thermal desorption measurements, and surface species are usually studied desorbing from ordered and relaxed surface arrays rather than a frozen random array. In field desorption, temperatures can be chosen such that surface diffusion does not occur and desorption becomes a stochastic process.

Field desorption is limited to a few selected systems, however. Not many materials of practical importance can be etched into emitter tips, and measurements may be disturbed by merely a few impurity atoms or molecules. The shortcomings of field desorption result form the required special shape of the field emitter, e.g., a tip or a wire with only a small surface area that can be investigated.

17.3 Investigations in the Field Ion Microscope

In an FIM, field desorption rates can be measured simply by counting the image spots of desorbed particles, either by using the desorbing ions directly for activating the image amplifier and luminescent screen or by observing desorption processes at the surface with the aid of an auxiliary imaging gas. The minimum electric field strength which is required to desorb a surface compound depends on the binding energy E_A to the surface. With increasing field strength different binding states with increasing E_A

will be successively desorbed. PANITZ [17.25] has described an isothermal ramped field desorption technique and demonstrated how different binding states of benzene on a tungsten surface can be characterized. The clean tungsten field emitter at 78 K was dosed with benzene. Subsequently a linearly increasing field strength was applied; the rate of field-induced desorption of surface species in the form of positive ions was registered; and spectra were obtained with remarkable similarity to those produced by thermal desorption at zero field. RENDULIC and LEISCH [17.26] used this technique to study the binding energy of common gases (H_2, N_2, CO_2, O_2) on metals like Ir on W. In a similar way, although with field pulses, COCKE et al. [17.27] were able to identify different binding states of oxygen and surface oxides on tungsten.

Binding states at surfaces are frequently characterized by the value of the desorption or evaporation field strength. The desorption field strength is defined by a certain desorption rate (for instance, 0.1 layer/s), which can be measured in the FIM. At the constant desorption rate, different corresponding values of temperature and field strength are evaluated. Since field desorption is a thermally activated process, desorption fields decrease with increasing temperature for each binding state. This can be demonstrated by one example. Recently, NISHIKAWA et al. [17.28,29] investigated the binding states of Ga and Sn on the surface of W. For Ga, for instance, they found by FIM structure determination, in accordance with LEED measurements, three different binding states: a pseudomorphic structure, a superstructure, and an overlayer. At a constant desorption rate (0.1 layer/s), measured in He imaging gas, the three different binding states are clearly identified (Fig.17.2). The highest binding energy is observed for the first layer directly at the W surface, where the W{111} surface displays higher binding energies (desorption fields) than W{112}. Next in decreasing binding energy is the superstructure at W{011}, followed by a second adsorption layer, the overlayer, at W{112} and {011}. This observation is in accord with the statement, that the Ga-W binding energy (at low coverage) exceeds the Ga-Ga binding energy in a second or higher surface layer. The temperature dependence, i.e., decreasing evaporation fields with increasing temperature, is as expected for this thermally activated desorption step.

In special cases, evaporation fields are determined by surface reactions rather than by the binding energies of the initial surface species. This behavior is similar to reactive thermal desorption [17.20]. One interesting recent example is the field desorption of surface layers from a Si{111} plane [17.30]. A {111}-oriented Si field emitter was imaged in hydrogen at 78 K.

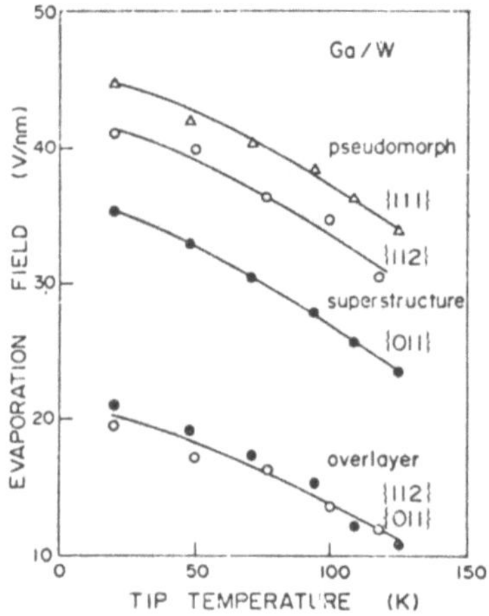

◄Fig.17.2. Temperature dependence of the evaporation field of Ga layers on the W surface. Open and solid circles are for {112} and {011} planes, respectively [17.29]

Fig.17.3. Temperature dependence of field desorption rate k_e of silicon {111}. Applied voltage and hydrogen pressure are kept constant, 11.0 kV and $6 \cdot 10^{-5}$ torr, respectively [17.30]

Fig.17.4. Temperature dependence of field evaporation voltage V_E. Field evaporation rate and hydrogen pressure are kept constant, 0.2±0.01 planes/s and $6 \cdot 10^{-5}$ torr, respectively. The experimental results are shown as solid circles. Calculated values are shown as lines A-F for A=10, B=11, C=12, D=13, E=14, and F=15 V/nm [17.30]

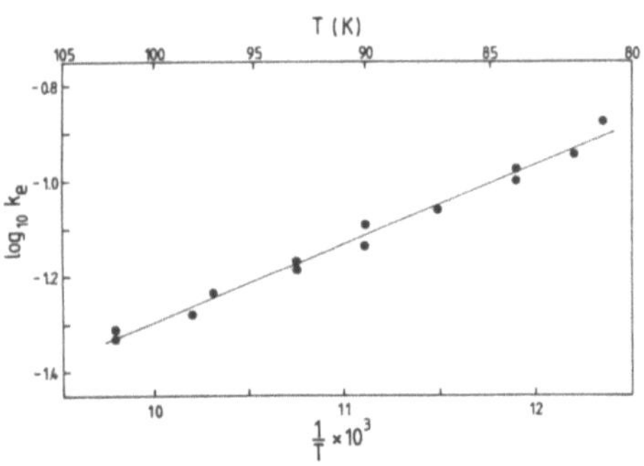

Fig.17.3

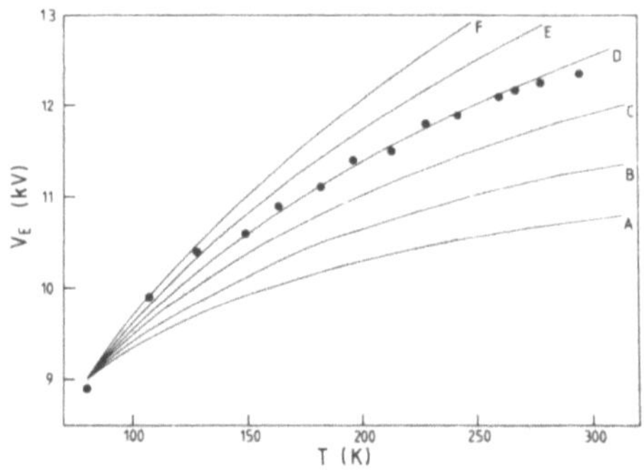

Fig.17.4

A field strength of about F = 11 V/nm, which is the best image field strength for hydrogen imaging, causes a continous evaporation or desorption of {111} planes. Applying a technique first described by BRANDON [17.31], desorption rates are measured by counting the number of planes which are removed during a certain time. The evaporate of a plane always starts at the outskirts. The plane gradually shrinks to a central spot; when it finally disappears, the next plane starts to shrink, and so forth. Thus the rate of disappearance of surface planes is simply measured with a stop watch.

An unexpected temperature dependence of the desorption rate k_e was observed for Si{111}. As shown in Fig.17.3, k_e decreases with increasing temperature. Between T = 80 K and T = 105 K, the log of k_e is inversely proportional to temperature, the slope corresponding to 0.034 eV. This unusual behavior is further confirmed by determining the field evaporation voltage V_E (which is linearly related to the field evaporation field strength) as a function of tip temperature. Keeping the evaporation rate (0.2 planes/s) and the hydrogen pressure constant, the temperature dependence of V_E behaves as indicated by the filled circles in Fig.17.4. The value of $V_E = 8.9$ kV at 80 K increases to $V_E = 12.3$ kV at 300 K. This is not the characteristic of a thermally activated, but of a thermally retarded process. However, the temperature effect can be readily understood if we consider the influence of the hydrogen pressure: k increases linearly with the H_2 pressure and with applied voltage, while the tip temperature is kept constant. At a voltage of 10.7 kV and 80 K, for instance, $k_e = 3.5 \cdot 10^{-2}$ planes/s at $p_{H_2} = 4 \cdot 10^{-6}$ torr, and $k_e = 0.7$ planes/s at $p_{H_2} = 8 \cdot 10^{-5}$ torr. The desorption of surface layers of the Si{111} plane is stimulated by hydrogen. Without an electric field, molecular hydrogen is not or is only weakly adsorbed on silicon at or above 80 K, while atomic hydrogen forms surface hydrides [17.32]. In the presence of the high electric field, however, hydrogen is field adsorbed in a molecular form. The calculated field binding energy ($\alpha/2 \cdot F^2$) at F = 13 V/nm is precisely 0.034 eV, i.e., as derived from the temperature dependence of k_e. The explanation for the unexpected temperature effect is given by the temperature dependence of the hydrogen adsorption. With increasing temperature the hydrogen coverage which is necessary to activate the field desorption process diminishes.

The following mechanism explains our observations. At the silicon surface field-adsorbed molecular hydrogen undergoes a field-induced dissociation reaction,

$$\mathrm{Si\{111\}} + H_2 \xrightarrow{F} \mathrm{Si\text{-}H} + H^+ + e^-_{(s)} \quad . \tag{17.4}$$

A process which is observed in space only at F > 36 V/nm [17.33] occurs on Si already at F = 13 V/nm. The dissociation is energetically favored at the surface for three reasons: Each hydrogen atom can saturate one of the dangling bonds of a Si{111} surface and gain the Si-H bond energy. During this process the electron also gains energy (given by the work function) by entering the solid ($e^-_{(s)}$). Furthermore, the field penetration depth into the semiconductor is enlarged (see below). Field penetration, however, means charge depletion and hence lowering of Si-Si bond energies. As described in more detail elsewhere [17.30], the concerted mechanism of electron tunneling, of field dissociation of hydrogen, and of surface hydride formation as in (17.4) can be calculated by applying the WKB approximation. These calculations, using the experimental rate at T = 80 K as reference, yield the solid curves A to F in Fig.17.4 for different assumed field strength values, with F = 13 V/nm in D as the most probable one (from radius determination a value of F = 11 V/nm was expected for D, in quite good agreement). Two other observations support the mechanism of (17.4):

1) Silicon hydrides are major desorption products, and

2) SiH_4 as the imaging gas dissociates and forms atomic hydrogen much easier than H_2. The desorption voltage V_E is reduced in SiH_4 to 58% of the value in H_2.

The field-induced formation of surface hydrides on silicon sheds further light on yet unexplained details in the mechanism of field evaporation of semiconductors [17.34]. Dangling bonds form surface states, which are occupied by electrons at a positive tip voltage. These occupied surface states compensate for most of the charge which is necessary to establish the high electric field. After removing the surface states by hydrogen atoms, the penetration depth of the electric field into the semiconductor increases (> 5 Å), causing Si-Si bond weakening beneath the top layer by charge depletion. This is one of the reasons for the unexpectedly low evaporation fields of semiconductors in hydrogen.

17.4 Surface Reactions Investigated by Field Pulse Techniques and Time-of-Flight Mass Spectrometry

Desorption fields of the order of 10 V/nm can be reached by field pulses within 10 to 100 ns. Although this is still a long time compared with that for many chemical reactions, which may occur within less than 1 picosecond, field pulses are often suitable for the instant analysis of surface composition. Following the early work of INGHRAM and GOMER [17.35], this field pulse technique has been used to identify intermediates of surface reactions and to determine kinetic parameters, like rate constants of individual reaction steps.

Examples for former work are the reaction of sulfur molecules on W, Pt, or Ag surfaces [17.36,37], the polymerization of acetone [17.38], or the reaction of dicyan at Pt surfaces [17.39]. We will discuss recent investigations of the formation of nickeltetracarbonyl and the intermediates thereof formed during the interaction of CO with Ni surfaces [17.40].

Field pulses, as schematically described in Fig.17.5a, are applied with a certain repetition frequency. During the time between pulses, known as the reaction time t_R, molecules are adsorbed from the continuously supplied

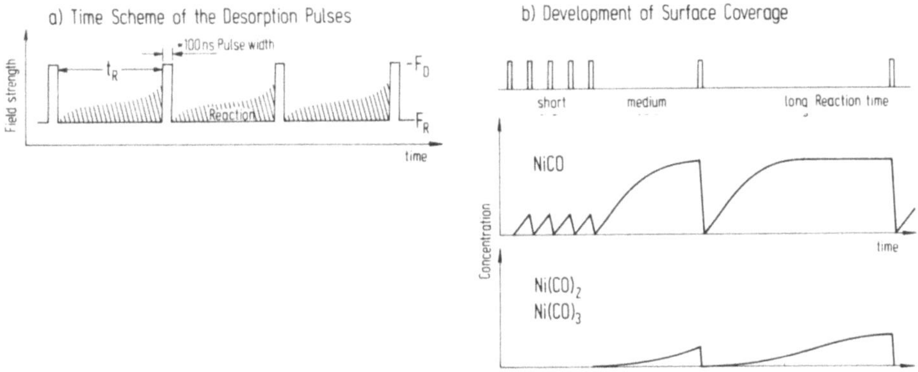

Fig.17.5. (a) Schematic presentation of field pulses in application for kinetic measurements for the system Ni/CO. (b) Different regimes with initial rates, medium reaction times and steady state condition

gas phase, and intermediates and products are formed at the surface. These species are then desorbed by the next pulse and mass analyzed by a TOF spectrometer. The reaction time t_R may be varied from 10^{-5} s to several seconds. Two requirements have to be fulfilled for using this field pulse desorption method as a quantitative tool for reaction kinetic studies [17.22]:

1) The reaction partners have to be desorbed quantitatively by each pulse.

2) During the pulse rise time any field-dependent alterations of the chemical composition of pertinant compounds have to be excluded.

The first requirement can often be tested by gradually increasing the desorption field amplitude. When ion intensities reach a certain level and become independent of the further increase of the field strength, quantitative field desorption can be assumed. The second requirement can be controlled by successively increasing the field strength (reaction field) during the reaction time t_R. Intensity alterations show whether a surface reaction is field sensitive or rather field insensitive. Another possibility for testing whether reaction products are due to alteration while the field pulse increases consists of successively changing the slope of the increasing pulse amplitude from fast pulses, which are desirable for kinetic measurements, to a slow pulse rise which is used in the field ramp method, described previously.

Kinetic parameters have been measured for the reaction of CD with Ni at 270 to 380 K and CO gas pressures up to 2×10^{-4} torr [17.40,41]. For short reaction times, only the initial stages of the surface reaction are analyzed. For example, in the Ni-CO system at 295 K shown in Fig.17.5b, when t_R is short the CO molecules are only in contact with the surface long enough to form NiCO before being desorbed by the next pulse. On increasing t_R, the concentration of desorbed $NiCO^+$ eventually reaches a steady state where further change in t_R has no effect. Also, as t_R is increased, the other intermediates are sequentially desorbed (Fig.17.5b). However, $Ni(CO)_4^+$ is not (or seldom) observed. This technique provides us with the steady-state concentrations of the intermediates and the time taken to reestablish this steady state after each pulse. From this we can gain information about the reaction mechanism and the rates of the various individual steps.

Chemisorbed CO is field desorbed as CO^+ or $NiCO^+$ as shown by the influence of CO pressure and field pulse amplitude on ion intensities. Obviously, the CO chemisorption bond must be strong enough so that a surface Ni atom can be removed together with a chemisorbed CO. Ni^+ and Ni^{++} ions also desorb during the field pulse so that the surface is continuously renewed during the experiment.

The results of the reaction time variation experiments (Fig.17.6) show an early increase in $NiCO^+$ intensity which saturates for $t_R \sim 2 \cdot 10^{-3}$ s. $Ni(CO)_2$ only begins to form at $t_R > 10^{-3}$ s, showing a noticeable delay after the formation of NiCO. After a further short delay $Ni(CO)_3$ is formed. Steady-state conditions for all species are attained after $\sim 5 \cdot 10^{-2}$ s. A detailed kinetic analysis [17.40] shows that the intermediate with the longest delay time not necessarily is the product of the rate determining step. Unambiguous conclusions are possible only after rates of the back reaction, the decomposition of $Ni(CO)_4$, are known.

Field ion microscopic observations have shown that, in the presence of high electric fields, the reaction of CO with a Ni surface to yield volatile Ni products occurs preferentially at kink sites. Under experimental conditions, we expect a near-saturated layer of CO molecules and it may be possible that some Ni atoms, particularity those at steps or kink sites (which are more exposed), may be bonded to two CO molecules. We then suggest that the subsequent slow step consists of the migration of an exposed Ni atom away from a kink site to form an adsorbed $Ni(CO)_2$ species which can then react relatively quickly with 2 more CO molecules to form the tetracarbonyl. The temperature dependence shows a maximum rate of intermediate formation at ≈ 340 K (Fig.17.7). Such a maximum has also been reported for studies with macroscopic crystals.

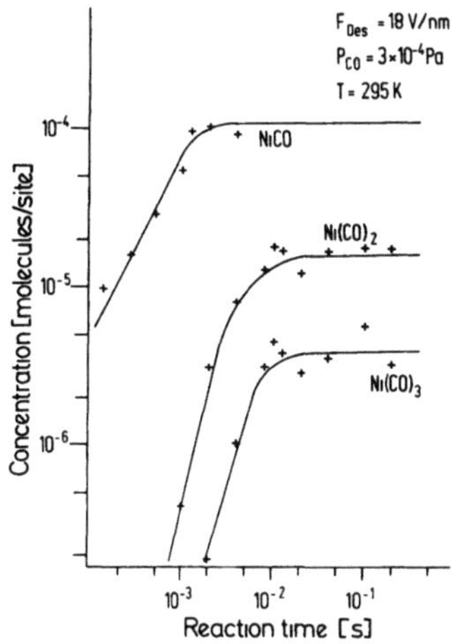

Fig.17.6. Intermediates of the reaction of CO with nickel as function of the reaction time

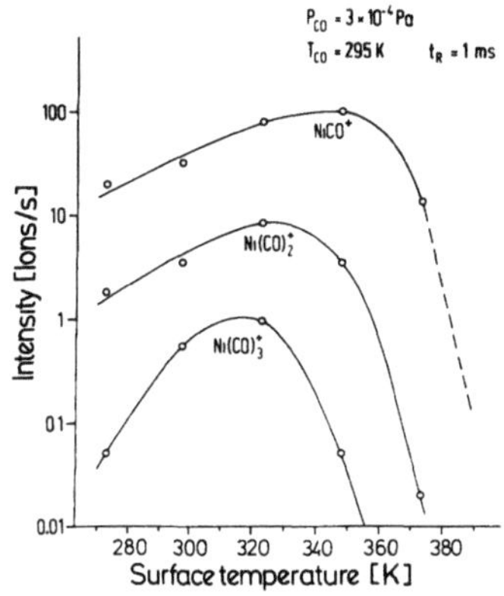

Fig.17.7. Temperature dependence of $Ni(CO)_x$ formation

A reaction rate of 10^{-2} molecules/site. 5e is derived from our data for $Ni(CO)_2$. This is roughly one order of magnitude higher than those values reported for the microscopic reaction at higher pressures and rough surfaces [17.42,43]. The importance of surface crystallography in the formation of $Ni(CO)_4$ can further be observed in alterations of the assemblage of forms of the Ni crystal during this reaction. De GROOT et al. [17.43] recently showed that Ni{100} and Ni{110} are transformed into Ni{111} facets during this reaction. The Ni{111} is the most stable plane.

17.5 Field Ion Appearance Spectroscopy

The electric field in front of the emitter surface reduces the potential barrier for ion desorption, as discussed in Fig.17.1. The binding energy can thus be compensated by electrostatic forces or, at elevated temperatures, partially by thermal activation. The activation energy of field desorption (17.2) is therefore field dependent. From Fig.17.1 we also can deduce the energy which the electric field has to provide for the ion desorption. As discussed in [17.45] this determines the appearance energy A_E

$$A_E = (I - or) + (E_A + P) - Q(F) \quad . \qquad (17.5)$$

Experimentally ΔE can be evaluated from the ion's energy distribution. In our laboratory the onset in the integral energy distribution is measured by a

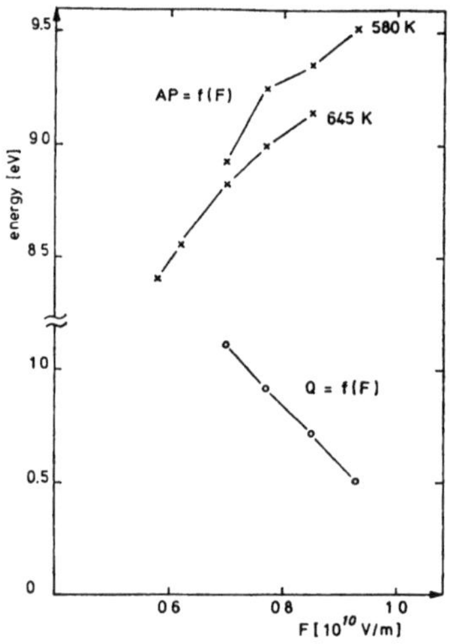

Fig.17.8. Field desorption of Ag^+ from a tungsten emitter. Experimental activation energies Q from an Arrhenius plot and AP($\equiv A_E$) values as function of the field strength F [17.45]

retarding field. This onset δ_0 has to be corrected by the work function ϕ_R of the retarder electrode in order to determine the appearance energy A_E in the integral energy distribution.

As an example, let us discuss the measurements of HUMMEL et al. [17.45] on the field desorption of Ag ions from a tungsten surface. Two kinds of investigations have been performed. The rate of field desorption of Ag^+ ions has been measured to evaluate Q as function of the field strength. Simultaneously, the onset of integral energy distribution of Ag^+ ions gave corresponding A_E data. The results are demonstrated in Fig.17.8. As to be expected from Fig.17.1, Q values decrease when the field strength is raised, while A_E values — identical with the notation appearance potential AP — increase. The sum A_E + Q is constant and only slightly temperature dependent. From these measurements, interaction energies of adsorbed layers can be derived. In another case, measurements of the field evaporation of Rhodium were used to determine the shape of the potential-energy curve and to evaluate the Surface Debye temperature [17.46].

17.6 Electron-Stimulated Field Desorption

Before we consider the influence of photons, a few experiments will be discussed which show how electrons may stimulate desorption processes. The general question is, whether high-energy electronic levels of an adparticle can be populated for sufficient time, such that field ionization and desorption of an electronically excited particle can occur at distances smaller than x_c, i.e., within the forbidden zone. Field ionization within the forbidden zone has already been observed by MÖLLER et al. [17.47,48].

Another fundamental question concerns the field dependence of ionization cross sections of adparticles. Without a high electric field these cross sections for electron impact ionization are smaller for adsorbed particles than for gaseous ones by orders of magnitude [17.49,50]. Deexcitation by the solid surface is significant and may be caused by a reneutralization process. Since the ionization potential of the desorbing particle usually exceeds the work function of the metal, the escaping ion has to pass a zone immediately in front of the surface where it is energetically unstable and where it may recapture an electron from the solid. The question is whether reneutralization may be influenced by an opposing electric field and whether ionization cross sections, consequently, would be field dependent.

Since it is impossible to focus an electron beam of sufficient density ($> 10^4$ A/cm^2) onto the apex of an emitter tip from an external conventional electron source without bombarding and heating the whole emitter system, electrons have to be provided by other means. An auxiliary gas with a lower ionization potential than the surface adparticles to be investigated is introduced. At the field strength of field desorption this auxiliary gas is ionized far away (> 100 nm) in the space in front of the emitter surface, where the field strength is just sufficient for ionization. The electrons released during this ionization step are accelerated and directed towards the tip surface. The energy distribution of the auxiliary gas ions is determined by field ion appearance spectroscopy (FIAPS), as described in detail elsewhere [17.51]. This gives the potential of the origin of electrons, i.e., the energy distribution of impinging electrons. Their density is derived from current measurements and tip radius determination.

Electron-impact-stimulated field desorption has been studied only for some rather simple systems of field-adsorped gas atoms by ERNST et al. [17.52]. He and Ne ions, field desorbed by electron impact from W, Ir, and Rh surfaces, were studied using hydrogen as the auxiliary gas. We will discuss results from the Ne-W system.

In Fig.17.9 onset curves of the integral energy distribution for singly charged ions of ^{20}Ne are shown. The potential of 17.20 V for the onset described by k_f, the rate of field-ionized Ne, means that these ions are not generated precisely at the tip potential (+ 3.000 kV, potential difference V = 0), but at a distance (x_c) where the potential is 17.20 V less than the tip potential. The appearance energy of ions is obtained when the work

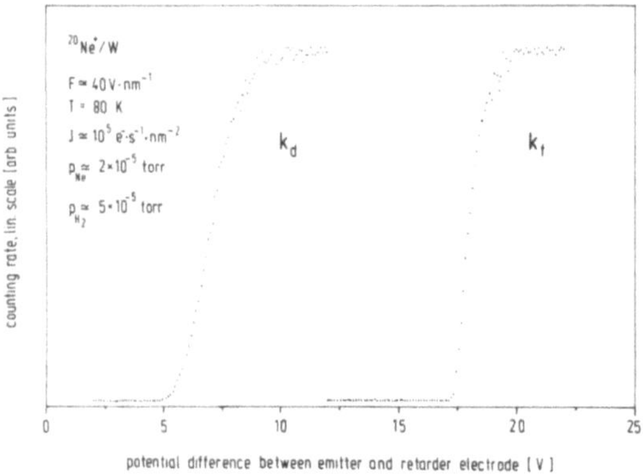

Fig.17.9. Electron-stimulated field desorption of Ne^+ from tungsten. Onset curves of integral energy distributions of field-desorbed Ne^+ (k_d, ions/s) and normally field-ionized $Ne^+(k_f)$ as measured with an electrostatic filter lens [17.52]

function of the retarder (4.36 V) is taken into account. As can be demonstrated in an energetic cycle [17.51], the onset potential is determined by the ionization potential of the rare-gas atom (I_{Ne} = 21.56 V) and the retarder work function (V = 17.20 = 21.56 - 4.36). The Ne-atoms are field ionized at distances $\gtrsim x_c$ with the counting rate k_f. A further onset curve, counting rate k_d (ions/s), comes from electron-stimulated field-desorbed species. The onset of these ions is 12.4 V smaller than that for field-ionized particles. When field ionization in a field of 40 V/nm occurs at $x_c \approx 4.2$ Å in front of the surface, field- desorbed ions are generated at a distance slightly more than 1 Å above the surface.

Following the treatments by TSONG and MÜLLER [17.53] and RENDULIC [17.54], the surface coverage Θ of Ne can be described by a Langmuir isotherm,

$$\Theta = \frac{bp}{1 + bp} \tag{17.6}$$

where b is dependent on temperature T, field strength F, and also the impingement rate J of electrons. If the field-dependent gas supply, including the sticking coefficient, is described by the factor a, we can write for b the ratio of adsorbing and desorbing particles,

$$b = \frac{a(FT)}{\nu\exp(-\Delta B/kT) + Q^+J} \tag{17.7}$$

where ΔB is a short-range binding energy and Q^+ a cross section. The desorption rate of ions is then $k_d = \Theta(T,F,J)Q^+J$, or combining (17.6) and (17.7)

$$k_d(T,\ F,\ J) = \frac{b(T,\ F,\ J)pQ^+J}{1 + b(T,\ F,\ J)p} \tag{17.8}$$

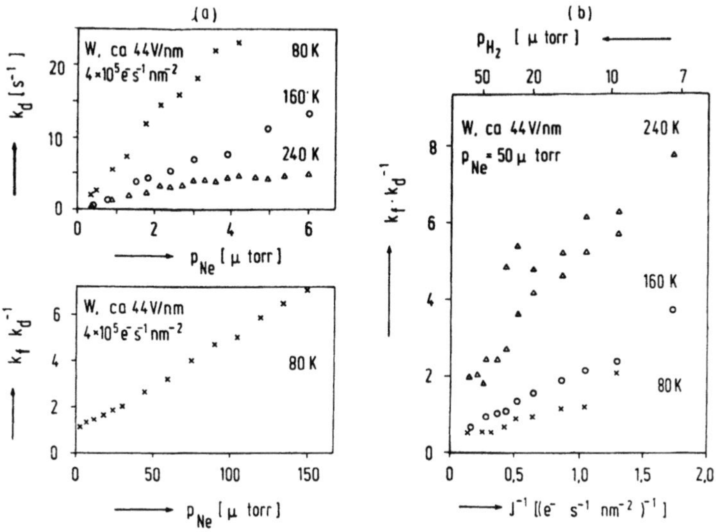

Fig.17.10a,b. Experimental results on absolute desorption rates k_d[(a) upper part] and relative rates $k_f k_d^{-1}$ [(a) lower diagram, and (b)] of electron-stimulated field-desorbed Ne^+ as function of Ne gas pressure and hydrogen partial pressure which determines the density J of impinging electrons [17.51]

If c is the ionization probability of normal field ionization, the rate k_f is $k_f = c(F, T)p$, and

$$\frac{k_f}{k_d} = \frac{c(F,T)}{Q^+J}\,p + \frac{c(F,T)}{b(F,T,J)Q^+J} \quad . \tag{17.9}$$

The cross section can be evaluated from the slope in the $(k_f/k_d)/J$ diagram.

The rates k_d and k_f have been measured as a function of the gas pressure of neon in order to change the surface coverage Θ_{Ne}. By changing the hydrogen pressure, the electron impingement rate J can be varied. In Fig.17.10a (upper part) the absolute rate k_d is seen to be nearly linear with respect to the neon partial pressure p_{Ne}. This indicates low surface coverage, i.e., the initial low-pressure slope of a Langmuir adsorption isotherm. From the rates at different temperatures the binding energy of field-adsorbed neon is determined to be 0.03 eV. Avoiding intensity fluctuation the ratios $k_f \cdot k_d^{-1}$ (Fig.17.10a, lower part) are more accurate for measurement of desorption cross sections. The electrons released during field ionization of hydrogen had a mean energy of 450 eV with a half width of ca. 220 eV. The $k_f \cdot k_d^{-1}$ relative rate dependence on the electron impingement rate is given for three different temperatures in Fig.17.10b. At constant p_{Ne} the relative rate increases with electron impingement. The slope increases with temperature as to be expected for decreasing surface coverage. A cross section Q^+ can be deduced from these measurements and gives a value of $Q^+ \approx 10^{-16}$ cm^2. This value is orders-of-magnitude higher than desorption without an electric field and approaches the value for gas-phase ionization.

One interesting new result, recently observed by ERNST [17.55], is the formation of doubly charged rare-gas ions such as He^{2+} and Ne^{2+} during electron-induced field desorption of field-adsorbed rare gases.

17.7 Photon-Induced Field Desorption

17.7.1 Instrumental Development

In early experiments [17.56] on current-voltage characteristics of field desorption it was found that ion intensities could be increased by photon impingement. This was verified in at least some cases, for instance for argon or acetone on an Al_2O_3-coated tungsten emitter. Detailed information about the influence of photons on field desorption was, however, only obtained after time-correlated TOF mass spectrometry was developed [17.57,58]. KELLOGG et al. later used a pulsed-laser atom probe for similar investigations [17.59].

The photon-induced field ionization (PIFI) mass spectrometer, as described in detail elsewhere [17.57,58], operates as follows (Fig.17.11): Short photon pulses (2.5 or 5 ns pulse length) impinge onto a field emitter from a laser light source which is the combination of a nitrogen laser and a dye laser yielding an energy density of $\gg 10^{-2}$ J/pulse/cm^2. Light pulses with a repetition rate of up to 60 s^{-1} generate ions which are superimposed onto ion intensities of normal field ionization currents. The ions are accelerated to a few kV before entering a field-free drift tube of $\approx$ 30-cm length. In the microscope mode, ions from about $5 \cdot 10^3$ surface sites are observed at the image intensifier with a luminescent screen. The TOF measurements are performed in such a manner that the ion-detecting systems are only activated for a short time ($\approx$ 10 μs) after the photon pulse has arrived. With this gating method background signals could be reduced and, during normal field ion emission with $\approx 10^6$ ions/s, an average PIFI rates of $\approx$ 10 ions/s still gave sufficient intensity for a TOF mass spectrum after accumulation of TOF data from 10^3 to 10^4 photon pulses into a multichannel analyser. For each light pulse the photon intensity (which fluctuates) was measured and stored together with the corresponding mass spectrum.

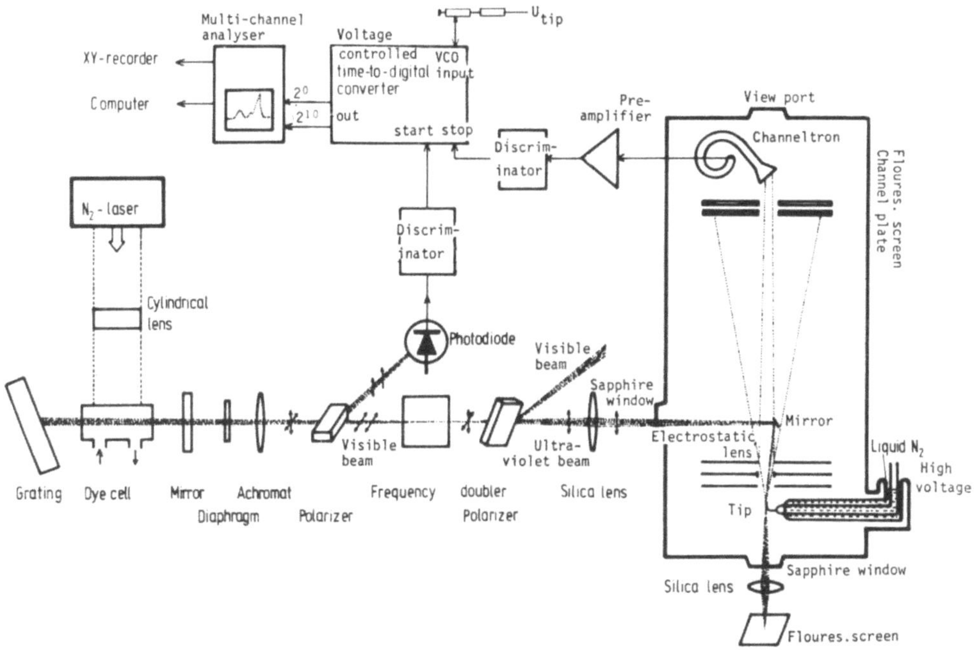

Fig.17.11. PIFIMS experimental arrangements [17.58]

17.7.2 Electronic Excitation of Adparticles

As discussed by LICHTMAN and SHAPIRA, cross sections of photodesorption (of neutral molecules and ions) from metal surfaces are extremely small ($< 10^{-22}$ cm^2), corresponding to a quantum efficiency of less than 10^{-8} molecules per photon [17.60]. Again the question arises whether these cross sections are influenced by external electric fields. Unfortunately the wavelength regions of our photons, from a frequency-doubled dye laser, reach only minimum values of ≈ 270 nm or maximum photon energies of ≈ 5 eV. These photon energies are not sufficiently high for performing experiments with field-adsorbed neon, as studied with electron impact.

In a series of experiments, true electronic excitation (quantum effects) in PIFI have only been observed in a few cases. The common influence of photon impact is of a thermal nature. Quantum effects in PIFI are identified by two requirements. First, electron excitation due to photons leads to ion desorption at field strength values which are not sufficiently high for normal field emission and, second, the number of released ions should be proportional to the number of impinging photons. Both of these requirements are fulfilled for PIFI of ethylene on Ag at low temperatures, $T < 100$ K. This is

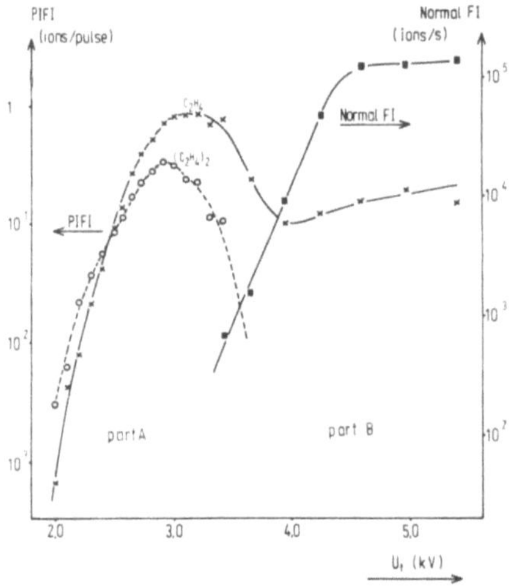

Fig.17.12. PIFI mass spectrum of $C_2H_4^+$ and $(C_2H_4)_2^+$ as a function of tip voltage (left-hand side) compared with normal field ion emission (FI) (right-hand side), Ag tip, 80 K, photons with λ=300 nm (4.1 eV) [17.58]

described in detail elsewhere [17.60]; only the most striking arguments for the quantum effect will be given here.

In Fig.17.12, the normal field ion emission (FI) intensity of $C_2H_4^+$ ions as a function of the tip voltage is given at the right-hand side of the diagram. At a tip voltage $U_t \approx 3.2$ kV, normal field ion emission starts to rise. This voltage corresponds to a field strength of 9.3 V/nm. The rising part of the normal field ion curve is determined by the increase of ionization probabilities with increasing field strength. At $U_t \approx 4.7$ kV a plateau is reached, which is caused by limiting particle supply, with a steady-state ion current of ca. $2 \cdot 10^5$ ions/s. When the tip surface is irradiated by photon pulses [λ = 300 nm, 60 pulses/s, ca. 1 photon $(\text{pulse})^{-1}$ $(\text{surface atom})^{-1}$] some surface molecules are immediately ionized even at a field strength $F < 9.3$ V/nm. These ions form a mass spectrum with sharp mass signals. The mass peaks in amu, at low fields, correspond to M 28 ($C_2H_4^+$), 29($C_2H_5^+$), 42($C_3H_6^+$), 56($C_4H_8^+$), 57($C_4H_9^+$), 70($C_5H_{10}^+$), 71($C_5H_{11}^+$), 84($C_6H_{12}^+$), and 85($C_6H_{13}^+$). The main intensities $C_2H_4^+$ and $C_4H_8^+$ are presented as functions of the tip voltage in the left-hand scale of Fig.17.12. The onset of PIFI starts at 5.8 V/nm and reaches a maximum when normal field ion emission begins. The signal of $C_2H_4^+$ has a second slightly increasing intensity at high tip voltages, when normal field ion emission is saturated. Under these higher fields $\approx 10^{-1}$ ions/s from a time correlated PIFI measurement are superimposed on $2 \cdot 10^5$ ions/s of normal field ion emission and still display a good TOF mass spectrum on top of the statistically measured ions of not time-correlated

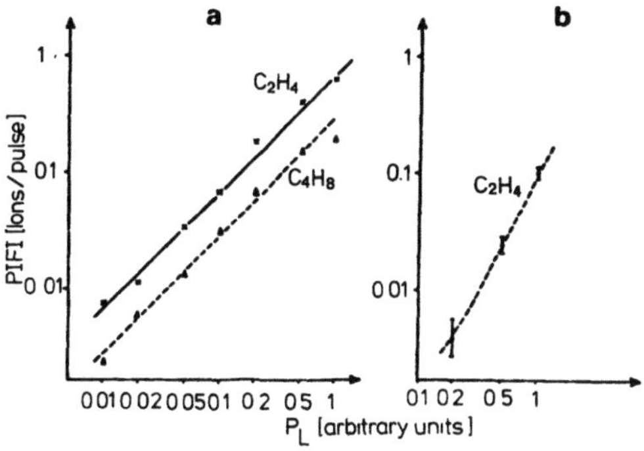

Fig.17.13,a,b. Ion intensities as a function of laser power P_L at constant photon energy (λ=300 nm), 60 Hz, T_t=78 K, $P_{C_2H_4}$=2 10^{-5} torr, (a) PIFI at U_t=3.199 kV, (b) at U_t=5.4 kV with predominating normal FI [17.60]

normal field ion emission. The mass spectrum at $F < 9.3$ V/nm shows that chemical reactions occur in an adsorption layer of ethylene on the Ag surface. In fact, a multilayer of C_2H_4 is formed at these low temperatures. This multilayer is necessary to find ions at $F < 9.3$ V/nm. When the tip is heated to $T \gtrsim 102$ K, the multilayer disappears and so do PIFI signals at $F < 9.3$ V/nm. At $F > 9.3$ V/nm PIFI signals are still observed at higher temperatures which are, as will be shown, of thermal origin.

The power of laser light P_L at constant wave length is proportional to the number of photons impinging onto the tip surface. The experiment of power variation should therefore be a test of whether PIFI is due to an excitation mechanism by quantum effects, since in this case ion intensity would be proportional to the power P_L. In Fig.17.13a the slope is ≈ 1.0, suggesting an excitation mechanism by single photons. The determination of quantum yield is intricate since details of photon absorption and of the excitation process are still unknown. If we assume that 10^3 photons/pulse are absorbed in the monitored area of the tip surface, we have to correlate this value with the maximum of $\approx$1/ion pulse in Fig.17.13a. From these experimental data an unexpectedly high value for the ion yield (of $\approx 10^{-3}$ ions/photon) must be deduced.

In Fig.17.13b the slope is much steeper. Since multiphoton effects have to be excluded at these low photon densities, this must be an indication for thermal effects, where an exponential dependence of ion intensities on P_L is expected.

PIFI at Ag has been obtained at $F < 9.3$ V/nm only for photons with λ = 300 nm wave length, not with λ = 440 or λ = 600 nm. Details of the excitation process are still obscure. Since there are no energy levels to be excited by 4.1 eV photons in the ethylene molecule, an indirect transition of an electron from the ethylene electronic ground level into the Fermi level of the Ag surface has been discussed. This involves a subsequent fast charge transport by electron hole conduction in a C_2H_4 multilayer. However, more sophisticated experiments have to be performed in order to confirm such a model.

17.7.3 Thermal Activation of Adparticles

The conversion of the absorbed photon energy into thermal energy in metals is a very fast process, with typical times of 10^{-13} to 10^{-14} s. This time is short compared to the photon pulse length (10^{-8} to 10^{-9} s) which is used in the author's laboratory. We, therefore, can assume immediate heat formation after light absorption and can use macroscopic heat conductivity data for estimating heat dissipation and local temperature rise. The photon pulse causes a simultaneous temperature pulse with identical time structure. Since the laser spot which shines onto the apex of the emitter tip is much larger than the tip diameter, the temperature rise at the tip surface occurs homogeneously.

Several facts complicate the precise evaluation of the temperature values while the laser pulse shines: variations in laser pulse power densities, deviation of optical absorption coefficients for small particles (tip radius $< \lambda$), and the reflectivity of the curved surface. A desirable method to measure surface temperatures of a tip emitter, based on the determination of the energy distribution of field-emitted electrons as demonstrated by KÖRMENDI [17.61], could not be applied so far. SULLIVAN et al., COWIN et al., and READY [17.62-64] considered different possibilities for estimating the temperature rise caused by laser radiation. For the experiments described in Figs.17.12,13, the temperature rise was of the order of only 10 K. In other cases (for instance, shining the light pulses of the nitrogen laser of Fig. 17.11 directly onto the emitter) a temperature increase of several hundred or even a thousand degrees could be achieved. With this technique, all the regimes between pure field desorption, with minor temperature rise, on the one side, up to high temperature rise, with pure thermal ionization, on the other side can be explored. The latter is known as laser microprobe mass analyzer (LAMMA) [17.65]. A recent review of laser applications in mass spectrometric analysis is given by COUZERNIUS and CAPELLEN [17.66]. Mass spectra of surface compounds may be very different for regimes with low- and with high-temperature rise.

As discussed before, the experiments described in Fig.17.12 are conducted with a rather low-temperature rise. Of particular interest is the origin of the PIFI intensities at $F > 9.3$ V/nm, where normal field ion emission prevails. At very high fields and saturation of field ion currents, PIFI can still be observed. This is due to the thermal activation of molecules adsorbed below the critical distance x_c. For ethylene with molecular dimensions of ≈ 30 nm along the C-C axis and ≈ 15 nm in the perpendicular direction, field adsorption is very likely. At the onset field of 9.3 V/nm the critical distance is approximately $x_c \approx 60$ nm. Even at the highest applied fields (with $x_c < 30$ nm), thermal activation in PIFI could still be observed. At a very high field strength, where field evaporation of the metal begins, complex ions $[Ag(C_2H_4)_n^+$ $(n \leqslant 3)]$ are obtained with a silver emitter.

The thermal activation process obviously increases the supply into the ionization zone above the surface. This can happen either by exciting vibrational modes directly at the apex of the emitter, with bond elongation in the anharmonic potential curve (Fig.17.1), or by supply from the near apex shank region. These diffusion processes can require more time than given by the pulse length. In TOF mass spectra this phenomenon causes signal broadening.

17.7.4 Surface Diffusion

The temperature pulse which is synchronous with the laser pulse enhances not only thermal desorption but also surface diffusion. Particle motion by surface diffusion may lead to a delayed supply into the ionization zone. This time delay is reflected in particular signal shapes of TOF mass spectra [17.67] and can be used to study surface diffusion.

In some PIFI mass spectra mass signals display a certain "tailing". One example with pronounced signal tailing is demonstrated in Fig.17.14 for a mass spectrum of argon with residual CO at an iridium field emitter. A normal field ion current of $8 \cdot 10^4$ ions/s is measured, and appears as background in the mass spectrum. This spectrum contains 58,050 signals, which are obtained in a sampling time of 1500 s. The first signal represents the peak due to the laser pulse because the channeltron is sensitive to stray photons. The sharpness of this signal also proves that signal broadening due to electronic jitter times is negligible. At a TOF of $t \approx 1.7$ μs a slightly tailing signal represents CO^+ molecular ions. The main signal ($t = 2$ μs) reaches a maximum at $m/z = 44$ and then displays a long signal tail. The peak maximum of Ar^+ is reached 110 ns after the laser pulse impinged, and the decay of the signal intensity can be approximated by an exponential law; the relaxation time is $\tau = 185$ ns (for 1/e of the initial intensity).

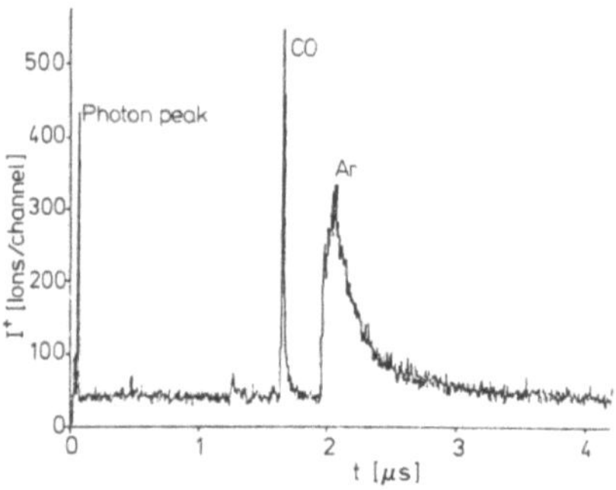

Fig.17.14. PIFI mass spectrum of argon (with CO residual gas) for an Ir emitter at 80 K, $p=2\cdot10^{-5}$ torr, $U_t=6.6$ kV, $F\approx42$ V/nm, $\lambda=300$ nm, pulse length 2.5 ns, $\approx10^{14}$ photons/cm^2/pulse and 60 pulses/s [17.67]

Signal tailing is much more pronounced for Ar than for CO. On the iridium surface CO forms a chemisorbed layer, which is rather immobile at 80 K. Argon, on the other hand, is mobile in a physisorbed layer. A temperature jump activates surface diffusion with remarkably long delay times. Extended investigations on the influence of field strength, laser power density and pulse repetition rate showed that different supply regimes are of importance in PIFI [17.62]. Besides desorption of neutral molecules and ions directly from the "visible" apex area, there is fast material supply from nearby regions by a field directed motion. From far-shank emitter regions, where the field strength is not sufficient for a field directed transport, statistical diffusion is combined with long delay times. The time for refilling an adsorbed layer after depletion by a single photon pulse could be measured and, thus, details could be evaluation of the particle supply regime by gas impact, and by surface migration during field desorption processes.

17.7.5 Formation of Complex Ions and Cluster Ions

At low temperatures and also small temperature pulses, cluster ions with low internal binding energies are found. Some examples can be demonstrated which have been observed during the field desorption of ethylene and water mixtures from emitters of IB metals [17.68]. At low light intensities a sequence of ions of ethylene and water is formed, as shown in Fig.17.15. The mass spectrum displays the parent molecular ion H_2O^+ and $C_2H_4^+$ and parent molecules with attached protons, H_3O^+ and $C_2H_4H^+$, dimers $(H_2O)_2^+$, $(H_2O)_2H^+$, $(C_2H_4)_2^+$, and $(C_2H_4)_2H^+$, water adducts of ethylene $(C_2H_4)(H_2O)_m^+$, $(m < 3)$, and a series of water clusters. $(H_2O)_mH^+$ with $m \leqslant 12$. Cluster ions of water have already been observed in field desorption mass spectrometry [17.69]. A single proton stabilizes a cluster of up to 12 water molecules. The binding energy of such a cluster must be minor. The fact that such an ensemble can desorb in ionic

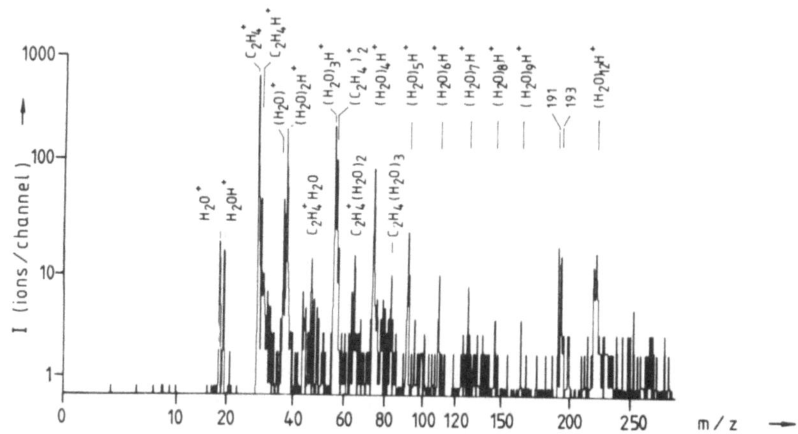

Fig.17.15. PIFI of coadsorbed C_2H_4 and H_2O from a silver emitter, T = 80 K, p = $2 \cdot 10^{-4}$ torr, λ = 290 nm, U_t = 2.1 kV, power intensity ≈ 0.1 j/pulse[1]/cm^2 [17.68]

form from the emitter surface proves that the energy transfer into the desorbing species is very small.

In earlier work [17.70,71] complex metal ions $Me^+(H_2O)_n$ (Me = Ag, Cu; n ⩽ 12) had already been observed under steady field conditions. Similarly with increasing light pulse intensity metal complexes are formed, with Cu, Ag, or Au emitters. With Au, for instance, complex ions with masses up to m/z ≈ 500 are formed like $Au_2(H_2O)_4^+$ and $Au_2(H_2O)_3\ C_2H_4^+$. The further increase of the laser power density (≈ 1 J/pulse/cm^2) causes an enlarged desorption of multiply charged metal ions and diminished intensities of complex ions. For gold, for instance, Au_3^{++}, Au_2^{++}, and Au^{++} are found besides Au_3^+, Au_2^+, and Au^+, and only small intensities of the complex ions $(AuC_2H_4)^+$ and $(AuH_2O)^+$ are left in the mass spectra under these conditions.

Extended measurements on copper [17.72] and nickel field emitters have shown that, at constant field strength, the proportion of doubly charged to singly charged ions sensitively increases with laser power density. This behavior gives further arguments for understanding details of the excitation mechanism during the ion generation processes at emitter surfaces [17.73,74].

After increasing the light power density by another order of magnitude a plasma can be formed in front of the emitter surface. Highly charged ions are then emitted with very broad energy distributions (up to ±100 eV with respect to the tip potential). Observation of the tip by built-in FIM revealed extensive surface reconstruction and often considerable damage to the emitter.

17.8 Summary

Field ion microscopy and field ion mass spectroscopy as well as photon-induced field ion mass spectrometry are now well established experimental tools for investigating fundamental problems in surface science. These techniques have a high lateral resolution in special events, with the ultimate sensitivity for individual surface sites and individual surface particles. These advantages are not offered by other methods in surface science. After years of instrumental development, these techniques are now sufficiently refined to provide research opportunities for solving fundamental questions about the physical chemistry of surfaces. In particular, the determination of the energy distribution of field desorption products provides information about the energy of surface interactions. The field pulse method has a proven ability to analyse reaction intermediates and products of surface reactions. Furthermore, with field pulse methods it is also possible to measure kinetic parameters which are otherwise not available. The photon pulse technique has only recently been developed and as yet has only been applied to a few selected systems. This technique has not yet been utilized for investigations of surface reactivity and photochemistry at interfaces. These avenues of research remain to be explored.

Acknowledgements. Thanks are due to P. Dowben and R. Driscoll for reading the manuscript and making valuable suggestions. Work reported from the author's laboratory has been performed in cooperations with G. Abend, G. Bozdech, D. Cocke, M. Domke, W. Drachsel, N. Ernst, O. Frank, E. Hummel, Th. Jentsch, N. Kruse, D.B. Liang, S. Nishigaki, T. Sakata, and W.A. Schmidt. The Deutsche Forschungsgemeinschaft supported this work in part by SFB 6/'81.

References

17.1 E.W. Müller: Phys. Rev. *102*, 618 (1956)
17.2 R. Gomer, L.W. Swanson: J. Chem. Phys. *38*, 1613 (1963)
17.3 E.W. Müller, T.T. Tsong: *Field Ion Microscopy, Principles and Applications* (Elsevier, New York 1969) p.230; Prog. Surf. Sci. *4*, 1 (1974)
17.4 H.D. Beckey: *Principles of Field Ionization and Field Desorption Mass Spectrometry* (Pergamon, Oxford 1977) pp.16ff
17.5 J.A. Panitz: Prog. Surf. Sci. *8*, 219 (1978)
17.6 P.A. Redhead: Vacuum *12*, 203 (1962)
17.7 G. Ehrlich: Adv. Catal. *14*, 255 (1963)
17.8 L.A. Petermann: Surf. Sci. *3*, 1 (1972)
17.9 D. Menzel: In *Interactions on Metal Surfaces*, ed. by R. Gomer (Springer, Berlin, Heidelberg, New York 1975) p.102
17.10 D.A. King: Surf. Sci. *47*, 384 (1975)
17.11 F.O. Goodman: Surf. Sci. *5*, 283 (1966)

17.12 S. Holloway, J.L. Beeby: J. Phys. C , 3531 (1975)
17.13 J.L. Beeby: In *Chemistry and Physics of Solid Surfaces II*, ed. by R. Vanselow (CRC, Boca Raton, FL 1979) p.73
17.14 I. Frenkel: Z. Phys. *26*, 117 (1924)
17.15 M. Drechsler: Angew. Chem. *79*, 987 (1967)
17.16 R. Gomer: J. Chem. Phys. *31*, 341 (1959)
17.17 R. Forbes: Prog. Surf. Sci. (in press)
17.18 G. Comsa, R. David: Chem. Phys. Lett. *49*, 512 (1977)
17.19 T.T. Tsong, E.W. Müller: J. Chem. Phys. *41*, 3279 (1964)
17.20 R.J. Madix: In *Chemistry and Physics of Solid Surfaces II*, ed. by R. Vanselow (CRC, Boca Raton, FL 1979) p.63
17.21 J.H. Block: In *Chemistry and Physics of Solid Surfaces I*, ed. by R. Vanselow, S.Y. Tong (CRC, Boca Raton, FL 1977) p.49
17.22 J.H. Block, A.W. Czanderna: In *Methods and Phenomena*, ed. by S.P. Wolsky, A.W. Czanderna (Elsevier, Amsterdam 1975) p.377
17.23 E.W. Müller, J.A. Panitz, S.B. McLane: Rev. Sci. Instrum. *39*, 83 (1968)
17.24 E.W. Müller, S.V. Krishnaswamy: Rev. Sci. Instrum. *45*, 1053 (1974)
17.25 J.A. Panitz: J. Vac. Sci. Technol. *16*, 868 (1979)
17.26 K.D. Rendulic, M. Leisch: Surf. Sci. *93*, 1 (1980)
17.27 D. Cocke, G. Abend, J.H. Block: Z. Phys. Chem. *108*, 247 (1977)
17.28 O. Nishikawa, M. Wada, M. Konishi: Surf. Sci. *97*, 16 (1980)
17.29 M. Wanda, M. Konishi, O. Nishikawa: Surf. Sci. *100*, 439 (1980)
17.30 T. Sakata, J.H. Block: Surf. Sci. Lett. (to be published)
17.31 D.G. Brandon: Brit. J. Appl. Phys. *16*, 683 (1965)
17.32 H. Ibach, J.E. Rowe: Surf. Sci. *43*, 481 (1974)
17.33 G.R. Hanson: J. Chem. Phys. *62*, 1161 (1975)
17.34 L. Ernst, J.H. Block: Surf. Sci. *49*, 293 (1975)
17.35 M. Inghram, R. Gomer: Z. Naturforsch. A*10*, 863 (1955)
17.36 D.L. Cocke, G. Abend, J.H. Block: Int. J. Mass Spectrom. Ion Phys. *24*, 271 (1977)
17.37 D.L. Cocke, G. Abend, J.H. Block: J. Phys. Chem. *80*, 524 (1976)
17.38 F.W. Röllgen, H.D. Beckey: Surf. Sci. *23*, 69 (1970); Surf. Sci. *26*, 100 (1971)
17.39 N. Kruse: Ph. D. Thesis, Technical University Berlin (1980)
17.40 D.B. Liang, N. Kruse, G. Abend, J.H. Block: Surf. Sci. (to be published)
17.41 D.B. Liang, G. Abend, J.H. Block, N. Kruse: In Proc. 28th Int. Field Emission Symp., Oregon, 1981, ed. by L.W. Swanson
17.42 D.E. Williams, J. Pritchard, K.W. Synkes: In Proc. 6th Int. Congress on Catalysis, ed. by G.C. Bond, P.B. Wells, F.C. Tompkins (The Chemical Society, London 1976) p.417
17.43 P. De Groot, M. Coulon, K. Dransfeld: Surf. Sci. *94*, 204 (1980)
17.44 T.T. Tsong, W.A. Schmidt, O. Frank: Surf. Sci. *65*, 109 (1977)
17.45 E. Hummel, M. Domke, J.H. Block: Z. Naturforsch. A*34*, 46 (1978)
17.46 N. Ernst, J.H. Block: Surf. Sci. *91*, L27 (1980)
17.47 E.W. Müller, S.V. Krishnaswamy: Surf. Sci. *36*, 29 (1973)
17.48 T. Sakurai, E.W. Müller: Surf. Sci. *49*, 497 (1975)
17.49 D. Menzel, R. Gomer: J. Chem. Phys. *41*, 3311 (1964); J. Phys. *42*, 886 (1964)
17.50 M.J. Drinkwine, D. Lichtman: Prog. Surf. Sci. *8*, 123 (1977)
17.51 N. Ernst, G. Bozdech, J.H. Block: Int. J. Mass Spectrom. Ion Phys. *28*, 33 (1978)
17.52 N. Ernst, G. Bozdech, J.H. Block: In Proc. 27th Int. Field Emission Symp., Tokyo, 1980, ed. by Y. Yashiro, N. Igata
17.53 T.T. Tsong, E.W. Müller: J. Chem. Phys. *55*, 2884 (1971)
17.54 K.D. Rendulic: Surf. Sci. *34*, 285 (1971)
17.55 N. Ernst: Phys. Rev. Lett. *45*, 1573 (1980)
17.56 T.T. Tsong, J.H. Block, M. Nagasaka, B. Viswanathan: J. Chem. Phys. *65*, 2469 (1976)

17.57 W. Drachsel, S. Nishigaki, J.H. Block: Int. J. Mass. Spectrom. Ion Phys. *32*, 333 (1980)
17.58 W. Drachsel, S. Nishigaki, J.H. Block: Adv. Mass Spectrom. *8*, 1037 (1980)
17.59 G.L. Kellogg, T.T. Tsong: J. Appl. Phys. *51*, 1184 (1980)
17.60 S. Nishigaki, W. Drachsel, J.H. Block: Surf. Sci. *87*, 389 (1977)
17.61 F. Körmendi: Private communication
17.62 T.E. Sullivan, P.H. Cutler, A.A. Lucas: Surf. Sci. *54*, 561 (1976)
17.63 J.P. Cowin, D.J. Auerbach, C. Becker, L. Wharton: Surf. Sci. *54*, 561 (1976)
17.64 J.F. Ready: J. Appl. Phys. *36*, 462 (1965)
17.65 F. Hillenkamp, E. Unsöld, R. Kaufmann, R. Nitsche: Appl. Phys. *8*, 341 (1975)
17.66 R.J. Couzemius, S.M. Capellen: Int. J. Mass Spectrom. Ion Phys. *34*, 197 (1980)
17.67 J.H. Block, Th. Jentsch, W. Drachsel: Int. J. Mass Spectrom. Ion Phys. *38*, 197 (1981)
17.68 W. Drachsel, Th. Jentsch, J.H. Block: In Proc. 28th Int. Field Emission Symp., Portland, Oregon, 1981, ed. by L. Swanson
17.69 H.D. Beckey: Z. Naturforsch. A*15*, 822 (1960)
17.70 W.A. Schmidt, O. Frank, J.H. Block: Surf. Sci. *44*, 185 (1974)
17.71 J.H. Block, W.A. Schmidt: In *The Physical Basis for Heterogenous Catalysis*, ed. by E. Dranglis, R.I. Jaffee (Plenum, New York 1975) p.383
17.72 Th. Jentsch, W. Drachsel, J.H. Block: Int. J. Mass Spectrom. Ion Phys. *38*, 215 (1981)
17.73 N. Ernst, J.H. Block: In Proc. 28th Int. Field Emission Symp., Portland, Oregon, 1981, ed. by L. Swanson
17.74 N. Ernst: Surf. Sci. *87*, 469 (1979)

18. Segregation and Ordering at Alloy Surfaces Studied by Low-Energy Ion Scattering

T. M. Buck

With 14 Figures

18.1 Introduction

Surface segregation in metal alloys, the equilibrium enrichment of the free surface by one of the components, has only recently been recognized as a very general phenomenon having important implications for catalysis, corrosion, microelectronics, and also grain-boundary segregation (Chap.19, [18.1]) with its influence on the strength of alloys. The term surface segregation, as it is usually employed and as it is used in this paper, refers to an equilibrium condition which is normally achieved at an elevated temperature to allow sufficient mobility of the atoms and in UHV to avoid interfering chemical reactions. Unwanted impurities in the bulk, e.g., S in Ni alloys, may also segregate to the surface and shift the equilibrium of interest. However, such impurities can usually be removed by ion bombardment at high temperature, above the range of the experimental measurements, leaving a depleted region deep enough so that impurity atoms cannot reach the surface during the measurements.

The phenomenon of surface segregation was suggested by catalysis research on Cu-Ni alloys using measurements of catalytic activity, hydrogen adsorption, and work function [18.2-4]. Essentially, it was found that a trace of copper in the bulk could produce a large effect on the catalytic behavior of Ni, suggesting that the copper had become enriched at the surface. Experiments with newly developed surface analytical techniques followed in considerable numbers. There have been several reviews, and briefer summaries of the literature accompanying new results in recent years [18.5-14] and since the field is still in a state of flux there will be no attempt in this paper at a definitive or comprehensive review. Rather, the approach will be to:
a) outline the significant features of some prominent theoretical models,
b) list the experimental techniques which have been used, together with some of their advantages and disadvantages, and then c) concentrate on the technique of low-energy ion scattering (LEIS or ISS) and results which have

been obtained by it, emphasizing those which shed light on critical features and predictions of the theoretical models. It will be shown that LEIS can provide information about the composition of the first three atomic layers of a segregated surface and about the atomic arrangement, and can also be used to investigate a closely related but opposing phenomenon, long-range order at the surface.

18.2 Principles of Surface Segregation

18.2.1 General Remarks

Present understanding of surface segregation in alloys has been based on the early writings of GIBBS [18.15-17] on the thermodynamics of solid and liquid surfaces; on the "regular solution" concept defined by HILDEBRAND [18.18] for liquid solutions and extended by GUGGENHEIM [18.19] and later applied to bulk phase equilibria in alloys [18.20,21] as the pair potential or quasi-chemical model; and, alternatively, on the relief of lattice strain brought about by removing a misfit atom from the bulk to the surface [18.22].

Surface segregation theory [18.9,10] based on the regular solution criteria and sometimes referred to as the bond-energy model predicts that the element with weaker bonds, as manifested in a lower heat of vaporization, will be enriched at the surface where bonds are missing. Less energy would be spent to create a surface of that element, by cleaving for example. However, results contrary to this rule are found. The other major driving force which has been recognized is lattice strain energy arising from differences in atom size, which was proposed originally by McLEAN [18.22] to explain grain-boundary segregation. In its simplest form this model would predict that the minority component, the solute, being the misfit atom if there is a significant size difference, will segregate to the surface in order to relieve strain. Since neither of these simple rules alone is in very good agreement with experimental results, although both of them appear to be significant, several combinations of bond-energy and lattice strain have been propsed, which will be outlined below.

18.2.2 Regular Solution Theory

Although this model has its failings it is derived from concepts and principles which appear in the more complete models and it is therefore instructive to examine it in some detail. The simple bond energy rule stated above would seem to predict complete coverage of the surface by the element with weaker bonds. However, it is the free energy which must be minimized at equilibrium. The free-energy difference between the two states for which $X_A^s = X_A^b$ (before segregation) and $X_A^s > X_A^b$ (after segregation), where X_A^s and X_A^b Are atom fractions of elements A at the surface and in the bulk, is given by

$$\Delta F_s = \Delta H_s - T\Delta S_s \quad . \tag{18.1}$$

This contains an enthalpy term ΔH favoring complete coverage, but also an entropy term $-T\Delta S$ which opposes the ordering effect of segregation; so that in general the coverage is not complete, although we will see that it may be very high in some cases.

A useful quantity, which enters into the definition of a regular solution and is sometimes called the regular solution parameter Ω, is defined for an alloy of A and B atoms by the equation:

$$H_{AB} = \Omega + \tfrac{1}{2}(H_{AA} + H_{BB}) \tag{18.2}$$

in which the H's are enthalpies (negative for an attractive interaction) of A-B, A-A, and B-B bonds which may be derived from experimental heats of solution and vaporization [18.9,23]. For an ideal solution $\Omega = 0$ and H_{AB} is just the arithmetic mean of the elemental bond enthalpies. However, for a regular solution, $\Omega \neq 0$ and may be positive or negative, but the entropy of mixing is that of an ideal solution,

$$\Delta S_m = -R(X_A \ln X_A + X_B \ln X_B) \tag{18.3}$$

in which R is the gas constant. This simplification, which assumes random mixing, is called the "zeroth approximation." A more realistic treatment which recognizes that $\Omega \neq 0$ implies preferential association of either like or unlike atoms is called the first-order approximation [18.19] and also the quasi-chemical approximation [18.20]. This refinement applied to the Ni-Cu case makes little difference [18.7] but could be significant in other cases.

The sign of Ω, positive of negative, gives useful physical insight into the segregation behavior. If $\Omega = H_{AB} - \tfrac{1}{2}(H_{AA} + H_{BB}) > 0$, demixing is favored, miscibility gaps appear in the bulk-phase diagram and segregation of the weaker bonding element is greater than for the ideal case. On the other hand, for $\Omega < 0$ mixing is favored, i.e., A-B bonds are preferred over A-A and B-B which can lead to bulk ordering for compositions having small integral ratios of the components. Segregation is expected to be weaker than for an ideal solution since it reduces the number of A-B interactions. The preference for A-B bonds also should lead to depletion of the segregating element in the second layer [18.9].

In the regular solution model surface composition is calculated by minimizing free energy with respect to surface-layer composition taking into account the difference in coordination number between an atom in the bulk and one at the surface. The equilibrium surface composition derived in this way [18.7,9,10] is given by

$$\ln\left(\frac{1 - X_s}{X_s}\right) = \frac{\Delta H_v}{RT} m - \frac{2\Omega}{RT} [\tfrac{1}{2} mZ - (m + \ell) Z X_b + \ell Z X_s] + \ln \frac{1 - X_b}{X_b} \qquad (18.4)$$

in which X_s and X_b are the surface and bulk concentrations of the element with higher heat of vaporization, ΔH_v the absolute difference between the pure element heats of vaporization, m the fraction of nearest neighbors of an atom which lie in a plane above or below it, ℓ the fraction in the plane, and Z the coordination number in the bulk. WILLIAMS and NASON [18.9] have also analyzed the case for segregation into layers 2, 3 and 4. The m factor in the first term represents the assumption that the surface energy is proportional to the heat of vaporization, with m as the fraction of nearest-neighbor bonds which are missing at the surface.

18.2.3 Influence of Atom Size Difference

As experimental results have accumulated it has become evident that a major shortcoming of the regular solution model is that it does not take into account atom size differences. For example, in the Cu- Au case segregation of Cu should be favored slightly by the bond-energy criterion whereas Au is the element which segregates [18.24] evidently because Au atoms are ~15% larger. Atom size difference was first invoked to explain grain-boundary segregation [18.22], in which case it was assumed that a misfit solute atom, whether too large or too small, would segregate to a grain-boundary where the strain would be eliminated entirely. WYNBLATT and KU [18.6] combined the driving forces of lattice strain and bond-energies, derived from bulk thermodynamic properties, as a sum of two independent terms, again assuming that solute strain energy is totally eliminated by the exchange of a bulk solute atom for a surface solvent atom. They used an expression of FRIEDEL [18.25] for the elastic strain due to the solute atom, rather than McLEAN'S. Measurement of segregation on gold-nickel alloys by AES agreed well with this model showing Au segregation at the Ni end and Ni at the Au-rich end; i.e., there was a reversal in segregation

Table 18.1. Thermodynamic values and atomic radii of some binary alloys

Element	ΔH_v [18.23] [Kcal/mol] [298 K]	$r_{at.}$ [Å] [18.81]	Integral heats of formation $0.5A + 0.5B = A_{0.5}A_{0.5}$		
			Alloy	Temp [K]	ΔH[cal/g at]
Cu	80.5	1.28	Au Cu	800	-1221
Ni	102.8	1.25	Au Ni	1150	+1807
Au	88	1.44	Cu Ni	973	+ 425
Pt	135	1.39	Ag Au	800	-1111
Ag	67.9	1.44	Ag Cu	1423	+1014

behavior which must be caused by the lattice strain effect since bond energies would favor gold segregation throughout (see Table 18.1). They also obtained good agreement between experimental and predicted heats of segregation.

ABRAHAM et al. [18.11,12] have also combined the bond-energy and lattice strain driving forces in their analysis and have mapped the combinations of bond-strength ratio ε^* and atom size ratio $\sigma^* = \sigma_2/\sigma_1$ where solvent or solute segregation should occur. In doing this they avoided continuum elasticity theory, the assumption of independent bond-breaking and elastic strain terms, and the use of bulk thermodynamic properties. Instead, they used a Monte Carlo method to find the boundaries between the segregation regions in the $\varepsilon^* - \sigma^*$ field, where the total driving force $\Delta Q = 0$. They used a Lennard-Jones interatomic potential and included relaxation of atoms around the solute atom in the bulk and at the surface. Thus, this model does not assume that elastic strain energy is eliminated for a solute atom at the surface, and furthermore it finds a larger driving force for segregation of large solute atoms than for small atoms since the lattice distortion is less pronounced for undersize atoms. The separate and combined effects of bond-energy and atom size differences are illustrated schematically in Fig.18.1 [18.12] where γ^* is the specific surface energy ratio γ_2/γ_1, taking into account the atomic surface area as well as bond energy of solute (2) and solvent (1), and σ^* is the atom size ratio. Panel 1 illustrates the bond-energy influence alone; there is no solute segregation, only solvent segregation, above the horizontal axis for $\gamma^* = 1$. Panels 2 and 3 illustrate effects of lattice strain alone, with a region on either slide of $\sigma^* = 1$ in which there is no segregation, and solute segregation everywhere else. This region is symmetrical about $\sigma^* = 1$ for linear elasticity [18.6] but extends further on the small solute atom side for nonlinear elasticity. In the lower panels the two driving forces are combined and if $\sigma^* \neq 1$ (for large or small solute atom) the solute segregation region is enlarged, symmetrically for linear elasticity, but less on the small-atom

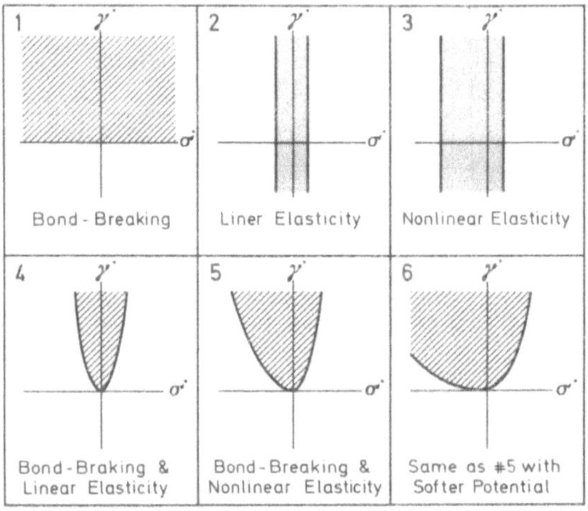

Fig.18.1. Schematic representation of surface segregation predictions for various theoretical models. The blank area is the solute segregation region, and the hatched area is the solvent segregation region. The shaded area denotes no segregation [18.12]

side for nonlinear elasticity and for a softer Lennard-Jones potential, which is more representative of metallic interactions. Predictions by this model were ~85% successful.

18.2.4 Miedema's Model

MIEDEMA [18.26] has proposed another combination of bond-energy and atom size effects which agrees well with experiment and also makes quantitative predictions of the degree of segregation. It is based on considerations of electron density and electronegativity, or chemical potential of electrons, at the boundaries of Wigner-Seitz atomic-cells, which lead to semi-empirical tabulations of surface energies of the pure metals [18.27,28] and bulk heats of solution for their alloy [18.29] from which the surface segregation predictions are derived. Surface energies are adjusted for atomic area $V_{at}^{2/3}$, and size mismatch energy is taken into account but is not given great weight, and is not assumed to be absent in the surface layer.

Table 18.2 shows a few examples, taken from a large table of Miedema [18.26] of the magnitude of the three contributions – heat of solution, surface energy, and size mismatch – to the segregation parameter $\ln(C_1^A/C_b^A)$, C_1^A being the surface concentration of the solute and C_b^A the bulk concentration. Of the three contributions surface energy is dominant. The predictions of these few examples and others in Miedema's table are all in qualitative

Table 18.2. Miedema predictions

A in B	I. (Heat of solution)	II. (Surface E)	III. (Size mismatch)	$\ln(C_u^A/C_b^A)$	Expt. info.	Technique	References
Ag in Au	-0.5	+2.1	0.0	+1.6	Y	L,A	[18.30,31]
Au in Ag	-0.6	-2.1	0.0	-2.7	N	A	[18.32]
Au in Cu	-0.8	+2.2	+1.0	+2.4	Y	L,A	[18.24,33]
Cu in Au	-0.6	-1.8	+1.0	-1.4	N	A	[18.24]
Cu in Ni	+0.7	+3.7	0.0	+4.3	Y	L,A	[18.7,34 34]
Ni in Cu	+0.7	-3.4	0.0	-2.7	N	L	[18.7]
Cu in Pt	-0.7	+4.1	+0.7	+4.1	Y	L,A	[18.7,35]
Au in Ni	+1.0	+6.8	+1.7	+9.5	Y	L,A	[18,6,14, 36,37]
Ni in Au				-4.2	N	L	[18.38-40]
					Y	A	[18.6]

agreement with experiment. Differences between LEIS (L) and AES (A) results on the case of Ni in Au will be discussed later.

CHELIKOWSKY [18.41] has recently extended the compilation to cover 2,550 combinations, including some III-V semiconductor compounds, following a modified version of Miedema's approach. Heats of segregation and also concentration factors are included. Only 2 out of 40 predictions of the segregating element disagreed with experimental results.

18.2.5 Bulk-Phase Diagram Rule

A simple rule for predicting surface segregation from bulk-phase diagrams has been proposed by BURTON and MACHLIN [18.42]. Observing that a solid surface differs from the bulk in characteristics similar to those of a liquid – lower symmetry, lower coordination, and no elastic strain – they stated that the solute should segregate if, and only if, it is also enriched in the liquid in equilibrium with the solid, and the degree of segregation should be roughly proportional to the separation between the liquidus and solidus curves. A list of experimental results for 16 alloys showed good agreement. However, the rule applied to Au(Ni) would predict Ni segregation which was observed by AES [18.6] but not by LEIS as will be discussed below.

18.2.6 Long-Range Order and Segregation

Most segregation results have been obtained on binary alloys which have a positive heat of solution and regular solution parameter Ω. However, as mentioned in Sect.18.2.2 some interesting differences in behavior may be expected when there is a negative heat of solution and negative Ω, especially when the two components are present in proportions of small integers. In this case ordering occurs in the bulk [18.20,21] below some critical temperature, i.e., the A and B atoms occupy preferred lattice sites in order to maximize the number of nearest neighbors of the opposite species. The Cu-Au alloys furnish some of the best-known examples. In Cu_3Au the {100} planes alternate between $Cu_{0.5}Au_{0.5}$ and pure Cu, below the critical temperature Tc = 390 C. This tendency may be expected to oppose surface segregation; for an atom at the surface the number of unlike nearest neighbors is reduced by the number in an adjacent plane parallel to the surface. Ordering and surface segregation have been investigated theoretically by MORÁN-LOPEZ and BENNEMAN [18.43] who used a Bragg-Williams [18.21] type of order-disorder analysis and the pair-bonding approach of regular solution theory with no atom size lattice strain term, in minimizing the free energy with respect to layer composition and the long-range order parameter which is a measure of the degree of order. In this way they calculated order parameters as functions of temperature for {100} and {111} surfaces, finding that segregation can cause the order parameter to disappear at temperatures below Tc, and that ordering may suppress segregation.

In another calculation [18.44] on the {110} surface of a bcc $A_{0.5}B_{0.5}$ alloy, both short- and long-range order parameters as well as layer compositions were calculated over a wide range of temperature, using KIKUCHI's method [18.45] to obtain the entropy. For an A-B type of first layer it was found that the concentration of one component should exhibit a maximum at the order-disorder critical temperature.

18.3 Surface Composition Analysis

Since the phenomena of interest occur in the first few layers, predominantly in the first, exceptional surface sensitivity is required in composition analysis. A comprehensive survey of techniques which have been used for surface and grain-boundary segregation has been published by JOSHI [18.46]. In the literature on surface segregation alone, three techniques appear most frequently—AES, LEIS, and the Atom-Probe Field-Ion Microscope, with AES as the most popular. This technique is convenient, widely available for surface analysis, has good elemental specificity, and good lateral resolution. Beam damage should be insignificant. On the other hand, quantitative interpretation is relatively complicated and depth resolution not easily defined. The atom-probe field ion microscope [18.47,48] has a single-atom-layer depth resolution over as many as 13 layers [18.47] and provides quantitative depth profiling, but with some difficulty in accumulation of adequate statistics. Preparation of the fine-tipped specimens requires skill and experience.

18.3.1 Low-Energy Ion Scattering

a) Noble Gas Ions with ESA. $E_0 \leq 3$ keV

Ion scattering at low energies (incident energy $\leq$ 10 keV) for surface analysis has now been performed in three different modes. In the most familiar, called ISS or LEIS, a well-collimated monoenergetic noble gas ion beam, usually He^+ or Ne^+ of $\leq$ 3-keV energy is incident on the target surface (Fig.18.2, left), and ions scattered from individual atoms in the surface through a particular scattering angle θ_L are collected in an electrostatic analyzer (ESA) which has a detector, usually a channel multiplier, to count the ions passing through the ESA at a given voltage setting on the plates. By scanning the voltage on the plates one obtains an energy distribution of the scattered ions, or more accurately an energy/charge distribution; e.g., 4-keV Ne^{++} would be passed at the same voltage setting as 2-keV Ne^+. However, this ambiguity is not a serious problem when a pure beam of singly charged ions is used since an increase to a higher charge state is very improbable in low-energy scattering. On the other hand, neutralization of noble gas ions in scattering is very probable, and the large fraction which become neutralized are, of course, not collected by an ESA. Neutralization is especially strong for ions which penetrate beyond the first atom layer. This neutralization effect and the large scattering cross sections which obtain at low energies account for the fact that this version of LEIS samples only the first atom layer of a surface [18.49-51]. Neutralization causes uncertainty in quantitative analysis since the probability depends on target atom identity [18.51], and neutralization factors have not been determined for many target atoms or energies. However, calibration with pure metal samples is usually a satisfactory means of overcoming this problem in surface segregation studies [18.7]. The original ESA design, illustrated in Fig.18.2 (left) uses a relatively high ion dose, up to $10^{15}/cm^2$, since ions in only a small energy window are collected at each voltage setting of the energy scan.

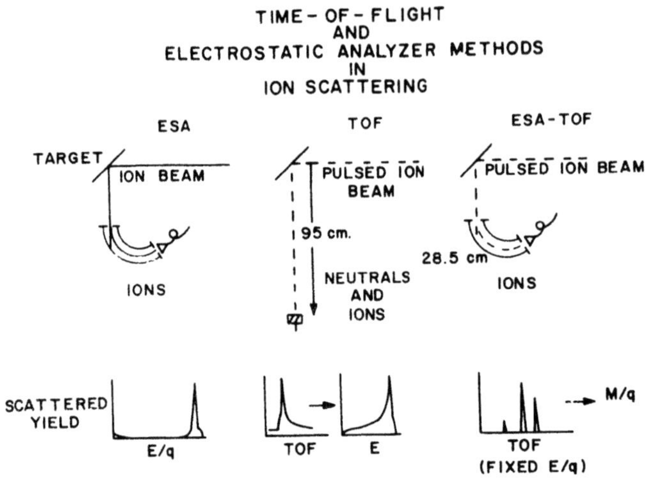

Fig.18.2. Schematic diagrams of electrostatic analyzer (ESA), and time-of-flight (TOF) techniques for low-energy ion scattering. The ESA-TOF method allows M/q identification of secondary ions removed by the beam

Substantial reductions in the required ion dose, by factors of $\sim 10^2$, have been achieved by the use of a clindrical mirror analyzer [18.7,52,53] which collects a cone of reflected ions instead of the usual small segment, but this design does not allow selection of a particular azimuthal angle, which is desirable in single crystal experiments.

b) LEIS (TOF)

The neutralization problem can be circumvented and the ion dose reduced by factors of 10^2-10^3 with respect to a parallel plate ESA by time-of-flight (TOF) analysis of the scattered ions and neutrals [18.54-56] illustrated in Fig.18.2 (center). The average pulsed ion beam current is $\sim$0.05 nA. Both the scattered ions and neutrals are detected by a Johnston MM1 multiplier at the end of the 95-cm flight path [18.54]. A timing system not shown collects a TOF spectrum which can be transformed to an energy spectrum. Since neutrals as well as ions are collected the scattering from one, two, or three layers, depending on crystal orientation, can be analyzed, thus increasing the information obtainable very considerably. The restriction to only a few layers is due to the large scattering cross sections which attenuate the penetrating and scattering beams. The ESA-TOF combination pictured in Fig.18.2 (right) can be used to identify recoil and sputtered particles coming off at a given energy setting by measuring their time-of-flight through the ESA [18.54a].

c) Alkali Ions with ESA

A third LEIS mode which has appeared quite recently and has not yet, so far as we know, been used for segregation studies, utilizes alkali ions which undergo little or no neutralization in scattering, with little or no depth dependence [18.57-60]. Energy spectra obtained with an ESA therefore provide information about layers beyond the first, and quantitative analysis is simplified as for the TOF method. An ESA technique is simpler in construction and maintenance than a TOF system, but if it is a single channel system it still requires a larger ion dose than does the multichannel TOF measurement.

d) Principles of Low Energy Ion Scattering

Kinematics — Identification by Peak Position. The ion-atom collisions are elastic within the energy resolution of most systems, and so from conservation of kinetic energy and momentum the energy E_1 retained by an ion of mass M_1 of incident energy E_0 after scattering from a target atom of mass M_2 through an angle θ_L is given by

$$\frac{E_1}{E_0} = \frac{M_1^2}{(M_1 + M_2)^2}\left[\cos\theta_L + \left(\frac{M_2^2}{M_1^2} - \sin^2\theta_L\right)^{\frac{1}{2}}\right]^2 . \tag{18.5}$$

For 90° scattering which is frequently used, this reduces to

$$\frac{E_1}{E_2} = \frac{M_2 - M_1}{M_2 + M_1} . \tag{18.6}$$

The energy scale can thus be calibrated as a mass scale for target atoms at the surface; with peaks for heavy atoms appearing at higher energies than those for light atoms.

Scattered Yield — Quantitative Analysis. The yield of scattered ions is proportional to N the density of target atoms exposed to the beam, σ the differential scattering cross section derived from a screened Coulomb potential, e.g., the Thomas-Fermi[1], E_1 the energy setting of the ESA which determines the width of the energy window, and P^+ the probability that the scattered particle remains ionized,

1 The computer program for cross sections used in examples of this paper was written by Yu Ssu Chen of Bell Labs and was based on the analysis of [18.61].

$$y^{+} \propto N\sigma E_1 P^{+} \quad . \tag{18.7}$$

For a binary alloy A-B then

$$\frac{N_A}{N_B} = \frac{y_A}{y_B} \times \frac{\sigma_B}{\sigma_A} \times \frac{E_B}{E_A} \times \frac{P_B^{+}}{P_A^{+}} \quad . \tag{18.8}$$

Usually, but not always, P_B^{+}/P_A^{+} is unknown. For TOF measurements these are simplified to

$$y \propto N\sigma \tag{18.9}$$

and

$$N_A/N_B = (y_A/y_B)(\sigma_B/\sigma_A) \tag{18.10}$$

which holds strictly only for first-layer scattering since a "focusing" effect must be considered in scattering from deeper layers. In the absence of foreign impurities, the concentration or atom fraction of A is given by

$$X_A^1 = \frac{N_A/N_B}{N_A/N_B + 1} = \frac{N_A}{N_A + N_B} \quad . \tag{18.11}$$

Shadow Cones and Focusing. Focusing occurs at the edge of a "shadow cone" [18.62] (Fig.18.3). Shadow cones are large at low energies; $r \approx 1$ Å for 5-keV Ne on Au as compared with ~0.1 Å for 1-MeV He. A shadow cone is the envelope of trajectories of ions which approach a target atom at random impact parameters (P). Ions coming in close, at $P \sim 0$ are scattered through large angles while large impact parameters produce smaller deflections, and for those close to S_0 there is a concentration of ion flux at the outer edge of the cone. An atom inside the cone receives no flux and is "shadowed" but one at the edge receives more than it would further away. This effect is called "focusing" and is somewhat analogous to the run-off at the edge of an umbrella in rain. "Blocking cones" [18.62] of atoms in or near the outgoing path also influence the scattering behavior. "Wedge focusing" [18.64] refers to the concentration of flux between the shadow cones of two atoms in the first layer onto an atom in the second.

In scattering from a single crystal surface one may, by proper choice of polar and azimuthal angles of incidence, scatter from the first layer only, or from the first and second layers, as illustrated in Fig.18.4 for an FCC{001} surface. Scattering from deeper layers will be discussed later. With reference to Fig.18.4, single scattering may be restricted to the first layer by scattering in a plane coincident with a <100> azimuth at an incidence

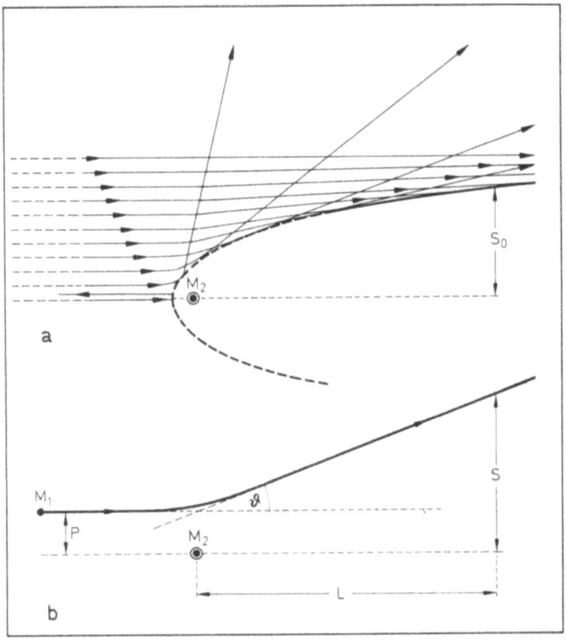

Fig.18.3

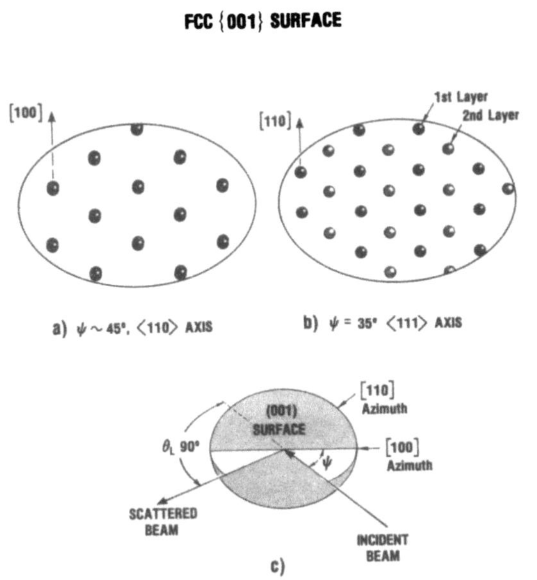

Fig.18.4

Fig.18.3a,b. Possible trajectories of a projectile M_1 scattered by a target atom M_2. In (a) the shadow cone behind M_2 is shown. In (b), S gives the position of the scattered particle for an impact parameter P and scattering angle θ, at a distance L behind the scattering center. The example shown is calculated for 5-keV Ne^+ ions on a Cu atom with impact parameters varying between 0.1 and 1.0 Å [18.63]

Fig.18.4a-c. Models of Fcc{001} surface atoms exposed to ion beam from two directions: (a) along <100> azimuth at Ψ = 45° incidence angle, and (b) along <110> azimuth at Ψ = 35°. Angles are defined in (c) [18.51] angle Ψ of 45°, corresponding to a <110> channeling axis, in which case atoms in the second and deeper layers are very effectively shadowed by first-layer atoms. Also, for a scattering angle θ_L of 90° there is very effective blocking of any ions which might otherwise emerge by sequential scattering from a first-layer atom to an atom in the second layer and out. Along the <110> azimuth second-layer atoms are exposed as well as first layer with deeper layers shadowed for $\Psi = 35^{\circ}$.

Considering an Fcc{001} alloy surface, the first layer would be analyzed by scattering on the <100> azimuth at $\Psi = 45^{\circ}$ and applying (18.11). For the second layer, the scattering yields along the <110> are measured and focusing factors introduced

$$f_A = \left(\frac{y_{1+2} - y_1}{y_1}\,\frac{X_1}{X_2}\right)^A \quad . \tag{18.12}$$

A similar relation holds for element B, and it can be shown that

$$X_2^B = F\left[\frac{(y_{1+2} - y_1)^A}{(y_{1+2} - y_1)^B}\,\frac{y_1^B}{y_1^A}\,\frac{X_1^A}{X_1^B} + F\right]^{-1} \tag{18.13}$$

in which $F = f_A/f_B$ and other quantities are measured. In the use of this relationship to analyze the Cu_3Au {001} surface which is described in Sect. 18.4.2, F = 1 is assumed without danger of a serious error. Computer simulations of ion trajectories should be useful in determining these factors when better accuracy is required.

18.4 Experimental Results—LEIS

18.4.1 Polycrystalline Alloy Surfaces

Experiments on Cu-Ni alloys [18.7,65] were performed with a CMA-type ESA system called NODUS [18.52]. Figure 18.5 shows scattered Ne^+ spectra obtained at 25° and $520^{\circ}C$ on a surface of a $Ni_{0.48}Cu_{0.52}$ (bulk) alloy sputter-cleaned before the $25^{\circ}C$ measurement. The clear resolution of the Ni and Cu peaks and the shoulders due to isotopes attest to the excellent energy resolution of the NODUS. The spectrum taken at $510^{\circ}C$ shows only a very small Ni peak. Surface compositions determined in this way are plotted against bulk composition in Fig.18.6. The straight diagonal line would be followed if there were no segregation. Instead the $500^{\circ}C$ points fall well above this line showing strong Cu segregation for the three bulk compositions used. The triangles

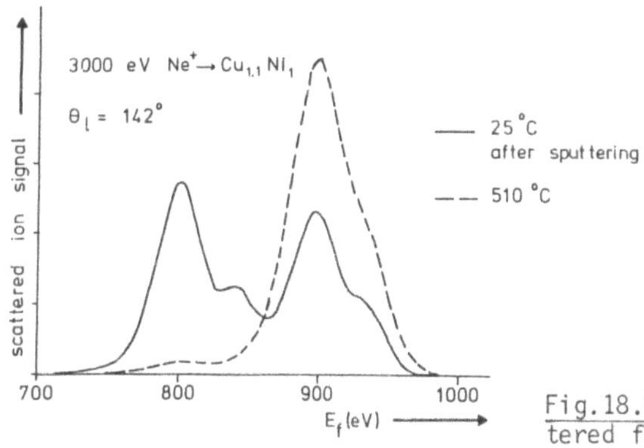

Fig.18.5. Energy spectra for Ne^+ scattered from Cu-Ni alloy of 52% Cu [18.7]

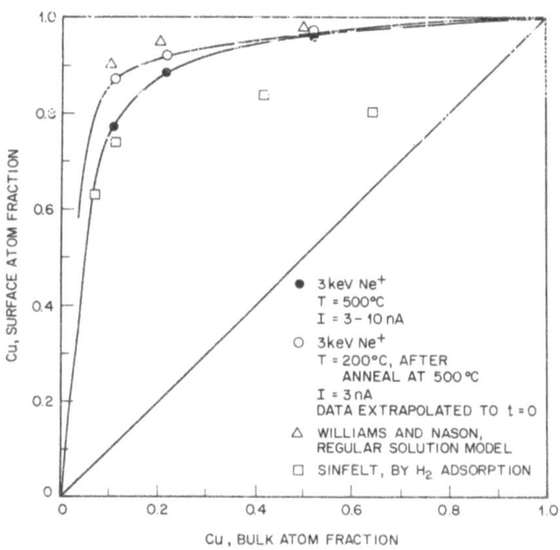

Fig.18.6. Surface concentration of Cu versus bulk concentration for Cu-Ni alloys. Ion scattering results compared with H_2 adsorption and calculations for an Fcc{111} surface. The open circles for 200°C do not represent an equilibrium condition but rather are estimated from diffusion rates to represent the 400°C concentration [18.7] (Sinfelt points from [18.3])

dot curve for the regular solution prediction for 500°C from (18.5) are even higher. The solid curve through the points also comes from a modification of (18,4) to include a "relaxation" correction $\sigma = 0.16$, assuming this fractional increase in enthalpies of bonds near the surface due to redistribution of electrons associated with the broken bonds. When this parameter is used to fit the data at one temperature and composition, results for other combinations are predicted correctly. This is a rather arbitrary but not unreasonable adjustment to make [18.66,67]. One might instead attribute the discrepancy between theory and experiment to some scattering from the second or third atom layers which have less copper, but TOF measurements of 2.4-keV Ne^+ scattering from a Ni{001} surface [18.51] showed a neutralization probability of ~100% on the second layer, so very little of the scattered ion yield is coming from the deeper layers. This should truly be a measurement of first-layer segregation and therefore comparable with the regular solution first-layer prediction, especially since there is no significant difference in atom sizes. The stronger segregation at 400°C is expected on the basis of entropy considerations. The Miedema factor (Table 18.2) corrected for temperature predicts much stronger segregation at the low Cu end.

This qualitative result, segregation of Cu rather than Ni is predicted by all segregation theories except the pure atom size lattice strain model which would have no prediction in this case. After some early AES experiments which found no segregation, all three experimental techniques mentioned in Sect.18.3 have agreed on Cu as the segregating element [18.7,34,47].

Au-Ag is another alloy system in which there is no significant difference in atom size, but a modest difference in bond energies favoring segregation of Ag. Two AES studies on this system [18.68,69], however, found no enrichment of either element at the surface, and BOUWMAN et al. [18.69] reasoned that the disagreement between this result and their regular solution calculation predicting Ag segregation might be due to a larger repulsive force between Au atoms [18.70] than between Ag atoms, which would favor Au segregation and balance out the bond-energy driving force. However this may be, NELSON [18.30] using a commercial ESA low-energy ion scattering system with both He^+ and Ne^+ beams found strong enrichment of Ag, as did a third AES measurement [18.71].

An alloy which has both a large binding-energy driving force and a large difference in atom sizes is Au-Ni. Strong gold segregation is predicted by the regular solution model throughout the composition range whereas, as mentioned in Sect.18.2.3, the pure lattice strain model, the formula of WYNBLATT and KU [18.6] including both bond-energies and atom size difference, and the bulk phase diagram rule [18.42] predict a reversal with Au segregating at the Ni-rich end and Ni at the Au-rich end, as was found by AES [18.6]. Between LEIS and AES there is no disagreement about the Ni-rich end. Several AES studies [18.6,14,36] as well as LEIS (TOF) measurements [18.37,38] have shown strong Au segregation, as should be expected. On the other hand, three LEIS measurements, two by ESA [18.39,40] and one by TOF [18.38], found Au segregating also at the Au-rich end, rather than Ni, contrary to the predictions and AES measurement cited above. The prediction by Miedema's formula [18.26,41] is Au segregation throughout the composition range.

18.4.2 Single Crystal Surfaces

Studies of segregation at single crystal surfaces can provide information on the more subtle aspects of segregation, such as individual layer composition, dependence on crystal face, and arrangement of segregated atoms in the first layer, which should be known for a complete understanding of th segregation phenomenon. Implicity or explicitly, it has generally been assumed that the segregated atoms occupy substitutional sites at the surface but, as BURTON et al. [18.72] have pointed out, this is rather unlikely, especially when there is a large difference in atom size. They investigated the matter using LEED and AES on a {111} surface of a Ni crystal containing 1% Au and obtained (6×6) and (13×13) hexagonal LEED patterns from Au segregated surfaces, indicating overlayer structures with the same symmetry and orientation as the substrate, but with large unit cell sizes. This they explained by a simple atom size mismatch argument, Au having ~15% larger diameter. They also predicted that Au segregated to the Ni{001} surface would form a hexagonal overlayer giving a $C(11\times2)$ LEED pattern. Other LEED studies [18.73] have suggested hexagonal Au overlayers on {001} surfaces of Cu and of Au itself. An implicit assumption here is that the segregated atoms are coplanar.

a) Ni{001} Au

The Ni{001} Au segregated surface has now been studied by LEIS (TOF) [18.37] and some of the results are shown in Figs.18.7-10. The experiments were performed on a Ni crystal doped with 1 at.% Au. Figure 18.7 shows an experimental energy spectrum derived from a TOF measurement, and a calculated spectrum, for Ne scattered from the first layer of the {001} surface after annealing for one hour at 600°C, and cooling to room temperature. The 5-keV beam was incident at $\Psi = 45^\circ$ along the <100> azimuth (<110> channeling axis) which leads to very efficient shadowing of atoms below the first layer (Fig. 18.4). The large Au and small Ni peaks, corresponding to ~11% Ni, evidently did not represent complete equilibrium since annealing overnight in a later experiment produced a first-layer spectrum in which there was no Ni detectable. The segregation is therefore very strong, as BURTON et al. found also on the {111} surface [18.72] and as Miedema's theory [18.26] predicts. The solid curves in Figs.18.7,8 were produced by a computer simulation which was

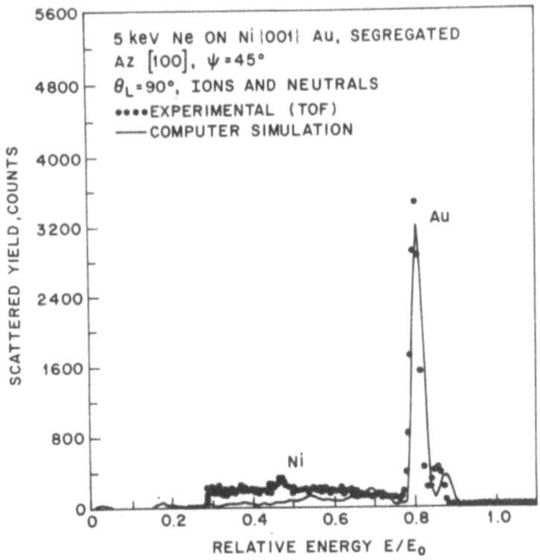

Fig.18.7. Experimental and simulated energy spectra of Ne scattered from Ni{001} Au (segregated) surface. Beam direction <100> 45. E_0 = 5 keV. Scattering from first layer only [18.37]

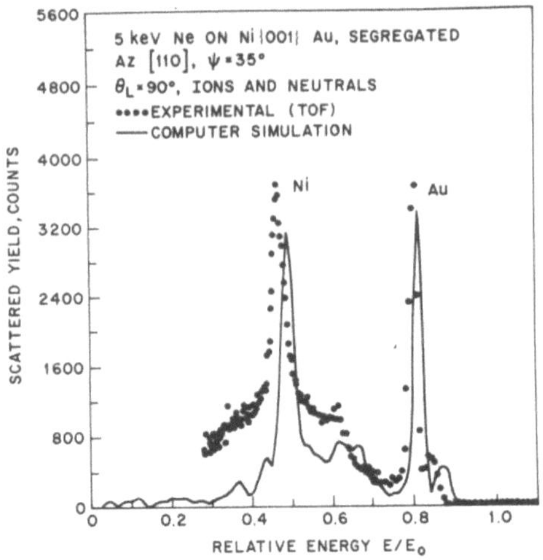

Fig.18.8. Experimental and simulated energy spectra of Ne scattered from first and second layers of Ni{001} Au (segregated) surface. Beam direction <110> 35. E_0 = 5 keV [18.37]

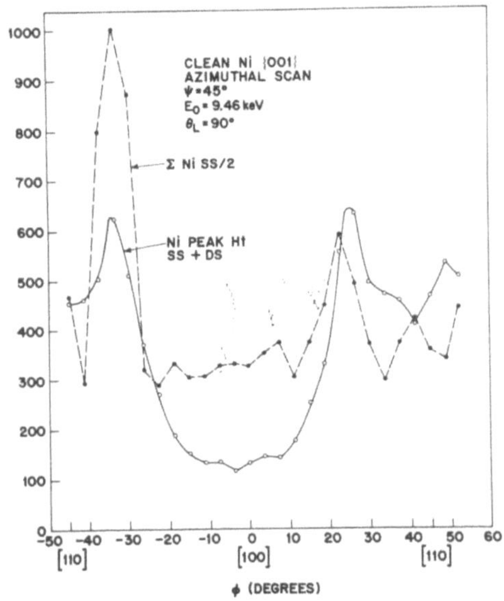

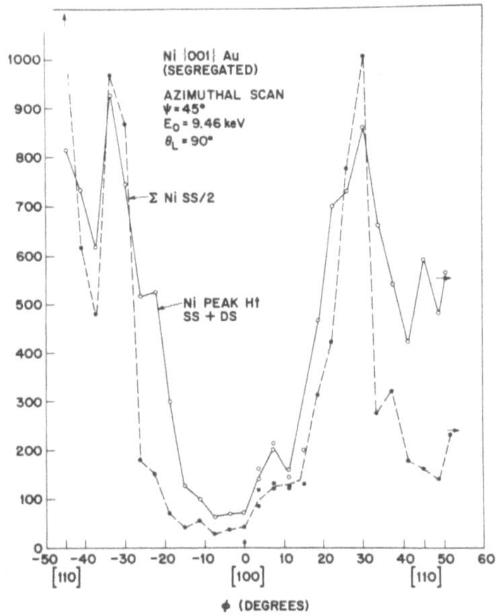

Fig.18.9. (a) Azimuthal dependence of Ni peak height and Ni single scattering yield on clean Ni{001} surface. Ne beam, E_0 = 9.46 keV. Ψ = 45°. (b) Same measurement on Ni{001} Au (segregated) surface [18.37]

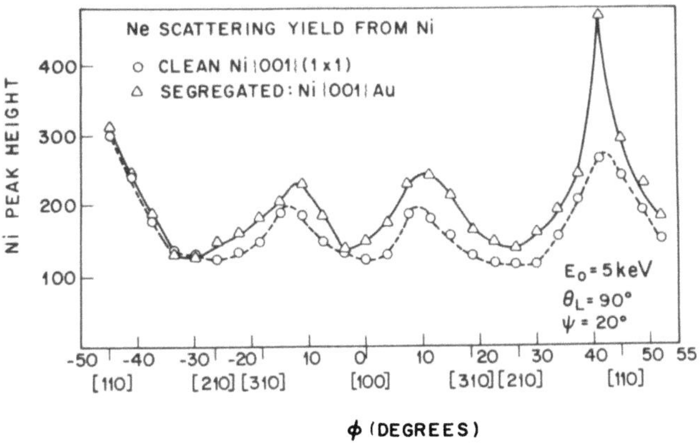

Fig.18.10. Azimuthal dependence of Ni peak height on Ni{001} and Ni{001} Au surfaces for Ψ = 20° [18.37]

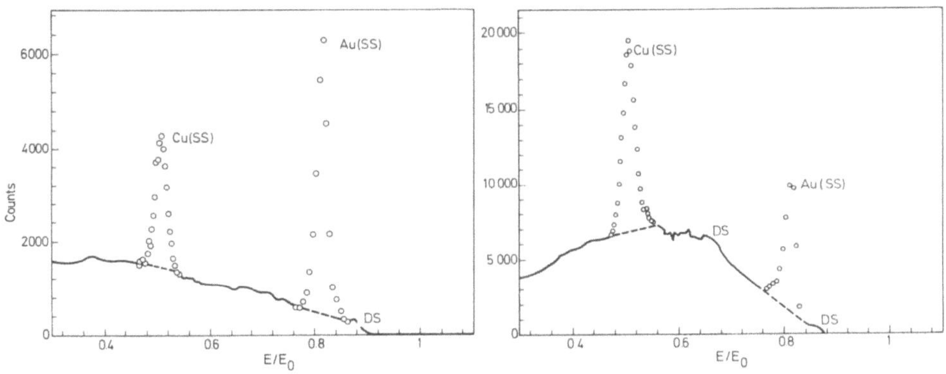

Fig.18.11. (a) Energy spectrum of Ne scattered from Cu_3Au\{001\} ordered surface, first layer only. E_0 = 5-keV, $\theta_L = 90^{\circ}$, <100> azimuth, $\Psi = 45^{\circ}$. Derived from TOF spectrum. (Expt. described [18.33]). (b) Scattering from first and second layers. <110> azimuth, $\Psi = 35^{\circ}$

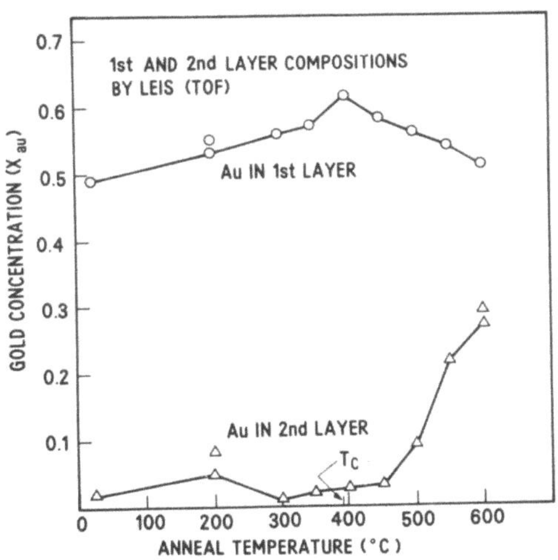

Fig.8.12. Compositions of first and second layers of Cu_3Au\{001\} surface as function of annealing temperature [18.33]

developed from one used for high-energy scattering [18.74]. It calculates the trajectories of ions in a fully three-dimensional vibrating lattice using sequential binary collisions. In this case it assumed 100% Au in the first layer with Au atoms on substitutional Ni sites, and no layer distance change.

If Au atoms are in substitutional Ni sites and in the first layer only, then one should expect a large increase in Ni yield along the <110> azimuth where second-layer atoms are exposed (Fig.18.4), and this is in fact observed in both the experimental and simulated spectra of Fig.18.8, where it can be seen that there is no significant increase in the gold peak, indicating little or no gold in the second layer. The large scattering yield peaked at $E/E_0 \sim 0.6$ is very similar to those obtained from a clean Ni{001} surface, for the same beam direction. It represents double scattering (DS) from two Ni atoms in <110> rows, mostly in the second layer in this case. This large DS yield is due to the close spacing of atoms in this direction, and the large separation between the Ni SS and DS peaks results from the large kinematic energy loss in the Ne → Ni collision.

In order to examine the symmetry of the Au overlayer more closely, azimuthal scans of the Ni scattering yield were performed, on both clean and segregated surfaces, by recording spectra as the target was rotated in 4-degree steps at a fixed incidence angle. Figure 18.9a shows the yield of Ne (neutrals and ions) scattered from a clean Ni{001} surface, plotted against azimuthal angle. The solid curve gives the Ni peak height which includes contributions from both single and double scattering while the dashed curve gives the sum of counts in the SS peak above the DS background, divided by 2 to keep it on the same scale. In general these will not have exactly the same shape. The broad maximum centered a few degrees left of $\phi = 0$, the nominal <100> azimuth, represents first-layer scattering only, the maxima at $\sim 30^{\circ}$ on either side of <100> are probably caused by strong focusing of Ne^+ onto second- and third-layer atoms by the first-layer atoms, and the small maximum a few degrees left of nominal <110> is probably due to wedge focusing on the second-layer atoms. More detailed computer simulation analysis and also shadow-cone analysis, which will be illustrated for Cu_3Au in the next section, should be performed in order to identify the scattering processes more accurately. However, the main object here is to show the similarity between the clean Ni{001} scan and the one for the segregated surface (Fig.18.9b). In the segregated case the curve represents Ni in the second and third layers as seen through open directions in the gold overlayer, except for the very low yield near <100> which comes from Ni in the first layer. The similarity is quite pronounced, especially in regard to the spikes at $\sim 30^{\circ}$. Scans for both the clean and segregated surfaces taken at $\Psi = 20^{\circ}$ instead of 45° are shown in Fig.18.10. At this low-incidence angle double and single scattering cannot be easily separated and so only the peak heights are plotted. The similarity here is very close: there are dips at the <100> planar channel and also for each pair of closely spaced <210> and <310> planar channels, but there is an increase in Ni yield for the <110> surface semichannel which is enhanced in the Au segregated case, presumably by stronger focusing.

Thus, the computer simulations which assume that the segregated Au atoms are on Ni lattice sites and which reproduce the experimental spectra closely, and the azimuthal scans of Ne scattered from Ni on clean and segregated surfaces which show similar symmetry, all seem to support the proposal that the Au atoms are on substitutional sites. However, LEED patterns taken on the segregated surfaces have consistently shown rows of extra spots with twofold symmetry which we index approximately as (2×6) rather than the (1×1) pattern to be expected of a substitutional overlayer. It is therefore suggested that the Au overlayer has Fcc{001} square symmetry with Au atoms on or above Ni sites, i.e., with alternate <100> rows raised slightly to accomodate the larger Au atoms, and with some buckling in the rows along a perferred orientation giving the higher order spots. The Au scattering yield variation in the azimuthal scans, which is not shown in these figures, is rather featureless but is slightly bowed, with a broad maximum centered at the <100> azimuth. This might result from a focusing action of the alternate raised rows on the lower rows of gold atoms. There is obviously more work to be done on this surface but is seems safe to say that the gold segregation occupies close to 100% of the first layer, that there is little or no gold in the nickel layer underneath and that the gold overlayer does not have the hexagonal structure which might be anticipated from earlier LEED studies of related surfaces.

b) Fe{001}SN

A more complicated system, Fe{001}Sn, studied by LEIS (TOF) and LEED has also shown strong segregation of the element having weaker bonds and larger atomic radius, i.e., Sn. In this case, however, there is evidence of Sn enrichment in the third layer as well as in the first [18.75].

c) $Cu_3Au\{001\}$. Ordering and Segregation

Cu_3Au is a classic ordering alloy, differing from the Au-Ni system on all the segregation criteria except the atom size disparity. It has negative enthalpy of solution and regular solution parameter Ω, and only a small difference in bond energies, favoring Cu segregation, while the atom size difference favors Au. Estensive studies of the bulk ordering by X-ray diffraction [18.76] have established that the atomic arrangement depicted in Fig.18.13a, with {100} layers alternating between pure Cu and 50-50 Au-Cu, exists at temperatures below T_c = 390 C. However, there is a first-order transition to the disordered state at this temperature, i.e., the long-range order parameter L drops abruptly from near 1 to 0. One may ask whether the ordered arrangement for $T < T_c$ extends to the surface and, if so, which layer is on top; whether antiphase domains, known to occur in the bulk, or steps, may result in a mixed surface; and whether surface segregation of either element occurs. Although the literature is not extensive, answers to some of these questions have appeared. SUNDARAM et al. [18.77], using LEED and AES, found that long-range order on the {001} surface, as manifested by LEED superlattice spots, decreased continuously with increasing temperature rather than sharply at T_c. POTTER and BLAKELY [18.78] using LEED found that the {001} and {111} surfaces ordered strongly but the {011} did not. VAN SANTEN and SACHTLER [18.79], applying the Bragg-Williams theory, predicted that Cu would segregate, but FAIN et al. [18.24] found Au enrichment by AES on polycrystalline Au-Cu films over a wide range of composition and also on a single crystal $Cu_3Au\{111\}$ surface. The Miedema model also predicts Au segregation. The predictions by MORAN-LOPEZ et al. [18.43,44] of competition between ordering and segregation have been mentioned in Sect.18.2.6. The first-layer type (Au-Cu or pure Cu) was not determined or predicted in any of these investigations.

In experiments on the $Cu_3Au\{001\}$ surface [18.33] with LEIS (TOF) supplemented by LEED and Auger it was found, in agreement with POTTER and BLAKELY [18.78], that LEED superlattice spots indicating long-range order appeared as the sample was cooled to room temperature after cleaning by heating and ion bombardment at 500°C or higher. However, the sample was usually annealed at 200°C overnight, or at room temperature for several days before taking LEIS measurements on the ordered surface. The question of first-layer type (Au-Cu or pure Cu) in the ordered state was answered by scattering along the <100> azimuth at $\Psi = 45^\circ$. A typical spectrum is shown in Fig.18.11a. Since there is a large Au peak the surface is obviously not pure Cu, and the

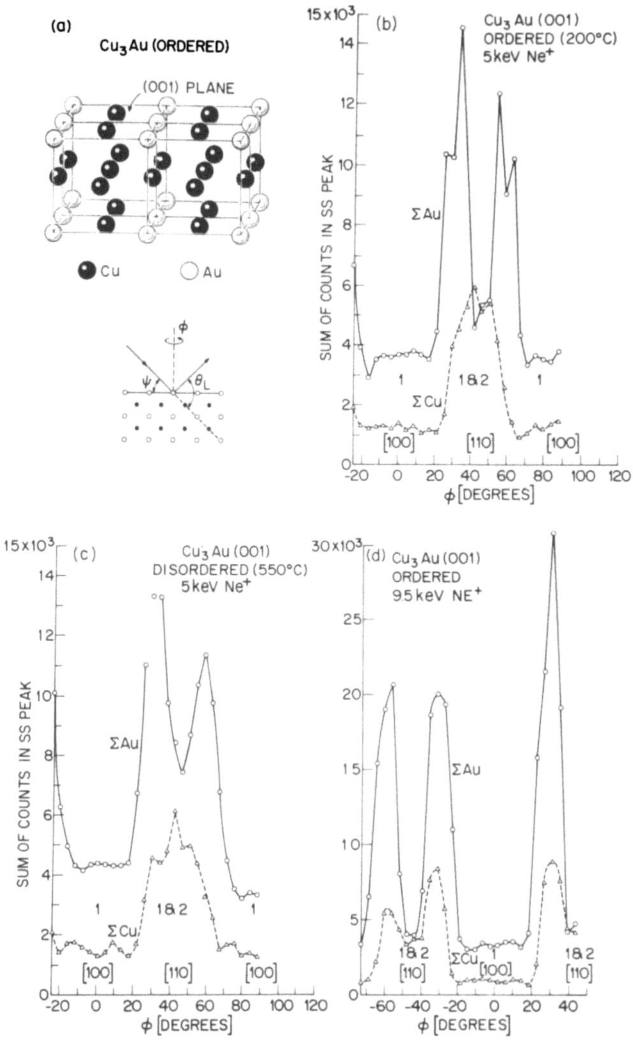

Fig.18.13. (a) Ordered arrangement of Au and Cu in the bulk, and (below) the scattering geometry. (b-d) Single scattering yield from Au and Cu as function of azimuthal angle ϕ. Incidence angle $\Psi = 45^{\circ}$. (b) 5-keV Ne on ordered surface. (c) 5-keV Ne on disordered surface. (d) 9.5-keV Ne on ordered surface. The crystal was removed and remounted between scans (c) and (d), shifting the scan limits by about 50° [18.33]

Cu and Au peak area ratio corresponds to $Cu_{0.47}Au_{0.53}$. (In practice, the sums of counts in the peaks of a TOF spectrum are more convenient than peak areas from the energy spectrum). The approximate 50-50 layer has been on top invariably after numerous sputter cleaning and annealing sequences. There is little if any evidence of antiphase boundaries or steps revealing the pure Cu layer, presumably because of the Au segregation tendency. In the spectrum of Fig.18.11b, taken along the <110> azimuth at 35° incidence where the second layer is exposed along with the first, there is a large increase in the Cu SS peak (and also in the DS background) but not much increase in the Au SS peak. A second-layer composition of $Cu_{0.94}Au_{0.06}$ is obtained by applying (18.13).

Compositions of the first and second layers are plotted as functions of annealing temperature in Fig.18.13. Annealing times were 16-40 hours at $25^{\circ}C$, 16 hours at $200^{\circ}C$, 2 hours at 300 and $350^{\circ}C$, and 1 hour at higher temperature. The first-layer behavior shows clear evidence of competition between ordering and segregation. The gold concentration increases with temperature, reaching $X_1^{Au} = 0.62$ at $400^{\circ}C$, and then decreases to 0.50 at $600^{\circ}C$, presumably approaching the average bulk value of 0.25 at sufficiently high temperature. In the absence of the ordering tendency, e.g., in the Au-Ni case, there should be no maximum in the curve, only a monotonic downward trend [18.6] in accord with the entropy term in the free energy of segregation, $\Delta F_s = \Delta H_s - T\Delta S_s$. However, in this case the entropy term favors segregation for $T < T_c$ but opposes it for $T > T_c$, since segregation represents less order and more order than the unsegregated conditions in the two cases, respectively. As mentioned in Sect.18.2.6, this type of behavior has been calculated for another surface, the {011} surface of $A_{0.5}B_{0.5}$ bcc ordering alloys [18.44]. The lattice-site ordering tendency suppresses the Au segregation but, as temperature is increased, the Au segregation progressively disrupts the order, as shown by the continuous weakening of the LEED superlattice spots [18.77] which was also observed in these experiments. Although the first layer does not assume the random bulk composition in the temperature range studied, the second layer does, but at a temperature above the bulk critical temperature, presumably because of the Au-Cu bonding preference and the lattice strain involved in placing more Au atoms in the second layer when the first-layer concentration is so high.

In addition to the compositions of the first two layers and the LEED patterns as evidence of long-range order at the surface, the angular dependence of scattering yields from Au and Cu, ($\sum Au$ and $\sum Cu$) in the azimuthal scans of Fig.18.14 gives further proof of lattice site order in the first three layers, especially when compared with calculations of shadowing and blocking for the ordered surface. In Fig.18.13b for 5-keV Ne^+ scattered from the ordered surface the broad minima centered about the <100> azimuth represent

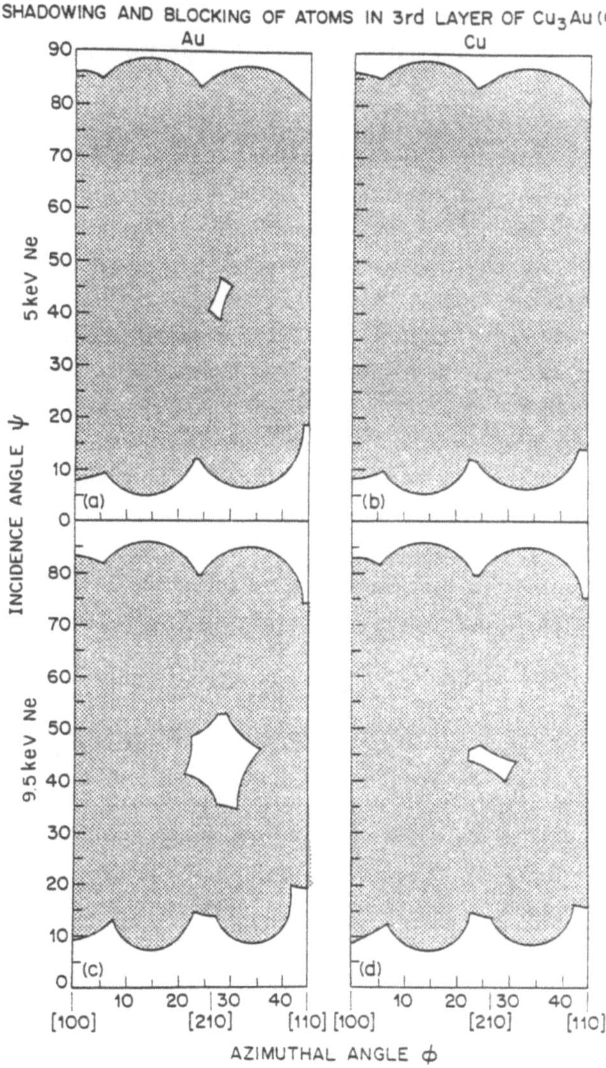

Fig.18.14a-d. Shaded areas map calculated combinations of Ψ and φ for which single scattering of Ne through 90°, by Cu or Au atoms in third layer, is prohibited by shadowing and/or blocking. Incident energy 5 keV in (a,b), 9.5 keV in (c,d) [18.33]

first-layer scattering, and the yield ratio, $\sum Au/\sum Cu = 2.93$, corresponds to the composition $Au_{0.55}Cu_{0.45}$. Starting at 20° to the right of the <100>, $\sum Au$ rises sharply to form a notched peak centered at ~30° from <100>, then drops to a minimum at <110>, showing that there is little gold in the second layer, and rises again to form another notched peak. Except for the broad minimum around the <100> azimuth, $\sum$Cu behaves quite differently, exhibiting no maximum at <100> + 30° or <100> + 60°, but peaking rather at the <100> azimuth where the Cu in the second layer is exposed.

The dramatic spikes in the gold yield can hardly be due to scattering from gold in the second layer, even with strong focusing, since measurements along the <110> azimuth have shown that there is very little gold in the second layer. Simple inspection of an Fcc crystal model suggests that third layer scattering must be involved, and this is supported by neutralization and energy loss of the Ne scattered from Au. However, this leaves the absence of corresponding features in the Cu curve to be explained, since the third layer is expected to be 50-50 Cu-Au like the first. The answer is found in an azimuthal scan at higher energy, 9.5 keV (Fig.18.13d) and in the shadow-cone analysis represented in Figs.18.14a-d. These diagrams are derived from a calculation [18.75] which determines the radius of a shadow cone (Sect.18.3.1d) in a plane normal to the beam direction and containing the target atom, and compares this radius with the distance in this plane from the cone axis to the target atom. Blocking cones of atoms in or near the outgoing path are also taken into account. This analysis is similar to one used recently [18.63] for scattering of Ne^+ ions from oxygen on Cu{110} in that an a/r^2 potential was used, but in this case the screening length was reduced by 20% and a more exact treatment was used for the shadow cone projection.

The experimental results of Fig.18.13 and the calculations of Fig.18.14 demonstrate several facts about the Cu_3Au{001} ordered and disordered surfaces: on the ordered surface Cu atoms in the third layer are shadowed at $\phi = 30^{\circ}$, $\psi = 45^{\circ}$, and 5 keV, but Au atoms are not. This difference arises from the fact that at this orientation each third-layer Cu atom is shadowed by two gold atoms in the first layer, one row over, whereas third-layer Au atoms are shadowed by Cu atoms. The Au shadow cones are larger, with a radius of 1.4 Å versus 1.03 Å for Cu, at a distance of 4.0 Å. Therefore there is no open space in Fig.18.14b for scattering from Cu at $\Psi = 45$, $\phi = 30$, but there is for Au in Fig.18.14a. For 9.5-keV Ne^+ the gold shadow cones are smaller and Cu atoms in the third layer are exposed to the beam, producing a maximum in $\sum$Cu at $\Psi = 30^{\circ}$ (Fig.18.13d) and a small opening centered at (43-27) in Fig.18.14d.

Not only is the third-layer Cu shadowed from 5-keV Ne at $\phi = 30^{\circ}$ and 60°, but so is half of the Cu in the second layer according to another shadow calculation not shown; this agrees approximately with the $\sum$Cu curve of

Fig.18.13b although it is a little unsymmetrical about $\phi = 45^\circ$. This partial shadowing involves the second-layer Cu atoms of <100> rows beneath Au rows. The other half under Cu rows are not hidden.

The most prominent difference between the 5-keV azimuthal scan taken at 550°C, well above T_c (Fig.18.13c), and that for the ordered surface in Fig. 18.13b is the increase in $\sum$Au at the <110> azimuth, indicating enrichment of Au in the second layer. Compositions derived in this case, using (18.10-13), are $Au_{0.54}Cu_{0.46}$ for the first layer and $Au_{0.22}Cu_{0.78}$ for the second. There seems to be evidence for short-range order persisting above T_c in the incipient peaks in $\sum$Cu at $\phi = 30^\circ$ and 50°, which must represent Cu atoms in Au sites of the third layer. These peaks should be more pronounced for random occupancy of sites and this suggests preferential arrangement of Au atoms over Cu atoms in accord with the fact that short-range order in the bulk is appreciable at this temperature [18.80].

It is interesting that the Cu_3Au\{011\} surface, in contrast with the \{001\}, does not exhibit long-range order according to LEED measurements [18.78]. The first- and second-layer compositions are the same, $\sim Cu_{0.75}Au_{0.25}$, after long annealing at 200°C, but gold enrichment of the first layer, to $X_1^{Au} \sim 0.65$, occurs upon annealing at 600°C and cooling to room temperature [18.33]. The absence of long-range order at the surface for $T \leq 200^\circ$C, even though it must be present in the bulk, presumably arises from the fact that a gold atom in an ordered first layer of the \{011\} surface would have only two Cu atoms as nearest neighbors whereas it has four on the \{001\}. On the other hand, the failure to segregate at 200°C is probably due to the existence of short-range order at the surface. Computer simulation of the scattering [18.37,59,60] will be needed for more complete understanding of both short- and long-range order effects, but the shadow-cone analysis is evidently a very useful first step.

18.5 Conclusions

It appears that while no current model of surface segregation is completely successful in either qualitative or quantitative predictions, a few do quite well and therefore can be useful in practical problems, especially in the absence of experimental results. The basic driving forces are recognized, and by adjusting their contributions properly it should be possible to increase the accuracy of predictions. Some aspects which deserve more attention in theory and experiment are the structure of the segregate overlayer, the heat of segregation, and the combined action of ordering and segregation.

Low-energy ion scattering offers first-layer analysis when noble gas ions are used with an electrostatic analyzer, and $E_0 \leq 3$ keV. For quantitative analysis in this case, calibration on the pure elements is necessary since neutralization efficiency varies with target element. Time-of-flight analysis permits collection of both ions and neutrals, thereby circumventing the neutralization problem and simplifying quantitative analysis considerably. Ion beam dose and resultant damage are reduced by factors of 10^2-10^3 by comparison with the conventional parallel plate ESA system. With a Ne^+ beam of 5-10 keV on a single crystal surface the sampling depth may be one, two, or three atom layers depending on crystal orientation with respect to the beam. Information on the structure of the overlayer and on relative positions of atoms in the first three layers has been obtained in this way.

Acknowledgments. The author is indebted to collaborators and co-authors of the several papers from which results have been taken for this review, and to L. Marchut and R.P. Frankenthal for useful discussions.

References

18.1 M.P. Seah, C. Lea: Philos. Mag. *31*, 627 (1975)
18.2 P. van der Plank, W.M.H. Sachtler: J. Catal. *12*, 35 (1968)
18.3 J.H. Sinfelt, J.L. Carter, D.J.C. Yates: J. Catal. *24*, 283 (1972)
18.4 V. Ponec, W.M.H. Sachtler: J. Catal. *24*, 250 (1972)
18.5 S.H. Overbury, P.A. Bertrand, G.A. Somorjai: Chem. Rev. *75*, 547 (1975)
18.6 P. Wynblatt, R.C. Ku: Surface Sci. *65*, 511 (1977); W.C. Johnson, J.M. Blakely (eds.): *Interfacial Segregation* (Am. Soc. Met., New York 1979) pp.115-136
18.7 H.H. Brongersma, M.J. Sparnaay, T.M. Buck: Surf. Sci. *71*, 657 (1978)
18.8 W.M.H. Sachtler, R.A. van Santen: Appl. Surf. Sci. *3*, 121 (1979)
18.9 F.L. Williams, D. Nason: Surf. Sci. *45*, 377 (1974)
18.10 R.A. Van Santen, M.A.M. Boersma: J. Catal. *34*, 13 (1974)
18.11 F.F. Abraham, N.H. Tsai, G.M. Pound: Surf. Sci. *83*, 406 (1979)
18.12 F.F. Abraham: Phys. Rev. Lett. *46*, 546 (1981)
18.13 M.P. Seah: J. Catal. *57*, 450 (1979)
18.14 J.J. Burton, C.R. Helms, R.S. Polizotti: J. Vac. Sci. Technol. *13*, 204 (1976)
18.15 J.W. Gibbs: *Collected Works* (Yale Univ. Press, New Haven 1957) Vol.1 [also (Dover, New York 1961)]
18.16 J.W. Cahn: In *Interfacial Segregation*, ed. by W.C. Johnson, J.M. Blakely (Am. Soc. Met., New York 1979) p.3
18.17 R. Defay, I. Prigogine, A. Bellemans: D.H. Everett: *Surface Tension and Adsorption* (Wiley, New York 1966)
18.18 J.H. Hildebrand: J. Am. Chem. Soc. *51*, 66 (1929)
18.19 E.A. Guggenheim: *Mixtures* (Clarendon, Oxford 1952)

18.20 R.A. Swalin: *Thermodynamics of Solids* (Wiley, New York 1966)
18.21 J.W. Christian: *The Theory of Transformations in Metals and Alloys*, 2nd ed. (Pergamon, New York 1975) Chap.6
18.22 D. McLean: *Grain Boundaries in Solids* (Clarendon, Oxford 1957)
18.23 R. Hultgren, P.D. Desai, D.T. Hawkins, M. Gleiser, K.K. Kelley: *Selected Values of Thermodynamic Properties of Binary Alloys* (Am. Soc. Met., New York 1973); a second volume: R. Hultgren, P.D. Desai, D.T. Hawkins, M. Gleiser, K.K. Kelley, D.G. Wagman: *Selected Values of Thermodynamic Properties of Binary Alloys and the Elements* (Am. Soc. Met., New York 1973)
18.24 J.M. McDavid, S.C. Fain, Jr.: Surf. Sci. *52*, 161 (1975); G.C. Shaw, S.C. Fain, Jr.: Proc. 7th Int. Vac. Congr. and 3rd. Int. Conf. Solid Surfaces, Vienna, 1977, p.2315
18.25 J. Friedel: Adv. Phys. *3*, 446 (1954)
18.26 A.R. Miedema: Z. Metallkd. *69*, 455 (1978)
18.27 A.R. Miedema, R. Boom: Z. Metallkd. *69*, 183 (1978)
18.28 A.R. Miedema: Z. Metallkd. *69*, 287 (1978)
18.29 A.R. Miedema, F.R. deBoer, R. Boom: CALPHAD, Vol.I, 341 (1977)
18.30 G.C. Nelson: Surf. Sci. *59*, 310 (1976)
18.31 B.J. Wood, H. Wise: Surf. Sci. *52*, 151 (1975)
18.32 G.A. Somorjai, S.H. Overbury: Surf. Sci. *55*, 209 (1976)
18.33 T.M. Buck, G.H. Wheatley, L. Marchut: To be published
18.34 C.R. Helms: J. Catal. *36*,114 (1975)
18.35 J.H. Anderson, P.J. Conn, S.G. Brandenberger: J. Catal. *16*, 404 (1970)
18.36 F.L. Williams, M. Boudart: J. Catal. *30*, 438 (1973)
18.37 T.M. Buck, I. Stensgaard, G.H. Wheatley, L. Marchut: Nucl. Instrum. Methods *170*, 519 (1980)
18.38 T.M. Buck, G.H. Wheatley, R. Seals, Y.-S. Chen: To be published
18.39 P. Biloen, R. Bouwman, R.A. van Santen, H.H. Brongersma: Appl. Surf. Sci. *2*, 532 (1979)
18.40 M.J. Kelley, P.W. Gilmour, D.G. Swartzfager: J. Vac. Sci. Technol. *17*, 634 (1980)
18.41 J.R. Chelikowsky: *Predictions for Segregation in 2550 Binary Intermetallic Alloys*, Exxon Research and Engineering Co. (Private communication)
18.42 J.J. Burton, E.S. Machlin: Phys. Rev. Lett. *37*, 1433 (1976)
18.43 J.L. Moran-Lopez, K.H. Benneman: Phys. Rev. *15*, 4769 (1977)
18.44 J.L. Moran-Lopez, L.M. Falicov: Phys. Rev. B*18*, 2542 (1978)
18.45 R. Kikuchi: Phys. Rev. *81*, 988 (1951)
18.46 A. Joshi: Interfacial Segregation: Am. Soc. Met. (1979), p.39
18.47 Y.S. Ng, T.T. Tsong, S.B. McLane, Jr.: Phys. Rev. Lett. *42*, 588 (1979)
18.48 H.W. Pickering, Y. Kuk, T. Sakurai: Appl. Phys. Lett. *36*, 902 (1980)
18.49 D.P. Smith: J. Appl. Phys. *38*, 340 (1967)
18.50 H.H. Brongersma, P.M. Mul: Chem. Phys. Lett. *14*, 380 (1972)
18.51 T.M. Buck, G.H. Wheatley, L.K. Verheij: Surf. Sci. *90*, 635 (1979)
18.52 H.H. Brongersma, N. Hazewindus, J.M. Niewland, A. Otten, A.W. Smets: Rev. Sci. Instrum. *49*, 707 (1978)
18.53 G.R. Sparrow, J.T. McKinney: New Mexico AVS Meeting, March 15-17, 1976
18.54a Y.-S. Chen, G.L. Miller, D.A.H. Robinson, G.H. Wheatley, T.M. Buck: Surf. Sci. *62*, 133 (1977)
18.54b T.M. Buck, G.H. Wheatley, G.L. Miller, D.A.H. Robinson, Y.-S. Chen: Nucl. Instrum. Methods. *149*, 591 (1978)
18.55 S.B. Luitjens, A.J. Algra, E.P.Th. Suurmeijer, A.L. Boers: Appl. Phys. *21*, 205 (1980)
18.56 S.A. Agamy, J.E. Robinson: Surf. Sci. *90*, 648 (1979)
18.57 I. Terzic, D. Ciric, B. Perovic: Surf. Sci. *85*, 149 (1979)
18.58 A.J. Algra, E. van Loenen, E.P,Th. Suurmeijer, A.L. Boers: Rad. Effects (In press)

18.59 E. Taglauer, W. Englert, W. Heiland, D.P. Jackson: Phys. Rev. Lett. *45*, 740 (1980)
18.60 S.H. Overbury, W. Heiland, D.M. Zehner, S. Datz, R.S. Thoe: Surf. Sci. *109*, 239 (1981)
18.61 M.T. Robinson, I.M. Torrens: Phys. Rev. *139*, 5008 (1974)
18.62 W. Turkenburg, W. Soszka, F.W. Saris, H.H. Kersten, B.G. Colenbrander: Nuc. Instrum. Methods *132*, 587 (1976)
18.63 A.G.J. Dewit, R.P.N. Bronckers, J.M. Fluit: Surf. Sci. *82*, 177 (1979)
18.64 H.H. Feijen, K.K. Verhey, E.P.Th. M. Suurmeijer, A.L. Boers: *Atomic Collisions in Solids*, Vol.2 (Plenum, New York 1975) p.573
18.65 H.H. Brongersma, T.M. Buck: Surf. Sci. *53*, 649 (1975)
18.66 R.A. Oriani: J. Chem. Phys. *18*, 575 (1950)
18.67 C. Herring: In *Metal Interfaces* (Am. Met. Soc., 1952)
18.68 S.C. Fain, J.M. McDavid: Phys. Rev. *39*, 5099 (1974)
18.69 R. Bouwman, L.H. Toneman, M.A.M. Boersma, R.A. van Santen: Surf. Sci. *59*, 72 (1976)
18.70 V. Heine: Surf. Sci. *53*, 408 (1975)
18.71 S.H. Overbury, S.J. Somorjai: Surf. Sci. *55*, 209 (1976)
18.72 J.J. Burton, C.R. Helms, R.S. Polizotti: Surf. Sci. *57*, 425 (1976)
18.73 P.W. Palmberg, T.N. Rhodin: J. Chem. Phys. *49*, 134 (1968); Phys. Rev. *161*, 586 (1967)
18.74 I. Stensgaard, L.C. Feldman, P.J. Silverman: Surf. Sci. *77*, 513 (1978)
18.75 L. Marchut: Ph. D. Thesis, University of Pennsylvania, Dept. of Materials Science and Engineering
18.76 D.T. Keating, B.E. Warren: J. Appl. Phys. *22*, 286 (1951)
18.77 V.S. Sundaram, R.S. Alben, W.D. Robertson: Surf. Sci. *46*, 653 (1974)
18.78 H.C. Potter, J.M. Blakely: J. Vac. Sci. Technol. *12*, 635 (1975)
18.79 R.A. van Santen, W.M.H. Sachtler: J. Catal. *33*, 202 (1974)
18.80 R. Smoluchowski: In *Handbook of Physics*, ed. by E.U. Gordon, H. Odishow, 2nd ed. (McGraw-Hill, New York 1967) pp.8-97
18.81 A.F. Wells: *Structural Inorganic Chemistry*, 3rd ed. (Clarendon, Oxford 1967) p.984

19. The Effects of Internal Surface Chemistry on Metallurgical Properties

C.L. Briant

With 22 Figures

19.1 Introduction

Research on interfacial phenomena is currently a very active field in metallurgy. However, in contrast to interfacial research in physics and chemistry, where the main emphasis is often on the gas-solid or vacuum-solid interface, most metallurgical studies of this type concern internal solid-solid interfaces. These are generally classed into two groups, grain boundaries and particle-matrix interfaces.

Most materials in nature are polycrystalline; they are made up of many small single crystals or grains, often less than 10 μm in diameter. These individual grains are usually randomly oriented with respect to one another. Since the orientations of the two neighboring grains are not the same, a grain boundary forms. An example of this is shown in the optical micrograph in Fig.19.1a.

In addition to being polycrystalline, most engineering alloys contain additional phases. These may be in the form of trapped inclusions such as oxides and sulfides in steels, or they may be purposely precipitated second phases such as carbides in steels or γ' (Ni_3Al) precipitates in nickel base superalloys. In both cases these second phases are discrete particles contained within a grain or at a grain boundary. Throughout this paper we will refer to the boundaries between the second phase and the primary phase as particle-matrix interfaces. Two examples are shown in Fig.19.1b,c.

The chemical composition of these internal interfaces can differ significantly from that of the bulk because of the segregation of elements from the bulk. This paper is concerned with this interfacial segregation and its effects on metallurgical properties. In Sect.18.2 segregation and the effects of the grain boundary structure on segregation will be discussed. Examples will be given where the segregation, in turn, has clearly affected the structure. In Sect.19.3 four metallurgical problems will be considered where the compositions of these interfaces are of utmost importance. These are

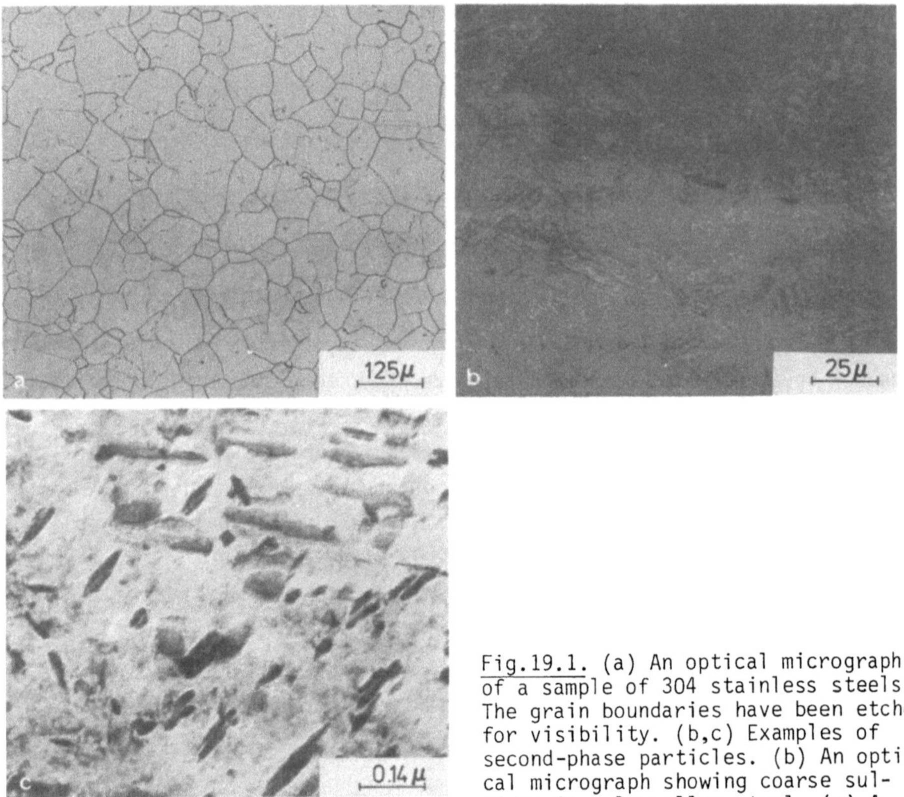

Fig.19.1. (a) An optical micrograph of a sample of 304 stainless steels. The grain boundaries have been etched for visibility. (b,c) Examples of second-phase particles. (b) An optical micrograph showing coarse sulfides in a low-alloy steel. (c) A transmission electron micrograph showing fine carbides in a low-alloy steel

temper embrittlement in steels, ductile fracture, intergranular corrosion of austenitic stainless steels, and grain boundary migration in iron -3% silicon.

Before the discussion of segregation a word should be said about the structure of these interfaces. At the atomic level grain boundaries are rather complex, and many researchers have attempted to model them [19.1-9]. In very simple grain boundaries, as shown schematically in Fig.19.2, a fraction of the atoms in the boundary occupy sites which are lattice sites in both grains. The atoms between these coincident sites are usually described by various structural units [19.1,18], many of which are distorted versions of the Bernal deltahedra shown in Fig.19.3. In the example shown in Fig.19.2 the interconnecting material is composed of distorted pentagonal bipyramids interlocked with capped trigonal prisms. Boundaries with a high density of coincident sites are actually quite rare, and the boundaries in most metals have a *very*

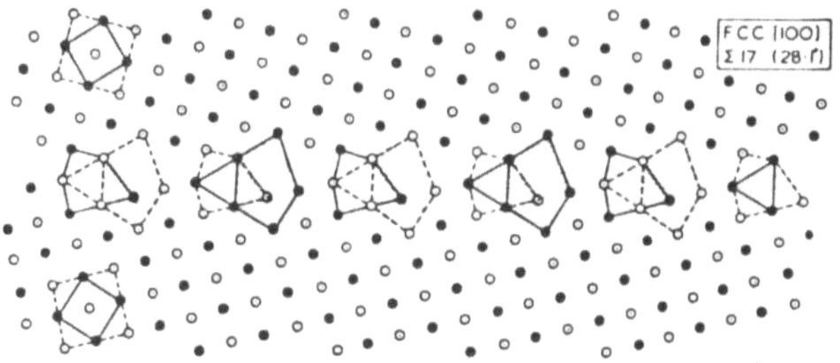

Fig.19.2. A schematic drawing of a grain boundary with a high density of coincident sites [19.1]

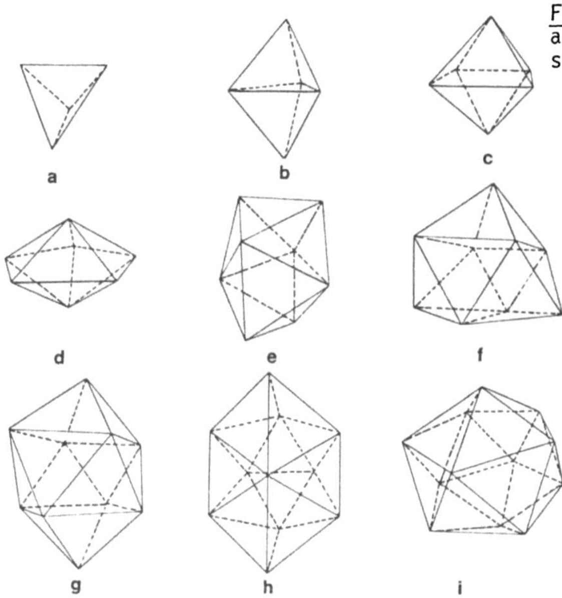

Fig.19.3. The Bernal deltahedra. Each face is composed of equilateral triangles. (a) Tetrahedron, (b) half-octahedron, (c) octahedron, (d) pentagonal bipyramid, (e) tetragonal dodecahedron, (f) capped trigonal prism, (g) Archimedian antiprism, (h) octadecahedron, and (i) regular icosahedron

Fig.19.4a-c. A schematic of the types of particle-matrix interfaces: (a) coherent, (b) semicoherent, and (c) noncoherent

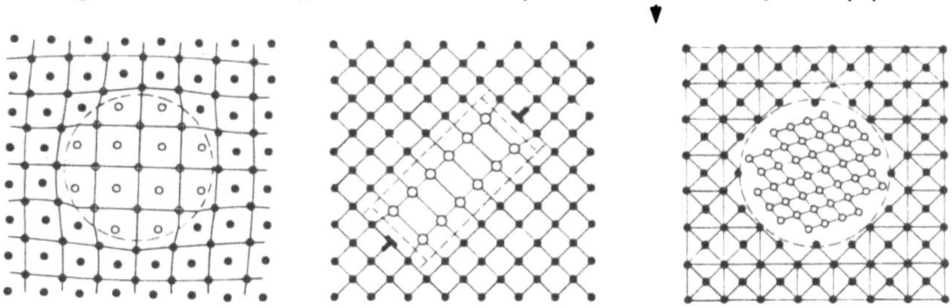

low density of coincident sites. However, even in these boundaries one can still find structural units such as the Bernal deltahedra which can be used to describe the boundary [19.1,2].

The structure of particle matrix interfaces have been studied in less detail. Generally, they are grouped into one of three categories: coherent, semicoherent, and noncoherent (Fig.19.4) [19.10]. In the coherent interface the lattice parameter of the particle closely matches that of the matrix. In the semicoherent interface the lattices can match with the aid of inserted dislocations. In the noncoherent interface there is no matching. In this paper I will only deal with noncoherent interfaces.

19.2 Segregation to Solid-Solid Interfaces

19.2.1 Grain Boundaries

The principal reason why metallurgists have been interested in segregation to grain boundaries is that impurity elements, which in the bulk have very low concentrations (e.g., 50-200 ppm) and are well below the limits of commercial control, segregate to grain boundaries. At the boundary their local concentration can become quite high (up to 5-10 at.%). Therefore, just as the addition of 5 to 10 at.% of an element to the bulk will usually change its properties, this segregation to the grain boundaries can also change the properties of the heavily segregated material which usually extends one or two atom layers on either side of the boundary. Elements which are not impurities but are intentionally added for alloying purposes may also segregate. However, it is almost always the impurities which cause the most damage and are most unexpected; therefore, they have been the most studied.

The following type of experiment is usually done to study grain boundary segregation. Samples are heat treated to produce the desired segregation. They are then placed in an Auger spectrometer and fractured under very high vacuum in the chamber. The Auger analysis, which gives the chemical composition of the fracture surface, is then performed on a freshly prepared and clean surface. If one is interested in grain boundary segregation, the fracture must proceed along the boundaries. Since many segregating species weaken the boundaries, these become the preferred fracture paths. In these cases the technique works well. If the boundaries are not inherently weak one must resort to other methods to try to achieve intergranular fracture [19.11,12]. The inability to achieve this in some situations has been the major limitation of these experiments.

By far the majority of experiments have been done on iron and steel, the discussion will begin with these alloys. Initially, one system will be considered in some detail, and then a brief review of the results of other studies will follow.

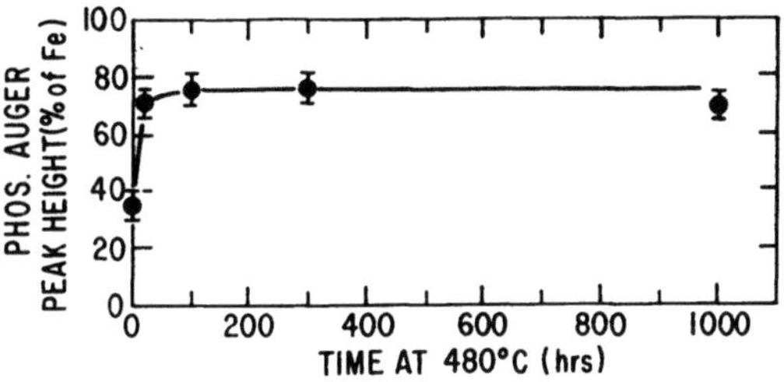

Fig.19.5

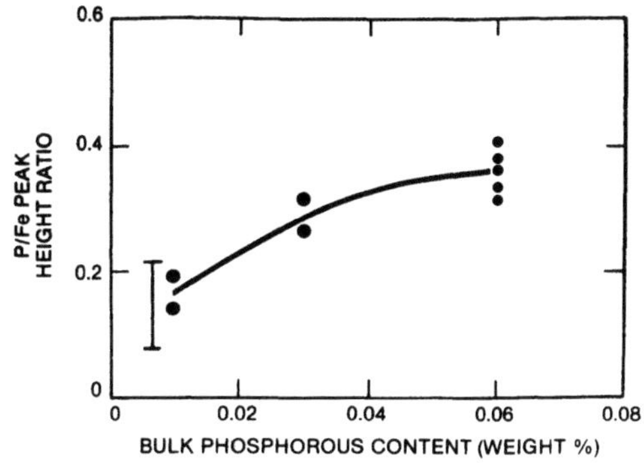

Fig.19.6

Fig.19.5. The kinetics of phosphorus segregation in a 3.5Ni, 1.7Cr, 0.3C, 0.06P steel aged at 480°C [19.18]

Fig.19.6. The equilibrium values of grain boundary phosphorus concentration at 480°C plotted as a function of bulk concentration. Each data point represents an average of over fifteen boundaries in one sample. The error bar to the left is typical of all data points. The full composition of these steels was 3.5Ni, 1.7Cr, 0.3C,P,Fe [19.13]

Phosphorus is a common embrittling segregant in steels [19.13-18]. When this segregation occurs the boundaries can be easily opened and studied by Auger electron spectroscopy. Figure 19.5 shows the kinetics of phosphorus segregation at 480°C in a 3.5Ni, 1.7Cr, 0.3C, 0.06P, balance Fe steel. (Throughout this paper all compositions are given in weight percent.) The phosphorus segregation reaches equilibrium within 100 hours of ageing at this temperature. Figure 19.6 shows equilibrium values of the phosphorus concentration plotted as a function of the bulk concentration. The curve rises rapidly for bulk concentrations up to about 0.03 wt.%. Beyond that the increase is much more gradual.

Previous work suggests that the amount of phosphorus which segregates depends on other alloying elements present in the steel [19.15,18-23]. In particular, it has been suggested that increasing the bulk concentrations of nickel, manganese, and chromium would also enhance phosphorus segregation even if the bulk phosphorus content was unchanged. Figure 7 shows the amount of phosphorus on the grain boundaries plotted as a function of bulk nickel, manganese, and chromium, respectively. Clearly, changing the concentration of these alloying elements does not change the amount of phosphorus segregation.

Most other studies of segregation in laboratory made steels have focused on tin and antimony [19.23-26]. These elements diffuse much more slowly than phosphorus so that even after thousands of hours equilibrium is often not

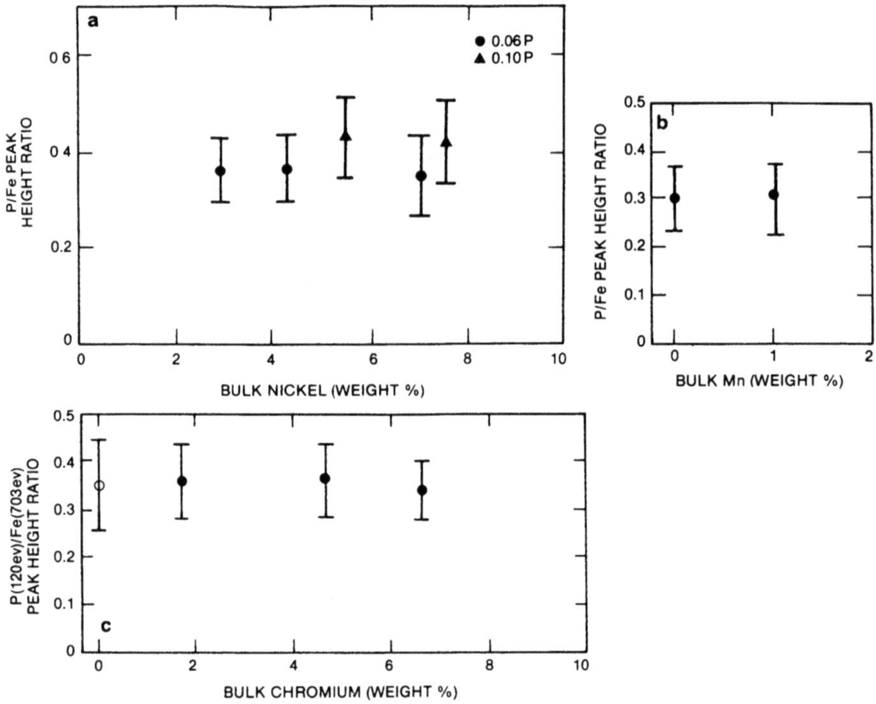

Fig.19.7. (a) The effect of nickel additions on phosphorus segregation in steel. The steels contained either 0.06 or 0.10P. (b) The effect of manganese on phosphorus segregation. The steel contained 0.03P. (c) The effect of chromium on phosphorus segregation. The data plotted with solid points is for steels containing 3.5Ni, 0.3C, 0.06P, Cr, Fe. The open data point is for a steel which also contains 0.6 Mn [19.13]

achieved (Fig.19.8). A number of studies have been performed on commercial steels. In these alloys the segregation chemistry is inevitably more complex [19.15,27-32]. Figure 19.9 shows the segregation kinetics in HY130 steel aged at 480°C. At short times P, N, Mn, Si, and Ni segregate. As ageing progresses, the phosphorus is displaced from the surface by silicon. Also, at very long ageing times tin begins to segregate.

Surprisingly little work has been done on high-purity Fe-binary alloys. It is clear that sulfur readily segregates to grain boundaries in iron [19.33-34]. (In steel S is usually precipiated as a sulfide and is not free to segregate). Phosphorus [19.33], nitrogen [19.33], selenium [19.35], tellurium [19.35,36], and tin [19.37] have also been observed to segregate in iron.

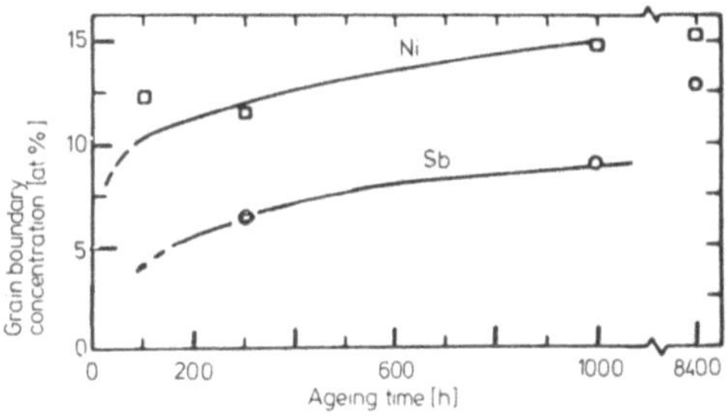

Fig.19.8

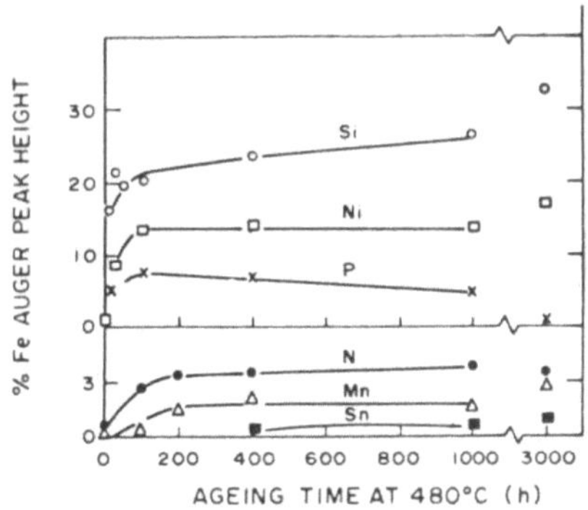

Fig.19.9

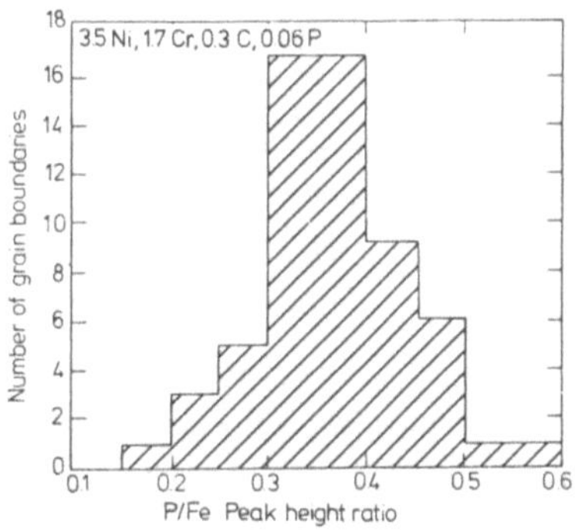

Fig.19.8. Segregation of nickel and antimony in a 3.5Ni, 1.7Cr, 0.3C, 0.06Sb steel. The ageing was done at 520°C [19.24]

Fig.19.9. Grain boundary segregation at 480°C in commercial HY130 steel [19.28]

Fig.19.10. A histogram of the values of phosphorus concentration on individual grain boundary facets in a 3.5Ni, 1.7Cr, 0.3C, 0.06P, balance Fe, steel

◄

Much less work has been done on nonferrous alloys. Bismuth segregation in copper [19.38] has been examined as has sulfur and tellurium segregation in nickel [19.39,40]. Some limited work on refractories has been reported [19.41,42].

The amount of impurity segregation varies significantly from grain boundary to grain boundary within each sample. The data for phosphorus segregation in Fig.19.6 are average values which are obtained by taking spectra from many different boundaries. The error bar in the figure indicates tremendous variation. This fact is seen more clearly in the histogram of phosphorus concentrations shown in Fig.19.10. The spectrum of values is quite large. Similar distributions have now been observed for many systems [19.21, 22,43-48].

The most likely cause for such variation in segregation is the result of differences in grain boundary structure. Furthermore, experiments have shown that this difference cannot be explained by deviations in the fracture path, which would cause an unequal distribution of the segregant from one side of the boundary or the other, or from the distribution of angles that the frac-

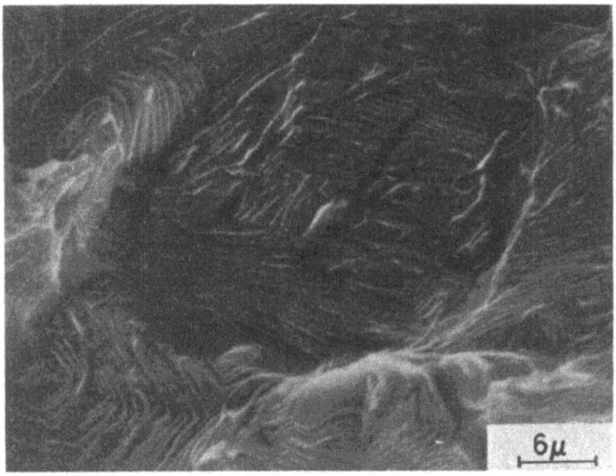

Fig.19.11. An example of facets in an Fe-Te alloy produced by segregation of Te

ture surface presents to the electron gun [19.47]. The relationship of grain boundary structure with segregation presents an interesting new area of study.

Just as the structure of the grain boundary must influence segregation to it, so the segregation that occurs can affect the structure. Many of the structural changes will only involve rearrangements of the nearest-neighbor atoms to the segregating species in the boundary. These will be very difficult to detect by most structural probes. However, in some cases much larger rearrangements are seen and these can be examined. A good example is the faceting which occurs in some systems as segregation progresses. Two examples are segregation of tellurium in iron [19.36] and bismuth in copper [19.49]. As the segregation progresses the boundaries change from generally smooth surfaces to severely faceted ones (Fig.19.11).

19.2.2 Particle-Matrix Interfaces

There have not been nearly as many studies of segregation to particle-matrix interfaces because of experimental difficulties. These interfaces occupy only a small fraction of any fracture surface. Most Auger spectrometers have too wide an electron beam to isolate these particle surfaces. Only the newest generation of Auger spectrometers is designed so that one can first operate in a scanning electron microscope mode and obtain an image of the surface and then perform Auger spectroscopy on very small regions of that surface. This technique is necessary for the study of particle-matrix interfaces.

JOHNSON et al. [19.50,51] have studied segregation to graphite nodules in ductile cast iron. They have found that many of the same elements that segregate to grain boundaries, e.g., S, O, Sn, and Sb, also segregate to these interfaces. BRIANT and BANERJI [19.46] have studied segregation to sulfide interfaces. They found that the sulfides were apparently coated with a layer of sulfur because of the intense sulfur Auger signal observed. They have also suggested from indirect etching results that phosphorus segregates to carbide interfaces [19.52].

19.2.3 Comparisoon with Surface Segregation

At the outset of this section the necessity of obtaining an intergranular fracture or a fracture along particle-matrix interfaces in order to obtain information about their composition was discussed. It has been suggested that that if this is not possible, one might employ surface segregation studies as a substitute [19.38,53]. In these experiments thermal treatments are performed on a small disc in the Auger spectrometer which resemble

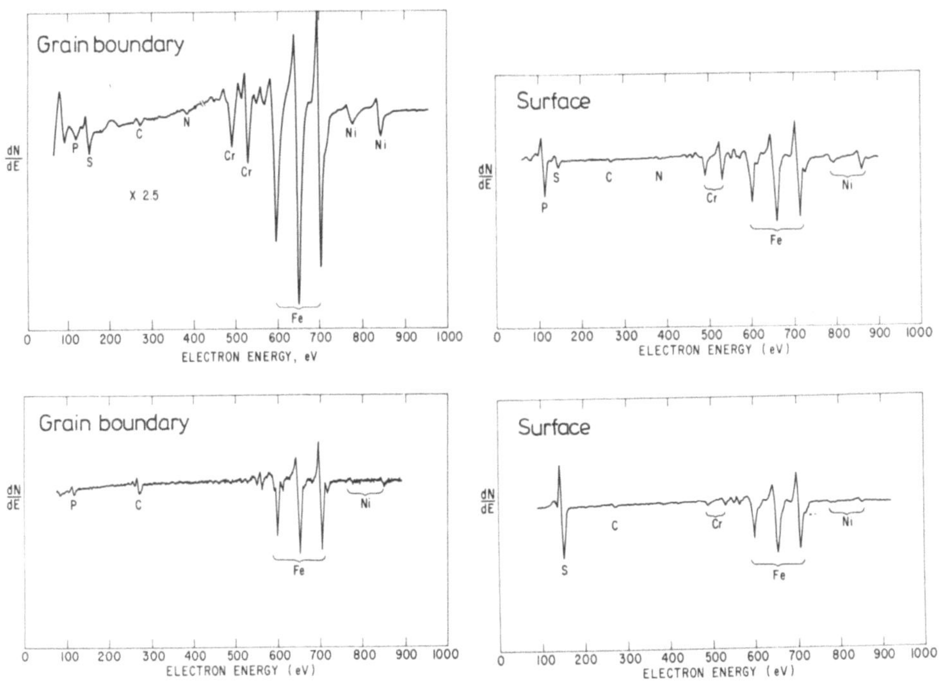

Fig.19.12. (a) A comparison of Auger spectra obtained from an intergranular fracture surface of 304 stainless steel and a surface. Both samples were solution annealed and then aged for 50 hours at 650°C. (b) A similar comparison for a sample of low-alloy steel heat treated at 850°C for 1 hour and quenched

those applied to the bulk sample. The segregation to the surface is monitored throughout these treatments. Figure 19.12 compares spectra obtained from grain boundaries and surfaces given identical heat treatments. In the two stainless steel samples heat treated for 50 hours at 650°C, the spectra look similar. However, for the low-alloy steel heat treated at 850°C the spectra are vastly different. Experiments in our laboratory have shown that site competition occurs much more readily in the surface than at grain boundaries. Therefore, it appears that at best only a qualitative picture of grain boundary segregation can be obtained from surface segregation studies.

19.3 Applications

In this section four metallurgical problems will be discussed where the compositions of internal interfaces are important.

19.3.1 Temper Embrittlement of Steels

When ferritic steels are mechanically tested over a range of temperatures, their fracture mode will change with increasing temperature from an easy, low-energy brittle fracture to a difficult high-energy ductile fracture. Therefore, when the steel is put into use, one wants the service temperature to be above the temperature range of brittle behavior.

If a typical low-alloy steel (Table 19.1) is properly heat treated, it will behave ductally at all temperatures above roughly -100 C. However, if it is not heat treated properly, its transition temperature from ductile to brittle behavior may increase to well above room temperature. As this change occurs the brittle fracture mode will change from transgranular cleavage (Fig.19.13a) to intergranular fracture (Fig.19.13b). This phenomenon is known as temper embrittlement.

The correct heat treatments which will avoid this problem usually involve tempering the steel at a temperature of 600-700°C and rapidly quenching it. The detrimental heat treatment involves either slow cooling after this tempering treatment or ageing the steel in the temperature range of 400° to 575°C. Both of these can occur in practice. The former is found when large forgings are quenched. Even though the steel is rapidly quenched on the surface the core may require hundreds of hours to cool. The latter type of heat treatment can occur if a steel is put in service in this temperature range.

It is now well established that temper embrittlement in steels occurs because impurities segregate to grain boundaries during these detrimental heat

Table 19.1. Typical compositions of steels discussed in this paper

Designation	Low-alloy steels										
	Ni	Cr	C	Mn	Mo	Si	Cu	Al	P	S	N	Fe
HY130	4.88	0.57	0.11	0.88	0.49	0.34	-	0.02	0.003	0.005	0.01	Balance
4340	1.50	0.85	0.38	0.75	0.20	0.20	0.14	0.05	0.006	0.004	0.007	Balance

Designation	Austenitic stainless steels								
	C	Mn	Ni	Cr	P	S	Si	Mo	Cu	Fe
316	0.057	1.67	12.77	17.14	0.035	0.025	0.54	2.21	0.31	Balance
304	0.078	1.12	8.49	18.1	0.025	0.027	0.41	-	0.21	Balance

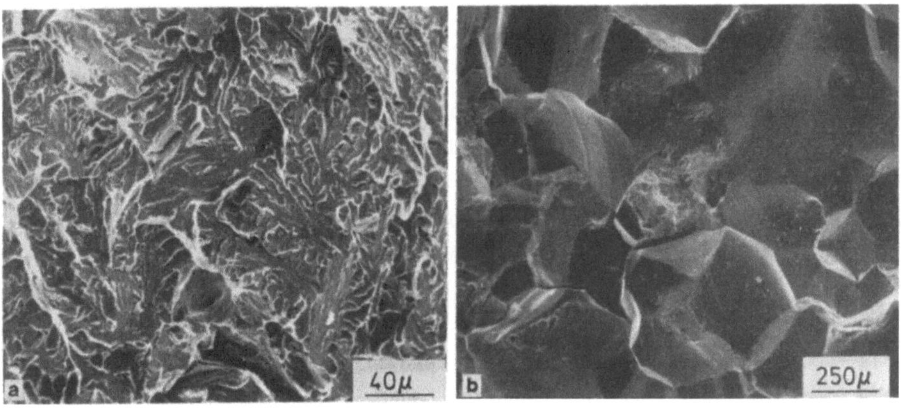

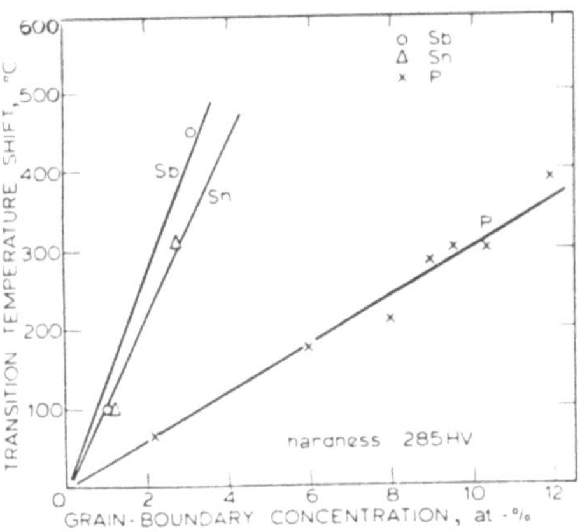

Fig.19.13. (a) Transgranular cleavage, the typical brittle fracture mode of an unembrittled bcc steel. (b) Intergranular fracture, the typical fracture mode of a temper-embrittled steel

Fig.19.14. The change in embrittlement as a function of grain boundary concentration of P, Sn, and Sb

treatments and weaken them [19.16]. Figure 19.14 shows the change in embrittlement as a function of grain boundary composition. It is quite clear that as phosphorus, tin, or antimony segregate to the grain boundaries, embrittlement increases.

Table 19.2. Elements known to cause embrittlement listed with their group identification from the periodic chart

Group I	IVA	VA	VIA
H		N	O
	Si	P	S
		As	Se
	Sn	Sb	Te

The impurities which most often cause temper embrittlement in steels are shown in Table 19.2. Of these, nitrogen, sulfur, and phosphorus are most commonly found in commercial steels. However, since nitrogen and sulfur are usally precipitated as nitrides and sulfides, respectively, phosphorus is the most common embrittler found in practice. The embrittling effectiveness, or potency, of phosphorus depends very much on the alloying elements surrounding it in the steel. For example, manganese and chromium enhance its potency [19.13,18,22,31]. Therefore, with a fixed phosphorus concentration on the grain boundaries, additions of either of these elements to the steel will increase the amount of embrittlement. In contrast, additions of molybdenum [19.21,54] and titanium [19.55] appear to counteract phosphorus-induced embrittlement.

One question which has remained unanswered through the years of research on this problem is why these segregated impurities cause embrittlement. One way the elements lower cohesion along the boundary is by changing the chemical bonding at the boundary. To investigate this question a technique must be used which can examine these bonds. Recently, BRIANT and MESSMER [19.56] applied the self-consistent-field X-α scattered-wave method developed by JOHNSON [19.57] and SLATER [19.58] to this problem. This technique allows one to calculate the molecular orbitals for a cluster of atoms and in that way examine the bonds which form. As discussed previously, the structure of grain boundaries can often be described in terms of distorted Bernal deltahedra. Therefore, these deltahedra can be used as the clusters for the X-α calculations. As an example of these calculations, consider the problem of sulfur-induced intergranular embrittlement in nickel. Figure 19.15 shows the cluster used for the calculations. In one calculation nickel was placed at the eight vertices of the cluster and in another calculation a sulfur atom was placed at the center of the Ni_8 cluster. Figure 19.16 shows a contour plot for the total valence charge density obtained for the two clusters. The plot is made in a plane that includes atoms #1, 4, 7, and 8 in Fig.19.15. Particular attention should be paid to contour number 4. In the Ni_8 cluster it clearly contributes to a bond between atoms 1 and 7, 4 and 8, and 7 and 8. However, in Ni_8S the

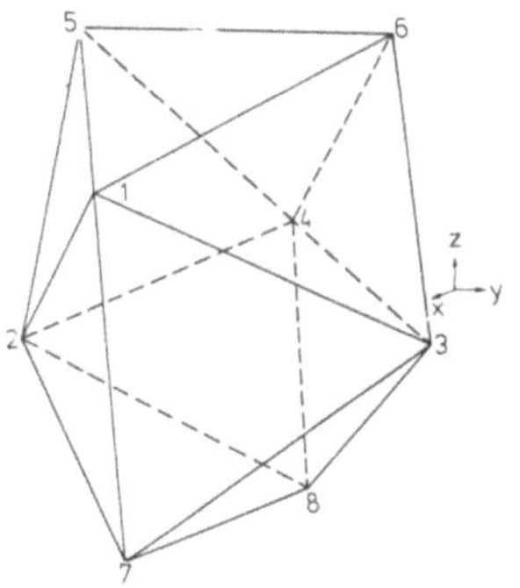

Fig.19.15. The tetragonal dodecahedron

Fig.19.16a,b. Contours of the total valence charge density in the X-Z plane of the cluster shown in Fig.19.15. (a) is for Ni_8 and (b) is for Ni_8S. Contour 4, which is drawn more darkly, differs most drastically between the two clusters [19.56]

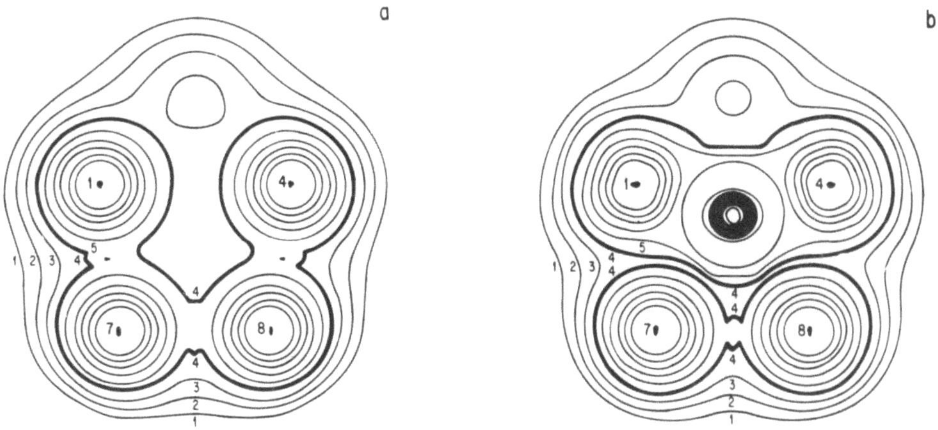

contour is completely changed. The contribution to bonds between atoms 1 and 7 and 4 and 8 is completely removed and now a strong bond is formed between the sulfur atom and the nickel atoms 1 and 4. In addition the Ni-Ni bond between atoms 7 and 8 is weakened. Of course, an identical picture would be obtained if the plane for the plot had been taken at 90° to this one so that it included atoms 2, 3, 5, and 6. The picture that emerges then is the following. Embrittling elements in the boundary form rather strong bonds with the metal atoms near to them and draw charge off of these metal atoms. The result of this bonding is that metal-metal bonds are weakened. These would then be easy to break and would give rise to easy brittle fracture. Calculations on other systems bear out the general conclusions of this argument. However, it should be recognized that distortions at the boundary simple caused by the size of the embrittling element or by changes in metal-metal bond lengths could also contribute to embrittlement.

19.3.2 Ductile Fracture

It has been mentioned above that ductile fracture was a high-energy fracture mode. However, the actual fracture energy is affected by segregation to particle-matrix interfaces. To prove this point we must first explain the mechanism of ductile fracture. When a material is deformed plastically, dislocations move through the grains and cause slip. When they encounter a hard particle their motion is impeded and additional stress will be required for them to continue their motion. While the dislocations are pinned at these particle-matrix interfaces, one of several events can take place. The particle can be sheared or cracked, or the particle-matrix interface can open up. Either of these last two events can give rise to voids in the material. As plastic deformation continues more and more voids will form until the alloy is only held together by the ligaments between them. With continued strain these ligaments will neck down and final fracture will occur. An example of a ductile fracture surface is shown in Fig.19.17. The voids around particles are quite obvious. This mode of ductile fracture is often referred to as microvoid coalescence.

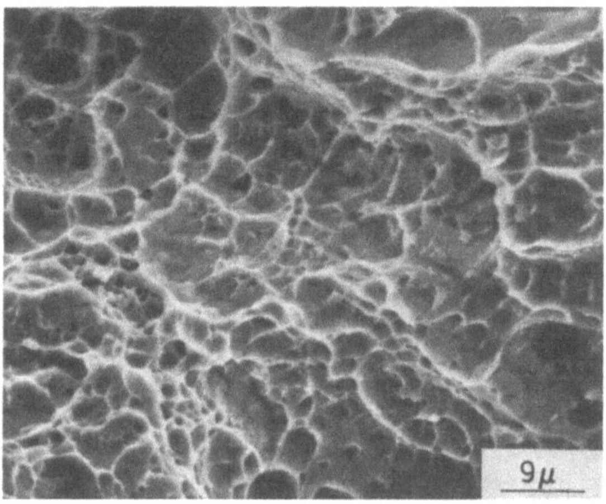

Fig.19.17. A ductile fracture surface. Note the voids formed around the particles

One of the factors which affects the total energy of the fracture process is the ease with which the particle-matrix interface opens up and allows voids to nucleate. If this interface is weakened by a segregated impurity, the fracture energy will decrease. KING and KNOTT [19.59] have examined the effects of impurity segregation to carbide-matrix interfaces in ductile fracture in steel. They found that segregation to these interfaces lowered the fracture energy. The sulfide-matrix interface is also often a site for easy void nucleation in steels. BRIANT and BANERJI [19.46] have shown than these interfaces are usually covered with a layer of sulfur which would make it easy to nucleate voids there.

19.3.3 Sensitization of Austenitic Stainless Steels

Austenitic (i.e., face-centered-cubic) stainless steels are primarily used because of their good resistance to corrosion, which is a result of the high chromium content of these alloys (Table 19.1). However, it has been known for many years that certain heat treatments make these steels quite susceptible to intergranular corrosion. The model proposed by BAIN [19.60] over forty years ago to explain this corrosion is still accepted today.

The basic idea of the model is as follows. If a steel is properly heat-treated by rapidly quenching it from a temperature between 1000° and 1200°C, it will be immune to intergranular corrosion. However, if it is slowly cooled from these temperatures or isothermally heated between approximately 550° and 750°C, the susceptibility to intergranular corrosion will greatly increase. An example is shown in Fig.19.18. BAIN proposed that this susceptibility increased because chromium carbides precipitate at the boundaries (Fig.19.18c). As these carbides grow, carbon can diffuse rapidly through the matrix and be available to enter the growing carbide. However, chromium will diffuse much more slowly. Therefore, the chromium that initially enters the carbide will be taken from a volume of material surrounding the precipitate. Because the carbides nucleate most easily at the grain boundaries, the material near the boundary will become depleted in chromium and lose its resistance to corrosion. With time, chromium will diffuse from the matrix to this region and the susceptibility to intergranular corrosion will decrease.

There are several differences between this example and the previous ones. First of all, we are concerned here with the depletion of an element rather than an enrichment. Secondly, the element involved is a major alloying element and not an impurity. Thirdly, this depletion can occur over a distance of several thousand angstroms on either side of the boundary in contrast to the segregation which only extends one or two atom layers into the grain. Because of this last factor most progress in studying this depletion has been made by using scanning transmission electron microscopy coupled with an X-ray energy analyzer. In these experiments the beam is moved across a boundary in the thin foil specimen and the composition profile is recorded. An example of such a profile is shown in Fig.19.19. One can observe the distinct depletion of chromium near the boundary.

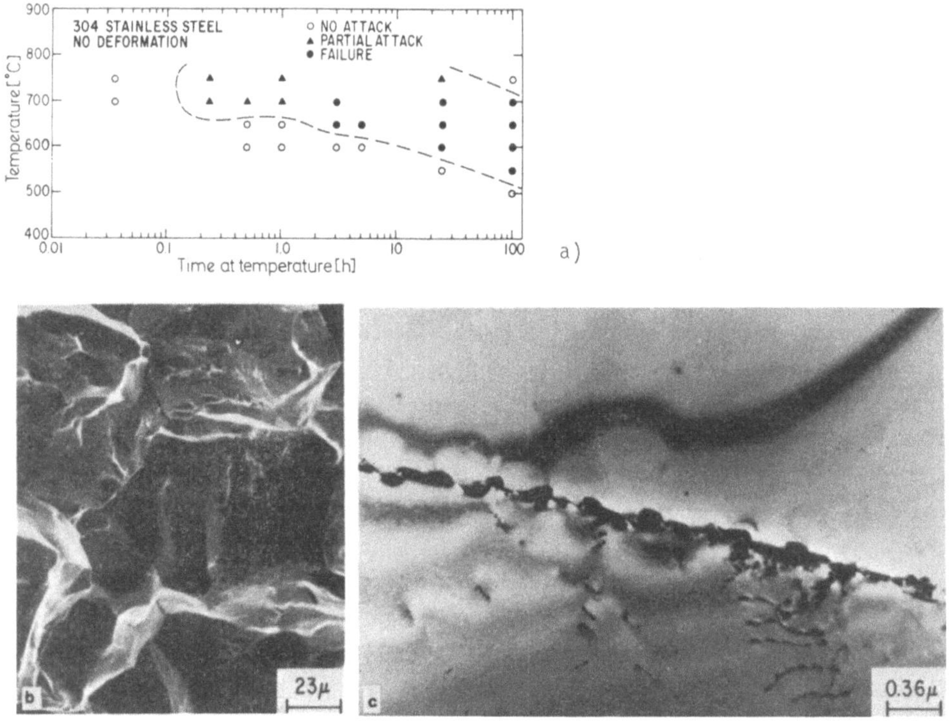

Fig.19.18. (a) The corrosion behavior of a type 304 stainless steel. No attack means no corrosion during the test. Partial attack means that corrosive attack was observed on the surface. Failure means that the corrosion penetrated completely through the sample. Each data point corresponds to a time and temperature of heat treating. (b) A scanning electron micrograph of an intergranular corrosion failure in these alloys. (c) A transmission electron micrograph showing the chromium carbides along the grain boundaries of a 316 stainless steel [19.11]

In recent years a number of advances have been made which can be used to avoid this problem. In many situations lowering the carbon content to less than 0.02 wt.% will alleviate sensitization. However, this removal has the disadvantage of lowering the strength of the steel. Another remedy is to add very strong carbide formers such as titanium or niobium to the steel. These elements will getter the carbon preferentially and prevent chromium carbides from forming.

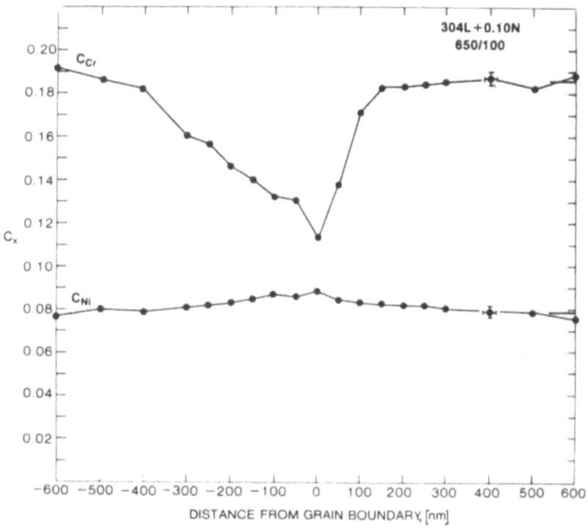

Fig.19.19. The composition of a 304 stainless steel as determined by X-ray energy analysis in scanning transmission electron microscopy. Note the depletion of chromium near the boundary. (Courtesy of E.L. Hall)

19.3.4 Grain Growth

In our discussion we have so far considered stationary grain boundaries. However, at high temperatures grain boundaries will migrate. The primary reason for this movement is that boundaries have energy associated with them, and the total energy of the solid will be lowered if they are removed. Indeed in some cases one grain can grow to consume most of the sample [19.61]. An example of this is shown in Fig.19.20.

Fig.19.20. Exaggerated grain growth in tungsten wire. Several grains are growing at the expense of all others [19.61]

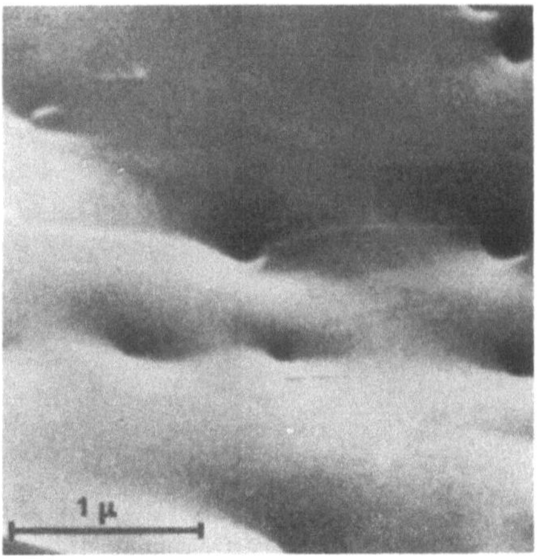

Fig.19.21. A moving grain boundary in Fe-3% Si that has been pinned by second-phase particles [19.62]

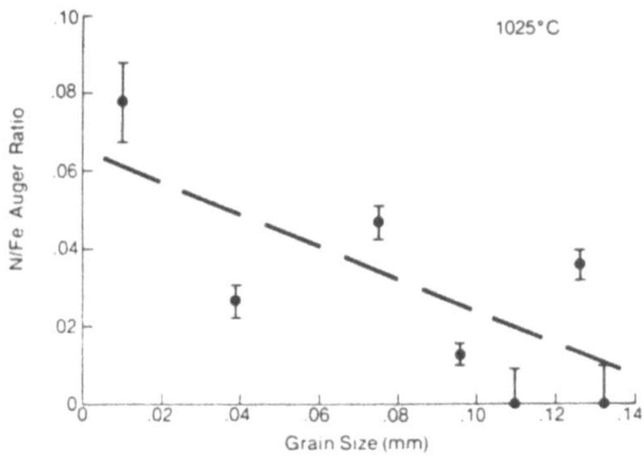

Fig.19.22. The N/Fe Peak height ration in Fe-3% Si plotted as a function of grain size [19.62]

It has been recognized for many years that second-phase particles can impede grain boundary motion. Figure 19.21 shows a scanning electron micrograph of a grain boundary in Fe-3% Si that has been broken open after a grain growth experiment. Note the cusps across the boundary when particles have pinned its migration.

However, it is also well established that segregated elements can affect grain growth. ROWE [19.62,63] has studied grain boundary migration in Fe-3% Si and shown that segregated nitrogen plays a role in retarding grain growth in these alloys. Figure 19.22 shows the N/Fe peak height ratio as measured by Auger electron spectroscopy as a function of grain size. Clearly when more nitrogen is present at the grain boundaries the resulting grain size is finer. As pointed out by ROWE [19.62,63], there is scatter in the data which he attributes to the presence of second-phase particles.

In another study of grain boundary migration, AUST found that most high-angle boundaries in lead-tin alloys had roughly equivalent velocities but that a few special, highly coincident boundaries moved much more rapidly [19.64]. He attributed this result to the fact that the special coincident boundaries have much less tin segregated to them in this alloy and thus would not be impeded by this segregation. The motion of the random boundaries would be impeded because of the segregants present.

If one considers the atomic processes that must be involved in grain boundary migration then it is not surprising that segregation would affect this process. Essentially, an atom must hop from one grain or lattice orientation across the boundary to another. The presence of a segregated atom will change the ease with which this hop can occur. Unless the segregated species can keep up with the moving boundary, it will be reincorporated into the bulk in an energetically unfavorable process.

19.4 Summary

This paper has examined the chemistry of solid-solid interfaces. It has been shown that many different elements will segregate to these interfaces during appropriate heat treatments. In particular, impurity elements which may have very low bulk concentrations can be concentrated at the interface and reach surprisingly high local concentrations. In specially made laboratory alloys the segregation of individual elements may be isolated and carefully studied. In commercial alloys the segregation may be quite complex with many elements segregating and some competing for sites at the interface. The amount of segregation varies greatly from grain boundary to grain boundary. This fact is almost certainly a reflection of the differences in the structure of the boundaries. This segregation plays an important role in many metallurgical processes. Examples have been given of grain boundary fracture, corrosion, and movement, as well as of ductile fracture which initiates at second-phase particles.

Our understanding of these problems from an engineering point of view is reasonably good although unexpected failures still occur. However, the scientific explanation of many of these phenomena is at best qualitative. The exciting areas of research in the future will, I feel, lie in finding these explanations.

Acknowledgments. The author would like to thank Jacqueline L. Briant for her helpful comments on the manuscript.

References

19.1 M.F. Ashby, F. Spaepen, S. Williams: Acta Metall. *26*, 1647 (1978)
19.2 A.P. Sutton, V. Vitek: Scr. Metall. *14*, 129 (1980)
19.3 G.H. Bishop, B. Chalmers: Scr. Metall. *2*, 133 (1968)
19.4 P.H. Pumphrey: Scr. Metall. *6*, 107 (1972)
19.5 M.J. Weiss, H. Gleiter, B. Chalmers: J. Appl. Phys. *42*, 2639 (1971)
19.6 P.D. Bristowe, S.L. Sass: Acta Metall. *28*, 575 (1980)
19.7 M.L. Kronberg, F.H. Wilson: Metall. Trans. *185*, 501 (1949)

19.8 H.J. Frost, M.F. Ashby, F. Spaepen: Scr. Metall. *14*, 1051 (1981)
19.9 A. Brokman, P.D. Bristowe, R.W. Balluffi: Scr. Metall. *15*, 201 (1981)
19.10 E. Hornbogen: In *Nucleation*, ed. by A.C. Zettlemeyer (Dekker, New York, 1969) p.309
19.11 C.L. Briant: Corrosion Houston *36*, 497 (1980)
19.12 J.Y. Park, S. Danyluk: Corrosion Houston *33*, 304 (1969)
19.13 C.L. Briant: Scr. Metall. *15*, 1013 (1980)
19.14 C.A. Hippsley, J.F. Knott, B.C. Edwards: Acta Metall. *28*, 869 (1980)
19.15 B.C. Edwards, H.E. Bishop, J.C. Riviere, B.L. Eyre: Acta Metall. *24*, 957 (1976)
19.16 C.L. Briant, S.K. Banerji: Int. Metall. Rev. *24*, 164 (1978)
19.17 C.L. Briant, S.K. Banerji: Metall. Trans. A*10*, 1729 (1979)
19.18 R.A. Mulford, C.J. McMahon, Jr., D.P. Pope, H.C. Feng: Metall. Trans. A*7*, 1183 (1976)
19.19 C.J. McMahon, Jr.: Mater. Sci. Eng. *25*, 233 (1976)
19.20 Ph. Lemble, A. Pineau, J.R. Castagne, Ph. Doumoulin, M. Guttmann: Mut. Sci. J. *13*, 496 (1979)
19.21 Jin Yu, C.J. McMahon, Jr.: Metall. Trans. A*11*, 277 (1980)
19.22 Jin Yu, C.J. McMahon, Jr.: Metall. Trans. A*11*, 291 (1980)
19.23 M. Guttmann: Surf. Sci. *53*, 213 (1975)
19.24 H. Ohtani, H.C. Feng, C.J. McMahon, Jr., R.A. Mulford: Metall. Trans. A*7*, 87 (1976)
19.25 H. Ohtani, H.C. Feng, C.J. McMahon, Jr.: Metall. Trans. A*7*, 1123 (1976)
19.26 A.K. Cianelli, H.C. Feng, A.H. Ucisik, C.J. McMahon, Jr.: Metall. Trans. A*8*, 1059 (1977)
19.27 M.P. Seah: Surf. Sci. *53*, 168 (1975)
19.28 C.L. Briant, H.C. Feng, C.J. McMahon, Jr.: Metall. Trans. A*9*, 625 (1978)
19.29 B.C. Edwards, B.L. Eyre, G. Gage: Acta Metall. *28*, 335 (1980)
19.30 J.P. Coad, J.C. Rivere, M. Guttmann, P. Krahe: Acta Metall. *25*, 161 (1977)
19.31 J.F. Smith, J.H. Reynolds, H.N. Southworth: Acta Metall. *28*, 1555 (1980)
19.32 A. Joshi, P.W. Palmberg, D.F. Stein: Metall. Trans. A*6*, 2160 (1975)
19.33 P.V. Ramasubramanian, D.F. Stein: Metall. Trans. *4*, 1735 (1973)
19.34 B.D. Powell, H.J. Westwood, D.M.R. Taplin, H. Mykura: Metall. Trans. *4*, 2357 (1973)
19.35 C. Pichard, M. Guttman, J. Rieu, C. Goux: J. Phys. *35*, 151 (1975)
19.36 J.R. Rellick, C.J. McMahon, Jr., H.L. Marcus, P.W. Palmberg: Metall. Trans. *2*, 1492 (1971)
19.37 M.P. Seah, C. Lea: Philos. Mag. *31*, 627 (1975)
19.38 E.D. Hondros, D. McLean: Philos. Mag. *29*, 771 (1974)
19.39 R.A. Mulford: To be published
19.40 H.L. Marcus, N.E. Paton: Metall. Trans. *5*, 2135 (1974)
19.41 A. Joshi, D.F. Stein: Metall. Trans. *1*, 2543 (1971)
19.42 C.L. White, R.E. Clausing, L. Heatherly: Metall. Trans. A*10*, 683 (1979)
19.43 J. Kameda, C.J. McMahon, Jr.: Metall. Trans. A*12*, 31 (1981)
19.44 T. Ogura, C.J. McMahon, Jr., H.C. Feng, V. Vitek: Acta Metall. *26*, 1317 (1978)
19.45 T. Watanabe, T. Murakami, S. Karashima: Scr. Metall. *12*, 361 (1978)
19.46 C.L. Briant, S.K. Banerji: Metall. Trans. A*12*, 309 (1981)
19.47 R.A. Mulford, C.L. Briant, R.G. Rowe: In *Scanning Electron Microscopy/1980/I*, ed. by O. Johari (SEM, Chicago, 1980) p.487
19.48 R.G. Rowe, C.L. Briant, F. Bacon: In *Eight Internat. Conf. on X-Ray Optics and Microanalysis*, ed. by D.R. Braman, R.E. Ogilvie, D.B. Wittry (Pendell, Midlands, MI 1980) p.393
19.49 A.M. Donald, A.J. Craven: Philos. Mag. A*39*, 1 (1979)
19.50 W.C. Johnson, W.B. Smartt: Metall. Trans. A*8*, 553 (1977)

19.51 W.C. Johnson, B.V. Kovacs: Metall. Trans. A*9*, 219 (1978)
19.52 C.L. Briant, S.K. Banerji: To be published
19.53 J.J. Burton, B.J. Berkowitz, R.D. Kane: Metall. Trans. A*10*, 677 (1979)
19.54 C.L. Briant, S.K. Banerji: Scr. Metall. *13*, 813 (1979)
19.55 H. Ohtani, H.C. Feng, C.J. McMahon, Jr.: Metall. Trans. *5*, 516 (1974)
19.56 C.L. Briant, R.P. Messmer: Philos. Mag. B*42*, 569 (1980)
19.57 K.H. Johnson: Adv. Quantum Chem. *7*, 143 (1973)
19.58 J.C. Slater: Adv. Quantum Chem. *6*, 1 (1972)
19.59 J.E. King, J.F. Knott: Metal Sci. *15*, 1 (1981)
19.60 E.C. Bain, R.H. Aborn, J.J.B. Rutherford: Trans. Am. Soc. Steel Treat. *21*, 489 (1933)
19.61 D.B. Snow: Metall. Trans. A*7*, 783 (1976)
19.62 R.G. Rowe: Metall. Trans. A*10*, 997 (1979)
19.63 R.G. Rowe: In *Recrystallization and Grain Growth of Multi-Phase and Particle-Containing Materials*, ed. by N. Hansen, A.R. Jones, T. Leffers (Riso National Laboratory, Roskilde 1980) p.243
19.64 K.T. Aust, J.W. Ritter: Trans. Metall. Soc. AIME *215*, 820 (1995)

Subject Index

MIX
Papier aus verantwortungsvollen Quellen
Paper from responsible sources
FSC® C105338

If you have any concerns about our products,
you can contact us on
ProductSafety@springernature.com

In case Publisher is established outside the EU,
the EU authorized representative is:
Springer Nature Customer Service Center GmbH
Europaplatz 3, 69115 Heidelberg, Germany

Printed by Libri Plureos GmbH
in Hamburg, Germany